Exotic Plant Pests and North American Agriculture

Contributors

L. A. Andres
J. A. Browning
C. O. Calkins
Alvin Keali'i Chock
W. E. Cooley
E. Crooks
L. E. Ehler
D. R. Forney
C. L. Foy
K. Havel
James G. Horsfall
Ke Chung Kim
John D. Lattin
J. W. Miller
Paul Oman
David T. Patterson
P. Lawrence Pusey
Reece I. Sailer
R. D. Schein
C. L. Schoulties
Ralph Scorza
C. P. Seymour
M. Shannon
R. D. Shrum
M. D. Simons
G. Snyder
T. Wallenmaier
Charles L. Wilson
C. E. Yarwood
Robert L. Zimdahl

Exotic Plant Pests and North American Agriculture

Edited by
CHARLES L. WILSON
Appalachian Fruit Research Station
Agricultural Research Service
United States Department of Agriculture
Kearneysville, West Virginia
CHARLES L. GRAHAM
Plant Disease Research Laboratory
Agricultural Research Service
United States Department of Agriculture
Frederick, Maryland

1983

ACADEMIC PRESS
A Subsidiary of Harcourt Brace Jovanovich, Publishers
New York London
Paris San Diego San Francisco São Paulo Sydney Tokyo Toronto

ACADEMIC PRESS, INC.
111 Fifth Avenue, New York, New York 10003

United Kingdom Edition published by
ACADEMIC PRESS, INC. (LONDON) LTD.
24/28 Oval Road, London NW1 7DX

Library of Congress Cataloging in Publication Data
Main entry under title:

Exotic plant pests and North American agriculture.

Includes index.
1. Pest introduction--United States.
2. Pest introduction. 3. Agricultural pests--United States. 4. Agricultural pests. I. Wilson, Charles L. II. Graham, Charles L.
SB990.5.U6E96 1983 632'.6'0973 83-6078
ISBN 0-12-757880-3

PRINTED IN THE UNITED STATES OF AMERICA

83 84 85 86 9 8 7 6 5 4 3 2 1

Contents

9. Stopping Pest Introductions

E. CROOKS, K. HAVEL, M. SHANNON, G. SNYDER, and T. WALLENMAIER

10. How to Detect and Combat Exotic Pests

KE CHUNG KIM

11. Research on Exotic Insects

C. O. CALKINS

Contributors

Numbers in parentheses indicate the pages on which the authors' contributions begin.

L. A. Andres (395), Biological Control of Weeds Laboratory, Agricultural Research Service, United States Department of Agriculture, Albany, California 94706

J. A. Browning (449), Department of Plant Sciences, Texas A&M University, College Station, Texas 77843

C. O. Calkins (321), Insect Attractants, Behavior, and Basic Biology Research Laboratory, Agricultural Research Service, United States Department of Agriculture, Gainesville, Florida 32604

Alvin Keali'i Chock (479), Region II, International Programs, Plant Protection and Quarantine, Animal and Plant Health Inspection Service, United States Department of Agriculture, 2585 The Hague, The Netherlands

W. E. Cooley (65), Department of Plant Pathology and Physiology, Virginia Polytechnic Institute and State University, Blacksburg, Virginia 24061

E. Crooks (239), Plant Protection and Quarantine, Animal and Plant Health Inspection Service, United States Department of Agriculture, Washington, D.C. 20250

L. E. Ehler (395), Department of Entomology, University of California, Davis, California 95616

D. R. Forney (65), Department of Plant Pathology and Physiology, Virginia Polytechnic Institute and State University, Blacksburg, Virginia 24061

C. L. Foy (65), Department of Plant Pathology and Physiology, Virginia Polytechnic Institute and State University, Blacksburg, Virginia 24061

K. Havel (239), Plant Protection and Quarantine, Animal and Plant Health Inspection Service, United States Department of Agriculture, Washington, D.C. 20250

James G. Horsfall (1), The Connecticut Agricultural Experiment Station, New Haven, Connecticut 06504

Ke Chung Kim (261), The Frost Entomological Museum, Department of Entomology, The Pennsylvania State University, University Park, Pennsylvania 16802

John D. Lattin (93), Systematic Entomology Laboratory, Department of Entomology, Oregon State University, Corvallis, Oregon 97331

J. W. Miller (139), Division of Plant Industry, Florida Department of Agriculture and Consumer Services, Gainesville, Florida 32602

Paul Oman (93), Systematic Entomology Laboratory, Department of Entomology, Oregon State University, Corvallis, Oregon 97331

David T. Patterson (381), Agricultural Research Service, United States Department of Agriculture, and Department of Botany, Duke University, Durham, North Carolina 27706

P. Lawrence Pusey (361), Southeastern Fruit and Tree Nut Research Laboratory, Agricultural Research Service, United States Department of Agriculture, Byron, Georgia 31008

Reece I. Sailer (15), Department of Entomology and Nematology, University of Florida, Gainesville, Florida 32611

R. D. Schein (419), Department of Plant Pathology, The Pennsylvania State University, University Park, Pennsylvania 16802

C. L. Schoulties (139), Division of Plant Industry, Florida Department of Agriculture and Consumer Services, Gainesville, Florida 32602

Ralph Scorza (219), Appalachian Fruit Research Station, Agricultural Research Service, United States Department of Agriculture, Kearneysville, West Virginia 25430

C. P. Seymour (139), Division of Plant Industry, Florida Department of Agriculture and Consumer Services, Gainesville, Florida 32602

M. Shannon (239), Plant Protection and Quarantine, Animal and Plant Health Inspection Service, United States Department of Agriculture, Washington, D.C. 20250

R. D. Shrum (419), Plant Disease Research Laboratory, Agricultural Research Service, United States Department of Agriculture, Frederick, Maryland 21701

M. D. Simons (449), Agricultural Research Service, United States Department of Agriculture, and Department of Plant Pathology, Iowa State University, Ames, Iowa 50011

G. Snyder (239), Plant Protection and Quarantine, Animal and Plant Health Inspection Service, United States Department of Agriculture, Washington, D.C. 20250

T. Wallenmaier (239), Plant Protection and Quarantine, Animal and Plant Health Inspection Service, United States Department of Agriculture, Washington, D.C. 20250

Charles L. Wilson (361), Appalachian Fruit Research Station, Agricultural Research Service, United States Department of Agriculture, Kearneysville, West Virginia 25430

C. E. Yarwood* (39), Plant Pathology Department, University of California, Berkeley, California 94720

Robert L. Zimdahl (183), Weed Research Laboratory, Department of Botany and Plant Pathology, Colorado State University, Fort Collins, Colorado 80523

*Deceased. Send requests for reprints to A. R. Weinhold, Chairman, Department of Plant Pathology, University of California, Berkeley, California 94720.

Preface

Controversy surrounds our handling of exotic pests. In 1981, we saw a Governor and a President pitted over control measures for the Mediterranean fruit fly in California. Even adversary groups in the U.S. Congress have called themselves "gypsy moths" and "boll weevils." Awareness at such political levels demonstrates the profound effect exotic pests have had, and continue to have, on our lives.

Responsibility for assessing potential foreign pest introductions has been placed primarily in regulatory agencies. Research to support their effort has often been "too little too late." Except for the efforts of the McGregor task force in 1973, no overall effort has been made to determine how our current knowledge measures up to the threat posed by potential immigrant pests.

Shockingly, we find no affirmation of our present quarantine system in the McGregor report. McGregor states, "Worldwide quarantine programs appear to be based on authority without scientific support or verification. Quarantine actions are a matter of public policy and the usefulness of these activities has not been verified."*

Scientists have not given the exotic pest problem the attention it demands. Consequently, we often find scanty knowledge upon which to devise and implement defensive strategies. The different scientific disciplines that have expertise to deal with this problem have just not been challenged to carry their full load. Hence, we find conflicting and confusing advice for combating exotics.

The contributors to this treatise were asked to look anew at the threat posed by exotic pest introductions and represent a variety of disciplines from the academic community, government research laboratories, and regulatory agencies. They have built us a higher platform of knowledge from which to view the exotic pest problem.

The treatise is divided into three major areas. We first take a look at the history

*McGregor, R. C. (1978). People-placed pathogens: The emigrant pests. *In* "Plant Disease," Vol. 2 (J. G. Horsfall and E. B. Cowling, eds.), p. 394. Academic Press, New York.

of past pest introductions (Chapters 1–4). Then an evaluation is made of where we can expect future introductions (Chapters 5–10). Chapters 11–17 discuss the research efforts and international cooperation that are needed. Dr. Horsfall starts us on this journey by looking at the impact that exotic pests have had on man's history.

Charles L. Wilson
Charles L. Graham

CHAPTER 1

Impact of Introduced Pests on Man

JAMES G. HORSFALL

The Connecticut Agricultural Experiment Station
New Haven, Connecticut

People have an uncanny facility for lugging their pests along as they move into new territory. The early invaders of the New World brought smallpox and measles to bedevil the American Indian and took back syphilis to bedevil their European confreres. The settlers of New Zealand took almost every apple pest they knew along with them when they went "down under."

The editors have asked me to discuss the impact of imported pests on man's social behavior. I shall leave the economic impacts to others. I shall not limit myself to the United States, however. Since sociological information is hard to come by, I shall discuss a few examples that I know personally or have researched considerably. Being a plant pathologist, I must begin with plant dis-

ISBN 0-12-757880-3

eases and then move on to weeds and then to insects, where the examples are plentiful and dramatic.

I. DISEASES OF CROP PLANTS

A. Red Rust of Wheat

Since wheat feeds more people than any other plant except rice, the red rust disease (as named by the Romans) has probably had more social impact than any other plant disease. Most historians agree that wheat and its concomitant rust disease evolved together in western Asia in the region of Persia and Turkestan. During its evolution, the wheat rust fungus learned to live a part of its life cycle on the barberry (*Berberis*) plant.

The first recorded social impact of wheat rust that I can find is described by Carefoot and Sprott (1967). About 1800 B.C., Jacob's son, Joseph, was sold by his brothers into slavery in Egypt. They told his father that Joseph had died. In Egypt Joseph became a favorite of the pharaoh, who had a famous dream of seven fat years followed by seven lean years. Joseph, being in charge of all farm land for the pharaoh, built and filled many granaries during the seven fat years. This tided the Egyptians over during the seven lean years when the wheat was heavily damaged with rust. When Joseph's father had to go to Egypt to buy grain, he found his long-lost son. Carefoot and Sprott show some evidence that 4000 years ago, Egypt's climate seemed damp enough for rust.

1. *Romans Create a Rust God, Robigus*

Wheat rust had an impact on the Romans (see Stakman and Harrar, 1957). About 700 B.C., it was so bad that the priests decided that the standard gods such as Jupiter and Ceres had become so overworked that they should create a new god. They named the new god Robigus after the red rust.

The red rust usually appeared on the Roman wheat in April, just after the rising of the red dog star, Sirius. *Ipso facto,* the dog star caused the wheat to become rusty. Accordingly, the priests of Rome staged the Robigalia, a ceremony to propitiate the red rust god, Robigus. When Sirius became visible in the night sky, the white-robed priests would light innumerable torches, begin to chant, and thus lead the citizenry out through the Catularia Gate of Rome, along the Via Claudia to an altar in a sacred holly grove. There the priests ascended the altar stairs and sacrificed to Robigus a red dog whose blood ran over the altar. Whereupon they added red wine to the gory mess, and burned wheat straw during the night until the red dog star set in the western sky.

The farmers were not quite sure that this ceremony could save their wheat from rust. Therefore, they took out insurance by planting laurel bushes among the wheat so that the rust would settle on the laurel instead.

As the Roman empire expanded west and north, the traders carried wheat with them on their ships and spread the crop throughout Europe and Britain. Since the rust fungus is not seed-borne, they did not carry it with the wheat. Nor did they take the barberry with them; and so, for centuries, the Europeans escaped famines due to wheat rust.

2. *Rust Invades Upper Europe, Changing Eating Habits*

It appears that during the last third of the first millennium, perhaps about A.D. 800, the barberry found its way north and west, and produced another social change. Central Europe, eastern France, and Germany had an ideal wheat rust climate—warm, damp weather in the spring. Therefore, rust grew to epidemic proportions and caused famines. As a result, the natives had to learn to eat rye bread because rye is less susceptible to rust than wheat.

3. *Rye Induces Ergotism, or St. Anthony's Fire*

The same moist weather that aided wheat rust also aided a disease of the rye grain called "ergot." This disease did not reduce the yield of rye, as the rust did on wheat. People eating ergot-contaminated grain came down with a dreadful disease, ergotism. The first recorded epidemic of ergotism occurred in the Rhine Valley in A.D. 957. Thousands of citizens experienced high fever, "heavenly light," dreadful hallucinations, and often death. This encouraged a Frenchman, Gaston de la Valoire, to build a hospital in the Rhine Valley to treat the sufferers of ergotism. He dedicated it to St. Anthony, and the disease came to be known as "St. Anthony's fire."

4. *Ergotism in Art*

This event calls to mind a book by Scheja (1969), "The Isenheim Altarpiece." On the altar in the church in the tiny Alsatian village of Isenheim is a painting of the agonies of St. Anthony's fire as symbolically suffered by St. Anthony himself. The tryptych was painted in about 1500 by Matthias Grünewald. Scheja says (p. 27), "On the right wing [of the tryptych] we have what is called 'The Temptation of St. Anthony,' but more correctly, the ordeal of the saint." Elsewhere, he says (p. 18), "Its nightmare scene is an appalling explosion of visionary fantasy. The saint lies sprawled in the foreground hard pressed by a hellish bestiary of demons . . . who menace him, strike him, bite him. The saint, however, seems unharmed, and above the landscape shimmers a heavenly light."[1] This is all depicted in living color.

Caporeal (1976) suggested, a few years ago, that ergotism could explain the witchcraft trials in Salem, Massachusetts, in 1692. This stirred my interest in the

[1]The quotations are reprinted with the kind permission of the publisher, Harry N. Abrams, Inc., New York.

subject because a few years before (1969), I had discovered a newspaper advertisement for 1832 that proclaimed that "Dr. Relf's Botanical Drops" would cure St. Anthony's fire, presumably caused by contaminated rye being advertised in the adjacent column by a grain dealer.

Recently, Matossian (1982) has reopened the subject of the relation between ergotism and the witchcraft trials. She makes a reasonably persuasive case based on the court records and contemporary diaries. One bit of convincing evidence is her assertion (p. 357) that "Beginning in 1590, the common people of England began to eat wheat instead of rye bread. The settlers in New England also preferred wheat bread but, troubled by wheat rust in the 1660s, they began to substitute the planting of rye for wheat. This dietary shift may explain why the witchcraft affair of 1692 occurred 47 years after the last epidemic of witch prosecution in England."

St. Anthony's fire is surely a dreadful side effect of wheat rust.

5. *Immigrants to Virginia Substitute Corn*

The fourth social impact of wheat rust that I know about occurred in the United States (Horsfall, 1958). The English brought wheat with them and settled two separate areas at first, New England and Virginia. They brought barberries, too, and therefore rust. Rust occurred on New England wheat but was not so extensive as to cause famine. Wheat did not do well in Virginia, because the warm, wet weather in the spring encouraged rust.

Like the Central Europeans before them, the Virginians had to turn to a different carbohydrate. Luckily for them, they did not have to choose rye. They selected instead a local carbohydrate that the Indians called "maize" and they called "corn." As a result, southerners eat corn bread with relish, along with grits and hominy.

These three impacts of wheat rust gave us three names for bread. Bread in Germany is "rye bread." Bread in England and New England is "wheat bread." Bread in the southern United States is "corn bread."

6. *Wheat Rust Triggers the First Environmental Protection Law*

Wheat rust had still another impact on history. For more than 200 years before De Bary explained it, farmers had been telling the intellectuals that barberries brought on wheat rust, but being yokels, they were ignored by the thinkers. Legislators, however, listened to their constituents, and not to the savants, and in 1660 in Rouen, France, they passed a law forcing people to destroy their barberry bushes. This was probably the first environmental protection law.

The New England settlers, having brought barberries, then attempted to eradicate them. Connecticut passed a barberry eradication law in 1725.

B. White Pine Blister Rust Gives Us the Quarantine Act

White pine blister rust is a traveling disease. It seems to have originated on the five-needle pines in Siberia. From there it moved to Europe and subsequently to the United States. It spread rapidly through the beautiful white pine forests of New England and the Adirondacks in New York State.

There it entered politics and created its social impact. Congressman Simmonds introduced a bill in Congress that became the Federal Quarantine Act of 1912 and ended a political impasse. For a dozen or more years, professional pest controllers, especially in California, had tried to get a national quarantine bill through Congress. They were successfully opposed by nurserymen using classic pressure group tactics. In those days, California was too far from Washington to have much political clout, but when the crowded easterners saw their beautiful pine forests going down, they got action through Congressman Simmonds of New York State.

As soon as the act was safely on the books, the quarantiners prohibited the importation of white pine seedlings. They had to do this for political reasons, although it was locking the barn door after the horse had been stolen.

Very soon, the social impact hit the ordinary citizen. Tourists returning happily from a trip abroad were shocked to find their luggage being pulled apart and examined for plant material that might harbor pests.

It was not only travelers to foreign countries who were subjected to the indignities of inspection, however. As the stay-at-homes traveled along their highways, they were stopped and examined. I am not sure that any bugs were stopped. I suspect that the Japanese beetles and corn borers laughed as they flew westward over the stalled road traffic below.

The white pine blister rust had a social impact on Roland Thaxter, the famed cryptogamic botanist of Harvard and an erstwhile plant pathologist at The Connecticut Agricultural Experiment Station. *Ribes* is the alternate host for blister rust, as barberry is for wheat rust. The federals set out to control the disease by eradicating *Ribes*. Thaxter had a garden on the Maine coast at Kittery Point. When the federals came to clean out his currants and gooseberries, he met them at the gate with his trusty old blunderbuss, thereby winning the ''Battle of Kittery Point.'' After millions of dollars had been spent throughout the nation, the project was abandoned.

Thaxter was vindicated, but only long after his death.

C. Dutch Elm Disease Overleaps the Quarantiners

Having swept through Holland and France, the elm disease reached the United States in 1930. Why didn't the quarantiners stop Dutch elm disease at the border ports?

In the 1920s, the elm disease was a boon to American veneer manufacturers, who happily imported the diseased logs for years. Like the foolish virgin, we went to the wedding without any oil in our lamp. We had the lamp, the quarantine act, but we had no oil. The port inspectors would confiscate some poor lady's houseplant and let tons of elm logs through. The law applied to living things. Logs were deemed nonliving, even though they were loaded with contaminated bark beetles that escaped to ravage living trees.

D. Chestnut Blight Quenches Two Industries

Chestnut blight, entering the United States from the Orient, also exerted an influence on our history. To this day, old-timers nostalgically remember from their childhood the safaris into the woods to collect chestnuts.

The white men who invaded North America found that chestnut wood was exceedingly resistant to decay, and so they made fence posts, railroad ties, and telephone poles out of it. Chestnut blight killed this industry. Settlers discovered, too, that chestnut bark contained tannins that could be used for tanning shoe leather. The blight killed this business too.

On the other hand, two new industries arose from the debris: creosoting and synthetic tanning.

E. Potato Blight Has Far-Reaching Effects

The *vade mecum* of all plant pathologists is the Irish famine, which made plant pathology a science. The potato blight is induced by the infamously famous fungus *Phytophthora infestans* (*Phytophthora* means "plant destroyer"). I will not summarize the starvation and emigration of the unfortunate Irish, but rather will discuss several other dramatic social impacts of the famine, one of which involves my family.

1. *It Makes Britain a Free Trade Nation*

The potato blight brought about the downfall of British Prime Minister Robert Peel, and the abandonment of the invidious corn laws that his party favored.

In 1815 the wealth of England was largely in the hands of the landholders. Using typical pressure group tactics that year, they persuaded the British Parliament to pass the so-called corn laws that prohibited the importation of corn (wheat), which would depress the price of British wheat. The landholders probably did not sense it, but their political clout was already in decline because the industrial revolution was shifting wealth and, hence, influence from the country to the city. This created a new pressure group comprised of manufacturers and ship builders who opposed the corn laws. They wanted free trade for their manufacturing, cargo for their ships, and cheap food for their laborers.

This gave me a historical connection with the episode. My great-grandfather, Thomas Berry Horsfall, inherited and expanded a large manufacturing and shipowning company in Liverpool. In the 1840s, as mayor of Liverpool, he pressured Parliament to repeal the corn laws, citing the starving Irish as evidence. A grateful Liverpool elected him to Parliament, but only after the law had been repealed.

Carefoot and Sprott (1967, p. 82) say that the repeal of the corn laws "appears to have been the first step toward the policy of free trade." This new policy enabled Britain to become a great center of trade and to build the biggest mercantile fleet in the world to service its world empire.

2. *It Provides Fertile Ground for Marx and Engels*

Carefoot and Sprott state further that the hunger on the mainland of Europe induced by the potato blight, although not as severe as in Ireland, generated widespread unrest. This, coupled with an economic depression, created fertile ground for the radical seeds being sown by Marx and Engels in the mid-nineteenth century. Governments got into trouble. King Louis Philippe abdicated in France. Lajos Kossuth created drastic reforms in Hungary. The Italians forced the Austrians out of Milan, and on and on.

3. *It Enhances Catholicism in the United States*

Prior to the Irish famine, the United States was a white, Anglo Saxon, Protestant country. Then, the overwhelming immigration of Irish Roman Catholics infiltrated the country, provided policemen for Boston and New York, built churches and parochial schools throughout the country, and eventually produced a Roman Catholic president.

4. *It Accelerates the German Surrender in World War I*

Carefoot and Sprott (1967) cite another striking social impact of the potato blight: In 1916 it destroyed the German potato crop. The military had scooped up all the copper in Germany for shell casings, electric wire, and (I suspect) brass buttons. The farmers, being low in the Prussian pecking order, lost the battle to get copper for their Bordeaux mixture to control the disease. The potatoes fell victim to an inordinately cool, wet summer, just as had occurred in Ireland in 1845.

Late blight rampaged over the potato fields, and *Phytophthora* destroyed again, producing hunger and outright starvation. The soldiers were fed, but their families at home were not. This affected the soldiers' will to win.

5. *Nematodes Affect the Sociology of Long Island*

Ever since colonial days, potatoes have thrived in the sandy loam soil of Long Island, that terminal glacial moraine off the shores of New England. Sometime prior to 1941, the golden nematode was imported, presumably from Europe, and the fat was in the fire. The quarantiners took over, slapping on quarantines and other regulations.

The disease became so serious and the regulations so restrictive that many disgusted farmers sold their land for houses, highways, and shopping centers. Those who remained turned to the nematocide Temik, which gave excellent control until the compound was discovered in the groundwater and then in the drinking water. This created a considerable upset among the population, which had been made chemophobic by the environmentalists.

II. WEEDS

Weeds, like sin, have always been with us. Unlike sin, however, no one likes a weed. Still, however far down the scale of dislikes weeds might be, Shakespeare finds something worse: ''Lilies that fester,'' says he, ''smell far worse than weeds.'' Whatever the psychological effect of weeds, it is difficult to find good examples of their sociological effects.

They do affect the economics of the farmer and, therefore, his sociology. Johnsongrass, for example, is one of the most pernicious weeds that was ever introduced. Within about 50 years, it invaded essentially all parts of the country, leaving in its wake many dull hoes and frustrated farmers. Throughout the cotton South, it kept many children home and out of school with hoe in hand. It generated a folk phrase among cotton farmers: ''The cotton is in the grass.'' Mind you, they did not say, ''The grass is in the cotton.'' In 1900, Johnsongrass led to the first federal appropriation specifically for weed control. It didn't seem to help me much, however, as I hoed johnsongrass out of cotton around 1918.

Witchweed (*Striga*), a parasite, was introduced in the 1950s. Its quarantine seriously restricts the movement of farm machinery and produce out of the affected area.

City people hate weeds, too. Ragweed pollen gives them hay fever. Crabgrass makes them crabby. Plantain makes them plaintive: ''Why does it pick on me?'' Robbins *et al.* (1952) say that the Europeans, moving westward in the United States, left so much plantain in their wake that the Indians called the weed the ''white man's footprint.'' Weeds such as crabgrass and plantain are to the city man what fleas are to a dog. They give him something to do.

III. INSECTS

A. The Gypsy Moth Has a Strong Social Impact

The gypsy moth, like the late blight of potatoes, has had a strong impact on our society. When it first appeared in Connecticut at the turn of the century, it stirred up a latent bug taboo. When DDT came along as a Nobel Prize-winning compound, it stirred up the poison taboo. Exploiting the poison taboo, Rachel Carson created a cult of chemophobic citizens and triggered the nascent environmental movement. Let us see how this came about.

For years, I was directly concerned with the gypsy moth because of my statutory responsibility as director of The Connecticut Agricultural Experiment Station. Hitchcock (1974, p. 87) has written, and I tend to agree, that "From its introduction into Connecticut in 1905, the gypsy moth has generated more worry, thought, and controversy than any other animal present in the State."

In 1869 the insect was deliberately imported into an area near Boston as a possible producer of silk. The only silk it ever created was a few dangling threads that often lodged in milady's hair.

1. *It Stirs Up the Bug Taboo*

During the century gone by, gypsy moth larvae have passed millions of acres of forest foliage through their alimentary canals, leaving the trees as bare in June as they were in February. Their crushed bodies have made the roads as slippery in June as the ice had done in February. Their droppings have fallen like rain onto picnic tables and into reservoirs, where they have tainted the water. The larvae have crawled up the sides of houses and into the parlor when doors were left ajar. People hated them.

2. *It Crawls into the Lawbooks*

In 1907 the gypsy moth crawled into the lawbooks of Connecticut when the legislators declared them a public nuisance. The bug taboo was in full cry. The gypsy moth was equated with Beelzebub, and the Agricultural Experiment Station was instructed to extirpate it. We tried—desperately.

We eradicated the pest the first two or three times it came into the State. Since the poison taboo was dormant, we were able to spray the trees with lead arsenate from the ground. We also treated the egg masses with creosote and scorched them with torches. Like the phoenix, however, the insects rose again from the ashes.

For 10 years prior to 1945, the damage was so negligible, however, that our state entomologist, R. B. Friend (1945), was lulled into a sense of false security. He wrote (p. 625), "The gypsy moth in Connecticut has attained the status of a native insect where the natural factors of control [have reduced it to] only sporadic local outbreaks . . . in forests." Would that he were right! The moth defoliated 1,400,000 acres in the state in 1981. Friend fell victim to the dictum that "Forecasting is a difficult business, especially forecasting the future."

Let us see what happened after Friend's paper appeared. Let us see how the gypsy moth ate its way deeper into the consciousness of the citizens of Connecticut and the other states in the Northeast.

Friend experienced only a lull before the storm, which lasted throughout the 1940s and early 1950s. The insects were still defoliating only 200 acres in 1951, but 2 years later (1953) the damage exploded tenfold to 2000 acres and even to 14,000 acres in 1954. The intensity of the bug taboo increased accordingly during the 1950s.

The citizens whose forests were defoliated and whose houses were invaded demanded that the State do something. They hid their antipathy to the bugs under the guise that the State had an obligation to "keep Connecticut green." The only option at the time was to spray the threatened forests with DDT from the air, making sure to obtain permission from the owners in every case. The governor financed the operation from his contingency account, insisting that the affected towns pay half the cost.

3. *It Stirs Up the Poison Taboo*

Looking back, I can see that the bug taboo and the poison taboo were on a collision course (Horsfall, 1965). As the 1950s wore along and the bugs and the spraying increased, so did both taboos. The people whose forests were not damaged greatly outnumbered those whose forests were. Hence, they could exert more political pressure to abandon spraying and let nature take its course.

In the meantime, and during the heyday of the bug taboo, the U.S. Department of Agriculture announced that it would eradicate the pesky bug with the new miracle compound, DDT. In 1955 they easily got a bill through Congress to provide the money. They proposed to start at the western edge of the infestation and squeeze the gypsy moth into the Atlantic Ocean.

In 1957 the Department of Agriculture blanketed New Jersey, Westchester County (adjacent to New York City), and Long Island with DDT. These were densely populated areas. Even so, the Department ignored the poison taboo and the rising antispray pressure from the conservationists (soon to be called environmentalists). They flew right over their objections, their houses, and their gardens, spraying DDT as they went.

The citizens could see the spray streaming down from the bellies of the spray planes, but they had no place to hide. My impression is that Rachel Carson's area was among those sprayed. No wonder she could write so feelingly about the poisonous rain from heaven. In any event, the episode hardened her resolve to write a book about the village where "no birds sang."

4. *It Triggers the Environmental Movement*

Rachel Carson's book, "Silent Spring," was on the bestseller lists for months in 1962. I was honored to serve on the panel appointed by President John F. Kennedy's science advisor to examine the impact of her book. We interviewed Miss Carson, who was intensely concerned with the poison taboo. She converted "pesticide" into a dirty word and created a whole new class of people, the chemophobes. Unquestionably, her book was the stimulus for the powerful environmental movement.

The advertising copywriters had a ball. Labels on consumer products blossomed with statements that they contained only "natural ingredients," no "chemical additives."

The gypsy moth has surely exerted a powerful influence on man's sociology.

B. The Codling Moth Runs into the Poison Taboo

The codling moth of apple was introduced into the North American continent in the late nineteenth century. It exerted a serious effect on the apple economy of the Annapolis Valley of Nova Scotia. The apple farmers there lived on their exports to Britain. In the 1930s, lead arsenate spray was the only known control for the codling moth. In using it, they ran into the poison taboo.

Just at that time, the British were discovering lead poisoning among the habitués of their famous pubs. The beer extracted lead from the lead pipes leading from the kegs below to the spigots above. When the British discovered that the Nova Scotians were using lead arsenate, they banned Nova Scotian apples and the apple industry faded.

Three thousand miles away, in the Ozarks of Arkansas and Missouri, the codling moth was so difficult to control that farmers pulled out the trees. After considerable dislocation of the people, the apple industry was replaced by a poultry economy.

C. The Boll Weevil Sends Farmers to Make Cars

The boll weevil invaded the cotton South from Mexico early in the twentieth century. It rapidly moved eastward, and by 1923 had "harvested more than half the cotton planted in the United States" (Metcalf and Flint, 1962, p. 9). In so doing, it tipped the economic balance in the worn-out hill farms of Arkansas and Mississippi so badly that the white farmers moved north in droves to make cars in Detroit. The boll weevil affected me too. In 1924 I got my first job in science, chasing boll weevils on a cotton plantation.

D. The Cottonycushion Scale Splits California Entomologists

The cottonycushion scale arrived in California in the late nineteenth century. It (1) threatened the young citrus industry in southern California, (2) resulted in the development of a method of biological control that has become an important tactic in the control of especially troublesome and difficult or costly-to-control pests, (3) divided entomological experts into two distinct and still warring biological and chemical factions within the University of California, and (4) resulted in more formalized, stricter quarantine regulations in the state of California and eventually in Washington.

E. The Mediterranean Fruit Fly Reverses California's Role

Until 1981, California strongly emphasized quarantines to protect its agriculture from bugs that might come from other states. As of July and August of 1981, however, other states have threatened to quarantine California produce to protect their agriculture from a bug that might come from that state, in this case the Mediterranean fruit fly.

The fruit fly was first discovered in 1980 near San Jose, California, the city made famous at the turn of the century by the arrival there of another pest that became known as the San Jose scale. The fly frightens farmers because it attacks a wide range of fruits and vegetables.

The episode put the nation into a very difficult position between the Scylla of the bug taboo and the Charybdis of the poison taboo. It badly needs the fruits and vegetables, but the citizens do not want to be sprayed. The newspapers, radio, and television have had a grand time pitting the chemophobes against farmers and entomologists. *Time* magazine featured it in its issue of August 31, 1981.

When the insect was discovered in 1980, the local regulators lined up on the side of the bug taboo to justify a little conflict with the poison taboo. They would spray with malathion, a compound so mild that it is registered for use on home gardens.

Governor Jerry Brown listened to the chemophobes and refused permission to spray, whereupon, according to *Time,* the farmers referred to the plague as ''Brown Rot.'' By mid-July 1981, the pressure caused by the bug taboo became too strong for the governor, and he reversed his position.

To counter the threat of quarantines by other states, California established an internal quarantine around the infested area. Within the quarantined area it labored strenuously to eradicate the fruit fly. This has succeeded so well that the quarantine was lifted in December 1982.

ACKNOWLEDGMENTS

Paul E. Waggoner has provided aid and comfort in the preparation of the chapter. I am grateful, also to the following for providing useful material: W. R. Horsfall, C. B. Huffaker, W. F. Mai, D. J. Patterson, R. J. Raski, and P. W. Santleman.

REFERENCES

Caporeal, L. R. (1976). Ergotism, the satan loosed in Salem. *Science* **192,** 21–26.
Carefoot, G. L., and Sprott, E. R. (1967). ''Famine on the Wind.'' Rand McNally, New York.

Friend, R. B. (1945). The gypsy moth in Connecticut. *Trans. Conn. Acad. Arts Sci.* **36,** 607–629.

Hitchcock, S. W. (1974). Early history of the gypsy moth in Connecticut. *In* "The 25th Anniversary of the Connecticut Entomological Society" (R. J. Beard, ed.), pp. 87–97. Conn. Agric. Exp. Stn., New Haven.

Horsfall, J. G. (1958). The fight with the fungi: Or the rusts and the rots that rob us, the blasts and the blights that beset us. *In* "Fifty Years of Botany" (W. C. Steere, ed.), pp. 50–60. McGraw-Hill, New York.

Horsfall, J. G. (1965). A socio-economic evaluation. *In* "Research in Pesticides" (C. O. Chichester, ed.), pp. 2–18. Academic Press, New York.

Matossian, M. K. (1982). Ergot and the Salem witchcraft. *Am. Sci.* **70,** 355–357.

Metcalf, C. L., and Flint, W. P. (1962). "Destructive and Useful Insects." McGraw-Hill, New York.

Robbins, W. J., Crafts, A. S., and Raynor, R. N. (1952). "Weed Control." McGraw-Hill, New York.

Scheja, G. (1969). "The Isenheim Altarpiece." Harry N. Abrams Inc., New York.

Stakman, E. C., and Harrar, J. G. (1957). "Principles of Plant Pathology." Ronald Press, New York.

CHAPTER 2

History of Insect Introductions

REECE I. SAILER

Department of Entomology and Nematology
University of Florida
Gainesville, Florida

I. INTRODUCTION

At a time in man's history when many people, as well as governmental agencies at national and international levels, worry about the extinction of animal and plant species, agriculturalists are concerned about additions to the fauna and flora of their agroecosystems. In each case, the source of worry is a consequence of the increasing presence of one highly adaptive and successful organism, *Homo sapiens* L.

ISBN 0-12-757880-3

After several million years of physical and cultural evolution, the human species has acquired the means of destroying a large portion of the earth biota. Although some species have been exploited to extinction, the greater danger is loss of habitats and food resources as the human population increases and requires ever more space for its agricultural and industrial enterprises. Nevertheless, although much of the world's biota is endangered by the growth and expansion of the human population, this same factor affords many plant and animal species the opportunity to surmount geographic and ecological barriers by which they were previously confined to their regions of evolutionary origin. These are the plants and animals that have been carried by commerce. Although a relatively small number are crop plants and domestic animals, a much larger number are ''camp followers''—the weeds, plant and animal pathogens, and insect pests. Despite the number of these pests that have already reached the United States, an equal or larger number await an opportunity to gain entry. In this chapter, I present a historical account of when and how immigrant insects and mites arrived in the United States and became part of our North American fauna. Factors influencing the sources and success of the immigrant species will be considered. Although emphasis is on the injurious species, note will also be taken of those that are of no known importance or are beneficial.

II. BIOGEOGRAPHIC CONSIDERATIONS

The question may be asked, why begin a history of insect introductions into the United States with a discussion of biogeography? After all, we are not primarily interested in those indigenous species whose presence predates the advent of commerce between North America and other land areas of the world. In fact, there would be little occasion for such a discussion if immigrant species could in all cases be recognized as such with certainty. Unfortunately, this is not the case.

From a biogeographic viewpoint, most if not all of the United States is included in the Nearctic Region. Only the southern tip of Florida and the lower Rio Grande Valley of Texas contain sufficient elements of the Neotropical fauna and flora to be included in that region by some authors (Muesebeck *et al.*, 1951). Although proximity to the Neotropical lands of Mexico and the Caribbean presents a problem in determining immigrant status when species from those regions are discovered in Florida or Texas, a much more serious problem is presented by the large number of species common to Europe and North America.

To understand the nature of the problem presented by the species found in both the United States and Europe, it is necessary to consider the close affinity of the Nearctic and Palearctic regions. Biogeographers generally treat these as subregions of a circumpolar Holarctic Region (Cox and Moore, 1980). Faunal and floral similarities are such that, in the case of insects, nearly all families represented in one are also found in the other. Most genera found in Europe will be represented by species in North America; more to the point, a considerable number of species exhibit Holarctic distribution.

To illustrate the magnitude and complexity of the problem presented by the Holarctic component of the Nearctic insect fauna, an examination of the North American order Diptera is informative. Although there is reason to believe that the Diptera may not be entirely representative of all insect groups, it is an order for which a relatively recent catalogue is available (Stone *et al.*, 1965). Also, a considerable number of dipterous species that have been treated as being of immigrant origin show distribution patterns that suggest indigenous origin.

Of the 16,130 species of Diptera recorded from North America, 1151, or approximately 7% of the total, have somehow managed to establish and maintain populations in both Eurasia and North America. This is in striking contrast to the 97 species of flies presently recorded as immigrants to the United States. Further examination of the North American distribution of the 1151 species reveals that 570 of them are recorded from Alaska, with 204 restricted to the Hudsonian Life Zone while 167 extend their range across the Canadian border through the Transition and into the Alleghanian Zone. Of the remainder, 163 reach even farther south into the Upper Austral Zone, with some being found in north Georgia.

This two-continent distribution seems best explained as the result of faunal exchange across the Bering land bridge. Most are cold-adapted species associated with the boreal forest, taiga, and tundra. Although widely distributed across Canada, their distribution in the United States tends to be limited to Maine, New Hampshire, Vermont, and an area including northern Michigan as well as portions of northern Wisconsin and Minnesota. In the West, they form an important component of the alpine Diptera. A relatively small number range from Alaska southward into the western United States and, although not known to occur in central Canada, they are present in the St. Lawrence Valley, Nova Scotia, and coastal areas of New England. Although most of these species are undoubtedly indigenous in the West, some if not most of the eastern populations are introduced.

More puzzling and interesting from a biogeographical viewpoint are the 581 species now restricted to areas of North America and Eurasia that were once characterized by temperate forest and are now highly modified by agriculture. The presence of such a large number of species having such widely disjunct populations raises many questions that are not easily answered. Is the pattern a result of faunal exchange across the Bering land bridge? A land bridge across the Atlantic? Or is it a consequence of continental drift, the populations having once been contiguous in the supercontinent Laurasia? And finally, what proportion, if not all, of the 581 species have been introduced into North America through commerce from Europe?

Serious objections can be raised to each of the possible explanations implied by the above questions. The possibility that appreciably more than the 97 species now listed as introduced are recent arrivals seems unlikely, for most of the remaining 495 species live in habitats or exhibit life history and behavior characteristics that minimize the possibility of movement through commerce.

Continental drift is an attractive explanation if we are willing to accept the status of these species as living fossils. Or, alternatively, it may be argued that any evolutionary changes affecting the species since their populations were isolated have progressed in a manner that has preserved their specific identities. In either event, it will be necessary to postulate a remarkable degree of evolutionary conservatism over a period of at least 65,000,000 years since the separation of North America from the Eurasian land mass (Matthews, 1973).

Repeated existence of the Bering land bridge is not questioned. It is believed to have existed during each of the four periods of ice accumulation in North America that occurred during the Pleistocene (Hopkins, 1959). However, pollen evidence (Colinvaux, 1964) and plant distributions (Hultén, 1937; Hustich, 1953) are purported to show that the central land bridge supported only tundra vegetation, across which the forests of Asia and America did not merge during the Pleistocene. If true, this would mean that any connection across the Bering bridge suitable to explain the present distribution of the 581 species of temperate climate flies common to Europe and North America would have to be moved back to the Oligocene. Again, this would mean that the species involved are remarkably old, having remained unchanged for 30,000,000 years.

A more plausible explanation is that evidence of the arctic or at least subarctic conditions of the Bering land bridge needs revision. Although data from pollen analysis obtained in northern Alaska (Colinvaux, 1964) do support the conclusion that subarctic to arctic conditions prevailed over the exposed Bering–Chukchi platform during the period of maximum Wisconsin glaciation 18,000 years ago, this does not mean that a relatively narrow band along the southern coast of the bridge could not have had a much more temperate climate. In a later paper, Colinvaux (1967) presents data from core samples taken from the Pribilof Islands. In this paper, he reports finding spruce pollen from a time beginning about 9500 years B.P. and continuing back to 35,000 B.P. He concludes that there was a real increase in the amount of spruce pollen falling on Saint Paul Island in late Wisconsin times. This is at least suggestive of climatic conditions along the southern shore of the Bering land bridge that would have allowed passage of boreal if not temperate species.

The presence of such a large number of temperate climate species having distributions now restricted to Europe and North America may be regarded as additional evidence of a period of temperate climate over at least a fairly wide southern band of the Bering land bridge during one or more of its several appearances during the Pleistocene. The alternatives are either to treat a remarkably large number of Diptera and other insect species now resident in the temperate United States as dating from the Oligocene or as recent immigrants. For purposes of this history of introductions they are regarded as indigenous, and the question of how they happen to be where they are remains open.

III. MODES OF ENTRY

Prior to the discovery of America in 1492, the land area of the 48 contiguous U.S. states was isolated by geographic and ecological barriers. After the retreat of the last ice sheet some 15,000 years ago, the fauna and flora of North America settled into biomes and life zones that characterized the continent when the first European colonists began to arrive on its eastern shores. Except for the exploitation of fire in clearing land and hunting game animals, the Indian peoples did little to disturb the natural patterns of the plant and animal communities. With the arrival of the Europeans, the long isolation of North America ended as the geographic and ecological barriers were breeched by commerce. The resulting changes in the North American biota have been as profound as those caused by the glaciers of the Pleistocene.

Where native forests once extended from the eastern seaboard well into the Mississippi Valley and merged into the western prairies, there are now millions of acres of cultivated land and pastures. Most of this land area is supporting plants and domestic animals that are foreign to North America. Although a few crop plants, such as corn, squash, and beans, were grown on a very small scale prior to the arrival of the Europeans, even these plants were originally of Mexican or Central American origin. For the most part, the pests that coevolved with the introduced crop plants and domestic animals did not arrive with their hosts, but as agriculture and commerce expanded, they found their way into the United States and are now estimated to cause 38–50% of all crop losses due to insects (McGregor, 1973). Because of the disproportionate importance of these foreign pests, agriculturists have reason to fear the consequences of invasion by additional foreign species. To reduce the danger of such invasions, Congress approved a Plant Quarantine Act in 1912. Since that date, all plants and most plant products entering the United States have been subject to inspection. When pests are found, appropriate action is taken to prevent their entry. Nonetheless, new foreign pests have been gaining entry with almost predictable regularity (Sailer, 1978).

In 1971 the U.S. Department of Agriculture established a task force to review the effectiveness of plant quarantines in preventing the entry of pests and, more specifically, to define and if possible quantify the risks from the entry of exotic pests and diseases (McGregor, 1973). As part of this effort, a list of immigrant insects and related arthropods was compiled. This included not only those of importance as pests but also those of no known economic importance and those that are beneficial. The list continues to grow as additional species are reported for the first time or found through a continuing literature search. Preparation of this list was undertaken not only as a means of learning what species were already in the United States but also for what might be revealed about their origin and means of arrival.

At the conclusion of the 1971–1972 study, 1115 species were recognized to be of foreign origin. By 1977 this number had increased to 1385 (Sailer, 1978). When the species were arranged in chronological order of the earliest known occurrence, it was evident that the proportional representation of the major orders had changed markedly during the period 1820–1977.

Based on species for which the earliest dates could be established with reasonable confidence, it was found that 90% of the species present in 1820 were of the order Coleoptera. Most of the remainder were mites. By 1840, the proportion of Coleoptera had dropped to about 60%, Lepidoptera had increased to 18%, and Heteroptera and other orders made up the remainder. In the period 1840–1860, Coleoptera contributed only 15% of the arrivals, whereas Lepidoptera had moved up to 40% and Homoptera was in second place with 20%. After 1860, the proportion of Homoptera increased until 1900, when 40% of the species belonged to this order. Subsequent to 1910, the number of Homoptera declined as the proportion of Hymenoptera increased. By 1960, Hymenoptera was contributing 40% of the recently established species, with about equal numbers of Coleoptera and Homoptera making up another 40%. Obviously, there were factors at play that influenced the proportional representation of the orders during the last 150 years.

These factors relate to technological changes affecting commerce, changes in commodities being moved in commerce, and the biological characteristics of the insect species. From examination of the dates of earliest record and from our knowledge of the biological characteristics of the immigrant species, four major modes of entry are discernible: (1) ships' ballast, (2) plant introductions, (3) range extension, and (4) beneficial insect introduction.

A. Role of Ship's Ballast

The early dominance of Coleoptera is readily explained as due to ballast traffic during the era of sailing ships. Although the role of ballast in the introduction of Coleoptera and other organisms commonly associated with soil was recognized by a number of earlier authors, the subject has been thoroughly researched by Lindroth (1957).

Lindroth was able to tabulate 638 species and subspecies of insects common to Europe and North America. Of these, he treated 242 as accidentally introduced to North America, whereas only 18 were moved in the other direction. He attributed this difference to the peculiar character of the ballast traffic. Ships sailed in ballast on their way west to Newfoundland, the Maritime Provinces of Canada, and New England. Toward the end of the sailing ship era, traffic extended around the Horn to the Puget Sound area of the Northwest.

From examination of old port records and other sources, Lindroth established that most ships engaged in the North American trade sailed from a small number of ports in southwestern England. One such port was Poole, from which, as an example of this traffic, 57 ships had sailed for North America in 1815. Of these, 17 were partly or fully loaded with ballast. Records showed that 1180 tons of ballast were loaded on these ships. For the most part, this ballast consisted of rubble and soil obtained from the vicinity of the wharves. At ports of call in America, this ballast was off-loaded as cargo was taken aboard. As a result of intensive collecting and study of local faunal lists, Lindroth was able to show that most of the species included in his list of insects introduced into North America from Europe occurred in southwestern England.

Based on the distribution of the carabid beetles, with which he had the greatest familiarity, Lindroth found two North American zones with the greatest concentration of European species. One of these included the St. Lawrence Valley, Nova Scotia, and Prince Edward Island. The second was centered on Puget Sound. No doubt the similarity of the climate in these areas to that of northern Europe contributed to the successful colonization of the species transported in ballast.

With the growth of industry in the United States and Canada, fewer ships left European ports without full cargoes. After 1880, there is little evidence that ballast contributed significantly to the introduction of additional species from Europe. However, in the period following the Civil War, there was increasing commerce with South American countries, and a number of important pests began to appear in the vicinities of the southeastern and Gulf Coast ports that engaged in the South American trade. Such pests as the fire ants, *Solenopsis invicta* Buren and *S. ritcheri* Forel, the mole crickets, *Scapteriscus vicinus* Scudder and *S. acletus* Rehn and Hebard, as well as the whitefringed beetles, *Graphognathus* spp., and vegetable weevil, *Listroderes costirostris obliquus* (Klug), are all native to the Rio de la Plata river basin of South America and, as soil insects, are believed to have arrived in ballast. The imported fire ant *S. invicta,* first discovered near Mobile, Alabama, in 1941, seems to have been the last important pest to arrive by the ballast route.

B. Role of Plant Introduction

A very large proportion of the immigrant insects now resident in the United States are species that gained entry on plants or plant seeds and fruit. More than one-fifth of all immigrant species are Homoptera (Fig. 2). These are the scale insects, whiteflies, and aphids that cannot survive or will soon die if removed from their host plants. No doubt some of these species were carried with plants during the era of the sailing ships; however, many would have failed to survive the long sea voyage. Also, most of the plants introduced during this period were carried as seeds. Certainly, the proportion of Homoptera among arriving species increased dramatically between 1840 and 1860 (Sailer, 1978, Fig. 2), a period during which the steamship assumed a prominent role in transatlantic commerce. Although the advent of the steamship undoubtedly shortened the time in transit and increased the likelihood that foreign species would reach the United States, there was also increased interest in introducing new kinds of useful plants.

As an agent of the colony of Pennsylvania from 1764 to 1775, Benjamin Franklin sent seeds and cuttings home from England. After gaining independence, the new nation began to establish diplomatic contacts and expand trade relations with other nations throughout the world. Consular officials were encouraged to follow Franklin's example (Powell, 1927). Recognizing the need to establish new crops, Congress as early as 1802 authorized expenditures to encourage the introduction and cultivation of grapes in Ohio. In 1838, land in Florida was made available to encourage the introduction of tropical plants (Powell, 1927), and in 1839, Congress appropriated $1000 to be used to obtain new and rare varieties of plants (Pieters, 1905). Foreign plants and seeds were exempted from customs duty in 1842.

In 1862, the U.S. Department of Agriculture was founded; it had as one of its objectives the introduction of foreign plants. Three years later, the department established a 40-acre garden in Washington, D.C., for propagation of plants and other purposes. About 1897, the Office of Plant Introduction was organized under the leadership of David Fairchild, and in succeeding years several plant introduction gardens were established principally around the perimeter of the United States. One of the largest and most important of these was established at Chico, California, about 1903. These gardens served as quarantine stations, as well as sites where the agricultural value of the crop plants could be tested. After his retirement, Fairchild (1945) wrote that the Office of Plant Introduction had brought nearly 200,000 named species and varieties of plants into the United States.

Also, following the Civil War, wealth accumulated in the industrial regions of the northern United States. As fine homes were built, there was an active demand for nursery stocks of ornamental trees and shrubs. European nurserymen found a ready market for their plants. These entered the United States with little or no safeguard against the introduction of pests until passage of the Plant Quarantine Act in 1912.

Although insects of most orders continued to arrive and become established in the United States during the period 1840–1920, Homoptera comprised 34% of the total number of species, and most if not all arrived at immature or mature stages on plants. Most of the 194 species became pests, among which the more important are the California red scale, *Aonidiella aurantii* (Maskell); Florida red scale, *Chrysomphalus aonidum* (L.); citricola scale, *Coccus pseudomagnoliarum* (Kuwana); citrus whitefly, *Dialeurodes citri* (Ashmead); cloudywinged whitefly, *D. citrifolii* (Morgan); cottonycushion scale, *Icerya purchasi* Maskell; purple scale, *Lepidosaphes beckii* (Newman); green peach aphid, *Myzus persicae* (Sulzer); citrus mealybug, *Planococcus citri* (Risso); white peach scale, *Pseudaulacaspis pentagona* (Targioni-Tozzetti); and greenbug, *Schizaphis graminum* (Rondani).

After enactment of the 1912 Plant Quarantine Act, there was no immediate decline in the number of new immigrant species recorded on an annual basis: however, beginning in 1920, a decline is noted. Where prior to 1920 annual numbers tended to increase exponentially, those following 1920 fall in a straight line, as shown in Fig. 1. The marked reduction in the number of species established each year following 1920 undoubtedly reflects the deterrent effect of the 1912 Plant Quarantine Act, with many of the species recorded between 1912 and 1920 having been established prior to 1912 but not discovered until a later date.

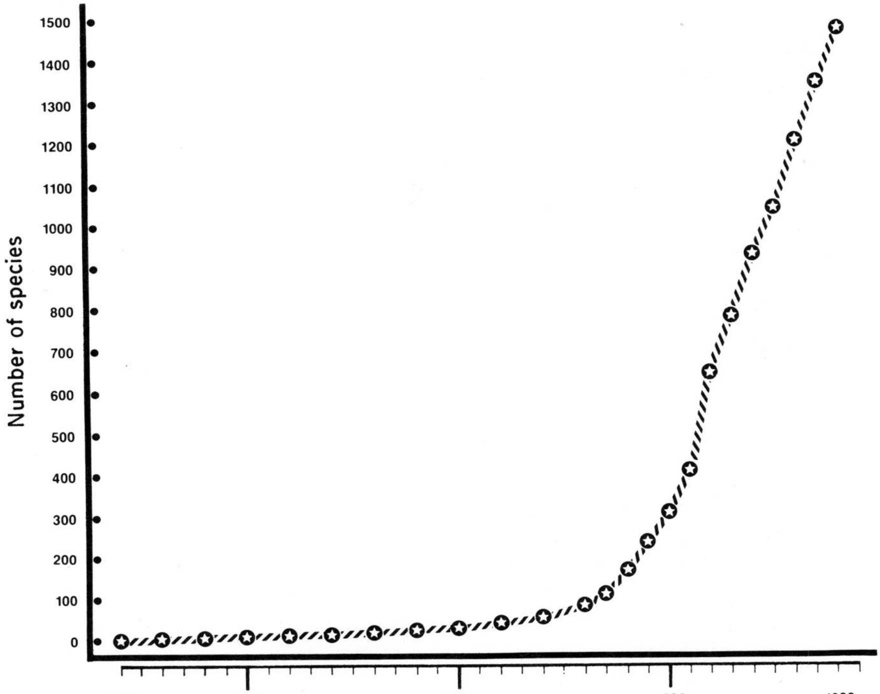

Fig. 1. Number of immigrant species of insects and mites resident in the 48 contiguous states at time intervals from 1640 to 1980. (Illustration by Limhout Nong.)

C. Beneficial Insect Introductions

Beginning about 1920, the number of Hymenoptera being reported for the first time as established in the United States began to increase, and from 1920 to 1970 constituted the dominant component of the annual increment of immigrant species. In large part, these represent intentional introductions. They are species that act as predators, or more often parasitoids, of insect pests that are themselves of foreign origin. They reflect the results of parasite exploration activities of the U.S. Department of Agriculture and the University of California, with a minor recent input from Florida, Texas, and Virginia. Although parasite introduction work is generally regarded as having started with importation of the vedalia beetle *Rodolia cardinalis* (Mulsant) for control of the cottonycushion scale, peak activity was during the period 1920–1940; however, substantial numbers continued to be introduced, and a number of those released prior to 1940 were not reported to be established until a later date. Thus, Hymenoptera remain a major contributor to the growing list of immigrant species.

It may surprise some biological control workers to learn that of 232 beneficial Hymenoptera of record in 1982, slightly more than 35% arrived through accident. Of the remaining 150, about 140 were introduced by the U.S. Department of Agriculture and the University of California, and 10 are species intentionally introduced into Canada that subsequently extended their range into the United States.

D. Range Extension

This is a troublesome category of immigrant species that have entered the United States from Canada, Mexico, and Cuba. Of those entering from Canada, most are European species introduced there at an earlier date. They are characteristically species that were moved in ballast and first established in the St. Lawrence Valley or the Maritime Provinces. Most are relatively innocuous or even beneficial beetles. However, the European crane fly, *Tipula paludosa* Meigen, now in Washington, is undoubtedly of this origin. The winter moth, *Operophtera brumata* (L.), also recently established in Washington and Oregon, is another European species that has entered the United States by way of Canada. Although there are a number of other such species, perhaps the most notable being the face fly, *Musca autumnalis* De Geer, most of the traffic has been in the other direction, with immigrant pest species moving from the United States to Canada.

On the border with Mexico and to a lesser degree south Florida, foreign species of two categories have invaded the United States. One of these consists of species not native to Mexico or Cuba but which, once established there, have extended their range into the United States. A particularly noteworthy example is the pink bollworm, *Pectinophora gossypiella* (Saunders). Originally an Indian species, it first invaded Mexico and from there entered Texas about 1920. The citrus blackfly, *Aleurocanthus woglumi* Ashby, is another such species. As a species of southeast Asian origin, it was first reported from Jamaica, Cuba, and later Mexico. In 1934, it entered the Florida Keys but was eradicated. In 1955, it moved into Texas from Mexico and was eradicated. However, it reinvaded Texas in 1969 and Florida in 1976. All infestations have been controlled by parasites introduced from India and southeast Asia.

The second range extension category involving Mexico and Cuba includes species native to those countries. Many of these are immigrant only in the sense that they are tropical forms having an extreme northern range that more or less coincides with south Texas or southern Florida. In favorable years, they extend their range into the United States. During years when conditions are less favorable, their range contracts. The Mexican fruit fly, *Anastrepha ludens* (Loew), seems to be such a species. A more troublesome category are those species that have been able to move into the United States and exploit a niche created by expansion of U.S. agriculture. Here we find several major pests, among which the cotton boll weevil, *Anthonomus grandis* Boheman, is perhaps the best documented. The Mexican bean beetle, *Epilachna varivestis* Mulsant, is a second example, having invaded the western United States from Mexico and gained entry to the eastern states through Mobile, Alabama. Supposedly of Central American origin, this beetle has shown little evidence of climatic limitation and has successfully invaded most areas of the United States where beans are grown. Another important pest of this category, although obviously more limited by climate, is the Caribbean fruit fly, *Anastrepha suspensa* (Loew). This species invaded Florida from Cuba in 1931 but did not become a pest of significance until the 1960s, when it spread from the Florida Keys into the peninsula. Seemingly, this is a case of climatic adaptation that has enabled the species to expand its range.

IV. INCREMENTAL INCREASE IN FOREIGN SPECIES, 1620–1980

A. Species Established before 1800

By 1800, the English colonies on the Atlantic Coast of North America from Maine to Georgia had consolidated into the United States and settlements were pushing westward into the Ohio Valley. At the beginning of the Revolutionary War, the population of this area amounted to little more than 3,200,000. In addition, approximately 54,000 English- and French-speaking people and 25,000 of Hispanic origin lived in areas not yet part of the United States. With such a sparse and widely scattered population, the growth of agriculture was slow and mostly of a subsistence nature. Apparently, the crops of these early colonists did not suffer seriously from insect pests. During the Revolutionary War, no more than six injurious insects had attracted attention. One of these was the Hessian fly, *Mayetiola destructor* (Say), first mentioned in 1778. A second was the codling moth, *Laspeyresia pomonella* (L.), and in North Carolina and Virginia there were reports of damage caused by the Angoumois grain moth, *Sitotroga cerealella* (Olivier). One homopteran, the oystershell scale, *Lepidosaphes ulmi* (L.), is also known to have been introduced as early as 1796 (Webster, 1892). The peachtree borer, *Anarsia lineatella* Zeller, is also known to have been in Pennsylvania before 1800 (Howard, 1930).

No doubt, if one were to read diaries of the period or search out old newspapers and almanacs, it would be evident that many more foreign insects and mites were already well established by 1800. Based on knowledge of their presence in England and the certainty that they would have been present in cargoes, ship stores, and personal effects of the people on the ships sailing for America, at least 28 foreign species would have been present in the port cities of the eastern seaboard. These would have included five mites, four of them pests of stored products, and one predacious mite that would have accompanied the grain mite, *Acarus siro* (L.). The oriental cockroach, *Blatta orientalis* (L.), is known to have been established on the eastern seaboard as early as 1758, and two Hymenoptera species that parasitize their eggs would have left the cockroach-infested ships and established themselves ashore as soon as their hosts became well established in the nearby taverns and inns. The bed bug, *Cimex lectularius* (L.), is reported to have appeared in England in 1503. By 1730, it was reported to be abundant in all seaport towns (Southwood and Leston, 1959). Ships leaving these ports for America would have soon transported the bugs to the colonial seaports. The honey bee, *Apis mellifera* (L.), was also an early arrival, having been reported to be in Virginia in 1622, and the town of Newbury, Massachusetts, established a municipal apiary in 1640 (Townsend and Crane, 1973). Among the Coleoptera, the pea weevil, *Bruchus pisorum* (L.), was reported as a pest in Pennsylvania as early as 1756, and at least two dung beetles, *Aphodius granarius* (L.) and *A. lividus* (Olivier), are believed to have arrived aboard ships carrying cattle from England to the colonies during the seventeenth century. Many additional species not reported until much later, in the nineteenth century, undoubtedly arrived in the eighteenth or even the seventeenth centuries but because of their innocuous character went unnoticed.

B. Species Established 1800–1860

Only 70 additional species have been listed as established between 1800 and 1860. Although 25 of these were Coleoptera, many are species that undoubtedly arrived much earlier and were reported only during this period. The more important pests belonging to this order were the bean weevil, *Acanthoscelides obtectus* (Say); cowpea weevil, *Callosobruchus maculatus* (F.); asparagus beetle, *Crioceris asparagi* (L.); Mexican bean beetle, *Epilachna varivestis* Mulsant; elm leaf beetle, *Pyrrhalta luteola* (Muller); strawberry root weevil, *Otiorhynchus ovatus* (L.); and black vine weevil, *O. sulcatus* (F.). Among the Diptera, the seed corn maggot, *Hylemya platura* (Meigen), was reported in 1865. Of the species of Homoptera reported from this period, four are major pests, including the purple scale, *Lepidosaphes beckii* (Newman); black pine aphid, *Pineus strobi* (Hartig); and pear psylla, *Psylla pyricola* Foerster. Among the Lepidoptera there was the imported cabbage worm, *Pieris rapae* (L.); and the eyespotted bud moth, *Spilonota ocellana* (Denis and Schiffermüller).

C. Species Established 1860–1910

As early as 1845, Asa Fitch, state entomologist for New York, recognized the importance of foreign species as crop pests and urged that natural enemies of the wheat midge, *Sitodiplosis mosellana* (Gehin), be imported from Europe (Howard, 1930). The continued discovery of new foreign insect pests and the increasing severity of insect pest problems undoubtedly contributed to the establishment of a Division of Entomology in the U.S. Department of Agriculture in 1863. Ten years earlier, Townsend Glover had received an appointment to the U.S. Patent Office as an "Expert for collecting statistics and other information on seeds, fruits, and insects in the United States." In 1865 he wrote that "it is well known that several of the insects most destructive to our crops are of European origin" and then urged that all foreign seeds and plants entering the country be examined and any insects found be destroyed (Howard, 1930).

Two pest problems commanded national attention in the first 2 decades of this period. Both were foreign to the areas affected, although they originated in the United States. During the 1850s, the Colorado potato beetle, *Leptinotarsa decemlineata* (Say), extended its range from Colorado across the western plains, and in the 1860s threatened potato production throughout the Middle West and the East. By 1870, the beetle was being controlled through the use of Paris green. As the consternation caused by the Colorado potato beetle began to subside, there was an extraordinary excursion of the Rocky Mountain locust, *Melanoplus spretus* (Walsh), which devastated crops in Kansas, Nebraska, Iowa, Texas, Oklahoma, and part of Missouri during the years 1874–1876. This resulted in establishment of a commission, with C. V. Riley as chairman, and his appointment as Entomologist to the U.S. Department of Agriculture in 1878, a position that expanded into the Bureau of Entomology, of which L. O. Howard became chief in 1894.

During the half-century following 1860, the United States experienced almost exponential growth. The population increased from 31,440,321 to 91,972,266.

Agricultural production quadrupled. Manufacturing increased sevenfold between 1863 and 1899. Foreign commerce increased at a similar or even greater rate. The number of immigrant insects established during the period shows a similar trend, with 4.5 times more species recorded in 1910 than in 1860. It is also significant that of the 475 species arriving in this 50-year period, 152, or 32%, were reported during the last decade.

Over the entire period, 187 species of Homoptera were added: of these, 135 were scale insects, whiteflies, and aphids of economic importance as pests. One of these was the cottonycushion scale, *Icerya purchasi* Maskell, first discovered in California in 1868. The highly successful control of this pest through introduction of the vedalia beetle, *R. cardinalis* (Mulsant), in 1888 marks the beginning of world interest in biological control.

The influx of so many foreign pests during this period, and the fact that they included such important species as the cotton boll weevil, *A. grandis* Boheman, 1892; the San Jose scale, *Quadraspidiotus perniciosus* (Comstock), 1879; and the gypsy moth, *Lymantria dispar* (L.), 1869, ensured continued public support for development of entomology in the United States. The number of foreign insect pests and evidence that their number increased annually caused L. O. Howard, chief of the Bureau of Entomology in the U.S. Department of Agriculture, and entomologists of several states to urge the establishment of a federal inspection and quarantine system. Discovery in 1910 of the European corn borer, *Ostrinia nubilalis* (Hubner), and an embarrassing situation arising in the same year from discovery of the oriental fruit moth, *Grapholitha molesta* (Busck), on flowering cherry trees sent as a gift to the American people from Japan created additional pressure for regulatory measures against foreign pests. This resulted in the enactment of the Federal Plant Quarantine Act in 1912 (Howard, 1930).

D. Species Established 1910–1980

Although passage of the 1912 Plant Quarantine Act serves as an important point of reference in a discussion of foreign insect pests in the United States, its effect on the statistics of newly arrived species is less than spectacular. As indicated in Fig. 1, no effect is discernible until the decade following 1920. This might be explained as resulting from the discovery of species that had actually become established prior to 1912. However, it should also be noted that the department's inspection and quarantine services were not fully operational until after World War I (Marlatt, 1920). Actually, several very serious pests became established during the decade 1910–1920, including the European corn borer, *Ostrinia nubilalis* (Hubner), 1917; Japanese beetle, *Popillia japonica* (Newmann, 1916; oriental fruit moth, *Grapholitha molesta* (Busck), 1913: and the European pine shoot moth, *Rhyacionia buoliana* (Denis and Schiffermüller), 1913. An astonishing number (12) of serious mite pests were all first reported in 1918, whereas 15 species of important aphids and scale insects are more evenly distributed throughout the decade. The northern cattle grub, *Hypoderma bovis* (DeGeer), 1912, is another important pest from this period.

Following the remarkable influx of 187 immigrant species in the period 1910–1920, there was a substantial drop in the number reported as becoming established on an annual basis; however, between 1920 and 1980, an additional 837 species were recorded. Approximately 10% of these are important pests and include such species as the pink bollworm, *Pectinophora gossypiella* (Saunders), 1920; vegetable weevil, *Listroderes costirostris obliquus* (Klug), 1922; white-fringed beetle complex, *Graphognathus* spp., 1936–1942; citrus blackfly, *Aleurocanthus woglumi* Ashby, 1934 (eradicated), 1955 (eradicated), 1971; Mediterranean fruit fly, *Ceratitis capitata* (Wiedemann), 1929 (eradicated), 1956 (eradicated), 1966 (eradicated), 1975 (eradicated), 1980–1981 (eradicated?); melon fly, *Dacus cucurbitae* Coquillett, 1956 (eradicated); oriental fruit fly, *Dacus dorsalis* Hendel, 1966 (eradicated); Egyptian alfalfa weevil, *Hypera brunneipennis* Boheman, 1939; eastern strain of the alfalfa weevil, *Hypera postica* (Gyllenhall), 1951; face fly, *Musca autumnalis* De Geer, 1954; khapra beetle, *Trogoderma granarium* Everts, 1953, eradicated after a long campaign, but again reported in 1982; and the cereal leaf beetle, *Oulema melanopus* (L.), 1962. These are but a few of the 80 important insect and mite pests that have arrived since 1920.

Although after 1920 the trend changed from nearly exponential increase to one characterized by approximately equal numbers in each successive decade (Fig. 1), new pests have continued to arrive with almost predictable regularity. Of about 14 species added annually, three have been intentionally introduced. Of the remaining 11, four can be expected to have little or no economic importance and seven will likely be injurious in some degree, although one or two might be beneficial in a given year. About every third year, a pest of major importance is included. Recently, these have tended to be repeaters, with the 1980 reinvasion of California by the Mediterranean fruit fly following eradication of earlier infestations being a prime example.

V. COMPOSITION OF PRESENT IMMIGRANT FAUNA

Sixteen insect orders, together with Acarina (mite) and Araneae (spiders), are represented among the 1683 species presently included in the file developed over the past 12 years (Fig. 2). Although this represents an increase of only 298 above the number reported in 1978 (Sailer, 1978), a much larger number has actually been added. The difference is made up of species removed for various reasons. Also, publications of two major reference works, *Introduced Parasites and Predators of Arthropod Pests and Weeds: A World Review* (Clausen, 1978) and the *Catalog of the Hymenoptera of North America, North of Mexico* (Krombein

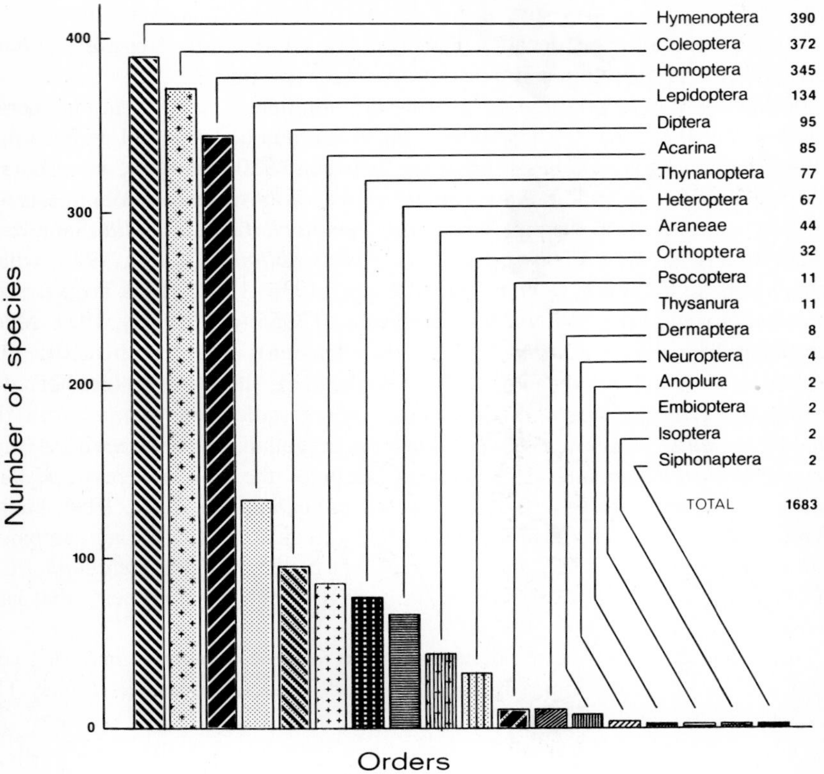

Fig. 2. Number of species in each of the orders represented in the immigrant arthropod fauna of the 48 contiguous states. (Illustration by Limhout Nong.)

et al., 1979), have made available much previously inaccessible information regarding both purposely introduced and adventitious species. The result has been the addition of a disproportionate number of Hymenoptera to the list and a shift of Homoptera from first position to third. Also, special attention to the spider literature has resulted in moving the Araneae from 15th to 9th place. Otherwise, there has been little change in the proportion of species, as shown by Sailer (1978). Three orders account for 66% of the total number of species. If intentionally introduced species are removed, Homoptera becomes the most numerous order (345), with Coleoptera a close second (342) and Hymenoptera falling to third place (233). Of the 134 Lepidoptera, nine are intentional introductions. As might be expected in view of the parasitic and predacious habits of many Diptera, almost 30% of the 95 immigrant species have been intentionally introduced. Of the remaining orders, only one species of Thysanoptera and one species of Heteroptera are intentional introductions.

Changes in Systematic Composition through Time

There have been pronounced changes in the kinds of foreign insects that have gained entry and become established during the past 360 years. This has already been discussed in Section III. These changes reflect changes in the technology of transport, with the time in transit reduced from several weeks to a few hours. Where once the most serious obstacle to colonization was the ability to survive a long sea voyage, today's insect traveler is more likely to find success limited by the availability of food or the ability to find a mate. From the early dominance of Coleoptera transported in ballast, and the shift toward predominance of Homoptera during the 50-year period of intensive unregulated plant introduction activity prior to passage of the 1912 Plant Quarantine Act, the order affinities of invading species have become much less predictable. Of those species not intentionally introduced, data for the period 1970–80 indicate that Coleoptera (25) and Homoptera (24) are still dominant orders; however, Heteroptera (13) has moved up to third place, closely followed by Lepidoptera (11). Most of the remaining immigrant species during this period are accounted for by Acarina (9) and Diptera (8).

VI. ECONOMIC STATUS OF IMMIGRANT INSECTS AND MITES

Schwartz and Klassen (1981) have reviewed the subject of losses caused by insects and mites to agricultural crops. They indicate that more than 10,000 species cause losses, but only about 600 are serious enough to warrant control measures each year. The 235 important pest species, as shown in Fig. 3, would be included in Schwartz and Klassen's 600.

Crop losses attributable to individual pest species are notoriously difficult to fix on a national scale. Attempts to establish standard reporting procedures have been defeated by the number and diversity of pests involved. Many of the pests attack several commodities, and the importance of many varies so much from place to place and from time to time that workers in different states cannot agree on what should be included in a list of important pests. Most often, losses are cited in terms of damage caused by a complex of species affecting a commodity, which in turn may be a composite, i.e., forage or stored products.

Despite the general view that foreign species are responsible for a disproportionate share of crop losses due to insects and mites, it is difficult to determine the actual extent of losses due to the foreign species. McGregor (1973) estimated that they were responsible for two-fifths of all losses caused by insects and mites. Sailer (1978) placed such losses at 50%.

In an effort to fix more firmly and document the proportion of losses due to foreign species, a series of summaries of "Estimated Damage and Crop Loss Caused by Insect/Mite Pests (1968–75)" prepared by the California Department

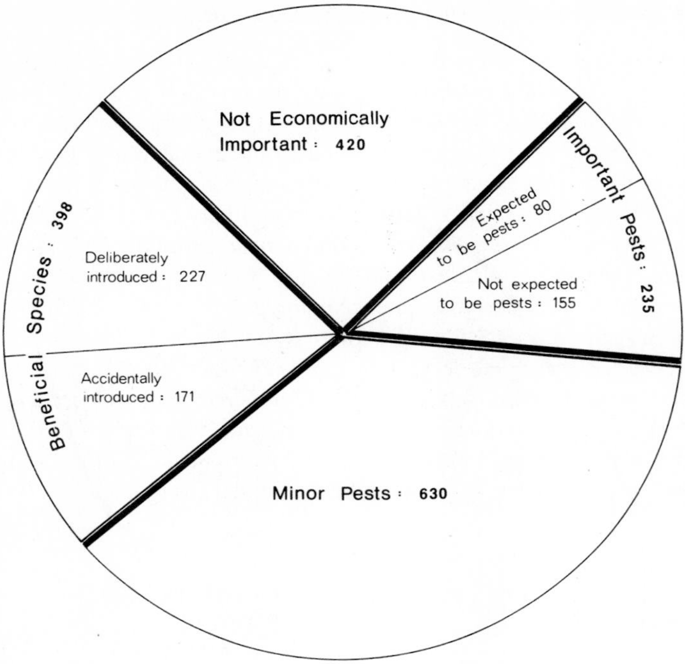

Fig. 3. Economic importance of immigrant arthropod species now resident in the 48 contiguous states, with the number of species in each economic category. (Illustration by Limhout Nong.)

of Agriculture have been examined. Of all the annual crop loss estimates prepared by various states, these provide the most complete data for individual pest species. Even here, some losses are attributed to "Aphids (nonspecific)" and "Mites (nonspecific)." However, by using losses credited to named species, it is possible to fix the amounts due to foreign and native species. Whether the proportion of losses due to foreign species obtained from the California data is applicable nationally is debatable. Although California is number one in agricultural production, the kinds of crops produced and the production practices followed suggest that foreign pest species may play a more important role here than elsewhere in the United States. However, 50% of the species for which losses were reported are foreign, and the reports show a remarkable year-to-year consistency. Approximately 38% of all losses were attributed to species of unquestioned foreign origin. This is increased to 50% if data for one species of questionable origin are included, the species in question being the corn earworm, *Heliothis zea* (Boddie). Although it is generally regarded as a native, certain lines of evidence strongly suggest that the corn earworm has extended its range into the United States from southern Mexico and Central America. In considering the California data, it should also be noted that several of the more important species, although native to the United States, were not originally found in California, i.e., the grapeleaf skeletonizer, *Harrisina brillians* Barnes and McDunnough.

When viewed nationally, foreign species (235) comprise 39% of approximately 600 important pest species. An additional 630 species have been added to the list as pests of lesser importance. Another 420, or almost 25% of the immigrant fauna, are species of no known importance, and the remaining 398 are in some degree beneficial. Of this last category, 171 (43%) are accidental introductions and include important pollinators, dung removal species, predators (particularly predacious mites), and several valuable parasites, such as *Encarsia perniciosi* (Tower). This aphelinid wasp was described from specimens reared from the San Jose scale, *Quadraspidiotus perniciosus* (Comstock), collected at Amherst, Massachusetts, in 1912 (Tower, 1913). The species has since been found to be widely distributed in the Eastern Palearctic and Oriental regions, and evidently gained entry to the United States on plants infested by the San Jose scale.

VII. GEOGRAPHIC ORIGINS OF IMMIGRANT FAUNA

On the basis of 1236 species whose origin was established in 1978 (Sailer, 1978), it has been determined that the Western Palearctic contributed 58%. As additional species have been added to the list, the effect has been to augment further the Western Palearctic component. This now comprises 66.2% of the present list of 1383 immigrant species (Fig. 4). Although most of the percentage increase has resulted from the addition of species found as the result of literature search, a number have been added at the expense of the Eastern Palearctic Region. These are species known to occur in both Europe and Japan. Where the earliest known occurrence in the United States preceded 1940, these species are treated as originating in Europe. After 1945, such species were more likely to enter from Japan or Korea.

Most of the species added since 1978 are considered minor pests or have no known economic importance. Of those that were accidentally introduced and beneficial, most are predacious mites added as a result of recent revisionary studies that revealed their presence and origin for the first time. Of the 467 species coming from regions other than the Western Palearctic, two-thirds are about equally divided between the Northern Neotropical and Oriental regions. However, it should be noted that some of our more serious pests, such as the mole crickets, *Scapteriscus vicinus* Scudder and *S. acletus* Rehn and Hebard, came from temperate South America, a region that has contributed only 1.7% of

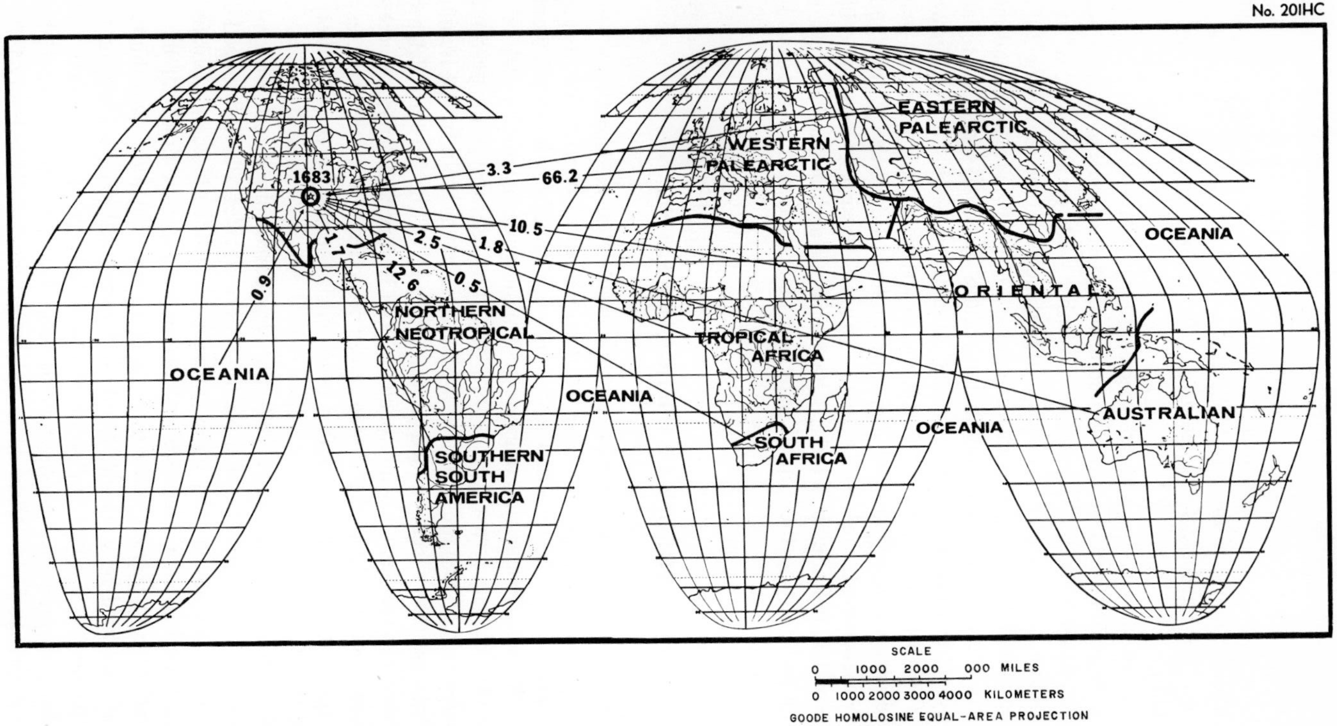

Fig. 4. Origins of the immigrant arthropod fauna, with the total number of recorded species and the percentage of species contributed by different world biogeographic regions. (Prepared by Limhout Nong.)

the species in the present immigrant file. Although now found in many tropical and subtropical countries, the Mediterranean fruit fly is believed to have originated in tropical (East) Africa, another region that has contributed few species to the immigrant fauna of the United States.

VIII. CONCLUSIONS

After more than 10 years of effort involving the collaboration of many people, a list has been compiled of 1683 species of insects and other arthropods of foreign origin now resident in the 48 contiguous states of the United States. At the rate that additional species are being found through a search of the literature, it seems likely that the list will eventually include about 2000 species. This will approach 2% of the total number of insects, mites, and spiders known to occur in the continental United States. About 600 species of insects and mites are of sufficient importance as pests to require annual application of some control measures at one or more locations in the United States. Of these, 39% are species of foreign origin, and depending on whether one very important pest, the corn earworm, *Heliothis zea* (Boddie), is included or excluded from the immigrant category, foreign species account for either 50% or 38% of all losses due to insect and mite pests. That this figure is not much higher is due to the activity of beneficial parasites and predators that have been intentionally or in some cases accidentally introduced.

Despite the deterrent effect of plant quarantine regulations that have been in force for more than 60 years, accidentally introduced foreign species continue to be added to the fauna of the 48 contiguous states at the rate of about 11 per year. Of the 11, seven are likely to be pests of some importance, and about every third year a pest of major importance is discovered. In recent years, the latter category has tended to be repeaters—species such as the Mediterranean fruit fly, which became established and subsequently eradicated on five occasions since 1929.

There remains a very large reservoir of foreign species of insects and mites that are potentially injurious to American agriculture. The exact number cannot be established. In part, this is because of the low predictability of the economic importance of species when moved from native to alien environments. Based on information available prior to the time when each of the 235 important foreign insect and mite pests were found in the United States, no more than one-third would have been expected to be serious pests (see Fig. 3). The Japanese beetle, *Popillia japonica* Newman, and the cereal leaf beetle, *Oulema melanopus* (L.), are examples of such species. Although some of these species may have found unusually favorable conditions of food and climate, more often their ability to increase to destructive numbers has been due to their escape from specialized natural enemies. The number of such species that have been controlled following introduction of parasitic or predacious species is evidence of the role of natural enemies in regulating the abundance of plant-feeding species in their countries of origin. These are also species that had no known economic importance prior to their establishment in the United States.

Thus, there are two categories of foreign insects and mites of potential danger to American agriculture. The first is a relatively small number of species having a known high potential for damage, such as the Mediterranean fruit fly and the khapra beetle, but also including a larger number of species that are known to be destructive to crop plants grown in the United States. There are about 600 such species (McGregor, 1973). They can be looked for and identified when found, and regulatory action taken for their exclusion. Unfortunately, the second category is in a sense more dangerous than the first, for it includes the species for which there are no records of economic importance. The number of such species is large, at least 10 times that of the known foreign species of potential importance. More disquieting is the knowledge that many of the 6000 suspect species will belong to poorly known faunas containing a high proportion of undescribed species.

At the current rate of establishment of additional foreign species, it is obvious that the probability of establishment of most of the known and suspect pest species in a given year is very low. At the same time, it is evident that except for the deterrent effect of plant quarantines, the number would be much greater. In the absence of regulatory measures, certain high-risk species, such as the Mediterranean fruit fly, khapra beetle, and screwworm fly, *Cochliomyia hominivorax* (Coquerel), would within a year invade the 48 contiguous states and quickly infest all areas climatically suited to the species.

Above all else, one thing is clear: If the hazard of invasion by potentially injurious foreign insects and mites is to be held to the present level, far more information is needed about the species that are associated with American crop plants growing in other parts of the world. This is particularly true of the Eastern Palearctic and Oriental regions, where the reservoir of suspect pest species for major crops such as the soybean is particularly large. This becomes even more serious as trade with China grows and increasing numbers of American tourists visit that country.

In addition to the need for more knowledge to assess the full magnitude of the problem posed by invading foreign insects and mites and the specific actions that can be taken to minimize the hazard, there is a need to ensure that the traveling public is fully aware of the problem. In this respect, the recent successful, but costly, campaign to eradicate the medfly from California has been of great public service. As a result of national news coverage, few Americans are now unaware of the threat of foreign insect and mite pests to American agriculture.

ACKNOWLEDGMENTS

Compilation of the data and information on which this discussion of arthropods immigrant to the 48 contiguous states is based was conducted in cooperation with The Insect Identification and Beneficial Insect Introduction Institute, ARS, USDA, with the *Florida Agricultural Experiment Station Journal,* and with partial support of funds provided by Specific Cooperative Agreement No. 58-32U4-2-357, USDA-ARS, Northeastern Region.

REFERENCES

Clausen, C. P., ed. (1978). "Introduced Parasites and Predators of Arthropod Pests and Weeds: A World Review," Agric. Handb. 480. U.S. Department of Agriculture, Washington, D.C.

Colinvaux, P. A. (1964). Origin of ice ages: Pollen evidence from arctic Alaska. *Science* **145,** 707–708.

Colinvaux, P. A. (1967). Bering land bridge: Evidence of spruce in late-Wisconsin times. *Science* **156,** 380–383.

Cox, C. B., and Moore, P. D. (1980). "Biogeography, An Ecological and Evolutionary Approach," 3rd ed. Wiley, New York.

Fairchild, D. G. (1945). "The World Was My Garden: Travels of a Plant Explorer." Scribner's, New York.

Hopkins, D. M. (1959). Cenozoic history of the Bering land bridge. *Science* **129,** 1519–1528.

Howard, L. O. (1930). A history of applied entomology. *Smithson. Misc. Collect.* **84,** Publ. 3065, 1–564.

Hultén, E. (1937). "Outline of the History of Arctic and Boreal Biota During the Quaternary Period." Bokoer. Aktieb., Thule, Stockholm.

Hustich, I. (1953). The boreal limits of conifers. *Arctic* **6,** 149–162.

Krombein, K. V., Hurd, P. D., and Smith, D. R., eds. (1979). "Catalog of Hymenoptera in America North of Mexico," 3 vols. Smithson. Inst. Press, Washington, D.C.

Lindroth, C. H. (1957). "The Faunal Connections Between Europe and North America." Almqvist & Wiksell, Stockholm.

McGregor, R. C. (1973). "The Emmigrant Pests," Report to Administration. APHIS, USDA, Hyattsville, Maryland.

Marlatt, C. L. (1920). Federal plant quarantine work and cooperation with state officials. *J. Econ. Entomol.* **13,** 179–180.

Matthews, S. W. (1973). The changing earth. *Natl. Geogr. Mag.* **143,** 1–37.

Muesebeck, C. F. W., Krombein, K. V., and Townes, H. K. (1951). "Hymenoptera of America North of Mexico, A Synoptic Catalogue," U.S. Dep. Agric., Agric. Monogr. 2. U.S. Govt. Printing Office, Washington, D.C.

Pieters, A. J. (1905). The business of seed and plant introduction and distribution. *U.S. Dep. Agric. Yearb.* pp. 291–306.

Powell, F. W. (1927). "The Bureau of Plant Industry; Its History, Activities, and Organization." Johns Hopkins Press, Baltimore, Maryland.

Sailer, R. I. (1978). Our immigrant insect fauna. *Bull. Entomol. Soc. Am.* **24,** 3–11.

Schwartz, P. H., and Klassen, W. (1981). Estimate of losses caused by insects and mites to agricultural crops. *In* "CRC Handbook of Pest Management in Agriculture" (D. Pimentel, ed.), Vol. I, pp. 15–77. CRC Press, Inc., Boca Raton, Florida.

Southwood, T. R. E., and Leston, D. (1959). "Land and Waterbugs of the British Isles." Frederick Warne & Co., Ltd., London.

Stone, A., Sabrosky, C. W., Wirth, W. W., Foote, R. H., and Coulson, J. R. (1965). "A Catalog of the Diptera of America North of Mexico," Agric. Handb. 376. U.S. Govt. Printing Office, Washington, D.C.

Tower, D. G. (1913). A new hymenopterous parasite on *Aspidiotus perniciosus* Comst. *Ann. Entomol. Soc. Am.* **6,** 125–126.

Townsend, G. F., and Crane, E. (1973). History of apiculture. *In* "History of Entomology" (R. F. Smith, T. E. Mittler, and C. N. Smith, eds.), pp. 387–406. Annual Reviews, Inc., Palo Alto, California.

Webster, F. M. (1892). Early published references to some of our injurious insects. *Insect Life* **4,** 262–265.

CHAPTER 3

History of Plant Pathogen Introductions

C. E. YARWOOD*

Plant Pathology Department
University of California
Berkeley, California

*Deceased. Send requests for reprints to A. R. Weinhold, Chairman, Department of Plant Pathology, University of California, Berkeley, California 94720.

ISBN 0-12-757880-3

I. HYPOTHESIS

In the United States and in all of North America, most living pathogens of agricultural crops have been or will be introduced from abroad by man. Documenting this hypothesis with logic and history will be the major objective of this chapter. Other hypotheses are that pathogens came from the native vegetation (Bennett, 1952) or from the environment; and/or that they entered by the natural movement of air, water, insects, or other vectors; or that they are primarily a result of a predisposition to disease caused by agricultural and industrial operations. All of these possibilities are real. Part of the problem is to estimate which of these and other possibilities are most likely in each situation. Pathogens, unlike insects, have no autonomous movement, or such movement is only a few centimeters per year.

This hypothesis has been advanced previously in other treatises (Orton, 1914; Smith *et al.,* 1933; Fairchild, 1938; McCubbin, 1946, 1954; McGregor, 1973; Cherrett and Sagan, 1977; Ebbels and King, 1979). It follows naturally from the facts that most food plants in the United States are introduced by man from abroad, that most pathogens have a relatively small host range, and that plants may carry pathogens.

II. LOGIC

There is a lack of hard evidence. With few exceptions, the time and mode of entrance of plant pathogens are unknown. The deliberate introduction of pests into new areas is rare, but does occur (Gregory, 1979). As stated by Gregory (p. 269), "History is concerned with the reconstruction and reasonable interpretation of past events from often fragmentary records." Much of the history of past introductions and most of the reasons for quarantine are based on logic, but much of the logic is based on reasonably well-established facts.

The source of pathogens and the time of their entry is logically based on their time of appearance at different places. We believe that wheat stem rust, Dutch elm disease, and potato wart came from Europe to America partly because they were recorded in Europe before they were recorded in the United States. In these examples, I believe that the conclusion based on the above and other evidence is sound, but it is dangerous to extend this logic.

Verticillium albo-atrum (or *V. dahlia,* because the two species are confused) was recorded in Europe in 1816 or 1879 but not clearly in North America until 1926, although California residents believe that a disease caused by this fungus was present by 1910 (Rudolph, 1931). However, the fact that this fungus has a wide host range (Rudolph, 1931) and occurs in virgin land (Ashworth and Zimmerman, 1976) indicates that it may be endemic. Similarly, tobacco necrosis was first reported in England (Smith and Bald, 1935), but has a wide host range and was recovered from virgin soils in the United States (Gold, 1960). In the two above cases, there is no firm basis for deciding whether these are introduced or endemic pathogens. They could have been reported first from Europe by chance or because plant pathology was more advanced in Europe than in the United States.

On the other hand, *Agrobacterium tumefaciens,* which causes crown gall, was recorded in the United States before it was recorded elsewhere. This could be because *A. tumefaciens* is endemic on native plants in the United States (Riker *et al.*, 1946; Brown and Evans, 1933) or because plant bacteriology developed in the United States before it developed in Europe. Riker *et al.* (1946) consider the origin of *A. tumefaciens* to be uncertain. *Pseudomonas campestris* (now *Xanthomonas campestris* pv *campestris*), which causes black rot of crucifers (Garman, 1894), was recorded in the United States before it was recorded in Europe. Because cabbage came from Europe to the United States (Bailey and Bailey, 1976), because the organism has a narrow host range (Walker, 1969), and because the infection is seed-borne (Harding *et al.*, 1904), it is likely that the disease came from Europe and was recognized first in the United States only because plant bacteriology in the United States was more advanced than that in Europe.

The difficulty of determining the source of pathogens is well illustrated by disagreements among investigators and even the change of opinion by the same investigator. For example, Salmon and Ware (1925) first believed that *Pseudoperonospora humili* came to England from Japan or America. Then they believed that it was native to England and had gone from nettle to hop. Finally (Salmon and Ware, 1926), they returned to the idea that it was an introduced disease. Bennett (1952) first believed that sugar beet curly top virus came from native vegetation in the western United States, but in 1971 (Bennett, 1971) stated that it came from the Mediterranean region. Clinton (1911) at first believed that chestnut blight was native to the United States. Shear (1911) first indicated that it came from Europe, but he later (Shear, 1913) retracted this statement. It is now (Heald, 1933) generally believed to have come from Asia.

My associates and I (Yarwood 1937, 1945; Yarwood and Gardner, 1974; Gardner *et al.*, 1970; Gardner and Yarwood, 1974) have reported more than 250 powdery mildews as new in the United States.Several could be introductions from other countries, but many are likely endemic, such as the one on *Eschscholzia californica,* whose host is endemic. Many may not be new—merely new hosts of powdery mildews present previously on other host species of the same genus or on other genera. Some are likely of long standing and became conspicuous as a result of tillage of the ground in which they were growing (Yarwood and Gardner, 1972).

Of these, only beet mildew has attracted national and international interest (Ruppel *et al.*, 1975). *Erysiphe polygoni* (syn. *E. betae*), powdery mildew on sugar beet (*Beta vulgaris*) was first observed in 1899 in Europe (Vanha, 1903). In the United States, it was first found in 1934 in California (Yarwood, 1937). This California infection could plausibly have come from Europe. The fungus remained quiescent and virtually absent from 1934 to 1974, when it was again found in California (Kontaxis *et al.*, 1974).

Eucalyptus is native to Australia, and powdery mildew on eucalyptus occurs there. The infection found in California on several species (Gardner *et al.*, 1970) might have come from Australia, but there is no hard evidence.

If the beet powdery mildew came from Europe to California and eucalyptus mildew came from Australia to California, why has hop powdery mildew (*Sphaerotheca humuli*) failed to move the much shorter distance from Utah or New York, where it occurs, to Idaho, Washington, Oregon, and California, where it does not yet occur? This situation gives further support to the idea that these pathogens were introduced by man, in which case distance is almost irrelevant.

I believe it is impossible to form a reliable opinion on the origin of most if not all of these powdery mildews, and the same is true of hundreds of other pathogens. Yet, the case for beet powdery mildew coming from Europe to the United States is basically the same as the case for grape powdery mildew going from the United States to Europe.

A. Introduction of United States Food Crops

Wheat probably originated and was domesticated in the Tigris–Euphrates area, rice and cucumber in India, potato in South America, coffee in Central Africa, etc. These and further details on the origin of crops are given by de Candolle (1885), Fairchild (1906), Vavilov (1950), and Darlington (1963). Naturally, authorities disagree. De Candolle believes that some of the origins given by Linnaeus are incorrect. According to Thresh (1980), only about 1% of the cultivated area of North America is planted with indigenous crops, whereas indigenous crops occupy about 80% of the cultivated areas of Southwest Asia. The first introduction of a crop to a new area may be that of incense trees from the land of Punt to northern Egypt in 1750 B.C. (Biset, 1925). The first permanent recognition of the value of an introduced plant may be the monument in Japan to the person who introduced citrus from China to Japan about A.D. 100 (Fairchild, 1906). Biset (1925) writes that when persons leave their homeland to establish a new life, they are likely to carry with them their favorite tree or ornamental plant.

Most crops are more successfully grown in areas of importation than in areas of origin. Wheat originated in Asia but is more successfully grown in the United States, Canada, and Australia. Rubber is a more successful crop in Malaysia, where most of the world's natural rubber is produced, than it is in Brazil, where

Hevea is native. Coffee originated in Central Africa, but it is more successfully grown in Brazil and Central America. Monterey pines grow as well or better in New Zealand, where they were introduced, than in California, where they originated. Eucalyptus grows as well if not better in California, where it was introduced, than in Australia, where it originated. A list of tropical crops which grow better at the site of introduction than at the site of origin is given by Purseglove (1963).

According to Anderson (1952), sunflower (*Helianthus annuus*) is the only food crop which originated and was domesticated in the United States, but Fairchild (1906) includes pumpkin, a few grapes, plums, and berries.

A major reason crops do better in areas remote from their place of origin is that when they were introduced, their pathogens, or some of them, were left behind. Purseglove (1963) has given other possible reasons. In addition, most timber trees of the United States, such as pine, spruce, redwood, oak, chestnut, and elm, are native.

B. Desirability of Introducing New Crops

Although the United States has a large complement of crops introduced from countries throughout the world by early settlers and plant explorers, new crops and new strains of old crops from abroad are still appearing. Although most of our present crops were developed in prehistory, it is logical to expect that there are still many species and strains of old species which would be an asset to U.S. agriculture. What we regard as a weed today may be eaten tomorrow (Crawford, 1922). Much of the current introduction is designed to secure strains of crops resistant to disease. The search for such strains is especially intense in areas where the crop originated, where the pathogen also originated, and where disease-resistant strains have developed by natural selection. Therefore, the search for potatoes resistant to late blight has centered on the Andes region of South America (Reddick, 1943), for *Ceratocystis*-resistant elms in Asia (Sinclair and Campana, 1978), and for *Sphaerotheca*-resistant cucurbits in Africa (Whitaker and Davis, 1962).

III. PLANT INTRODUCTION AS A GOVERNMENT ACTIVITY

The idea of plant introduction as a government activity started before the nature of plant disease was understood. Following the example of Benjamin Franklin, who served in England from 1764 to 1775 as an agent of the colony of Pennsylvania, American consular officials adopted the practice of sending home seeds and cuttings of foreign plants (Powell, 1927). In 1802, the U.S. Congress authorized expenditures to encourage the introduction and cultivation of grapes in Ohio. In 1838, land in Florida was made available to encourage the introduction of tropical plants (Powell, 1927). In 1838, Congress appropriated $1000 to be used for the purchase of new and rare varieties of plants (Pieters, 1905). Foreign plants and seeds were exempted from customs duty in 1842. In 1862, the United States Department of Agriculture (USDA) was founded, and one of its objectives was to introduce foreign plants to America. In 1865, a garden of 40 acres for the propagation of foreign plants and other purposes was started in Washington, D.C. An Office of Plant Introduction within the USDA was founded about 1897, and was directed by David Fairchild.

One of the first activities of this office was the sponsorship of a trip to Russia by Mark Carleton to search for rust-resistant wheats. The spectacular success of the durum wheats he introduced focused public attention on the value of plant introduction. Much of the early history of the Office of Plant Introduction is given in Fairchild's "The World Was My Garden" (1938). To facilitate plant introduction, several plant introduction gardens were established, principally around the periphery of the United States (Dorsett, 1916). One of the largest and most important of these was the one at Chico, California, started about 1903. These gardens serve as quarantine stations, as well as sites where the agricultural value of the new crop can be tested.

The Bureau of Plant Industry was started in 1902. The Federal Quarantine Act was passed in 1912. Until then, the United States had been the only major nation not protected against the indiscriminate importation of plants carrying pests. To carry out the quarantine work previously done by the Federal Horticultural Board, the Plant Quarantine and Control Administration was established in 1928 (Weber, 1930).

Crawford (1922) estimated that 50,000 kinds of plants had been introduced by the USDA up to 1923.

IV. ENTRANCE OF PATHOGENS WITH HOSTS

Pathogens usually evolve with their hosts (Eshed and Dinoor, 1980) and enter new areas with them, but this is not easy to document. Easier to establish is the fact that specific pathogens occur only where their hosts occur. Because of its narrow host range, *Phytophthora infestans* will be found only with potatoes and tomatoes. *Puccinia graminis* will be found only on cereals and grasses. *Plasmodiophora brassicae* will be found only on crucifers, etc. Further evidence for this is the steep gradient of decreasing incidence of a pathogen as one moves farther away from diseased plants (Gregory, 1961).

Normally, pathogens effectively enter an area with their hosts. This is primarily because the pathogen is concentrated by the host (see above) and secondarily because the host protects the pathogen from the environment. Pathogen propagules produce an infection or die. If conidia of *E. polygoni* from bean are placed on a dry slide in a normal glasshouse environment, they will be dead in about 1 day, whereas if they are placed on a bean leaf in the same environment, they will live as long as the leaf lives.

Another approach is to compare the production and survival of propagules. All infections produce a tremendous excess of propagules. Representative doubling times for pathogens under favorable conditions are: tobacco mosaic virus, 60 minutes; *Xanthomonas phaseoli* (now *X. campestris* pv *phaseoli*), 300 minutes; *E. polygoni,* 500 minutes (Yarwood, 1956); *Uromyces phaseoli,* 1 day (Yarwood, 1961). Representative doubling times for diseases under favorable field conditions are: potato late blight, 3 days; wheat leaf rust, 4 days; tobacco mosaic virus, 16 days; tristeza virus, 240 days (calculations from data of others presented by van der Plank, 1959). The average doubling time for the above pathogens is 0.4 day and for the diseases is 66 days. Although admittedly imperfect, these data emphasize that pathogens in nature do not attain their biotic potential for increase. This difference in rate of increase of the pathogen and rate of increase of disease represents the loss of the propagules to the environment and, indirectly, the protective effect of the host on the pathogen.

McCubbin (1954) estimates that imported nursery stock and other plants and seeds have been the source of introduction of 90% of the pests which have come to us from abroad. Seeds are the most common form of plant introduction. Seeds are usually free of fungi (Spaulding, 1911) and viruses (Heald, 1933), but wheat seeds can carry 55 species of pathogens (McCubbin, 1954). On the other hand, date palm seeds carry none (Carpenter, 1977). Seeds carrying known pathogens can usually be disinfected by heat or chemicals.

Not all pathogens come from abroad. *Erwinia amylovora,* the cause of fireblight, probably originated in native crab apple in the United States (Anderson, 1956); its principal host, pear, originated in Europe and China (Bailey, 1935). Tomato ringspot virus, the cause of peach yellow bud mosaic, is endemic to coastal California (Frazier *et al.*, 1961); peach, one of its many hosts, originated in China (de Candolle, 1885). *Phymatotrichum omnivorum* and *Armillaria mellea* are other examples of pathogens native to the United States which have caused serious diseases on introduced crops.

V. RETURN OF NATIVE PLANTS FROM ABROAD

When a crop is grown in a new area, it may acquire pathogens of that area which may be introduced to the native area if the crop is returned. A likely example is maize (*Zea mays*), which was introduced to the Philippines and elsewhere and returned to the United States with *Sclerospora* from the Philippines (Frederickson and Renfro, 1977). Another example is white pine, *Pinus strobus,* which is native to the United States. It was introduced to Europe, where it became infected with *Cronartium ribicola* of Europe. When infected white pine seedlings were brought back to the United States for reforestation, it is believed that they introduced *Cronartium ribicola* to this country.

VI. ENTRANCE OF PATHOGENS WITHOUT HOSTS

All pathogens can enter new areas without man and without their hosts, even though they usually do not. Fungi, bacteria, and viruses vary greatly in their adaptation to natural dissemination. Uredospores of rusts, teliospores of smuts, conidia of powdery mildews, and sporangia of downy mildews are borne free and aboveground, and are specially adapted to aerial dissemination (Craigie, 1942; Stakman and Christensen, 1946; Gregory, 1961). Uredospores of *P. graminis* may live for several weeks in normal environments (Maneval, 1924). They are regularly blown north from Mexico to infect wheat in the great plains of the United States and Canada, probably in several successive showers. Such spores were found in air 800 miles beyond wheat culture in Canada but were relatively rare over oceans (Bisby, 1935; Craigie, 1942). Spores have been collected at an elevation of 14,000 ft (Stakman and Christensen, 1946) and at 36,000 ft (McCubbin, 1954). If spores at 36,000 ft (9300 m) fall at 1 cm/sec (Gregory, 1961) in a 20-mile-per-hour wind, they could drift for about 5000 miles and could account for transoceanic travel.

Clouds of chlamydospores (teliospores) may disseminate smuts for many miles (McCubbin, 1946). *Erysiphe polygoni* moved from southern California to Nebraska at a rate of about 300 km per month (Yarwood, 1978).

Conidia of *Endothia parasitica* and bacteria of *E. amylovora* are also borne aboveground, in a matrix mass, and are thus less well adapted to aerial dissemination. Conidia of powdery mildews and basidiospores of rusts are borne singly and aerially but have thin walls and a high water content. As they lose their water in a dry environment, their capacity for germination decreases. But powdery mildews are effectively disseminated over great distances, and basidiospores rarely are. Sclerotia of *Sclerotinia sclerotiorum* and of *Claviceps purpurea* are long-lived but are usually soil-borne and not adaptable to rapid aerial dissemination. Bacteria are probably never borne free and aerially, and are commonly limited to dissemination over short distances by splashing rain or insect vectors. Viruses are always produced intracellularly; therefore, they are rarely free, and are commonly limited to dissemination by vectors or by host tissue. However, they can be carried as host-free suspensions.

When pathogens are separated from their hosts, their longevity is determined by the nature of the pathogen and of the environment. But when they are associated with their hosts in successive generations, their longevity is potentially unlimited. All these pathogens with such varied natural dissemination are easily but not necessarily equally subject to dissemination by man.

VII. ENTRANCE OF HOSTS WITHOUT PATHOGENS

In any given area, including those where pathogens occur, most plants are free of most pathogens, even those which are limited to the host in question. A random plant chosen for introduction to a new area or, in particular, a seed from that plant, would be unlikely to carry a pathogen which would jeopardize that or related plants in the new area. This is supported by many field inspections and by logic. Many routine surveys for plant disease have been made and reported in the "Plant Disease Reporter." For example, Tyler (1944) found that the incidence of six pathogens on carrot ranged from a trace to 2% and for seven pathogens on onions from 0.1% to 10%. Logically, if this generalization were not true, pest control would be more difficult than it is and pathogens would have spread throughout the world more rapidly than they have.

For example, potatoes must have been introduced to Europe and the United States several times from the 1600s on, but late blight did not appear until about 1830. However, potato tubers from a *Verticillium*-infested area usually carry the fungus (Wilhelm *et al.*, 1974). Coffee was introduced to Brazil in 1727, probably from an area where coffee rust was present, but *Hemileia vastatrix,* the coffee rust pathogen, did not appear until shortly before 1970 (Wellman, 1970). Rubber was introduced to Malaya from Brazil, where *Microcyclus ulei* was endemic, but the fungus has not yet arrived in Malaya.

California is isolated from most of the United States by mountains, desert, and oceans. Agriculture probably started with the missions about 1769 (Chapman *et al.*, 1970), but as late as 1850, Smith (1946) considered California to be relatively free of plant disease. Smith attributes this freedom partly to the fact that most plants were introduced as seeds, which are less likely to carry pathogens than are entire plants or cuttings (McCubbin, 1954; Wallace *et al.*, 1956). By 1870, following the heavy influx of people as a result of the gold rush and the expansion of agriculture, the plant disease situation in California had become alarming (Smith, 1946).

The fact that many potential pathogens are still not found in the United States after many years of unrestricted importation of plants before the Quarantine Act of 1912 (Stevenson, 1926; Hunt, 1946) is strong supporting evidence that most importations do not effectively introduce pathogens.

VIII. THE IMPORTANCE OF ALTERNATE HOSTS

With some rusts in some situations, an alternate host is essential for disease. Introducing the alternate host may initiate disease in almost the same way as introducing the pathogen. With wheat stem rust, the pycnial and aecial stages are produced on barberry and the uredinial, telial, and basidial stages are produced on wheat. Wheat is necessary for the disease on barberry, but barberry is not necessary for the disease on wheat, since the uredinial repeating stage occurs on wheat. In areas where the uredinial stage does not over-winter or blow in from other areas, barberry can be a source of new strains of the fungus and of localized epidemics. With some rusts, the repeating stage may be on the less economic host, as with *Cronartium ribicola* on *Ribes*. With other rusts, there may be no repeating stage, as with *Gymnosporangium juniperi-virginianae* on apple and cedar. These situations could have important implications for the introduction of pathogens.

Alternate hosts may also be important in pathogen introduction. Many pathogens have several hosts, including weeds, in addition to their primary crop hosts. Pathogens such as *V. albo-atrum, S. sclerotiorum,* and *Fusarium solani* have a wide host range, including weeds. These pathogens could have been introduced on an alternate host as well as on one of the important economic hosts.

IX. THE IMPORTANCE OF VECTORS

Insects and other arthropods, fungi, dodders, birds, and other animals may transport pathogens. Of these, insects are most important; they may function as passive carriers or as alternate hosts. Some insects actively migrate for long distances, and others may be windborne (Johnson, 1957). Birds may migrate for thousands of miles and may carry pathogens (Heald and Studhalter, 1914).

Disease may arise from the introduction of vectors without a host or pathogen. One possible example is aster yellows; the aster and the mycoplasm were presumably already present in the United States (Walker, 1969) when the vector *Macrosteles* appeared (Neave, 1940). Although vectors may be important in the introduction of pathogens, there is apparently no record of any important disease having been introduced in that way, although Dutch elm disease started from the simultaneous introduction of the pathogen and the vector (Beattie, 1933).

X. INTRODUCTION OF PATHOGENS BY TRAVELERS

In the early history of the United States, it was necessary to introduce plants in order to establish agriculture. Now this practice is usually illegal. Many immigrants and other travelers deliberately or accidentally introduce plants and other materials which may carry plant pathogens. Laws restricting the importation of plant products are usually respected, but also commonly disrespected. As one who grew up on the border between the United States and Canada, I have personally observed and learned of situations in which persons intentionally and illegally introduced plants and plant products from both countries without declaring them. The speed of modern travel has greatly increased the chance of survival of plants and pathogens in international travel. It has been suggested that in the initial introduction of potatoes from South America to Europe, the fragile *P. infestans* did not survive the passage through the hot tropics by sailboat, but with the faster steamboats, *Phytophthora* did survive. Interceptions and confiscations of diseased plants at points of entry into the United States average about 1000 per year (Hunt, 1958). With airplane travel, the chance of survival of delicate pathogens on long journeys is even greater.

XI. VARIABLE HOST RANGE

In terms of absolute numbers of host species and varieties, most pathogens have a wide host range, but in terms of the potential number of host species, most pathogens have a narrow host range. Expressed differently, most plants are resistant to most pathogens. As of 1960, the largest number of living pathogens on any host in the United States was 164 on apple (U.S. Department of Agriculture, 1960). Yet, apple is resistant to 10,000 species of plant pathogens, and most host species are resistant to an even greater number. From this point of view, it is easy to control any disease of any crop by growing a different crop. For example, *Venturia inaequalis,* an important pathogen of apple, can be eliminated by growing any of several hundred crop species resistant to *V. inaequalis*. No one crop is essential to society, although at present, many people depend heavily on wheat, corn, rice, millet, etc. The point is that a random pathogen, when introduced, has a very low chance of meeting a susceptible host under environmental conditions favorable for inspection, and that in an adequately organized society, even the introduction of an extremely virulent pathogen need not be disastrous. One reason *P. infestans* was so disastrous to Ireland is that the Irish at that time depended largely on a diet of potatoes, and loss of that crop caused a famine. In the United States, with a wide range of food crops, many of which are used as feed for livestock as well as for humans, the loss of any one crop due to disease or any other cause would not bring about a disaster, as it did in Ireland.

XII. CHANCE OF ESTABLISHMENT

Even if a specific pathogen is introduced, the chances of its becoming established are small. Many parasites of insects have been intentionally introduced for purposes of biological control, but most have failed to become established (Smith *et al.,* 1933). Citrus canker has often been intercepted at border inspection stations and has probably been introduced without being detected even more often, yet it has not become established since the eradication campaign was completed. Smith *et al.* (1933) list the following conditions necessary after successful introduction for a pathogen to become established: (1) contact with special tissue or wounds, (2) arrival at the right season, (3) proper synchronization with the development of the host, (4) adequate dosage, (5) existence of the proper vector, (6) existence of an alternate host if obligately heteroecious.

XIII. CHANGING VIRULENCE OF PATHOGENS

Each species of pathogen exists as several biologic forms, many of which, if not most, are created by man. These biologic forms range from virulent to nonvirulent, which in a given situation means ranging from being a pathogen to being a nonpathogen. For example, a single infection of ordinary cucumber mosaic virus (CMV) causes only a necrotic ± 1 mm lesion on cowpea, which causes no apparent damage to the host. But if CMV is passed through cowpea, this same virus may become systemic and cause severe stunting of cowpea (Yarwood, 1978). Other changes in pathogenicity are brought about by hybridization of pathogens. Changes in pathogenicity have led to hundreds of strains of pathogens of great economic importance, for example, *Puccinia graminis* (Walker, 1969). Naturally, the strain of the pathogen will determine its potential danger as an introduced pest. Dutch elm disease swept through western Europe and destroyed many trees starting in 1918 (Peace, 1962), but reached a peak in 1936–37 and declined in intensity. Then a new strain of the pathogen was introduced about 1970, and another epidemic of the disease developed (Brasier and Gibbs, 1973).

Verticillium albo-atrum is a long-established pathogen in the United States and Europe. A distinctive strain attacks alfalfa (*Medicago sativa*). The United States was free of this strain for many years. After a considerable history in Europe, it was found in Canada in 1964 and Washington state in 1976, but whether the U.S. infection came from Canada, Europe, or elsewhere is not clear (Heale *et al.*, 1979).

Pathogens may appear to change in virulence because of a change in environment. When a pest is introduced into a new area, it may be released from constraints which kept it suppressed in its original environment. These constraints may be unfavorable temperature, moisture, host resistance, predators, and hyperparasites.

XIV. CHANGING SUSCEPTIBILITY OF HOSTS

Each species of crop plant exists as varieties or strains, many of which are created by man. This is especially true of disease susceptibility: Varieties resistant to disease are being continually created, many of which are later discarded because they lose their resistance. Transient characters not selected for may be lost. This may explain the high susceptibility of potatoes to late blight (*P. infestans*) after 200 years of culture in the absence of the pathogen in Europe and the United States (Burdon and Shattock, 1980). This changed susceptibility could affect the threat of many other pathogens, but the subject has been only slightly studied. A change in susceptibility of the host may be confused with a change in virulence of the pathogen.

XV. PATHOGENS INTRODUCED INTO THE UNITED STATES

Taphrina deformans, Venturia inaequalis, Plasmodiophora brassicae, and hundreds of other pathogens were presumably introduced to the United States before they were recognized as such or before their role in disease was understood. Therefore, there is no solid record of their time or method of arrival, although they were likely introduced on plants or seeds by early settlers. Only those which were introduced after about 1870 can be documented, even though inadequately.

The following list, taken generally from McCubbin (1946), who took most of it from an unpublished manuscript by J. A. Stevenson (1919), gives the estimated dates of introduction of several well-known diseases.

Onion downy mildew	*Peronospora destructor*	1872
Grape anthracnose	*Elsinoe ampelina*	1879
Sorghum smut	*Sorosporium reilianum*	1880
Celery late blight	*Septoria apii*	1885
Hollyhock rust	*Puccinia malvacearum*	1885
Gooseberry leaf spot	*Mycosphaerella grossularia*	1886
White pine blister rust	*Cronartium ribicola*	1890
Cherry leaf spot	*Coccomyces hiemalis*	1890
Olive knot	*Pseudomonas savastanoi*	1890
Wheat stripe rust	*Puccinia glumarum*	1890
Corn brown spot	*Physoderma zeae-maydis* (now *Physioderma maydis*)	1890
Chrysanthemum rust	*Puccinia chrysanthemi*	1895
Rice smut	*Tilletia horridula*	1895
Cucurbit downy mildew	*Pseudoperonospora cubensis*	1895
Chestnut blight	*Endothia parasitica*	1900
Potato blackleg	*Erwinia phytophthora* now *E. carotovora* subsp. *atroseptica*	1900
Tomato leaf spot	*Septoria lycopersici*	1900
Crucifer black rot	*Xanthomonas campestris* now *X. campestris* pv *campestris*	1900

(continued)

Crucifer black leg	*Phoma lingam*	1905
Potato powdery scab	*Spongospora subterranea*	1905
Hop downy mildew	*Pseudoperonospora humuli*	1909
Citrus canker	*Pseudomonas citri* now *Xanthomonas campestris* pv *citri*	1910
Potato wart	*Synchtrium endobioticum*	1911
Wheat flag smut	*Urocystis tritici*	1919
Dutch elm disease	*Ceratocystis ulmi*	1930
Potato ring rot	*Corynebacterium sepedonicum*	1932
Strawberry red stele	*Phytophthora fragariae*	1935
Beet powdery mildew	*Erysiphe polygoni*	1934
Geranium rust	*Puccinia pelargoni-zonale*	1967
Carrot powdery mildew	*Erysiphe polygoni*	1975
Tomato powdery mildew	*Leveillula taurica*	1978

Some of these and others will be documented in more detail.

A. Wheat Stem Rust

Wheat was introduced to America at least as early as 1621 (Bidwell and Falconer, 1925). Wheat stem rust caused by *P. graminis* (stages 0 and I on *Berberis* sp.; stages II, III, and IV on *Triticum* and related species) is generally believed to have been present in Roman times in Italy (Heald, 1933). It was probably present in America as early as 1660 (Bidwell and Falconer, 1925), on the basis of symptoms on wheat; more certainly was present in 1726, on the basis of the injurious effect of barberry on wheat culture (Fullung, 1943); and was conclusively present in 1832, on the basis of finding the fungus on wheat straw of that date (Hendry and Hansen, 1934).

There are some 175 species of *Berberis* native to America (Bailey, 1935), many of which are susceptible to wheat stem rust. The species most commonly grown as an ornamental in the United States, and which has escaped from cultivation and become naturalized in many places, is *Berberis vulgaris*, the common barberry; it is a native of Europe. *Berberis vulgaris* is also considered to be the most important alternate host of *P. graminis*.

English colonists brought barberry to America in the 1600s, "along with it they imported the rust fungus, and it was inevitable that sooner or later the combined presence of wheat barberry and rust should cause trouble" (Fullung, 1943). Since *P. graminis* was hardly recognized as a distinct entity at that time, this conclusion by Fullung seems tenuous. Certainly, rust can be important without barberries, as in Australia (Butler, 1918), or in the wheat belt of the United States after the eradication of barberry. But the introduction of *P. graminis* with barberry is certainly possible. At any rate, a law requiring the eradication of barberries as a means of controlling wheat stem rust in Connecticut was passed in 1726 (Fullung, 1943). This was probably influenced by the passage of barberry eradication laws in France in 1660, but also by the observations in Connecticut that the proximity of barberries was injurious to wheat. This is one of the few cases in which observations of growers were correct in the face of the logic of scientists. An editorial in the "Gardener's Chronicle" (Anonymous, 1849) reads "and many a vulgar error springing from the same source, as for instance, that of the barberry in producing mildew (rust) in wheat, a notion dependent doubtless on the circumstance that both wheat and barberry are extremely subject to mildew though there is as great a difference as can be imagined between the intimate structures of the parasites by which the diseases are respectively produced."

Another possibility is that *P. graminis* existed as a mycoplasm (Ericksson, 1930) or mycelium in wheat seed. The mycoplasm theory seems largely discredited, but the presence of mycelium in seed is confirmed by Butler (1918). Still another possibility is that the fungus blew from Europe to America as uredospores, as hypothesized for coffee rust from Africa to Brazil by Bowden *et al.* (1971). The long-distance spread of wheat stem rust is discussed by Craigie (1942). It could also have been introduced as teliospores on straw. The possibility that it was here on native grasses before European settlers arrived cannot be excluded, but seems unlikely because of the relatively narrow host range of the pathogen. The actual method by which wheat stem rust was introduced to the United States and the time of its introduction will probably never be known.

B. Dutch Elm Disease

Dutch elm disease, caused by *Ceratocystis ulmi,* was recognized as a new and destructive disease in Holland in 1919 (Beattie, 1937; Peace, 1962), although the causal fungus was not correctly identified until 1922 (Sinclair and Campana, 1978). In the United States, it was first found in Cleveland, Ohio, in 1930 (May, 1930) and then in New Jersey in 1933 (Sinclair and Campana, 1978). It spread throughout the United States at the rate of about 6 km/month, which is much slower than that of clover powdery mildew, beet powdery mildew, or wheat stem rust (Yarwood, 1978). Dutch elm disease was clearly introduced to the United States in veneer logs from Europe (Beattie, 1933). The logs from trees killed by the European epidemic were distributed to many places in the eastern United States, carrying the living fungus and the living vector, *Scolytus multistriatus.* The origin of the European infections is unknown, but an Asian source is suspected since Asiatic elm species are resistant (Sinclair and Campana, 1978).

C. Citrus Canker

Citrus canker caused by *Xanthomonas citri* (now *Xanthomonas campestris* pv *citri*) was found in Florida in 1913 (Berger, 1914). It is reliably believed to have been introduced on Satsuma and Trifoliate orange stock from Japan, where it was a common disease (Tanaka, 1918). It was originally diagnosed as caused by *Cladosporium citri* (Wolf, 1916), but was shown by Hasse (1915) to be caused by a bacterium (Wolf, 1916). It soon spread to seven Gulf Coast states. Eradication was started in 1915 (Kellerman, 1927). Millions of orchard, nursery, escaped, and abandoned trees were destroyed (Gaddis, 1937). The last case was found in Texas in 1943 (McCubbin, 1950). This is the outstanding example of successful eradication and subsequent exclusion of plant disease in the United States.

D. Chestnut Blight

Chestnut blight caused by *Endothia parasitica* on *Castanea dentata* may have been present in the United States since 1893 (Metcalf and Collins, 1911), but the first record of the disease as caused by a species of *Cytospora* is from New York City (Merkel, 1906). Murrell (1906) named the fungus *Diaporthe parasitica*. Anderson and Anderson (1913) changed it to the genus *Endothia*. The controversy over its identity was analyzed by Farlow (1912).

Clinton (1911) at first believed that it was a native fungus which became virulent to chestnut subjected to severe winter stress. The resistance of Japanese chestnuts (Metcalf, 1908) and the finding that the fungus was apparently endemic in China (Fairchild, 1913) convinced most investigators that the fungus was introduced from Asia in planting stock.

The rapid spread and severe damage caused by the fungus led to the establishment of the Pennsylvania Chestnut Tree Blight Commission in 1911. Eradication failed, and by 1936 (Walker, 1969) the disease had destroyed most of the chestnuts in the Appalachian region. Anderson and Rankin (1914) stated that "the completeness of destruction is without parallel in the annals of plant pathology."

The discovery of the disease in Agassiz, British Columbia, in 1910 (Faull and Graham, 1914), with the suggestion that it may have been introduced in 1890, and the finding of *E. parasitica* in Oregon in 1929, Washington in 1932, and California in 1934 (Gravatt, 1935) may indicate that the United States West Coast introductions came from Canada and were unrelated to the discovery in New York. However, Gravatt (1935) believed that the Oregon infection came from Pennsylvania.

E. White Pine Blister Rust

American white pines (*Pinus strobus*) were introduced to Europe in the eighteenth century and were very successful there (Granhall, 1963). White pine blister rust (*Cronartium ribicola* stages 0, I on *Pinus;* stages II, III, IV on *Ribes*) was first described by Dietel from Russia in 1854. By 1900 it had spread throughout most of Europe and was regarded as a threat to pines everywhere. In 1898 Schenk warned against the danger of the rust in America, but in spite of this, white pine seedlings were purchased in Europe for planting in the United States (Hirt, 1956). Such importation continued until passage of the Quarantine Law of 1912, which was directed specifically against white pine blister rust.

The first record of the fungus in America is as the uredinial stage on *Ribes aureum* in Kansas in 1892 (Spaulding, 1911). There were no known white pines nearby and no recorded spread of the infection. In 1906 the fungus was found on *Ribes* at Geneva, New York (Stewart, 1906). This entire *Ribes* plantation was pulled out and burned and the disease apparently eradicated, but later findings indicated that the disease was present in the northeastern states as early as 1898 (Mielke, 1943). In 1909, the first infection on pine was found, and the disease was present in seven eastern states.

In 1921, *C. ribicola* was found in *Ribes* and *Pinus* in British Columbia. It is believed that these infections had occurred as early as 1910. It is likely that the New York and British Columbia infections started from infections on imported nursery stock, but were otherwise independent of each other. In 1921, infection was found on pines in Washington state only a few miles from the finding in British Columbia. Within the next few years, the rust spread south into Oregon and California, in spite of quarantine and in spite of the destruction of black currants in Washington near the northern border (Fullung, 1943).

F. Hop Downy Mildew

Hop downy mildew caused by *Pseudoperonospora humuli* on *Humulus lupulus* was first recorded in Japan in 1905 (Miyabe and Takahashi, 1907), then in Wisconsin in 1909 (Davis, 1910), in England in 1920 (Wormald, 1939), British Columbia in 1928 (Newton and Yarwood, 1930), New York in 1928 (Magie, 1942), Oregon in 1931 (Barss, 1931), and California in 1934 (Scott and Thomas, 1934). It is usually considered to have originated in Japan, where Miyabe and Takahashi believed that it was indigenous on wild hops and had moved to the cultivated hops which were introduced to Japan from America and Europe. Except for the 2-year priority of occurrence in Japan, there is no good reason to believe that the Wisconsin infection came from Japan. Hops are indigenous to the United States (Bailey, 1935), and the infections might have possibly originated in Wisconsin or British Columbia as well as in Japan.

Wild (or escape) hops are common in British Columbia. I observed infected wild hops about 200 miles north of commercial hops in 1929, but whether they were infected from the commercial hops or vice versa is unknown. In 1929, I observed infection on wild hops near Burlington, Washington. The sequence of discovery, British Columbia—Washington—Oregon—California, is impressive evidence that the U.S. infection on the Pacific Coast came from Canada. But the situation is complicated by the occurrence of the same fungus or a similar one on

Urtica in England (Salmon and Ware, 1925), British Columbia (Jones, 1932), and Oregon (Hoerner, 1940), since infection of nettles could easily be overlooked.

XVI. PATHOGENS EXPORTED FROM THE UNITED STATES

The conidial stage of powdery mildew (*Uncinula necator*) of grape (*Vitis vinifera*) appeared in glasshouses in England in 1845, and became widespread and disastrous in France about 1850 (Bulit and Lafon, 1978). Because a similar fungus had been previously found as an innocuous parasite of wild grapes in the United States (deSchweinitz, 1834), after some dispute about the identity of the two fungi, it is believed (Yarwood, 1978) that the European fungus came from the United States, although Hesler and Whetzel (1917) believe it came from Japan.

Plasmopara viticola, the cause of grape downy mildew, was first observed in the United States as an innocuous parasite on wild grapes. In 1878, an epidemic of grape downy mildew developed in France; it is believed that the fungus had been present in France since 1870. It is logically believed that the fungus was introduced to France on cuttings brought from the United States to control *Phylloxera* (Hesler and Whitzel, 1917).

Erwinia amylovora, the cause of fireblight of pome fruits, probably originated in the United States and has spread throughout the world (Van der Zwet, 1968).

XVII. OTHER PATHOGENS OF WORLD INTEREST

A. *Microcyclus ulei*

The South American leaf disease of rubber (*Microcyclus ulei* on *Hevea brasiliensis*) is probably the best example of a disease whose exclusion is responsible for a world industry. The elasticity of the coagulated sap of *Hevea* was discovered by Columbus and others (Wong, 1970), and many uses for this rubber, principally auto tires, were gradually discovered. As natural supplies in Brazil became depleted, rubber plantations were started in Brazil, but failed (Holliday, 1970), presumably because *M. ulei* was much more severe in monocultures than in mixed forests. At the same time, entrepreneurs realized that rubber might be a highly profitable crop in other parts of the world. The first two attempted introductions of rubber plants from Brazil to Malaysia were unsuccessful, but the third, based primarily on the efforts of Sir Henry Wickham, was successful (Rao, 1973) and was the start of the present natural rubber industry. It is generally believed that the reason for the success of the Malaysian plantations is their freedom from *M. ulei.* The history of this introduction has several versions (Wycherley, 1968). Whether Wickham and others realized that *M. ulei* or some other pathogen was responsible for the failure of the plantations in Brazil, and that by excluding *M. ulei* plantations could be successful in Malaysia, is not clear. Also, reports conflict as to whether Brazil permitted the export or attempted to prevent it. At any rate, 70,000 *Hevea* seeds from the Rio Trapagas region were collected by Wickham, packed between banana leaves, transported to England, and planted in Kew Gardens glasshouses in 1876.

Most of the seeds died, but in 1917 plants were dispatched by boat in 38 Wardian cases (miniature portable glasshouses) to Malaysia. They were the start of plantations there that now supply most of the world's natural rubber. Whether the danger of introducing *M. ulei* was responsible for the "quarantine" in England, or even whether the fungus could be carried on the seed, is not clear. But it does seem clear that the planned or accidental exclusion of *M. ulei* from Malaysia was responsible for the success of the rubber industry there. There is some likelihood of a correlation of high yield and susceptibility to blight in collections from Brazil (Chu, 1977).

B. *Hemileia vastatrix*

Hemileia vastatrix, cause of the rust of coffee (*Coffea arabica*), is another pathogen whose exclusion was responsible for a world industry. Coffee rust probably originated with the coffee plant in East Africa (Wellman, 1961: Waller, 1972). It spread slowly from east to west, although Large (1940) suggests that it may be a native of Ceylon. Coffee was introduced to Ceylon in 1658 (Wilson, 1970), presumably in the absence of rust, and these plantings were responsible for much of the world's supply of coffee for several years. Coffee rust reached Ceylon about 1869, and its movement throughout the world until 1954 is chronicled by Rayner (1960): by 1890, nearly 90% of the coffee area was abandoned in Ceylon.

Coffee was introduced to Brazil about 1727 (Wilson, 1970), and for many years up to the present, Brazil was the major source of the world's coffee. Coffee rust appeared in Puerto Rico about 1902 but was eradicated (Stevens, 1917). In 1970, coffee rust was reported to be well established in Brazil (Wellman, 1970); how it got there is not clear. Bowden *et al.* (1971) gave plausible reasons to believe that it may have been carried from Africa by prevailing winds in 5–7 days.

XVIII. DISCUSSION

The entrance of plant diseases into a new area is primarily a consequence of the necessity of introducing crops and the occurrence of pathogens on their propagules. Most plant pathogens in the United States were probably introduced before the nature of the disease or the methods of dissemination of pathogens was known, so that there was no basis for trying to exclude them. The primary reasons for believing that most pathogens of U.S. food crops came from outside the United States is that most crops came from other countries and that most pathogens are limited to one or a few crop species.

The primary reasons for believing that most pathogens were introduced by man are that most crops were introduced by man; natural dispersal of most pathogens is not extensive enough to account for their arrival here: all pathogens are readily carried by man; and entry of a few pathogens has been specifically traced to man.

Even after modern concepts of plant disease were developed and precautions to exclude pathogens were instituted, pathogens continued to enter. It would be difficult to show that increased knowledge or increased efforts to exclude them have decreased the rate of entry. However, the anticipated and demonstrated financial return from the exclusion of pathogens and the promise of improved methods of exclusion (Anonymous, 1980; McGregor, 1973) are a logical basis for continued efforts.

REFERENCES

Anderson, F. (1952). "Plants, Man and Life." Little, Brown, Boston, Massachusetts.

Anderson, H. W (1956). "Diseases of Fruit Crops." McGraw-Hill, New York.

Anderson, P. J., and Anderson, H. W. (1913). The chestnut blight fungus and a related saprophyte. *Pa. Chestnut Blight Comm. Bull.* **41,** 1–26.

Anderson, P. J., and Rankin, W. H. (1914). *Endothia* canker of chestnut. *Bull.—N.Y., Agric. Exp. Stn. (Ithaca)* **347,** 533–618.

Anonymous (1849). Editorial. *Gard. Chron.* **9,** 7–24.

Anonymous (1980). USDA funds research for anti-smuggling device. *BioScience.* **30,** 783.

Ashworth, L. J., and Zimmerman, G. (1976). *Verticillium* wilt of the Pistachio nut tree: Occurrence and control by soil fumigation. *Phytopathology* **66,** 1447–1451.

Bailey, L. H. (1935). "The Standard Encyclopedia of Horticulture." Macmillan, New York.

Bailey, L. H., and Bailey, L. Z. (1976). "Hortus Third." Macmillan, New York.

Barss, H. P. (1931). Hop downy mildew in Oregon. *Oreg., Agric. Exp. Stn., Circ. Inf.* **53,** 1–7.

Beattie, R. K. (1933). How the Dutch elm disease reached America. *Proc.—Int. Shade Tree Conf.* **9,** 101–105.

Beattie, R. K. (1937). The Dutch elm disease in Europe. *Am. For.* **43,** 159–161.

Bennett, C. W. (1952). Origin and distribution of new or little-known virus diseases. *Plant Dis. Rep., Suppl.* **211,** 43–46.

Bennett, C. W. (1971). The curly top disease of sugar beet and other plants. *Am. Phytopathol. Soc. Mon.* **7,** 1–81.

Berger, W. W. (1914). History of citrus canker. *Bull.—Fla., Agric. Exp. Stn.* **124,** 27–30.

Bidwell, P. W., and Falconer, J. I. (1925). History of agriculture in northern United States. *Carnegie Inst. Washington. Publ.* **358,** 1–512.

Bisby, G. R. (1935). Are living spores to be found over the (Atlantic) ocean? *Mycologia* **27,** 84–85.

Biset, P. (1925). Plant introduction. *Parks Recreation* **8,** 316–319.
Bowden, J., Gregory, P. H., and Johnson, C. P. (1971). Possible wind transport of coffee rust across the Atlantic ocean. *Nature (London)* **229,** 500–501.
Brasier, C. M., and Gibbs, J. N. (1973). Origin of the Dutch elm disease epidemic in Britain. *Nature (London)* **242,** 607–609.
Brown, J. G., and Evans, M. M. (1933). The natural occurrence of crown gall in the cactus *Carnegia gigantea. Science* **78,** 167–168.
Bulit, J., and Lafon, R. (1978). Powdery mildew of the vine. *In* "The Powdery Mildews" (D. M. Spencer, ed.), pp. 525–548. Academic Press, New York.
Burdon, J. J., and Shattock, R. C. (1980). Disease in plant communities. *Appl. Biol.* **5,** 145–219.
Butler, E. J. (1918). "Fungi and Disease in Plants." Thacker, Spink, Calcutta.
Carpenter, J. B. (1977). Date palm (*Phoenix dactylipera*). *In* "Plant Health and Quarantine" (W. B. Hewitt and L. Chirappa, eds.), pp. 155–162. CRC Press, Cleveland, Ohio.
Chapman, C. E., Cutter, D. C., Shields, P. H., and Baur, J. F. (1970). California. *Encycl. Br.* **4,** 631–641.
Chu, K. H. (1977). Combatting South American leaf blight of *Hevea* by plant breeding and other measures. *Planter* **53,** 287–296.
Cherrett, J. M., and Sagar, G. R. (1977). "Origin of Pest, Parasite and Disease, and Weed Problems." Blackwell, Oxford.
Clinton, G. P. (1911). Notes on plant diseases in Connecticut. *Conn., Agric. Exp. Stn., Rep.* **33,34,** 713–738.
Craigie, J. H. (1942). Aerial dissemination of plant pathogens. *Proc. Pac. Sci. Congr., 6th, 1939* Vol. 4, pp. 753–757.
Crawford, R. P. (1922). World crops for America. *Sci. Am.* **126,** 226–227.
Darlington, C. D. (1963). "Chromosome Botany and the Origin of Cultivated Plants." Hafner, New York.
Davis, J. J. (1910). A new hop mildew. *Science* **31,** 752.
de Candolle, A. (1885). "Origin of Cultivated Plants." Appleton, New York.
de Schweinitz, L. D. (1834). Synopsis Fungorum in America Boreali media degentium. *Trans. Am. Philos. Soc.* **4,** 141–317.
Dorsett, P. H. (1916). The plant introduction gardens of the Department of Agriculture. *U.S. Dep. Agric. Yearb.* pp. 135–144.
Ebbels, D. L., and King, J. E., eds. (1979). "Plant Health." Blackwell, Oxford.
Ericksson, J. (1930). "Fungus Diseases of Plants." Thomas, Springfield, Illinois.
Eshed, N., and Dinoor, A. (1980). Genetics of pathogenicity in *Puccinia coronata. Phytopathology* **70,** 1042–1046.
Fairchild, D. (1906). Our plant immigrants. *Natl. Geogr. Mag.* **17,** 179–201.
Fairchild, D. (1913). The discovery of the chestnut bark disease in China. *Science* **38,** 297–299.
Fairchild, D. (1938). "The World Was My Garden." Scribner's, New York.
Farlow, W. G. (1912). The fungus of the chestnut tree blight. *Science* **35,** 717–722.
Faull, J. H., and Graham, G. H. (1914). Bark diseases of the chestnut on British Columbia. *For. Q.* **12,** 201–203.
Frazier, N. W., Yarwood, C. E., and Gold, A. H. (1961). Yellow bud virus endemic along California coast. *Plant Dis. Rep.* **45,** 649–651.
Frederickson, R. A., and Renfro, B. L. (1977). Global status of maize downy mildew. *Annu. Rev. Phytopathol.* **15,** 249–275.
Fullung, E. H. (1943). Plant life and the law of man. *Bot. Rev.* **19,** 481–592.
Gaddis, B. M. (1937). Citrus canker eradication. *Plant Dis. Rep., Suppl.* **99,** 41–44.
Gardner, M. W., and Yarwood, C. E. (1974). A list of powdery mildews of California. *Univ. Calif. Agric. Exp. Stn., Ext. Serv.* pp. 1–31.

Gardner, M. W., Yarwood, C. E., and Raabe, R. D. (1970). Unreported powdery mildews. IV. *Plant Dis. Rep.* **54,** 399–404.

Garman, H. (1894). A bacterial disease of cabbage. *Rep.—Ky., Agric. Exp. Stn.* **3,** 43–46.

Gold, A. H. (1960). A tobacco-necrosis-like virus isolated from potato-tuber lesions and California soils. *Phytopathology* **50,** 84.

Granhill, I. (1963). Plant introduction and quarantine. *Genet. Agrar.* **17,** 537–542.

Gravatt, G. F. (1935). Chestnut blight in California. *Calif., Dep. Agric. Mon. Bull.* **24,** 173–177.

Gregory, P. H. (1961). "The Microbiology of the Atmosphere." Leonard Hill, London.

Gregory, P. H. (1979). Movement of diseases between neighboring states: Some South American examples. *In* "Plant Health" (D. L. Ebbels and J. E. King, eds.), pp. 269–273. Blackwell, Oxford.

Harding, H. A., Stewart, F. C., and Prucha, M. J. (1904). Vitality of cabbage black rot germ on cabbage seed. *Bull.—N.Y., Agric. Exp. Stn. (Ithaca)* **251,** 177–194.

Hasse, C. H. (1915). *Pseudomonas citri,* the cause of citrus canker. *J. Agric. Res. (Washington, D.C.)* **4,** 97–100.

Heald, F. D. (1933). "Manual of Plant Diseases." McGraw-Hill, New York.

Heald, F. D., and Studhalter, R. A. (1914). Birds as carriers of chestnut blight fungus. *J. Agric. Res. (Washington, D.C.)* **2,** 405–422.

Heale, J. B., Isaac, I., and Milton, J. M. (1979). The administrative control of *Verticillium* wilt. *In* "Plant Health" (D. E. Ebbels and J. E. King, eds.), pp. 71–78. Blackwell, Oxford.

Hendry, G. W., and Hansen, H. N. (1934). The antiquity of *Puccinia graminis* and *Ustilago tritici* in California. *Phytopathology* **24,** 1313–1314.

Hesler, L. R., and Whetzel, H. H. (1917). "Manual of Fruit Diseases." Macmillan, New York.

Hirt, R. R. (1956). Fifty years of white pine blister rust in the northeast. *J. For.* **54,** 435–438.

Hoerner, G. R. (1940). The infection capabilities of hop downy mildew. *J. Agric. Res. (Washington, D.C.)* **61,** 332–334.

Holliday, P. (1970). South American leaf blight (*Microcyclus ulei*) of *Hevea brasiliensis. Commonw. Mycol. Inst. Phytopathol., Pap.* **12,** 1–31.

Hunt, J. (1958). List of intercepted plant pests, 1957. *USDA Bur. Entomol. Plant Quarantine Ser., 44th Rep.*

Hunt, N. R. (1946). Destructive plant diseases not yet established in North America. *Bot. Rev.* **12,** 593–627.

Johnson, B. (1957). Studies on the dispersal by upper winds of *Aphis craccivora* in New South Wales. *Proc. Linn. Soc. N. S. W.* **82,** 191–198.

Jenes, W. (1932). Downy mildew infection of hop and nettle seedlings in British Columbia. *J. Inst. Brew.* **38,** 194–196.

Kellerman, K. F. (1927). Citrus canker under control and final eradication expected. *U.S. Dep. Agric. Yearb.* pp. 183–184.

Kontaxis, D. G., Muster, H. G., and Sharma, R. K. (1974). Powdery mildew epiphytotic on sugarbeets. *Plant Dis. Rep.* **58,** 904–905.

Large, E. C. (1940). "The Advance of the Fungi." Holt, New York.

McCubbin, W. A. (1946). Preventing plant disease introduction. *Bot. Rev.* **12,** 101–139.

McCubbin, W. A. (1950). Plant pathology in relation to federal and domestic plant quarantines. *Plant Dis. Rep., Suppl.* **191,** 87–91.

McCubbin, W. A. (1954). "The Plant Quarantine Problem." Munksgaard, Copenhagen.

McGregor, R. C. (1973). "The Emigrant Pests." Mimeo report to Animal and Plant Health Inspection Service, Hyattsville, Maryland.

Magie, R. O. (1942). The epidemiology and control of downy mildew of hops. *Bull.—N.Y. Agric. Exp. Stn. (Ithaca)* **267,** 1–48.

Maneval, W. E. (1924). The viability of uredospores. *Phytopathology* **14,** 403–407.

May, C. (1930). Dutch elm disease in Ohio. *Science* **72,** 142–143.
Merkel, H. W. (1906). A deadly fungus on the American chestnut. *N.Y. Zool. Soc. Annu. Rep.* **10,** 97–103.
Metcalf, H. (1908). The immunity of the Japanese chestnut to bark disease. *U.S. Bur. Plant Ind. Bull.* **121,** 55–56.
Metcalf, H., and Collins, J. F. (1911). The control of chestnut bark disease. *Farmers' Bull.* **467,** 1–24.
Mielke, J. L. (1943). White pine blister rust in Western North America. *Yale Sch. For. Bull.* **52,** 1–155.
Miyabe, K., and Takahashi, Y. (1907). A new disease of the hop vine caused by *Peronoplasmopara Humuli* n sp. *Sapporo Nat. Hist. Soc. Trans.* **1,** 149–155.
Murrell, W. A. (1906). A new chestnut disease. *Torreya* **6,** 186–189.
Neave, S. A. (1940). "Nomenclator Zooligicus." Zool. Soc. London, London.
Newton, W., and Yarwood, C. (1930). The downy mildew of the hop in British Columbia. *Sci. Agric. (Ottawa)* **10,** 508–512.
Nyland, C., Raju, B. C., and Purcell, A. H. (1980). Peach yellow leaf roll. *Plant Dis.* **64,** 903.
Orton, W. A. (1914). The biological basis of international phytopathology. *Phytopathology* **4,** 11–19.
Peace, T. R. (1962). "Pathology of Trees and Shrubs." Oxford Univ. Press (Clarendon), London and New York.
Pieters, A. J. (1905). Business of seed and plant introduction and distribution. *U.S. Dep. Agric. Yearb.* pp. 291–306.
Powell, F. W. (1927). "The Bureau of Plant Industry." Johns Hopkins Press, Baltimore, Maryland.
Purseglove, J. W. (1963). Some problems of the origin and distribution of tropical crops. *Genet. Agric.* **17,** 105–122.
Rao, B. S. (1973). South American leaf blight. *Q. J.—Rubber Res. Inst. Sri Lanka* **50,** 218–222.
Rayner, R. W. (1960). Rust disease of coffee. *World Crops* **12,** 187, 222, 261, 301.
Reddick, D. (1943). Development of blight immune varieties. *Am. Potato J.* **20,** 118–126.
Riker, A. J., Spoerl, E., and Gutsche,A. E. (1946). Some comparison of bacterial plant galls and their causal agents. *Bot. Rev.* **12,** 57–100.
Rudolph, B. A. (1931). *Verticillium hadromycoses. Hilgardia* **5,** 201–356.
Ruppel, E. G., Hills, E. J., and Mumford, D. L. (1975). Epidemiological observations in the sugarbeet powdery mildew epiphytotic in Western U.S.A. in 1974. *Plant Dis. Rep.* **59,** 283–286.
Salmon, E. S., and Ware, W. M. (1925). The downy mildew of the hop. *G. B. Minist. Agric. J.* **31,** 1144–1151.
Salmon, E. S., and Ware, W. M. (1926). The downy mildew or "spike disease" of the hop in 1925. *G. B. Minist. Agric. J.* **33,** 149–161.
Scott, C. E., and Thomas, H. E. (1934). Downy mildew of the hop in California. *Phytopathology* **24,** 1146.
Shear, C. L. (1911). The chestnut blight fungus. *Phytopathology* **2,** 211–212.
Shear, C. L. (1913). *Endothia radicalis* (Schw). *Phytopathology* **3,** 61.
Sinclair, W. A., and Campana, R. J. (1978). Dutch elm disease. *Search: Agr.* **8,** 1–52.
Smith, H. S., Essig, E. O., Fawcett, H. S., Peterson, G. M., Quayle, H. J., Smith, R. E., and Tolley, H. R. (1933). The efficacy and economic effects of plant quarantine in California. *Bull.—Calif. Agric. Exp. Stn.* **533,** 1–276.
Smith, K. M., and Bald, J. F. (1935). A description of a necrotic disease affecting tobacco and other plants. *Parasitology* **27,** 231–245.
Smith, R. E. (1946). Protecting plants from their enemies. *In* "California Agriculture" (C. B. Hutchison, ed.), pp. 239–315. Univ. of California Press, Berkeley.

Spaulding, P. (1911). The blister rust of white pine. *U.S., Dep. Agric., Bull.* **206.**
Stakman, E. C., and Christensen, C. M. (1946). Aerobiology in relation to plant disease. *Bot. Rev.* **12,** 205–253.
Stakman, E. C., Henry, A. W., Curran, G. C., and Christopher, W. N. (1923). Spores in the upper air. *J. Agric. Res. (Washington, D.C.)* **24,** 599–605.
Stevens, F. L. (1917). Noteworthy Puerto Rican plant diseases. *Phytopathology* **7,** 130–134.
Stevenson, J. A. (1926). "Foreign Plant Diseases." U.S. Govt. Printing Office, Washington, D.C.
Stewart, F. C. (1906). An outbreak of the European currant rust. (*Cronartium ribicola* Dietr.) *N.Y., Agric. Exp. Stn., Geneva, Tech. Bull.* **2,** 61–74.
Tanaka, T. (1918). A brief history of the discovery of citrus canker in Japan and experiments in its control. *Fla. Plant Board Quarantine Bull.* **3,** 1–15.
Thresh, J. M. (1980). The origin and epidemiology of some important plant virus diseases. *Appl. Biol.* **5,** 1–65.
Tyler, L. J. (1944). Vegetable storage disease in New York. *Plant Dis. Rep.* **28,** 143–155.
U.S. Department of Agriculture (1960). "Index of Plant Disease in the United States." USDA, Washington, D.C.
van der Plank, J. E. (1959). Some epidemiological consequences of systemic infections. *In* "Plant Pathology—Problems and Progress 1908–1958" (C. S. Holten *et al.*, eds., pp. 566–573. Univ. of Wisconsin Press, Madison.
Van der Zwet, T. (1968). Recent spread and present distribution of fireblight in the world. *Plant Dis. Rep.* **52,** 698–702.
Vanha, J. (1903). Eine neue Blattkrankheit der Pube. Der echte Mehltau der Rube. *Microsphaera betae* (nova species). *Z. Zuckerind. Boehmen* **27,** 180–186.
Vavilov, N. I. (1950). The origin, variation, immunity and breeding of cultivated plants. *Chron. Bot.* **13,** 1–364.
Walker, J. C. (1969). "Plant Pathology." McGraw-Hill, New York.
Wallace, J. M., Oberholzer, P. C. J., and Hofmeyer, J. D. J. (1956). Distribution of viruses of tristeza and other diseases in propagative material. *Plant Dis. Rep.* **40,** 310.
Waller, J. M. (1972). Coffee rust in Latin America. *PANS* **18,** 402–408.
Weber, G. A. (1930). The plant quarantine and control administration. *Brookings Inst. Serv. Monogr.* **59.**
Wellman, F. L. (1961). "Coffee." Wiley (Interscience), New York.
Wellman, F. L. (1970). The rust *Hemileia vastatrix* now firmly established on coffee in Brazil. *Plant Dis. Rep.* **54,** 539–541.
Whitaker, T. W., and Davis, G. N. (1962). "Cucurbits." Leonard Hill, London.
Wilhelm, S., Sagen, J. E., and Tietz, H. (1974). Resistance to *Verticillium* wilt in cotton sources, techniques of identification, inheritance trends and the resistance potential of multiline cultivars. *Phytopathology* **64,** 924–931.
Wilson, T. C. (1970). Coffee. *Encycl. Br.* **6,** 26–30.
Wolf, F. A. (1916). Citrus canker. *J. Agric. Res. (Washington, D.C.)* **6,** 69–100.
Wong, W. (1970). Rubber. *Encycl. Br.* **19,** 680–693.
Wormald, H. (1939). "Diseases of Fruits and Hops." Crosby Lockwood, London.
Wycherley, P. R. (1968). Introduction of *Hevea* to the Orient. *Planter* **44,** 127–137.
Yarwood, C. E. (1937). Unreported powdery mildews. *Plant Dis. Rep.* **21,** 180–182.
Yarwood, C. E. (1945). Unreported powdery mildews. II. *Plant Dis. Rep.* **29,** 698–699.
Yarwood, C. E. (1956). Generation time and the biological nature of viruses. *Am. Nat.* **90,** 97–102.
Yarwood, C. E. (1961). Uredospore production by *Uromyces phasioli. Phytopathology* **51,** 22–27.
Yarwood, C. E. (1970). Reversible host adaptation in cucumber mosaic virus. *Phytopathology* **60,** 1117–1119.
Yarwood, C. E. (1978). History and taxonomy of powdery mildews. *In* "The Powdery Mildews" (D. M. Spencer, ed.), pp. 1–37. Academic Press, New York.

Yarwood, C. E., and Gardner, M. W. (1972). Powdery mildews favored by agriculture. *Phytopathology* **62,** 799.
Yarwood, C. E., and Gardner, M. W. (1974). Unreported powdery mildews. III. *Plant Dis. Rep.* **48,** 310.

CHAPTER 4

History of Weed Introductions

C. L. FOY, D. R. FORNEY, and W. E. COOLEY

Department of Plant Pathology and Physiology
Virginia Polytechnic Institute and State University
Blacksburg, Virginia

I. INTRODUCTION

When man began the shift from a nomadic, hunting–gathering existence to an agrarian-centered lifestyle some 10,000 or more years ago (Baker, 1974; Bronowski, 1974), he had already begun to fulfill his destiny as the dominant force shaping the patterns of the earth's biota. At first simply harvesting the hybrid grains which evolved naturally soon after the last ice age (Bronowski, 1974), man next began hand-pulling competing vegetation, and gradually learned to

EXOTIC PLANT PESTS AND
NORTH AMERICAN AGRICULTURE

ISBN 0-12-757880-3

grow food plants in areas of his own choosing (Baker, 1974; Shear, 1983). His cultural activities, and the consequent disturbance of natural equilibria, increased geometrically (Rousseau, 1966). Ground clearing and related operations simulated natural disturbances, such as glaciers, avalanches and forest fires, to which certain types of plants had adapted (Bates, 1955), and at every opportunity, these plants sprang up to interfere and compete with the plant communities of man's design. Even up to the present, with a multitude of machines, chemicals, biocontrol organisms, and integrated strategies available for dealing with pest problems, agriculture is still described as "man's controversy with weeds."

In North America, vast expanses of virgin vegetation were relatively recently converted to a heterogeneous complex of disturbed areas, varying in size and in the extent of alteration. This was followed by an invasion of exotic weeds, native to other parts of the world where their habitat was similar to that produced by the removal of the native vegetation (Muenscher, 1949; Mulligan, 1965). The situation is perhaps analogous to that more recently documented on the island of Java (van der Pijl, 1969). There, the opening of arable land released few native weedy species; most plants were shade-adapted. More than 300 foreign species are now established there. New Zealand is similar; beginning in 1979, roughly 1000 alien species became naturalized, transforming the vegetation of the island.

The introduction of the vast majority of plants to new areas is attributable to man. He purposefully spreads food, fiber, and other plant types, and transports weed-contaminated ballast, wool, seed, straw, soil, and countless other minor sources. The most likely contaminants of such sources are specialized species, such as ruderals and crop-associated weeds (Harper, 1965). Thus, in addition to his frequent and widespread disturbance of natural equilibria (i.e., creation of weed habitats) and his lack of ability to control fully the subsequent weed invasions, man is also solely responsible for the introduction of many of the most troublesome species.

The history of weed introductions is the story of when, where, how, and why many of our serious weed problems began and, to some extent, the results of these problems. The story reflects the power of man to create problems inadvertently while advancing his culture. It also reflects the remarkable capacity of certain species to establish, perpetuate, and spread themselves opportunistically over new habitats into which they are introduced.

In this chapter, we will begin the story by establishing the importance of introduced weeds in North America. Next, we will discuss the nature of weeds: the traits which render these plants undesirable; the biological and physiological characteristics which contribute to their ability to invade and survive; and the environmental pressures that lead to the evolution of "weediness." Third, we will summarize the various means by which plants are brought to new areas, with special regard to the dramatic rate at which international travel and commerce have increased since the colonization of North America. Fourth, we will discuss a number of introduced weed species in relation to North American history, touching upon their introduction, spread, and importance. Last, we will review the history of legal sanctions designed to arrest the introduction and spread of weeds, from early seed laws through the 1974 Federal Noxious Weed Act, and the efforts of the Animal and Plant Health Inspection Service (APHIS) in this struggle.

II. THE IMPORTANCE OF INTRODUCED WEEDS IN NORTH AMERICA

A. Influence of Weeds on Human Affairs

Weeds profoundly influence human affairs in all walks of life. A major division of the agricultural industry, which feeds, clothes, and shelters our society, is occupied with the control of weeds, and the livelihood of countless farmers is diminished and threatened each year. The combined annual costs and losses attributable to weeds were estimated to exceed $5 billion in 1965 (U.S. Department of Agriculture, 1968), $11 billion in 1978 (Foy, 1978), and, more recently, $18 billion (Sanders, 1981), making weeds the costliest group of all agricultural pests. In addition, millions of hay fever sufferers are directly affected each year, and controlling weeds is a major ongoing endeavor of highway departments, railroads, utilities, industries, homeowners, and gardeners across the land.

B. Foreign Origins of Weeds

Of all the flowering plants and ferns found in the central and northeastern United States and adjacent Canada, more than 12.5%—1051 species—have been introduced (Table I). Many of these introduced species have gone on to become troublesome weeds (Fogg, 1966). Indeed, most important weed species found in

TABLE I

Places of Origin of the Introduced Flowering Plants and Ferns in the Central and Northeastern United States and Adjacent Canada[a]

Place of origin	Number of introduced species
Europe	692
Eastern Asia	121
Eurasia	111
Tropical America	66
Central and Western North America	44
Others (Africa, South America, etc.)	7
Total	1051

[a] Fogg (1966).

North America are not indigenous (Table II). Among the world's 18 most serious weed species, as ranked by Holm *et al.* (1977), only 3 are possibly native to North America. Of the additional 56 weed species included among the world's worst, only 5 are North American natives. Among 29 species cited as the most prevalent problem weeds in croplands of the United States by a United States Department of Agriculture survey in 1965, 19 can be traced to foreign origins (U.S. Department of Agriculture, 1968). In "Weeds Today," a popular periodical, it was noted that about half of the common weeds found in the United States have been introduced, and that 13 out of the top 15 originated abroad (Anonymous, 1972). Hitchcock (1951) reported that of the 169 numbered genera and 1398 numbered species of grasses known to grow in the continental United States, 46 genera and 156 species have been introduced, mostly from the Eastern Hemisphere. A disproportionate number of these introduced grasses are among our most troublesome weeds.

Most authors addressing weeds comment extensively on the importance of "aliens" among our weed populations (Allan, 1978; Anderson, 1977; Baker, 1972; Brenchley, 1920; Crafts, 1975; Crafts and Robbins, 1962; Dewey, 1896; Fogg, 1956; Haughton, 1978; King, 1966; McNeill and Adams, 1978; Muenscher, 1949; Muzik, 1970; Reed and Hughes, 1970; Robbins *et al.*, 1952).

Humans have migrated throughout the world since before the advent of crop husbandry (Bronowski, 1974; McNeill and Adams, 1978), and our efficiency as agents of plant dispersal has resulted in the cosmopolitan character of many weeds (Salisbury, 1961). Also, man's activities, such as agriculture, have en-

TABLE II

Source of 500 Weeds of the Northern United States[a]

	Number of species	Percentage of species
Native to North America	196	39.2
Widespread	51	10.2
Eastern	95	19
Western	42	8.4
Southern United States	8	1.6
Tropical America	15	3
North America and Europe	13	2.6
North America and Eurasia	16	3.2
Europe	177	35.4
Asia	12	2.4
Eurasia	66	13.2
Africa and Eurasia	3	0.6
Doubtful	2	0.4
Total	500	100

[a] Muenscher (1949).

hanced the evolution of weeds (DeWet and Harlan, 1975; Young and Evans, 1976) and promoted the survival and spread of many species that otherwise would not have survived (Anderson, 1969). It has been stated that the history of weeds is the history of man (Anderson, 1969; Bunting, 1960). The United States is a nation of immigrants (McNeill and Adams, 1978), and immigrant plants have a sure place in North American history (Haughton, 1978), especially as we struggle toward optimum crop production and maintenance of a pleasant and safe landscape and environment.

III. THE NATURE OF WEEDS

A. Definitions

What is a weed? Numerous plant scientists, botanists, and agriculturists have offered definitions (Baker, 1965b; Brenchley, 1920; Hitchcock, 1951; Weed Science Society of America, 1975). Most often, the primary characteristic involves unwantedness or interference with human activities. Weeds interfere with crop production by competing with crop plants for water, nutrients, light, air, and space; by blocking irrigation systems; by impeding planting and harvest; by harboring other pests, such as disease organisms, insects, and rodents; and by reducing the quality of produce. In addition, they are often directly harmful to man and his animals by the production of toxins and immunogens. They interfere with water utilization for transportation, recreation, and power generation. They are unsightly and interfere with our aesthetic enjoyment of the landscape.

Because definitions of a weed are so often based on human concerns, Stearn (1956) suggested that weeds as a whole are less a botanical than a human psychological category within the plant kingdom. This artificiality, and exclusive association with man, of what constitutes a weed has generally been well recognized by authors addressing the problem. Most, however, also refer to certain biological, physiological, and ecological characteristics (Baker, 1965b; Harlan and DeWet, 1965; Muenscher, 1949) which may be used to assess more objectively the weediness of a given species.

Baker (1965b) has listed 14 characteristics which favor weediness in a given taxon. Most are also characteristics of pioneer plants, in the sense of ecological succession, and contribute to the likelihood of a plant's being introduced and succeeding as a weed in a new area. For example, special adaptations for long-distance and short-distance dispersal, discontinuous germination and great longevity of seed, lack of special environmental requirements for germination, and rapid seedling growth all may assist a plant in being transported to and getting started in a new place. Tolerance of climatic and edaphic variations, special competitive ability, rapid onset of flowering, lack of special pollination requirements (and nonobligatory self-compatibility), some seed production in a wide range of environmental circumstances, with continuous, high output of seed under favorable growing conditions, and vigorous vegetative growth and reproduction for perennial species all favor the survival and spread of plants introduced to new areas. Baker (1965b) suggests that a plant having all of the listed characteristics would probably take over the world. His definition of a weed, however, as a plant whose "populations grow entirely or predominantly in situations markedly disturbed by man" (p. 147) returns us to the idea that without human concern, there would be no weeds.

Recently, J. A. Duke (personal communication, 1981) submitted a questionnaire to approximately 100 scientists in an attempt to predict the weediness potential of exotic plants on the basis of about 60 selected botanical or other attributes. Opinions varied widely, as anticipated, thus illustrating the complexity of the task when attempting to generalize across all weed–crop situations. Although no significant correlations were found in this broad survey, the preliminary results were considered inconclusive, primarily because the various ecological systems (e.g., rangeland, aquatic, row crop) were not differentiated in the questionnaire. This approach may still prove fruitful in future studies, with appropriate refinements in the survey methods.

B. Success of Alien Weeds

Why are alien weed species so successful and important? As indicated above, they arrive in new areas where natural equilibria have been disrupted, and where the native species are not best adapted to take advantage of the changed conditions. As a class of plants, they possess certain characteristics which endow them with good colonizing ability. Many of the species possess a general-purpose, opportunistic genotype (Baker, 1974; Young and Evans, 1976) with sufficient plasticity to allow for a rapid development of weediness when the chance for aggressive colonization arises. These characteristics result from the origin and evolution of weeds along with agriculture in the Old World, a consideration first pointed out by the noted American botanist Asa Gray (Baker, 1974; Gray, 1879; McNeil, 1976).

The dynamic nature of agroecosystems, with the recurrent manipulations necessary for agricultural production, promotes rapid evolution and selection of competitive genotypes (Baker, 1974). Thus, in addition to facultative weeds, which may grow in wild as well as cultivated habitats, we also have obligate weeds (Zohary, 1962), such as field bindweed (*Convolvulus arvensis* L.) (Muzik, 1970) and pineappleweed [*Matricaria matricarioides* (Less.) Porter] (Baker, 1965a), which are never found in the wild state. The origins of these weeds are unclear, as with many of our most common plants (Anderson, 1969; Baker, 1965b), and it is probable that they evolved under man's influence.

Muzik (1970) describes, along with facultative and obligate weeds, a third evolutionary class—mimics of crop plants. These weeds may utilize a strategy involving seed production with such close similarity to that of the crop that dispersal and reseeding are accomplished in the harvesting and replanting of the crop, as with large-seed false flax [*Camelina sativa* var. *linicola* (L.) Crantz], mimic of flax (*Linum usitatissimum* L.) (Baker, 1974). In light of man's dominance of the earth's vegetation, this has been considered to be the most successful form of weed adaptation (Wellington, 1960). Another form of crop mimicry is evidenced by shattercane [*Sorghum bicolor* (L.) Moench] (Fawcett, 1980b) and wild oat (*Avena fatua* L.) (Holm *et al.*, 1977), in which vegetative similarity to crops makes control exceedingly difficult, and shattering and variable dormancy ensure survival of the seed.

Along with genetic adaptation and exploitation of new niches, absence of natural control agents, such as diseases and insects, is a major factor contributing to the success of introduced weeds (Baker, 1965b). Especially as a result of the long-distance transport provided by human dispersal mechanisms, natural control agents may be left behind, and an introduced plant is restricted only by climatic factors.

C. Weed Dispersal

Seeds and fruits are, of course, the most frequent means by which the range of plants is expanded, and by which plants are introduced to new areas by natural or human agents of dispersal. As implied by Baker (1965b), weedy plants are prolific seed producers, and this has been quantified and documented by a number of workers (Muenscher, 1949; Salisbury, 1942; Stevens, 1957). Extreme examples include hedge mustard [*Sisymbrium altissimum,* now *officinale* (L.) Scop.] (Muenscher, 1949) and witchweed (*Striga* spp.) (Pavlista, 1980), each of which may produce more than 500,000 seeds per plant. The most dramatic case, however, involves tumbleweed (*Amaranthus albus* L.), a single plant of which yielded, upon maturity, more than 3,000,000 seeds (Gleason and Cronquist, 1964).

The dispersal of plants concerns scientists in various disciplines, but Ridley's text (1930) seems to be the authoritative work on the role of natural forces (wind, water, animals, and birds) in dispersing plants. He provides great detail on specialized and nonspecialized species and mechanisms. Carlquist (1981) cites Ridley on this topic, and makes the point that chance events involving physical forces in concert with biological mechanisms have been responsible for the long-distance dispersal of organisms, resulting in transoceanic plant introductions. Even Ridley (1930) and Carlquist (1981), however, with their major emphases on natural phenomena, point out the greater influence of man, through a gradually increasing action, in the intercontinental transport of plants and the alteration of floras. Certainly, plants have evolved with a multitude of mechanisms for dispersal, including winged and plumed fruits and seeds for riding the wind; floating fruits and seeds for dispersal by water; colored fruits, foliage, and branches for attracting feeding birds; and hairs, plumes, lobes, spines, barbs, and viscid exudates for adhesion of fruits and seeds to passing animals. Transport by these mechanisms without the assistance of human activities is, however, generally very local in scope. The effect of human activities is especially evident in light of the worldwide wanderings of people and, within the last 200–300 years, the extent of long-distance commerce and commodities transport.

In addition to seeds as units of plant dispersal, vegetative portions of perennial species have become increasingly important under man's influence. Kudzu [*Pueraria lobata* Willd. (Ohwi)], a perennial legume, was introduced into North America from Japan in the early 1900s for erosion control; between 1935 and 1942, some 85,000,000 crowns were distributed by the United States Soil Conservation Service (Dickens, 1974). The plant now causes somewhat of a weed problem in various situations throughout the South. Hydrilla (*Hydrilla verticillata* Royle), a perennial submersed aquatic weed, was introduced as an ornamental around 1960, and has spread exclusively by vegetative means (Haller, 1976). Hydrilla now causes millions of dollars worth of damage annually in a number of subtropical areas of North America.

IV. MODES OF WEED INTRODUCTION

The means by which plants have been introduced into North America by man can be divided into two broad categories: intentional and unintentional.

A. Intentional

Humans have intentionally transported plants to new areas for a number of reasons, and plant transport and introduction have always been a major enterprise of explorers, travelers, and colonizers (Brockway, 1979). Williams (1980) reported on 36 species of purposefully introduced plants that have become weeds in the United States (Tables III and IV). Some are poisonous (Table III), whereas others are objectionable because of other weedy characteristics (Table IV). Under the heading "Purpose of Introduction," the most frequent entry by far is "ornamental," which appears for 21 of the weeds. Another seven were introduced as forage plants, and the rest for such purposes as herbs, medicines, and dyes.

Great numbers of weed species were brought to North America as garden plants by colonizers from Europe in the seventeenth century. Common dandelion (*Taraxacum officinale* Weber), common chickweed [*Stellaria media* (L.)

TABLE III

Purposeful Plant Introductions That Have Become Poisonous Weeds in the United States[a]

Species	Purpose of introduction	Weed type	Problem areas	Toxic principle
Belladonna *Atropa bella-donna* L.	Herb	Annual	Roadsides, waste areas	Atropine
Bouncingbet *Saponaria officinalis* L.	Ornamental	Perennial	Pastures	Saponins
Brazilian peppertree *Schinus terebinthifolius* Raddi	Ornamental	Perennial	Parks, forests, yards	Irritants
Buckthorn *Rhamnus* spp.	Ornamental	Perennial	Grazing areas	Anthraquinone
Chinaberry *Melia azedarach* L.	Ornamental	Perennial	Grazing areas	Unknown
Corn cockle *Agrostemma githago* L.	Ornamental	Annual	Wheat, grasslands	Saponins
Crotalaria *Crotalaria spectabilis* Roth *Crotalaria retusa* L.	Green manure, hay	Annual to perennial	Ranges, waste areas, soybeans	Monocrotaline
Foxglove *Digitalis purpurea* L.	Ornamental	Biennial	Pastures, waste areas	Aglycones
Goatsrue *Galega officinalis* L.	Forage	Perennial	Pastures, canal banks	Galegin
Hemp *Cannabis sativa* L.	Fiber, medicine	Annual	Pastures, waste areas	Tetrahydrocannabinol
Henbane *Hyoscyamus niger* L.	Medicine	Annual or biennial	Roadsides, waste areas	Atropine, scopolamine, hyoscyamine
Jimsonweed *Datura stramonium* L.	Ornamental	Annual	Pastures, cropland	Solanaceous alkaloids
Johnsongrass *Sorghum halepense* (L.) Pers.	Forage	Perennial	Cropland, pastures	Cyanogenic glycosides
Lantana *Lantana camara* L.	Ornamental	Perennial	Fence rows, ditchbanks, fields	Lantadene
Melaleuca *Melaleuca leucadendron* (L.) L.	Ornamental	Perennial	Swamps, cities	Respiratory and skin irritants
Precatory bean *Abrus precatorius* L.	Ornamental	Perennial	Fence rows, roadsides	Abrin
Sicklepod milkvetch *Astragalus falcatus* Lam.	Forage	Perennial	Rangeland	Nitro compounds
Tansy *Tanacetum vulgare* L.	Herb	Perennial	Old gardens, roadsides	Unknown

[a] Williams (1980).

TABLE IV

Purposeful Plant Introductions, Other Than Poisonous Plants, That Have Become Weeds in the United States[a]

Species	Purpose of introduction	Weed type	Problem areas
Bermudagrass *Cynodon dactylon* (L.) Pers.	Forage	Perennial	Pastures
Dalmatian toadflax *Linaria genistifolia* (L.) Mill. subsp. *dalmatica* (L.) Maire and Petitmengin	Ornamental	Perennial	Rangeland
Dyers woad *Isatis tinctoria* L.	Dyes	Biennial	Rangeland, crops
Hydrilla *Hydrilla verticillata* (L.f.) Royle	Aquarium trade	Perennial	Lakes, reservoirs, waterways
Cogongrass *Imperata cylindrica* (L.) Beauv.	Forage	Perennial	Southern farms
Japanese honeysuckle *Lonicera japonica* Thunb.	Ornamental	Perennial	Wooded areas, pastures
Japanese knotweed *Polygonum cuspidatum* Sieb. and Zucc.	Ornamental	Perennial	Lowlands, homesites
Kochia *Kochia scoparia* (L.) Schrad.	Forage, ornamental	Annual	Widespread
Kudzu *Pueraria lobata* (Willd.) Ohwi	Ornamental, erosion control, forage	Perennial	Forests, rights of way, field borders
Macartney rose *Rosa bracteata* Wendl.	Ornamental	Perennial	Pastures
Multiflora rose *Rosa multiflora* Thunb. ex. Murr.	Windbreaks, cover plantings	Perennial	Pastures
Reed canarygrass *Phalaris arundinacea* L.	Forage	Perennial	Canal and ditch banks
French tamarisk *Tamarix gallica* L.	Ornamental	Perennial	Pastures, flood plains, waterways
Strangler vine *Morrenia odorata* (Hook. and Arn.) Lindl.	Ornamental	Perennial	Citrus
Water fern *Salvinia auriculata* Aubl.	Ornamental	Perennial	Canals and waterways
Water hyacinth *Eichhornia crassipes* (Mart.) Solms	Ornamental	Perennial	Canals, lakes, waterways
Wild melon *Cucumis melo* L.	For observation	Annual	Imperial Valley, cropland
Yellow toadflax *Linaria vulgaris* Mill.	Ornamental	Perennial	Rangelands

[a] Williams (1980).

Cyrillo], plantains (*Plantago* spp.), and others were brought in as pot herbs (Haughton, 1978). Common mullein [*Verbascum thapsus* (L.)], yellow toadflax (*Linaria vulgaris* Hill), and common yarrow (*Achillea millefolium* L.) were brought in as medicinal plants. Redroot pigweed (*Amaranthus retroflexus* L.) was brought to the New World tropics from Spain by the conquistadors as a food plant; from there it spread to North America, where it is now one of our most common weeds (Muenscher, 1949).

Large crabgrass [*Digitaria sanguinalis* (L.) Scop.] and johnsongrass [*Sorghum halepense* (L.) Pers.], two major North American weed pests, got their start here as forage crops, as did a number of others mentioned by Williams (1980). Crabgrass, one of the first crops to be cultivated for grain, was introduced in 1849 by the United States Patent Office (Haughton, 1978). It was not widely used here, however, until it was reintroduced around 1900 by many Central European immigrants. They brought it as a "sure" crop that could be counted on to provide some grain even if planted late in the season or on poor ground. They quickly abandoned it for more productive and more easily produced crops such as corn and wheat, and crabgrass stayed on to flourish as a weed.

Johnsongrass was introduced into the southern United States as a forage crop, probably by a number of independent farmers, in the early 1800s, and began to cause serious problems as a weed by 1850 (McWhorter, 1971). Its more recent introduction into Canada was also the result of farmers' experimenting with it as a forage crop (Alex *et al.*, 1979); it had been advertised in a farm journal as "an amazing perennial forage." Johnsongrass is acknowledged to be the sixth worst weed in the world (Holm *et al.*, 1977).

Among plants introduced as ornamentals that have become serious weeds, water hyacinth [*Eichhornia crassipes* (Mart.) Solms] is a striking example (Williams, 1980). Thought to be a native of South America (Monsod, 1975), it has long been cultivated for the beauty of its flowers. It was brought to New Orleans from Brazil in 1884 as part of a horticultural display and taken from there to Florida by admirers. Within 13 years, it had begun to cause severe navigational problems (Williams, 1980), and had spread north to Virginia and west to California (Monsod, 1975). Now water hyacinth is the scourge of waterways throughout tropical and subtropical regions of the world, and is considered the world's eighth worst weed (Holm *et al.*, 1977).

Multiflora rose (*Rosa multiflora* Thunb.), a perennial shrub, was introduced into North America from the Orient as an ornamental some time prior to 1811 (Zimmerman, 1955). By 1900 it was widely used as rootstock material for grafting and budding of other roses (Moore, 1979), a use that continues today. By the 1930s, it was being widely spread as a living fence, for soil conservation through wind erosion control, as a highway safety barrier, and as a wildlife habitat (Anderson and Edminster, 1949). By about 1950, its potential as a weed was recognized (Rosene, 1950) because infestations were already severe. Multiflora rose spreads naturally by its seeds, which are dispersed by birds, and causes serious problems in permanent pastures and other noncultivated areas (Fawcett, 1980a).

B. Unintentional

The unintentional transport of plants by human activities is in part due to the strategies that plants have evolved for seed dispersal, and in part to man's extensive mobility and transportation systems. A prime characteristic of weediness is prolific seed production; this alone in areas of human activity increases the probability that some seeds will be transported inadvertently. In addition, all of the various mechanisms that apparently evolved for natural dispersal, previously mentioned, and covered by Ridley (1930), enhance the likelihood that weed seeds will contaminate commodities or cargo being moved by man. The fact that man's mobility is continuously increasing is a simple consequence of the evolution of society. Walker (1917) has stated that civilization is measured in terms of transportation, in the extent to which man has surmounted the time and space limitations of nature. Thus, inventions that increase human mobility, such as water craft, wheels, powered vehicles, and the airplane are milestones in the history of civilization. All have contributed incidentally to the worldwide dispersal of plants.

The most common methods by which weeds have been spread inadvertently by man involve the transport of contaminated materials associated with agriculture (Ridley, 1930). Crop seed has probably been the most important source. The problem has been recognized since biblical times, as evidenced in the Bible: "Thou shalt not sow thy field with mingled seed" (Leviticus 19:19). The dependability of crop seed contamination was demonstrated by Stapledon (1916). He proved it possible to trace the countries from which samples of commercial oats originated by determining the quantities and varieties of weed seed they contained. Turkestan alfalfa has also been a consistent source of contamination (Groh, 1940), and has provided North America with Russian knapweed (*Centaurea repens* L.) and others.

Despite increased concern, international cooperation and legislation, and advanced technology, it remains impossible to remove weed seed completely from contaminated crop seed, due to the similarity of seed characters (Hitchings, 1960). Thus, modern efforts to alleviate the problem concentrate on avoiding contamination by maintaining seed-crop fields free from weeds.

Among weeds which owe their existence in North America to arrival along with crop seed, wild oat is a classic example. Considered to be among the world's worst weeds (Holm *et al.*, 1977), it has been associated with the cultivation of oats and other cereals since at least 500–700 B.C. It is presumed to have been transported to North America in the crop seed of the earliest settlers. In agricultural fields where wild oat populations are established, infestations of more than 150 plants per square meter are not uncommon, and crop yield reductions may be as high as 80% (Nalewaja, 1970).

Other stowaways to arrive in crop seed have been hedge and field bindweed (*Convolvulus sepium* L. and *C. arvensis* L.), which came with colonial settlers (Haughton, 1978), Russian thistle (*Salsola kali* var. *tenuifolia* Tausch), brought to the northern plains by Russian settlers around 1870 in their flax and wheat seed (Mitich, 1980), and Russian pigweed (*Axyris amaranthoides* L.), which entered Canada in the late 1800s as a grain contaminant and continues to enter and spread in this way (Blackwell, 1978).

Another source of unintentional weed introductions is the wool of sheep and the hair of other animals, along with their excreta and fodder, which are necessarily transported along with them (Ridley, 1930). Cockleburs (*Xanthium* spp.) arrived in America in this way with the first loads of goats and sheep (Haughton, 1978). An interesting example involves the filaree [*Erodium obtusiplicatum* (Maire, Weiller and Wilczak)] (J. T. Howell), which now is a common weed of California rangelands (Baker, 1972). This native of the Mediterranean region became a weed in Australia after its transport there with sheep. It was later recorded as a weed in England in fields near woolen mills where Australian wool was being cleaned.

Ships' ballast was a common source of foreign plants in sailing days. Soil was carried as ballast when ships were not fully loaded, and was dumped in ports when cargos were taken on. The soil used was naturally infested with seed from a variety of plants, especially ruderals. Nelson (1917) studied the flora of a ballast heap at Linnton, Oregon. He found 32 species indigenous to the Pacific Coast, 88 species that were considered to be introduced but also found in other parts of the state, and 93 species that could be found only in the ballast area, and not elsewhere in the state.

Fogg (1966) states that hundreds of European species were introduced into North America in ballast. He names many, including chickweed, bindweed, dandelion, mullein, plantain, and crabgrass, that are also cited by others in reference to other modes of introduction. Thus, the most common weeds, those that occur most widely in association with man, have been nurtured, carried along, and distributed by man in a variety of ways throughout human history.

Other sources of stowaway weeds are numerous. Packing materials, such as straw, are frequently contaminated, as well as the soil associated with plants being transported. Soil, along with the weed seed it may contain, may stick to the surface or lodge in cracks and crevices of practically any object, incuding packages, tires, and shoes. Allan (1978) reported that a ball of earth adhering to the leg of a partridge yielded 82 plants when broken up and watered after 3 years in storage. Also, because of previously mentioned adaptations for dispersal, many weed seeds need no additional help to stick to surfaces: for example, puncture vine (*Tribulus terrestris* L.) has sharply pointed seeds that stick to tires (Crafts and Robbins, 1962). A native of the Sahara Desert, it presumably arrived in North America on the tires of military planes returning from that region, and has been widely spread here via aircraft and automobile traffic.

V. A CHRONICLE OF WEED INTRODUCTIONS IN NORTH AMERICA

We now present a few selected examples of weed introductions within the framework of an abridged history of North America.

Since human activity is clearly the overriding factor affecting the spread of weeds, the primary concern in considering the establishment of our alien weed flora is the immigration and subsequent activities of man in North America.

Little information is available regarding weed introductions in prehistoric times (King, 1966). We know that the peopling of North America most likely began with the movement of Eurasian migrants across the Bering causeway between 30,000 and 10,000 years ago (Jennings, 1974). Although it seems likely that some plant species accompanied the nomads, there is no evidence to suggest this, and Rousseau (1966) states that they brought only the dog.

North America remained isolated for quite some time after its initial colonization. Beginning about 6700 B.C. with a shift from a meat-dominated to a plant-dominated diet among the nomads, a sophisticated agriculture evolved in the New World, with its center of origin in the Tehuacán Valley of southern Mexico (MacNeish, 1964). Not much is known about the weed problems encountered by the earliest farmers. It seems that weeds should have evolved along with agriculture in the New World, as, according to Gray (1879) and others (McNeil, 1976), they did in the Old World. Several amaranths, including *Amaranthus hybridus* L., *A. powellii* S. Wats., and *A. spinosus* L., are apparently examples (Holm *et al.*, 1977; King, 1966; Muenscher, 1949), but among the flowering plants and ferns introduced into the central and northeastern United States and Canada, disproportionately few are New World natives (Table I).

After the early arrival of America's first inhabitants, the next intercontinental contact between the Old and New Worlds probably occurred when Scandinavians (Vikings) made a series of sporadic visits to Newfoundland in the eleventh or twelfth century A.D. (Stewart, 1973). Their physical and cultural impact was minimal, however, and it is doubtful that many weeds arrived at that time.

The influx of European plants to the New World began with the voyages of Christopher Columbus, starting in 1492. On his second trip to Española, he brought 17 ships, 1200 men, and seeds and cuttings for the production of a variety of fruits and vegetables (Crosby, 1972). Although no specific information regarding weeds is available, we can safely assume that some weeds must have come on this voyage, and Crosby (1972) states that most of the plants that crossed the Atlantic from 1492 to 1600 were brought inadvertently as weeds. In addition, as previously mentioned, the conquistadors brought redroot pigweed potherb; it subsequently escaped to become one of our more ubiquitous weeds. The Spaniards also brought hemp (*Cannabis sativa* L.) to the New World in about 1545 (Walton, 1938). Their use of the plant was as a fiber crop; it was introduced as a recreational drug plant by slaves at some later time. *Cannabis* is now considered a serious weed by most, although not all, segments of our society.

The greatest numbers of weedy plants began to arrive in North America with the English colonists of the seventeenth century. Haughton (1978) includes chickweed, mullein, plantains, wild carrot (*Daucus carota* L.), thistles (*Carduus* spp.), dandelion, and yarrow among "essential herbs" brought for Puritan gardens, and buttercups (*Ranunculus* spp.), dodders (*Cuscuta* spp.), and cocklebures (*Xanthium* spp.) as stowaways in the first shipments of seed, hay, and livestock. Josselyn (1672) reported that couchgrass [*Agropyron repens* (L.) Beauv.], shepherdspurse [*Capsella bursa-pastoris* (L.) Medic.], dandelion, common groundsel (*Senecio vulgaris* L.), sowthistle (*Sonchus* spp.), nightshade (*Solanum* spp.), stinging nettle (*Urtica* spp.), mallows (*Malva* spp.), plantains, knotweed (*Polygonum aviculare* L.), mayweed (*Anthemis cotula* L.), and mullein were among the weeds that had "sprung up" in New England "since the English planted and kept cattle there" (p. 85). It must be reaffirmed here that many of these common plants, which had long been associated with human activity in Europe, were afforded a variety of chances to colonize the New World, being brought both intentionally and unintentionally by early travelers. In addition, in tracing the origin of weeds, the historical record is simply not clear. For example, Gray (1889) points out that common purslane (*Portulaca oleracea* L.), now considered one of the world's worst weeds (Holm *et al.*, 1977), was generally assumed to have been introduced into North America in colonial times. This species, however, was reported by several botanists to occur indigenously in a number of locations in the United States. Likewise, common yarrow was brought by colonists, according to Haughton (1978); yet, this species has also been considered to be a New World native (Muenscher, 1949) and appeared in an Aztec herbal (King, 1966). Of course, the two theories (native vs. introduced) are not mutually exclusive, and the possibility exists that hybridization has resulted in weedier types.

By 1700, the American colonies were firmly established, with a population of some 250,000 (Ver Steeg, 1964). Wherever the new inhabitants roamed, their weeds followed, as evidenced by the American Indians' name for plantain—"white man's foot" (Rousseau, 1966). Expansionism in the eighteenth century involved immigration of English and non-English alike. For example, more than

200,000 Germans settled chiefly in Pennsylvania and the Carolinas between 1700 and 1770. They apparently brought wild garlic (*Allium vineale* L.) as a spice (Dewey, 1896); it is now a serious weed in wheat. As the colonies expanded, so did American agriculture. From Philadelphia, 350,000 bushels of wheat and 18,000 tons of flour were exported in 1765, a spectacular increase over the limited amounts exported in 1700 (Ver Steeg, 1964). The greater proportion of seeds sown in North America continued to come from Europe: thus, a host of European weed species continued to be sown (Dewey, 1896). In addition, trade and commerce flourished, increasing the opportunities for exotic weeds to be transported in cargos and ballast.

When the British defeated France in the Great War for the empire in 1760, the region west of the Appalachians was opened to exploitation by the colonials (Ver Steeg, 1964). The Indians, having lost their strategic advantage of playing off the English against the French, were swept aside by the growing English–American settlement. After the liberation of the colonies in 1776, and especially the Louisiana Purchase in 1803, the stage was set for fulfillment of what later became known as "manifest destiny," the coast-to-coast expansion of the United States.

In the 1800s, transportation systems evolved rapidly (Walker, 1917). Between 1800 and 1826, more than $150,000,000 was expended by federal and state governments in canal construction. Hay and straw brought in as feed and bedding for beasts of burden employed in this effort afforded a continuous supply of weed seed. By 1847, the legislature of New York had passed a special law requiring destruction of noxious weeds along the banks and sides of canals (Dewey, 1896). In 1807, Fulton invented the steamboat (Walker, 1917), and when the botanist William Baldwin visited the Midwest in 1819, he found every boat landing on the Mississippi and Missouri rivers already populated with European weeds (Gleason and Cronquist, 1964). Railroads were also a primary avenue for weed transport. Many weeds migrated westward with the Canadian Pacific Railroad, almost keeping pace with construction (Dewey, 1896). Russian thistle, now a pest in all western states (Evans and Young, 1974), appeared first along railways in 16 out of 21 states and territories into which it was introduced by 1896 (Dewey, 1896). This weed had come into North America in 1873 with flax seed from Russia (Crafts and Robbins, 1962).

The importation of foreign crop seeds throughout the 1800s and into the 1900s marked a continuing effort to optimize American agriculture. Johnsongrass was brought intentionally as a forage crop around 1830 (McWhorter, 1971) and crabgrass around 1849 (Haughton, 1978). In the drought years of the early 1930s, millet seed was imported from China, and with it came giant foxtail (*Setaria faberii* Herrm.) (Knake, 1977). Native to the Szechwan Province of China, where the soils and climate are similar to those in the Midwest, giant foxtail spread rapidly and became "the most serious annual grass weed in the

Midwest'' by 1977 (Knake, 1977). Also arriving from Asia in the 1930s was halogeton, which invaded western rangeland, and has been responsible for numerous disastrous sheep kills due to its accumulation of toxic amounts of oxalic acid (Williams, 1973).

In spite of concern and legislation regarding foreign weed introductions (see Section VI), alien weeds continue to be spread to new areas from outside and from within North America. Dearborn (1959) reported on a number of weeds being introduced into Alaska during the agricultural development of that state in recent years (Crafts and Robbins, 1962). Elsewhere in this book, current efforts to generate knowledge about potential weed invaders are described, and several of the prime conditions are discussed. We can only hope that McNeil (1976), although perhaps too optimistic, is correct in his statement: ''Although existing weeds are still being . . . spread around the world . . . the process is probably approaching completion in temperate regions'' (p. 408). His logic is that not only must a weed be introduced into a new area, it must find a niche to which it is better adapted than species already present. The problem is that man, in continuing to develop agricultural technology, is continuing to create new niches.

VI. LEGAL CONSIDERATIONS

A. General Introduction

The idea of laws designed to promote the control of weeds is not new. In the thirteenth century, Alexander II of Scotland ruled that a tenant should be hanged for allowing ''guilds'' (possibly a species of *Chrysanthemum*) to grow on his land. Moreover, a bond servant could be fined one sheep per weed for the same offense (Smith and Secoy, 1981). In the eighteenth century, Danish farmers were required by law to keep their land free from corn marigolds (Smith and Secoy, 1981). During this period, farmers in France could also be sued by their neighbors for not removing thistles from their fields (Smith and Secoy, 1981). From these examples, we can see that our ancestors realized a need for legislation to encourage weed control. With modern means of transportation increasing the possibilities of weed introductions, and with the cost of controlling weeds increasing each year, we can see that such laws are needed today more than ever.

The effectiveness of these laws is exemplified by the Kansas Noxious Weed Law, which was enacted in 1937 and supplemented by later amendments (Day, 1968). Provisions of this law govern both the prevention and control of noxious weeds on all lands in the state. The law includes regulations involving movement of weed seeds, infested crop seeds, and feeds into and within the state; prescribes approved control methods; establishes the duties of state and county officials in administration of the law and fixes penalties for noncompliance by landowners (Day, 1968). A very successful program for noxious weed control has been developed in Kansas as a result of this legislation (Day, 1968).

However, if not properly designed, legislation may become a hindrance to weed control. The Manitoba Noxious Weeds Act, which was passed in Canada in 1895 at the request of farmers, is an example of a law which became outmoded and caused a problem. Since the use of herbicides was not a stipulated procedure in this law, authorities found that they could not be used as a weed control measure. The law was later amended (Day, 1968).

Laws governing the control and spread of weeds can be divided into two basic types—seed laws and weed laws. Seed laws regulate the sale and transport of contaminated crop seeds. Weed laws may regulate a number of aspects of weed control, including (1) accepted methods of control, (2) transport of weeds or weed parts, including seeds, and (3) quarantine of recently introduced weed species that are potential problems. A discussion of seed laws is presented first, followed by a discussion of the Federal Noxious Weed Act of 1974.

B. Seed Laws

The need for measures to improve crop seeds and thereby retard the dissemination of weeds has long been recognized. Connecticut passed the first seed law in the United States in 1821 to prohibit the sale of forage-grass seed containing seeds of Canada thistle (Day, 1968). E. H. Jenkins later opened the first seed-testing laboratory in the United States at the Connecticut Agricultural Experiment Station in 1876 (Day, 1968).

All states in the United States, as well as the federal government, have now put seed laws into effect. The uniformity which exists among these laws is a result of a model seed law which was recommended by a joint committee of the International Crop Improvement Association, the Association of American Seed Control Officials, the Association of Official Seed Analysts, and the American Seed Trade Association (Day, 1968). These laws are important in reducing the spread of weed species into and within the United States (Day, 1968).

In 1939, the Federal Seed Act was passed. This law protects purchasers from contaminated seed by requiring that information on seed purity be given on all crop seed imported into the United States or shipped in interstate commerce (Day, 1968; U.S. Congress, 1939). In part, this label information must include the percent by weight of all agricultural seeds, the percent by weight of all weed seeds, and the kinds and rate of occurrence of weeds considered noxious by the state and/or the Secretary of Agriculture (U.S. Congress, 1939).

Verification of the label information is accomplished by the United States Department of Agriculture in cooperation with state agencies (Day, 1968; U.S. Congress, 1939). Seeds shipped into a particular state are analyzed in a state seed laboratory by state officials. Seeds to be imported are sampled by customs officials and analyzed by the United States Department of Agriculture, which may approve or refuse importation (Day, 1968). Noxious weed species which may not enter the United States as contaminants of imported seeds include whitetop (*Cardaria* spp.), Canada thistle [*Cirsium arvense* (L.) Scop.], dodder (*Cuscuta* spp.), quackgrass, johnsongrass, field bindweed, Russian knapweed (*Centaurea repens* L.), perennial sowthistle, and leafy spurge (*Euphorbia esula* L.) (U.S. Congress, 1939). Seeds are also prohibited entry if they contain more than 2% of weed seeds of all kinds (Day, 1968).

As mentioned above, all states in the United States have relatively uniform seed laws which are similar to the federal seed law. Some problems may arise, however, in interstate seed shipment due to the fact that the list of noxious weeds may differ among states. Another limitation to the effectiveness of state seed laws arises from the fact that 80% of small grain seeds planted are produced by individual farmers (Day, 1968). Drill-box surveys indicate that such seeds often have a weed-seed contamination level which is higher than would be permitted in commerce (Day, 1968).

C. The Federal Noxious Weed Act of 1974

Although the need for such legislation had long been recognized, there was no federal law prohibiting the importation of any plant or plant part, including seeds—with the exception of plants which were parasitic or a host to some pest—until the Federal Noxious Weed Act of 1974 was passed (Rodgers, 1974; U.S. Congress, 1976). Concern for this matter had been expressed by some members of the Weed Science Society of America since the late 1950s (Rodgers, 1974). At its annual meeting in 1963, the National Association of State Departments of Agriculture adopted a resolution calling for a federal weed law (Rodgers, 1974). Representatives from the United States Departments of Agriculture, Defense, Interior, and Health, Education and Welfare united as a Federal Interagency Ad Hoc Committee on Preventive Weed Control in 1968. An in-depth study by this committee concluded that existing federal or state legislation did not provide effective protection against importation of noxious weeds, weed parts, and seeds (Rodgers, 1974).

The cries from these groups finally reached the ears of legislators, and on April 19, 1973, H.R. 7278, a proposed federal noxious weed act, was introduced in the House of Representatives, by Congressman Louis Frey, Jr., from Florida. Although this bill was endorsed by representatives from the United States Department of Agriculture, a representative of the Weed Science Society of America, and several other organizations and individuals, a lack of understanding of its intent regarding regulation of the interstate movement of seeds was the cause of some dissent. The bill was subsequently rewritten to delete most significant sections other than the requirements for inspection for noxious weeds at points of importation (Rodgers, 1974). Meanwhile, a bill similar to Frey's original bill was introduced into the Senate on November 20, 1973, by Senator Herman E. Talmage of Georgia (Rodgers, 1974). This bill, eventually passed and signed into law by President Gerald R. Ford on January 3, 1975, became known as the Federal Weed Act of 1974 (Buchanan, 1975; U.S. Congress, 1976).

This act prohibits the movement of noxious weeds into or within the United States except as authorized by the Secretary of Agriculture under a general or specific permit. Under this law, it is also illegal to receive, give, sell, barter, or deliver for transport such a weed (U.S. Congress, 1976). The term "noxious weed," as defined by Section 3(c) of the act, "means any living state (including but not limited to, seeds and reproductive parts) of any parasitic or other plant. . . which is of foreign origin, is new to or not widely prevalent in the United States, and can directly or indirectly injure crops, other useful plants, livestock or poultry or other interests of agriculture, including irrigation, or navigation or the fish or wildlife resources of the United States or the public health" (U.S. Congress, 1976). Specific weeds considered to be noxious are identified by the Secretary of Agriculture (U.S. Congress, 1976).

The act also authorizes the Secretary of Agriculture to issue quarantines and/or other regulations to prevent the dissemination of noxious weeds into or through the United States. He may call for the inspection of any product, article, or means of conveyance as a condition to their movement into or through the United States. Also, given reason to believe that an infestation of noxious weeds exists in a state, territory, or portion thereof, the Secretary may issue a quarantine restricting or prohibiting the interstate movement from that area of any product, article, or means of conveyance (U.S. Congress, 1976).

Under certain circumstances, the act allows for products or articles to be destroyed. As an emergency measure, the Secretary of Agriculture may order the seizure, destruction, quarantine, treatment, or other means of disposal of any item which he has reason to believe is infested by noxious weeds or is otherwise in violation of this act. The owner of any product, article, or means of conveyance which is in violation of the act may be ordered by a U.S. district court to destroy that item at his own expense (U.S. Congress, 1976). However, such action is to be taken only when, in the opinion of the Secretary of Agriculture, no less drastic action would be adequate to prevent the dissemination of the infecting weed (U.S. Congress, 1976). If he feels that the action taken was not authorized by this act, the owner of the product destroyed by such action may, within 1 year, challenge the action in the U.S. district court for the District of Columbia to receive compensation for his losses (U.S. Congress, 1976).

Sanctions for noncompliance are also provided under this act. A person violating the Noxious Weed Act may be convicted of a misdemeanor and is subject to a fine of up to $5000 and/or 1 year in prison (U.S. Congress, 1976).

D. The Role of APHIS

Although the Federal Noxious Weed Act was signed into law on January 3, 1975, Congress did not provide funds for its implementation until October 1, 1978 (Spears, 1979). No funding for the act was included in the original budget which President Jimmy Carter submitted to the Congress for fiscal year 1979.

TABLE V

Plants Designated as Noxious Weeds by the United States Department of Agriculture[a]

Aquatic weeds
Eichhornia azurea (Sw.) Kunth (anchored water hyacinth)
Hydrilla verticillata F. Muell. (hydrilla)
Monochoria hastata (L.) Solms
Monochoria vaginalis (Burm. f.) Presl
Salvinia molesta D. S. Mitchell
Sparganium erèctum L.
Stratiotes aloides L.
Parasitic weeds
Orobanche aegyptiaca Persoon
Orobanche cernua Loeft.
Orobanche lutea Baumg.
Orobanche major L.
Striga spp. (witchweed)
Terrestrial weeds
Avena ludoviciana Dur.
Carthamus oxycantha M. B.
Commelina benghalensis L. (tropical spiderwort)
Crupina vulgaris Cassini (bearded creeper)
Digitaria scalarum (Schweinf.) Chiov. (blue couch)
Drymaria arenarioides H. B. K. (alfombrilla)
Galega officinalis L. (goatsrue)
Imperata brasiliensis Trin.
Ischaemum rugosum Salisb. (saromaccagrass)
Leptochloa chinensis (L.) Ness
Mikania cordata (Brum. F.) B. L. Robinson (mile-a-minute)
Mikania micrantha Kunth
Mimosa invisa Martius (great sensitive plant)
Oryza longistaminata A. Chev. and Roehr.
Oryza punctata Kotschy ex Steud.
Oryza rufipogon Griff (red rice)
Pennisetum clandestinum Hochst. (kikuyugrass)
Pennisetum pedicellatum Trinius (kyasumagrass)
Pennisetum polystachion (L.) Schult. (missiongrass)
Prosopis ruscifolia Griseb. (mesquite)
Rottboellia exaltata L. f. (itchgrass, raoulgrass)
Saccharum spontaneum L. (wild sugarcane)

[a] Westbrooks (1981).

However, an additional $1.5 million was included by the Senate Agricultural Appropriations Committee, and after the bill went to conference, an agreement was reached to include $550,000 for activities under the Federal Noxious Weed Act (Spears, 1979).

The responsibility for implementing the act, and thus also for spending the money allocated by Congress for this purpose, was given to the Animal and Plant Health Inspection Service (APHIS) of the United States Department of Agriculture (Spears, 1979, Westbrooks, 1981). This money is spent to (1) develop a survey on candidate weed species, (2) enforce appropriate quarantines, (3) apply control or eradication measures, (4) coordinate work with agencies that have authority for weed control, and (5) strengthen inspection for foreign noxious weeds at ports of entry (Spears, 1979).

Recommendations for the Federal Noxious Weed List are made by a Technical Committee to Evaluate Noxious Weeds. This branch of APHIS recommends for the list those weeds which are potentially serious problems. Presently, this list (Table V) includes 33 species representing 24 genera plus all species of the genus *Striga* (witchweed) (Westbrooks, 1981). A total of 125 other species (Table VI) are presently being investigated as possible candidates for the list (Anonymous, 1981; Westbrooks, 1981). Although most of the designated species are not found in the United States, a few of these weeds have been introduced. However, in meeting the criteria set forth by the Noxious Weed Act's definition of a noxious weed, these introduced species have very limited distribution. Among others, these plants include water hyacinth, hydrilla, and witchweed (Westbrooks, 1981).

Quarantine and eradication programs have been initiated for certain weeds, including witchweed and hydrilla (Westbrooks, 1981).

After being discovered in the Carolinas in 1956, witchweed has been under federal and state quarantines since 1957 (Westbrooks, 1981). Since the Federal Noxious Weed Act was not in effect at that time, the federal quarantine was possible because witchweed is parasitic and can therefore be classified as a pathogen (Day, 1968). The eradication of this plant has involved three approaches: (1) prevention of seed production, (2) reduction of seed population in the soil, and (3) prevention of seed transport (Westbrooks, 1981).

Programs to control hydrilla in irrigated citrus have been initiated at LaBelle, Florida, and in the Imperial Valley Irrigation District of California (Westbrooks, 1981). These APHIS-funded programs to control this aquatic plant involve detailed treatments with various herbicide combinations (Westbrooks, 1981).

APHIS inspectors have been involved for many years with port of entry inspections for insects, nematodes, and plant diseases; and since the Federal Noxious Weed Act has been passed, these inspectors now have authority to prevent entry of noxious weeds (Spears, 1979). Inspection is made for plant propagules which may be harbored by various commodities (Westbrooks, 1981). Part 360 of the Noxious Weed Regulations used by APHIS classifies commodity

TABLE VI

United States Department of Agriculture's Proposed Additions to the Noxious Foreign Weed List[a]

Scientific name	Common name(s)
Aquatic weeds	
Azolla pinnata R. Brown	Mosquito fern, water velvet
Hygrophila polysperma T. Anderson	Miramar weed
Ipomoea aquatica Forskal	Water spinach, swamp morning-glory
Limnophila sessiliflora (Vahl) Blume	Ambulia
Sagittaria sagittifolia L.	Arrowhead
Parasitic weeds	
Cuscuta spp.	Dodders, other than those found in the United States
Terrestrial weeds	
Alternanthera sessilis (L.) R. Brown ex de Candolle	Sessile joyweed
Avena sterilis L. (including *Avena ludoviciana* Dur.)	Animated oat, wild oat
Borreria alata (Aublet) de Candolle	
Chrysopogon aciculatus (Retzius) *Trinius*	Pilipiliula
Crupina vulgaris Cassini	Common crupina
Digitaria velutina Forskal Palisot de Beauvois	Velvet fingergrass
Euphorbia prunifolia Jacquin	Painted euphorbia
Galega officinalis L.	Goatsrue
Heracleum mantegazzianum Somier and Levier	Giant hogweed
Ipomoea triloba L.	Little bell, aiea morning-glory
Mikania cordata (Burman f.) B. L. Robinson	Mile-a-minute
Mimosa invisa Martius	Giant sensitive plant
Pennisetum clandestinum Nochstetter ex Chiovenda	Kikuyugrass
Pennisetum pedicellatum Trinius	Kyasumagrass
Pennisetum polystachion (L.) Schult.	Missiongrass, thin napiergrass
Prosopis spp.	Mesquite (25 species listed)
Saccharum spontaneum L.	Wild sugarcane
Solanum torvum Swartz	Turkeyberry
Tridax procumbens L.	Coat buttons

[a] Anonymous (1981).

types which may harbor these plants into two groups—high risk and moderate risk (U.S. Department of Agriculture, 1976). Included in the high-risk category are such things as wool, cotton, soil or debris contamination, containers, and bagging, and plant collections, dried flowers, etc. Commodities listed as moderate risks include pallets and dunnage, animal hides, woody plant seeds, commercial aircraft, and residue from permit cargo (U.S. Department of Agriculture, 1976). Surveys are also taken by some officers to detect new weeds around ports of entry and other high-hazard areas (Westbrooks, 1981).

Various options are available to APHIS officers in dealing with products which are found to be contaminated with noxious weeds. These options include various forms of cleaning, such as steam cleaning, chemical treatment including fumigation, and mechanical removal of seeds, and refusal of entry (U.S. Department of Agriculture, 1976; Westbrooks, 1981).

Augmenting all these efforts are Exotic Weed Surveys which are conducted in cooperation with APHIS by universities and state departments of agriculture in 25 states and the Commonwealth of Puerto Rico (Westbrooks, 1981).

VII. SYNOPSIS AND CONCLUSIONS

Introduced weeds are a problem of man, created by man, and to be solved by man. This chapter has pointed out the magnitude of the problem. Weeds affect people in all walks of life, and control becomes more costly each year. The factors which enable many foreign plants to become weeds were discussed. This chapter has also emphasized how man, throughout the course of American history, has been responsible for the introduction of many of our most troublesome weeds. Then we looked at man's response to this problem—the legal measures enacted and implemented to prevent weed introduction and spread. In assessing the present status of this problem, two questions should be answered: (1) have these laws been successful? (2) what more can be done to prevent future weed introductions?

With a problem such as weed introductions, citing failures is much easier than documenting successes. The failures are our weed problems, which are here for all to see, but determining the problems which have been avoided is very difficult. The quarantine of witchweed, however, presents an excellent example of a legal measure which has enabled us to prevent the spread of a weed. Although there are no clear natural limitations to its spread, witchweed—a plant with both a capacity for and a history of spreading—has remained along the Eastern Coast of the United States since its introduction more than 25 years ago (Pavlista, 1980). Certainly, success has been realized with respect to weed and seed laws, and this success will continue and even increase if these laws are properly amended and updated in response to future developments in weed science. These laws fall short, however, in one key respect—they prevent the entry only of plants known to be weedy, overlooking intentional introductions of plants which may later become weeds.

Thomas Jefferson has said, ''The greatest service which can be rendered to any country is to add a useful plant to its culture.'' Introduction of new plant species is essential if the United States is to maintain its preeminence in world agriculture (Williams, 1980). Most of the important food and fiber crops so widely distributed throughout the world today actually evolved in a rather restricted geographic area (Hodge and Erlanson, 1956; Leppick, 1970). The need for new germ plasm in the United States and in most countries is as great as ever (Waterworth and White, 1982). However, when foreign pests, including plants which later become weeds, are carelessly introduced to new areas of the world, serious losses can result. As mentioned above, many intentionally introduced plants have become noxious weeds. Williams (1980) has suggested that future research efforts should identify factors which may predict the ability of plants to become weedy upon introduction. This research would be designed to identify and intercept those plants which would become weedy only after they encounter their new environment. Thus, both plant introductions and quarantine are necessary (Waterworth and White, 1982).

This is one example of how the problem of weed introductions may be handled more efficiently in the future. Surely, many ideas are yet to emerge from the creative minds of weed scientists as they are faced with the challenges of the future.

REFERENCES

Alex, J. F., McLaren, R. D., and Hamill, A. S. (1979). Occurrence and winter survival of johnsongrass (*Sorghum halepense*) in Ontario. *Can. J. Plant Sci.* **59,** 1173–1176.

Allan, M. (1978). ''Weeds: The Unbidden Guests in Our Gardens.'' Viking Press, New York.

Anderson, E. (1969). ''Plants, Man, and Life.'' Univ. of California Press, Berkeley.

Anderson, W. L., and Edminster, F. C. (1949). Multiflora rose for living fences and wildlife cover. *U.S. Dep. Agric., Leafl.* **256,** 1–8.

Anderson, W. P. (1977). ''Weed Science: Principles.'' West Publ. Co., New York.

Anonymous (1972). Weed facts. *Weeds Today* **3**(3), 17.

Anonymous (1981). USDA proposes additions to noxious weed list. *Agrichem. Age* **25**(10), 18.

Baker, H. G. (1965a). Discussion of paper by Dr. Malligan. *In* ''The Genetics of Colonizing Species'' (H. G. Baker and G. L. Stebbins, eds.), pp. 144–146. Academic Press, New York.

Baker, H. G. (1965b). Characteristics and modes of origin of weeds. *In* ''The Genetics of Colonizing Species'' (H. G. Baker and G. L. Stebbins, eds.), pp. 147–172. Academic Press, New York.

Baker, H. G. (1972). Migrations of weeds. *In* ''Taxonomy, Phytogeography, and Evolution'' (D. H. Valentine, ed.), pp. 327–347. Academic Press, New York.

Baker, H. G. (1974). The evolution of weeds. *Annu. Rev. Ecol. Syst.* **5,** 1–24.

Bates, G. H. (1955). ''Weed Control.'' Spon, London.

Blackwell, W. H. (1978). The history of Russian pigweed, *Axyris amaranthoides* (Chenopodiaceae, Atripliceae), in North America. *Weed Sci.* **26,** 82–83.

Brenchley, W. E. (1920). ''Weeds of Farmland.'' Longmans, Green, New York.

Brockway, L. H. (1979). ''Science and Colonial Expansion.'' Academic Press, New York.

Bronowski, J. (1974). ''The Ascent of Man.'' Little, Brown, Boston, Massachusetts.

Buchanan, G. (1975). Editorial: Applaud lawmakers. *Weeds Today* **6**(1), 4.

Bunting, A. H. (1960). Some reflections on the ecology of weeds. *In* "The Biology of Weeds" (J. L. Harper, ed.), pp. 11–26. Blackwell, Oxford.

Carlquist, S. (1981). Chance dispersal. *Am. Sci.* **69,** 509–515.

Crafts, A. S. (1975). "Modern Weed Control." Univ. of California Press, Berkeley.

Crafts, A. S., and Robbins, W. W. (1962). "Weed Control: A Textbook and Manual." McGraw-Hill, New York.

Crosby, A. W., Jr. (1972). "The Columbian Exchange: Biological and Cultural Consequences of 1492." Greenwood Press, Westport, Connecticut.

Day, B. E., chairman (1968). "Principles of Plant and Animal Pest Control," Vol. 2. Natl. Acad. Sci., Washington, D.C.

Dearborn, C. H. (1959). Weeds in Alaska, and some aspects of their control. *Weeds* **7,** 265–270.

DeWet, J. M. J., and Harlan, J. R. (1975). Weeds and domesticates: Evolution in the man-made habitat. *Econ. Bot.* **29,** 99–107.

Dewey, L. H. (1896). Migration of weeds. *U.S. Dep. Agric. Yearb.* pp. 263–286.

Dickens, R. (1974). Kudzu—friend or foe? *Weeds Today* **5**(3), 8.

Duke, J. A. (1981). Chief, Econ. Bot. Lab., Agricultural Research, Science and Education Administration, U.S. Department of Agriculture, Beltsville, Maryland (personal communication).

Evans, R. A., and Young, J. A. (1974). Today's weed—Russian thistle. *Weeds Today* **5**(4), 18.

Fawcett, R. S. (1980a). Today's weed: Multiflora rose. *Weeds Today* **11**(1), 22–23.

Fawcett, R. S. (1980b). Today's weed: Shattercane. *Weeds Today* **12**(1), 11–14.

Fogg, J. M., Jr. (1956). "Weeds of Lawn and Garden." Univ. of Pennsylvania Press, Philadelphia.

Fogg, J. M., Jr. (1966). The silent travelers, handbook on weed control. *Brooklyn Bot. Gard. Rec.* **22,** 4–7.

Foy, C. L. (1978). Weed science today. *Weeds Today* **9**(3), 16–17.

Gleason, H. A., and Cronquist, A. (1964). "The Natural Geography of Plants." Columbia Univ. Press, New York.

Gray, A. (1879). The predominance and pertinacity of weeds. *Am. J. Sci.* **118,** 161–167 (cited in McNeil, 1976).

Gray, A. (1889). "Scientific Papers of Asa Gray, Selected by Charles Sprague Sargent," Vol. I. Houghton, Boston, Massachusetts.

Groh, H. (1940). Turkestan alfalfa as a medium of weed introduction. *Sci. Agric. (Ottawa)* **21,** 36–43.

Haller, W. T. (1976). Hydrilla a new and rapidly spreading aquatic weed problem. *Circ. Univ. Fla., Inst. Food Agric. Sci.* **S-245,** 1–13.

Harlan, J. R., and DeWet, J. M. J. (1965). Some thoughts about weeds. *Econ. Bot.* **19,** 16–24.

Harper, J. L. (1965). Establishment, aggression, and cohabitation in weedy species. *In* "The Genetics of Colonizing Species" (H. B. Baker and G. L. Stebbins, eds.), pp. 243–268. Academic Press, New York.

Haughton, C. S. (1978). "Green Immigrants." Harcourt Brace Jovanovich, New York.

Hitchcock, A. S. (1951). Manual of grasses of the United States. *Misc. Publ.—U.S., Dep. Agric.* **200,** 1–1051.

Hitchings, S. (1960). The control and eradication of weeds in seed crops. *In* "The Biology of Weeds" (J. L. Harper, ed.), pp. 108–115. Blackwell, Oxford.

Hodge, W. H., and Erlanson, C. O. (1956). Federal plant introduction—a review. *Econ. Bot.* **10,** 299–334.

Holm, L. G., Plucknett, D. L., Pancho, J. V., and Herberger, J. P. (1977). "The World's Worst Weeds." Univ. Press of Hawaii, Honolulu.

Jennings, J. D. (1974). "Prehistory of North America." McGraw-Hill, New York.

Josselyn, J. (1672). "New-Englands Rarities Discovered." London (facsimile ed. by Massachusetts Historical Society, Boston, 1972).

King, L. J. (1966). "Weeds of the World: Biology and Control." Wiley (Interscience), New York.
Knake, E. L. (1977). Giant foxtail—the most serious annual grass weed in the Midwest. *Weeds Today* **9**(1), 19–20.
Leppick, E. E. (1970). Gene centers of plants as sources of disease resistance. *Annu. Rev. Phytopathol.* **8,** 323–344.
McNeil, J. (1976). The taxonomy and evolution of weeds. *Weed Res.* **16,** 399–413.
McNeill, W. H., and Adams, R. S. (1978). "Human Migration: Patterns and Policies." Indiana Univ. Press, Bloomington.
McWhorter, C. G. (1971). Introduction and spread of johnsongrass in the United States. *Weed Sci.* **19,** 496–500.
MacNeish, R. S. (1964). The origins of New World civilization. *Sci. Am.* **211**(5), 29–37.
Mitich, L. W. (1980). The intriguing world of weeds. *Weeds Today* **12**(1), 10–11.
Monsod, G. G., Jr. (1975). "Man and the Water Hyacinth." Vantage Press, New York.
Moore, R. S. (1979). Cutting propagation of roses. *Int. Plant Propagators Soc.* **28,** 170–175.
Muenscher, C. W. (1949). "Weeds." Macmillan, New York.
Mulligan, G. A. (1965). Recent colonization by herbaceous plants in Canada. *In* "The Genetics of Colonizing Species" (H. G. Baker and G. L. Stebbins, eds.), pp. 127–146. Academic Press, New York.
Muzik, T. J. (1970). "Weed Biology and Control." McGraw-Hill, New York.
Nalewaja, J. D. (1970). Wild oat: A persistent and competitive weed. *Weeds Today* **1**(2), 10–13.
Nelson, J. C. (1917). The introduction of foreign weeds in ballast, as illustrated by the ballast plants at Linnton, Oregon. *Torreya* **17,** 151–160 (cited in Brenchley, 1920).
Pavlista, A. D. (1980). Why hasn't witchweed spread in the United States? *Weeds Today* **12**(1), 16–19.
Reed, C. F., and Hughes, R. O. (1970). Selected weeds of the United States. *U.S. Dep. Agric., Agric. Handb.* **366,** 1–463.
Ridley, H. N. (1930). "The Dispersal of Plants Throughout the World." L. Reeve & Co., Ltd., Ashford, Great Britain.
Robbins, W.W., Crafts, A. S., and Raynor, R. N. (1952). "Weed Control: A Textbook and Manual." McGraw-Hill, New York.
Rodgers, E. G. (1974). Needed: A federal noxious weed act. *Weeds Today* **5**(4), 9–10.
Rosene, W., Jr. (1950). Spreading tendency of multiflora rose in the southeast. *J. Wildl. Manage.* **14,** 315–319.
Rousseau, J. (1966). Movements of plants under the influence of man. *In* "The Evolution of Canada's Flora" (R. L. Taylor and R. A. Ludwig, eds.), pp. 81–89. Univ. of Toronto Press, Toronto, Ontario.
Salisbury, E. (1942). "The Reproductive Capacity of Plants." Bell, London (cited in Stevens, 1957).
Salisbury, E. (1961). "Weeds and Aliens." Collins, London (cited in Harlan and deWet, 1965).
Sanders, H. J. (1981). Herbicides. *Chem. Eng. News* **59**(31), 20–35.
Shear, G. M. (1983). Introduction and history of limited tillage. *Weed Sci. Soc. Am. Monog. Ser.* (in press).
Smith, A. E., and Secoy, D. M. (1981). Weed control through the ages. *Weeds Today* **5**(4), 9–10.
Spears, J. F. (1979). Federal noxious weed act founded. *Weeds Today* **10**(2), 11.
Stapledon, R. G. (1916). Identification of the country of origin of commercial oats. *J. Board Agric. (G.B.)* **23**(1), 105–116 (cited in Brenchley, 1920).
Stearn, W. T. (1956). Book review of "Weeds" by W. C. Muencher. *J. R. Hortic. Soc.* **81,** 285–286 (cited in King, 1966).
Stevens, O. A. (1957). Weights of seeds and numbers per plant. *Weeds* **5,** 46–55.
Stewart, T. D. (1973). "The People of North America." Scribner's, New York.

U.S. Congress (1939). United States Statutes at Large. 76th Congress, 1st Sessions. U.S. Govt. Printing Office, Washington, D.C.

U.S. Congress (1976). United States Statutes at Large. 93rd Congress, 2nd Session. U.S. Govt. Printing Office, Washington, D.C.

U.S. Department of Agriculture (1968). Extent and cost of weed control with herbicides and an evaluation of important weeds, 1965. *U.S. Agric. Res. Serv., ARS* **34-102.**

U.S. Department of Agriculture (1976). Animal and Plant Health Inspection Service. Part 360. Noxious weed regulations. *Fed. Regist.* **41**(220), 49987–49988.

van der Pijl, L. (1969). "Principles of Dispersal in Higher Plants." Springer-Verlag, Berlin and New York.

Ver Steeg, C. L. (1964). "The Formative Years." Hill & Wang, New York.

Walker, G. M. (1917). "The Measure of Civilization." Guy M. Walker, New York.

Walton, R. P. (1938). "Marijuana, America's New Drug Problem." Lippincott, Philadelphia, Pennsylvania.

Waterworth, H. E., and White, G. A. (1982). Plant introductions and quarantine: The need for both. *Plant Dis.* **66,** 87–90.

Weed Science Society of America (1975). "Herbicide Handbook of the Weed Science Society of America," 4th ed. WSSA, Champaign, Illinois.

Wellington, P. S. (1960). Assessment and control of the dissemination of weeds by crop seeds. *In* "The Biology of Weeds" (J. L. Harper, ed.), pp. 94–107. Blackwell, Oxford.

Westbrooks, R. G. (1981). Introduction of foreign noxious plants into the United States. *Weeds Today* **12**(3), 16–17.

Williams, M. C. (1973). Halogeton . . . sheep killer in the West. *Weeds Today* **4**(3), 10–11, 22.

Williams, M. C. (1980). Purposefully introduced plants that have become noxious or poisonous weeds. *Weed Sci.* **28,** 300–305.

Young, J. A., and Evans, R. A. (1976). Responses of weed populations to human manipulations of the natural environment. *Weed Sci.* **24,** 186–190.

Zimmerman, F. R. (1955). Multiflora rose: Living fence. *Wis. Conserv. Bull.* **20**(4), 1–8.

Zohary, M. (1962). "Plant Life of Palestine." Ronald Press, New York (cited in Muzik, 1970), p. 2).

CHAPTER 5

Where Are the Exotic Insect Threats?

JOHN D. LATTIN and PAUL OMAN

Systematic Entomology Laboratory
Department of Entomology
Oregon State University
Corvallis, Oregon

I. INTRODUCTION

The past is prologue. Any speculation regarding geographic sources of potential pest arthropods from abroad should begin with the excellent summarization of the past record, as presented by Sailer (1978). It is abundantly clear that the immigrant arthropod fauna now present in North America originated for the most

EXOTIC PLANT PESTS AND
NORTH AMERICAN AGRICULTURE

ISBN 0-12-757880-3

part in the western Palearctic Region (Europe and the Mediterranean Basin). The same generalization holds for the complex of agriculturally important species. Next in importance, numerically, are the Orient (South Asia) and the eastern Palearctic (north Asia). South and Central America, Africa (the Sahara and southward), Australia and the outlying major islands, and the remainder of the earth's land surfaces have contributed much less generously to our immigrant fauna.

The reasons for the disparate source pattern are varied, but understandable, as analyzed by Sailer (Chapter 2, this volume). Quite obviously, as evidenced by the changes in the kinds of immigrants over time, the situation is fluid and likely to continue so for the foreseeable future. The ''saturation point'' for the arthropod fauna of North America is surely many eons away. Several factors suggest that although the general pattern of immigration may continue, some adjustments in relative numbers will probably occur.

In the context of discussions in this chapter, ''exotic'' organisms are construed as those organisms not indigenous to the Nearctic Region. Only the discussion of home-grown and cryptic hazards concerns indigenous species that may become crop pests.

If we accept as valid the idea that movement across latitudes has a greater chance of success than south–north movements, we should anticipate a continuing stream of immigrants from the western Palearctic. But surely some, and perhaps most, will be repeaters, i.e., species that have already breached the Atlantic Ocean barrier and are now established in North America. Although perhaps not so great a hazard as the initial invaders, these repeaters, by supplementing the range of genetic materials available in North American populations, increase the possibility of the development of more capable pest biotypes. Hazards of this sort will serve to keep attention focused upon the agricultural crop pests that occur in the western Palearctic Region.

Of course, North America has not been the sole recipient of emigrant pests, other major land masses have also been invaded by exotic pest species. This means that the country of origin is not necessarily the place from which a pest will depart for invasion of our continent. A case in point is the 1981 invasion of the Mediterranean fruit fly. The present known world distribution of the pest clearly shows many potential sources from which this recent infestation in California might have originated.

As noted, the record of immigrant species tends to support the hypothesis that exotic arthropods with the potential to damage crop plants in the United States are most likely to be found in climatically similar regions of the world. But climatic similarity alone is insufficient to identify regions where the greatest hazards likely occur; regions where floral or agroclimatic analogues occur should also be considered (Nuttonson, 1965). For example, the vegetative diversity of the southern and southeastern United States includes grasses and other herbaceous plants, broadleaf deciduous trees, and needleleaf evergreen trees. Global analogues of such a climatic and vegetative region include virtually all of western Europe, and also an extensive band across central Asia, including Manchuria, the far eastern USSR, and northern Japan (Espenshade, 1979). Thus, attention needs to be directed toward *regions* of the world, rather than specific countries, except as commerce and tourism may serve to facilitate movement of pest species. One of the best sources of information about the distribution of major insect pests of agricultural crops, worldwide, is a series of maps prepared and periodically updated by the Commonwealth Institute of Entomology (Anonymous, 1951–1980).

No reliable inventory of the world's crop pests exists. The number of species that are known to reduce yields is legion and includes 10,000 species of insects (Klassen *et al.*, 1975, p. 275). Even less certain is the number of phytophagous species in the world from which more pest species may emerge. There are some 55,000 known (i.e., described and named) species of the phytophagous beetle family Curculionidae, many of which are known pests. Yet, this number is believed to represent only about 10–20% of the world's species of that family (Batra *et al.*, 1977). These discouraging data are probably reasonably representative of the phytophagous insect fauna in general, although reliable inventories of described species are available for very few groups that contain numbers of known pest species.

Some comprehension of the limitations on our knowledge of insects of the world may be attained by scrutiny of data pertaining to some groups for which the published records have been summarized. One such is the homopterous family Cicadellidae (= Cicadelloidea of Metcalf). Metcalf's catalogues (1962–1968, incl.) covered the literature through 1955. He recorded 9078 species assigned to 1074 genera. In the 27-year period since 1955, the number of recognized genera has more than doubled, accompanied by an enormous increase in the number of recognized species. This explosive expansion of knowledge of the group involves all faunal regions, but is primarily due to increased attention to previously little known areas. Even so, an approximate inventory of many regions is as yet far from attained. For example, an analysis of the cicadellid fauna of Chile (Linnavuori and DeLong, 1977) revealed only 119 species. Of these, 43 were first described in the aforementioned report, and another 41 had been described since the cutoff date (1955) of Metcalf's catalogue. In brief, the as yet very limited survey of the Chilean leafhopper fauna increased the known number of species by more than 200% during the past quarter-century.

II. RECOGNITION OF INSECT PEST SPECIES

A. Status of Present Knowledge of Insect Pests Worldwide

Our knowledge of insect pests is best in the North Temperate areas of the world, particularly Western Europe and North America. The temperate regions of eastern Asia are less well known. This general knowledge is tied to the development of agriculture, coupled with a relatively high level of taxonomic activity supporting applied entomological investigations. The resurgence of interest in the entire insect fauna of agroecosystems, including beneficial species, has increased the taxonomic load considerably since the time when main target species received most attention and control relied heavily on chemicals only.

Information on the subtropical and tropical pests is limited largely to the major pests of main crops, and even here there are taxonomic problems of great magnitude. Detailed information on lesser pests and pests of lesser crops is still in short supply. The same is true of the vast array of beneficials. The general paucity of resident systematists in these regions is partly responsible for this condition. So much information is needed, and so few people are able to provide it. A strong effort should be made to establish priorities for investigations. Most of the systematic expertise lies outside the region, further delaying accurate identifications within a reasonable period of time. Well-organized Systematic Service and Research Centers are urgently needed (Lattin, 1980; Lattin and Knutson, 1982) to assist in the assemblage of the required data bases and to participate in the creation of new knowledge in appropriate languages. Quality investigations of the total insect fauna of crops in virtually every area, coupled with the preservation of adequate reference materials, is badly needed. It is anticipated that the production of literature will gradually shift from North Temperate systematists to resident systematists as the expertise is acquired. Greater utilization of existing centers and museums is needed, together with the establishment of new centers at appropriate sites. A more organized system of providing the proper education and training of resident applied systematists is needed to ensure the supply of an ever-increasing cadre of individuals capable of providing their own identification and training their own personnel.

B. Factors Complicating Species Recognition

The need for identification of the organisms involved is generally understood in dealing with pest situations. Unfortunately, in entomology, our ability to make precise identifications is severely handicapped by the multiplicity of phytophagous arthropods with which we must deal. Many of the species are polymorphic, and most of them occur in several quite different-appearing developmental stages, making identification of population samples a very complex task. The problem is further compounded by the fact that many insect "species" appear to be composed of a bewildering array of sibling species, host races, biotypes, or other infraspecific taxa that may interact differentially with different varieties of crop plants (Cronquist, 1977).

The ability of wild insects to adapt to changing environmental conditions is well established (Hendrick *et al.*, 1976), and the existence of genetic diversity in insects is amply illustrated among the pests of such major crop plants as rice, wheat, maize, and alfalfa. Among cereal insect pests, there are three biotypes of the greenbug [*Schizaphis graminum* (Rondani)] that differ in their ability to live on and damage varieties of wheat, barley, and sorghum. Another pest of sorghum, the corn leaf aphid [*Rhopalosiphum maidis* (Fitch)], has four or more biotypes that differ in fecundity or survival capacity on different host plants. And the Hessian fly [*Mayetiola destructor* (Say)] is now known to have eight biotypes which interact differentially with varieties of wheat that have different gene combinations (Gallun *et al.*, 1975). Biotypes of pest species in other major insect taxa are also known.

The noctuid genus *Spodoptera* (Lepidoptera) includes important pests of many crops in almost all parts of the world. Ashley (1981) mentions 23 species; the exact number is uncertain. The exotic species most frequently mentioned in the literature are *Spodoptera exempta* (Walker), *S. maurita* (Boisduval), *S. littoralis* (Boisduval), and *S. litura* (Fabricius). Unfortunately, which distribution record is applicable to which nominal species is far from clear. Under the name "*Prodenia litura,*" the Egyptian cottonworm, *S. litura* was indicated as having a very wide distribution (Anonymous, 1957, No. 20, pp. 385–386). Rivnay (1962, p. 104) stated that the species "is reported from all continents except the American. However, not everywhere does it occur or cause damage in equal measure." Subsequently, Hill (1975, p. 307) considered the cotton leafworm of the Mediterranean basin to be *S. littoralis,* with *S. litura* absent from Africa but occurring in Asia and Australia. The five species of *Spodoptera* considered to be important agricultural pests in the tropics (*exempta, exigua* (Hubner), *littoralis, litura,* and *mauritia* (Hill, 1975, pp. 303–308) show considerable overlapping of distributions. Correct allocation of past records is clouded by time and may never be satisfactorily resolved. Situations such as this are very suggestive of species complexes, with the expectation that intergradation and possibly hybridization may occur.

Although biochemical methods promise to provide means for discrimination among biotypes, sibling species, etc. of a few intensively studied groups of insects (Berlocher, 1979, 1980), for pest species in general the state of our knowledge is such that we are still unable to make satisfactory predictions as to which species, species groups, or genera are likely to be of agricultural importance (Batra *et al.*, 1977, p. 289).

C. How Lacunae in Available Information Affect Determination of Pest Status or Potential

Among the most difficult questions to answer are these: What will be the status of an insect species detected for the first time in the United States? What will be the evaluation of an insect species likely to be introduced? Well-known species (e.g., the winter moth) offer fewer problems simply because we have considerable information on them. Even here, it is difficult to judge how any particular species will act in a new environment. The species cited above is a good example. The winter moth is not considered a pest in England, but we have a great deal of information on it provided by Varley and Gradwell (1968). Once introduced into eastern Canada (Embree, 1965), it took some time before it was apparent that it might reach pest status. The same situation occurred when it was detected in British Columbia (Gillespie *et al.*, 1978) and, somewhat later, in the northwestern United States (Ferguson, 1978; Miller and Cronhardt, 1982). Extensive information was available on the species in Europe, including detailed data on the wide array of parasites. Some of these were introduced into Canada and, later, into the United States. We are not so fortunate in most instances.

When an exotic species is first detected and its native home is determined to be Southeast Asia or the dry coast of South America or southern Brazil, we may not be so fortunate. Detailed knowledge of most exotic species—except for those coming from areas whose fauna is not only well known but for which extensive information is available on food plants, members of generations, and/or parasites and predators—is often fragmentary or not known at all. The extensive searches for biological control agents throughout the world provide ample evidence of that fact. Many of these searches must start with educated guesses and may involve years of work in order to narrow down the possibilities and gather sufficient data to permit intelligent decisions to be made. True parasite–host relationships may be obscured because of incomplete taxonomic knowledge. The lacunae in our knowledge of most of the insects of the world are embarrassingly large. The sheer numbers of species, combined with our limited biological information on them, makes the task exceedingly difficult. It is a tribute to the comparatively few people working in entomology (compared to the number of insect species) that we have done so well, and yet we are constantly surprised by our ignorance. Shortcuts must be taken to extrapolate from available information because the number of people able to acquire such data is so limited. The unknown factors involved in determining in advance just how a species will behave in a new environment are still unacceptably large. Some of the complicating factors are discussed in Section III.

III. CHARACTERISTICS OF ORGANISMS THAT INFLUENCE PEST STATUS

A. Introduction

A natural reaction to the invasion and temporary establishment of any exotic arthropod, other than those considered as beneficial, is one of concern. If the invader is a known agricultural pest, the level of concern is immediately high; if the pest potential of the immigrant is unknown or uncertain, steps to determine it should be initiated. Unfortunately, our understanding of the fundamental biological processes that predispose some species to become pests is still far from adequate to permit accurate predictions, although the probability of a high degree of genetic plasticity may reasonably be assumed if the immigrant is one of a complex of biological races, semispecies, or subspecies, some of which are known pests.

But invasion and establishment are not necessarily equated with pest status. A considerable proportion of the immigrants are likely to be of no economic importance, or at most minor pests (Sailer, 1978), due in part to the different environmental conditions encountered. And even the pest species are usually ''fine-tuned'' to environmental conditions, so that their population levels tend to fluctuate from year to year. A major factor in such population fluctuations is the impact of natural enemies—pathogens, parasites, and predators—which, unfortunately, seldom accompany immigrants to their new homes. But, of course, other factors, such as agricultural practices, which may vary considerably from one region to another even in ecologically similar areas, may influence the impact of pest species. The value of *a priori* knowledge of the biology and behavior of a pest becomes evident during the decision-making period that follows detection of a new immigrant.

Environmental extremes that affect crop plants, especially temperature and humidity, are also critical factors in pest population dynamics. Species that are pests in environments lacking in temperature extremes and low humidities may not survive high summer or low winter temperatures and low humidity during periods of dormancy, even though tolerant of ambient temperatures during their active stages. Some pests, intolerant of either summer or winter climatic stresses, cause extensive damage in cropping areas to which they migrate seasonally. Such species, even though capable of invading different major land masses, may not always find suitable seasonal refugia that permit population survival during noncrop periods. Thus, available knowledge of a pest's lack of climatic tolerance will influence the level of concern accorded an invader. For example, the periodic occurrence of *Sogatodes orizicola* (Muir), and the outbreaks of hoja blanca disease of rice in Florida, Louisiana, and Mississippi during the late 1950s and early 1960s (Everett and Lamey, 1969), are presumed to indicate that the vector is unable to tolerate the winter conditions that usually prevail in the Gulf Coast region. Consequently, hoja blanca currently receives little attention in the United States, although it is a major rice disease in more southerly regions of the Western Hemisphere.

Food availability is obviously a very important factor in pest population dynamics. A species that depends primarily upon a single crop plant as a food source may go unnoticed, or be considered only a minor pest, when the crop plant is not routinely available. However, the same species may have the capacity to build up to damaging levels through succeeding generations if its food plant becomes available year after year through repetitious planting, conditions often dictated by economic considerations under our highly specialized crop production methods.

Although we cannot afford to be indifferent to the arrival of immigrant pests, it is important to remember that neither agriculture (Jones, 1973) nor the tactics and strategies for dealing with crop pests are static (Knipling, 1979). Pest management methods are constantly being improved, and these developments alter attitudes toward invading pests. The citrus blackfly (*Aleurocanthus woglumi* Ashby) caused a great deal of concern when infestations occurred in Texas and Florida during the 1960s and early 1970s. Nevertheless, the widespread occurrence of the pest in Florida in the late 1970s, considered in light of the success of a biological control program against it in Mexico (Smith *et al.,* 1964) and elsewhere, became less objectionable, leading to the decision to live with it in the southeastern citrus-growing region.

The concept and practice of coordinated control programs against pests in other parts of the world (Brader, 1979) will produce much useful information in dealing with potential invaders. Information about the natural enemies of those pests that may become permanent residents in our agroecosystem is especially important (U.S. Department of Agriculture, 1978). A great deal of useful information in this and other areas of pest management tactics in numerous countries has been acquired since the early 1960s through the U.S. Department of Agriculture's P. L. 480 program. Examples of the types of information available are contained in the following technical reports (Ghani and Muzzaffar, 1974; Grover and Prasad, 1976; Hamid, 1976; Ibrahim, 1979; Isa, 1979; Israel and Padmanabhan, 1976; Lipa, 1975; Niemczyk *et al.,* 1976; Shaikh, 1978; Tao and Chiu, 1971; Wiackowski *et al.,* 1976; Yen, 1973).

B. Factors Influencing the Potential for Translocation

The various avenues for dispersal or translocation of any potential pest are important considerations. Natural spread from adjacent geographic areas via prevailing winds might allow the establishment of a pest from a nearby country, i.e., island hopping in the Caribbean to Florida, or movement south on prevailing winds from western Canada or north into the Southwest from Mexico. Naturally, these same avenues for translocation may go, at times, in the opposite direction. Successful establishment will depend upon the survival of the vagile stage, whatever that stage might be.

Spread by human activity is probably a more common mode of translocation. According to Sailer (1978), approximately 750 of some 1300 immigrant species in the United States had become established by 1930. Another 150 or so were added by 1940. The greatly increased air traffic subsequent to 1940 may have contributed to the 400 or so species added since that time. Certainly, the rapid transit time, together with the increased ability to move foodstuffs over long distances, enhances the opportunity for successful movement of pest species from one area to another.

The phenology of an insect or its placement in time in association with its hosts may determine the potential for movement from one site to another. The coincidence of the life stage with suitable sustenance once in place in a new locality may determine successful colonization. Although phenological coordination with site is important in all such cases, it is particularly true for organisms moving from one hemisphere to the other. The reversal of seasons is an obvious barrier of considerable magnitude for most organisms. Comparatively few of the species introduced into United States have come from the Southern Hemisphere (Sailer, 1978).

The duration of the vagile stage influences the dispersal capabilities of any species. The longer the stage, the greater the opportunity for accidental or intentional dispersal. Although such statements are meant here to apply to accidental introductions, certain of these are relevant to persons interested in the purposeful introduction of biological control agents. Indeed, in the last instance, artificial prolongation of the vagile stage might be considered.

The specific niche association of the mobile or vagile stage will influence the potential for dispersal. Insects whose eggs are inserted into plant tissues or larvae or adults that are borers are more likely to go undetected than a larva on a leaf or a pupa attached to the trunk of a sapling. Control methods applied to such plant materials will have a greater degree of success with stages more vulnerable to those methods. Insects laying eggs on leaves are unlikely to be dispersed through commerce if the sapling is shipped as a leafless, bare-rooted whip or as a seed.

The degree to which individuals of a particular species are sensitive to control procedures applied prior to movement, i.e., fumigation or spraying of plant materials, will influence the potential success for movement. Normal resistance to the materials used would of course apply, but so would improper application of such lethal doses. Similarly, any control tactic applied at the point of entry may also influence the chance for survival essentially to the same conditions. The location of the life stage being moved is important. Eggs embedded in the tissue of plants may be much less susceptible to control than those on the surface of the plant.

Insects with narrow temperature tolerances present a much smaller "window" for potential transport than those whose movable stage shows great tolerance to temperature fluctuations. The latitude of origin may influence the transport potential since wider temperature fluctuations normally would be experienced in the mid-latitudes compared to the tropics. An egg or pupa in deep diapause in response to higher latitudinal conditions might provide more time for transport.

C. Factors Influencing Establishment, Survival, and Subsequent Spread

The type of reproduction possessed by a potential pest and/or beneficial species may influence the ease with which the organism becomes established. Bisexual species normally require both sexes, although, of course, that problem may be avoided if a female is fertilized prior to the introduction. Once introduced, the necessity of locating a mate in order to produce a generation may reduce the chances of a successful colonization.

Species able to reproduce parthenogenetically would seem to have a distinct advantage in successful establishment, other things being equal. Indeed, a number of pest species already introduced are parthenogenetic, e.g., the white-fringed weevil and many aphids. The life stage involved in transport would exert some influence, particularly if adults were required rather than eggs. Short-lived adults would require rapid transport. Parthenogenetic reproduction may permit very rapid spread once establishment occurs, e.g., the blue alfalfa aphid, *Acyrthosiphon kondoi* Shinji. Some of the flightless weevils may spread far more slowly. Hermaphroditism is rare in insects, but there is at least one example of a hermaphroditic pest species: the cottonycushion scale, *Icerya purchasi* Maskell.

The number of generations passed by any species may influence the success or failure of establishment. Given suitable host material, it would seem advantageous to have multiple generations simply to provide the possibility of building up the establishing population. On the other hand, if the species normally undergoes multiple generations, the absence of suitable host material for a second or third generation may prevent the establishment of the species. The synchrony of a single-generation species may be easier than that of a species with multiple generations.

A number of grass-feeding species of plant bugs (Hemiptera: Miridae) have been introduced into the United States from Western Europe: *Leptopterna dolobrata* (Fallén), *Megaloceroea recticornis* (Geoffrey), *Lopus decolor* (Fallén), etc. All have a single generation per year. Several very common grass-inhabiting mirids of Western Europe—*Notostira elongata* (Geoffrey), *N. erratica* (Linnaeus), and *Stenodema* (*Brachytropis*) *calcaratum* (Fallén)—have not yet been detected in North America. All have two generations per year, and all overwinter as adults. *Stenodema* (*Brachytropis*) *trispinosum* (Reuter) has two generations and also overwinters as an adult, but is considered a true Holarctic species, occurring naturally in both the Old and the New World (Wagner and Weber, 1964; J. D. Lattin, unpublished). It may be that multiple generations compound the problems of successful colonization.

Assuming that an adequate number of individuals arrived as an introduction, the generation time could exert a significant influence upon the success of establishment. Short generation time, combined with multiple generations, could enable an invading population to build up significant numbers to create a population large enough to ensure survival. Further, should the invaders arrive near the end of the season of the preferred host plants, there still might be time for a generation to complete its development. Species with long generation times require greater synchrony with the new environment if successful colonization is to occur since adequate resources must be available throughout the life cycle.

Even if close synchrony exists between the phenology of the invading species and the new physical environment, the absence of a suitable primary or secondary host plant would doom attempted colonizations to failure unless a dramatic host plant shift occurred. The possibility of such a host shift actually occurring might indeed be greater in a totally new environment, where host plants not previously encountered might be found. Naturally, the nutritional value of such plants may play an important role in survival.

Another factor to be considered is the accessibility of the host plants to the invader. Barriers of distance from the point of introduction and the occurrence of the host plant are important. The likelihood of successfully locating suitable host material is a function of distance.

Successful colonization depends upon the ability of an invading species to adapt to an environment different from that of its native home. Further, the points of introduction are likely to be sites where the environment has undergone considerable alteration, usually reducing the number or diversity of available host plants. Species arriving from areas that have already undergone extensive alteration (e.g., Western Europe) may have been exposed to selective pressures inherent in such environments and thus may show some preadaptation to similarly disturbed environments in new locations. This may account for the unusually large number of successful introductions from Europe.

Any invading species must be able to cope with the new environment as it exists at the time of arrival, regardless of the life stage of the invader. An active immature or adult stage would seem to face the greatest difficulty simply because food normally would be an immediate concern. The egg or pupal stage would face similar problems once that stage had ended.

Competition with existing native species is a form of environmental resistance that might make successful establishment difficult. An invading species coming into a balanced fauna where most of the habitats already are occupied might have a more difficult time than a species entering an imbalanced fauna associated with heavily disturbed environments. The latter are far more likely to be found around most points of introduction.

Species requiring both sexes for successful establishment of a colony face yet another barrier—that of locating the opposite sex. The great diversity of mate-finding tactics known for most species makes it difficult to generalize. Those species whose systems are the most highly developed are likely to be favored. Species requiring highly specific conditions for mate location may not become established if these conditions are not available. Where an immigrant flora has preceded the insect (whether weed or crop), the particular requirements may be met. Absence of the appropriate plants or conditions that may facilitate mate location would be a deterrent to successful establishment.

Although range extensions depend upon dispersal, such dispersal does not necessarily result in range extension. The biological and behavioral traits of species (populations) that make them successful usually enable them to expand their ranges rapidly once they are established in North America or some comparable land mass. Those that were initially dependent upon human activities for their successful invasion will still be largely dependent upon humans for further dispersal, barring certain atmospheric conditions to be mentioned later. The utility of regulations designed to prevent domestic translocations of pests in general is limited to pests of this category.

Vagile pest species that disperse or migrate by flight pose quite different problems (Johnson, 1969). Such species will presumably spread to the limits of their bioclimatic tolerance, with corresponding gradients in pest states. Rapidity of spread by such species usually depends upon the direction and magnitude of prevailing air mass movements during times of population dispersal and the availability of suitable hosts along dispersal pathways. A few examples of successful invaders will illustrate some of the conditions that favor spread of pests.

The boll weevil (*Anthonomus grandis* Boheman), an exotic species in the sense of not being native to the United States, became a pest when extension of the cotton culture made a favored plant readily accessible to it. Even so, the rate of spread was not especially rapid. In contrast, the spotted alfalfa aphid (*Therioaphis maculata* Buckton), a highly vagile species with a great reproductive capacity, was able to expand its occupied territory to the limit within a few growing seasons. This rapid development was aided by parthenogenetic reproduction, *a priori* climatization, and the existence of the favored host plant, alfalfa, at virtually all places where airlifted immigrants landed. In both cases, the direction of spread was from southwest to northeast, and was aided by the prevailing movement of summer air.

In contrast to the boll weevil–spotted alfalfa aphid pictures, the gypsy moth [*Porthetria dispar* (Linnaeus)] made relatively little progress on its own in westward extension of its range, major population extensions being attributable to transport by humans. Within and immediately adjacent to areas where major infestations of the species occur, some extension of the occupied areas occurs by "ballooning" by early instars.

Other contrasts that should be noted are those exemplified by pests established in the Northern Hemisphere that spread northward, and those that spread southward from their initial points of establishment. Elsewhere, mention is made of the periodic northward movement of essentially tropical insect pests that may become established temporarily, either for a single season or for two or more seasons during periods of mild winters. Such pests, intolerant of the usual climatic conditions in the continental United States, are but occasional visitors, unlikely to develop biotypes capable of becoming permanent resident pests. Their invasions are presumably possible because of favorable atmospheric conditions at a suitable time. Thus, the range of these northward-moving pests may be visualized as a sort of accordion effect, expanding and contracting with climatic conditions, with permanent survival in the southern refugia.

The pattern of north-to-south spread of invading pests is somewhat different. Prevailing winds during periods of population dispersal seldom favor long-range movements of migrants, so the extension of infested areas is a more gradual process. However, once a southward expansion of range is accomplished, populations appear to persist. Examples are as follows:

Macropsis fuscula (Zetterstedt), a leafhopper vector of the *Rubus* stunt mycoplasma in Europe, was first discovered in North America in British Columbia in 1952 (Bierne, 1954). In 1969, it was found to occur in northwest Washington on wild blackberry, and by 1981 was well established southward at numerous sites in the Willamette Valley of Oregon, most abundantly on wild *Rubus*. A continued southward spread in the western mountains and valleys where *Rubus* is abundant is to be expected.

Eight of the species listed in this section as potential new pests are already present in North America. If the foregoing assumptions of dispersal and range extension are valid, those known to be established in Canada (*Acleris comariana, Incurvaria rubiella, Psylla mali,* and *Psylloides chrysocephala*) will eventually arrive within our borders. Range extension of *Acleris comariana* southward along the west coast will be aided by prevailing summer air movement. If *Chaetocnema concinna* proves to be established in Massachusetts, a gradual range extension will occur. *Stenoma catenifer,* present in Central America and Mexico, is likely to invade northern areas where its host (avocado) is grown. The two beetles that occur in Mexico and Central America (*Diabrotica speciosa* and *Epicaerus cognatus*) probably pose less of an immediate threat because of the bioclimatic conditions that will be encountered as they move northward.

Some species require a dispersal flight prior to the development of the ovaries. For such species, at least three major factors are involved in successful dispersal and colonization: (1) the dispersal itself, (2) the location of a mate, and (3) the location of a suitable habitat. The second step would not be of concern to a parthenogenetic form. Species whose ovaries develop prior to dispersal might have two dispersal states—unfertilized and fertilized. The timing of the dispersal flight relative to the development of the eggs could influence the duration and distance of the dispersal flight.

Any consideration of a potential pest should include the potential of that organism for the development of biotypes to fit particular habitats. This type of information may not be readily available but must be surmised from available evidence. Does the species occur over a wide geographic area in its homeland, or does it have a restricted range? Does the species occur in a wide range of habitats, or is it limited to one or several restricted habitats? Does the species show considerable morphological variation, or is it rather uniform in appearance? Is the species known to be associated with a wide range of hosts, or is it quite host specific? Does the species have multiple generations, or is it univoltine? Does the species have more than one form—i.e., is it parthenogenetic, is there a migratory phase, or is it polymorphic? The problem may be less difficult if the species is from a well-known fauna, so that answers to some or all of the above questions may already be known. Species originating in countries where the fauna is poorly known may make predictions difficult or impossible. Detailed, systematic knowledge may make it possible to extrapolate from closely related taxa whose habitats are better known. Recent systematic revisions may be of enormous assistance in the evaluation of the potential pest status, e.g., the recent revision of *Epilachna* in the Coccinellidae by Gordon (1975).

Introduced species may have the capacity to become established in a new environment and effectively replace an existing species, whether the second species is itself an introduced species or a native. This replacement may be due to the ability of the newly introduced taxon to overcome the normal environmental resistance and compete with a species already in place. The resident species may not have been exposed to vigorous competition from another species, or it may be occupying a marginal environment to which it is poorly adapted but able to survive in the absence of a competitor. A habitat that has been perturbed but not yet exposed to invasion is an example of such a condition. Any preadaptation that may have occurred in the invading species, such as adapting to disturbed environments in its own original home, may make it a superior competitor in a new environment on this continent. Species introduced from Europe, where completely natural environments are comparatively rare, may be those in which some preselection has taken place. A variety of grass-inhabiting plant bugs (Hemiptera: Miridae) that occur today on grasses in disturbed situations in Europe are found in similar environments in the United States, often replacing native U.S. species. *Leptopterna dolobrata* (Fallén) from Europe replaces *L. ferrugata* (Fallén), a Holarctic species common in the United States, in the Willamette Valley of western Oregon (J. D. Lattin, unpublished) in disturbed habitats. The winter moth, *Operophtera brumata* (Linnaeus), may have replaced the native species, *O. occidentalis* (Hulst), in the northern portion of the Willamette Valley, western Oregon, in the urban area of Portland and vicinity (Miller and Cronhardt, 1982).

The distribution and availability of host plants suitable for the successful establishment of a newly arrived pest species are of great importance. Acceptable plant species that are commonly encountered, particularly at points of introduction, and whose occurrence is continuous or almost so, in contrast to those with a very patchy distribution, should enhance the likelihood of contact, establishment, and subsequent spread. The vagility of the invading species plays an important role, but the even distribution of some plants, e.g., those commonly occurring along roadsides or found commonly in disturbed habitats or other types of agroecosystems, may enhance the successful location of a suitable host. Pure or almost pure stands of many western tree species, coupled with current harvesting and reforestation plans, may make these habitats vulnerable to successful colonization and subsequent spread of introduced forest insect pests. Once established, the wide range occupied by some of these tree species would make it comparatively easy for the spread of a new species of pest. Thus far, the western forests are remarkably free of such pest species. Increased trade with eastern Asia, where, for example, the closest relatives to the Douglas fir (*Pseudotsuga menziesii*) are found, may change the present situation.

Invading species combining the characteristics of comparatively easy detection (e.g., pheromone traps) and slow buildup of an adequate population to permit successful dispersal may lend themselves to containment or eradication. Early detection of invasion and the availability of acceptable eradication methods may allow successful spot treatment. Species for which no readily available assay method has been developed, or species whose successful control and possible eradication may involve practices unacceptable for a variety of practical and/or environmental reasons, present an entirely different problem. For example, although assay methods are available for the gypsy moth, environmental–political factors may complicate the implementation of an eradication program. Assay methods are unavailable, at present, for a wide variety of pest species either because of undeveloped methodology or lack of interest.

It is axiomatic that any exotic phytophagous pest species that invades our domain concerns American agriculture. Yet invasion, even if followed by establishment, does not necessarily result in crop damage. In viewing the array of exotic pests, it becomes evident that there is a wide range of potential for damage, as well as for invasion. Thus, the degree of concern generated by an invader is likely to be tempered by the realization that some pest management strategies and tactics against it have already been tested. The unintentional widespread dispersal of agricultural pests and the intentional global movement of crop plants have brought many pest species into contact with new and different hosts. The increasing implementation of integrated pest control programs (Brader, 1979), exploitation of genetic diversity in crop plants to counter pests (Chiang, 1978; Gallun *et al.*, 1975; Kiritani, 1979; Singh and van Emden, 1979; Radcliffe, 1982), and ongoing international cooperative research programs such as those conducted at International Agricultural Research Centers and by the U.S. Department of Agriculture under Public Law 480, produce a wealth of useful information about the pest potential of species in many parts of the world and about methods for dealing with them should they arrive in North America. Pest management strategies are constantly being improved, making less distasteful the sometimes inevitable decision to live with an immigrant pest.

D. Potential New Pests

The following list of exotic pests is far from exhaustive and provides only examples of the kinds of arthropods of concern to American agriculture. The names are arranged alphabetically by order, family, genus, and species. Some general comments regarding the potential for certain taxa to invade or establish are given for each order. Entries following each pest species listed are, in sequence: common name, abbreviated summary of hosts attacked, summary of distribution, notation concerning biology if deemed significant to emigration potential, and source(s) of additional detailed information.

Acarina

Agriculturally important mites usually infest a variety of hosts, primarily broad-leaved plants. That trait, together with their small size and the fact that females customarily overwinter as adults, make them well adapted for dispersal in commerce. Four species deserve consideration as potential immigrants. Mite–plant associations are discussed in an evolutionary context by Krantz and Lindquist (1979).

EUPODIDAE

Halotydeus destructor Tucker. Red-legged earth mite. Vegetables. South Africa, Australia, New Zealand (Anonymous, 1958, No. 16, pp. 313–314).

TENUIPALPIDAE

Cenopalpus pulcher (Canestrini and Fanzago). Flat red mite. Pome and stone fruit. North Africa, Western Europe, and Middle East (Anonymous, 1968, No. 51, pp. 1143–1144).

TETRANYCHIDAE

Eutetranychus orientalis (Klein). Oriental red mite. Citrus and general. Northeast Africa, South Africa, Middle East, and India (Anonymous, 1969, No. 34, pp. 673–674).

Tetranychus viennensis Zacher. Fruit-tree spider mite. General. Europe (Anonymous, 1963, No. 31, pp. 897–898).

Collembola

Springtails are not usually considered crop pests, but one species is a troublesome pest of clover and alfalfa in the moister parts of Australia.

SMINTHURIDAE

Sminthurus viridis Lubbock. Lucerne flea. Legumes and other field crops. Throughout most of Europe, including Britain, northern coast of Africa, Middle East, Japan, Iceland, China, Australia, and New Zealand (Anonymous, 1958, No. 13, pp. 253–254).

Orthoptera

Although migratory locusts and some other saltatorial Orthoptera are notorious pests in many parts of the world, they are unlikely candidates for translocation to North America because of their reproductive requirements, large size, and conspicuous appearance. One species of mole cricket appears to have a high capacity for emigration.

GRYLLOTALPIDAE

Gryllotalpa africana (Beauvois). African mole cricket. Potatoes, turf, and a great variety of crop plants. Africa, southern Spain, Australia, South and Southeast Asia, Philippines, Japan, Guam, and Hawaii (Anonymous, 1974, No. 5, pp. 41–43).

Hemiptera/Heteroptera

The potential for invasion by true bug pests seems greatest for those species that either overwinter as adults (Lygaeidae, Pentatomidae, Piesmidae, Tingidae) or oviposit in plant tissue (Miridae).

LYGAEIDAE

Dimorphopterus pilosus (Barber). Rice chinch bug. Upland rice; also maize and several soft-stemmed grasses, not a pest of paddy rice. Papua, New Guinea (Anonymous, 1976, Nos. 48–52, pp. 897–899).

Macchiademus diplopterus (Distant). South Africa grain bug. Graminaceous plants and fruit. South Africa, but intercepted in North America (Anonymous, 1973, No. 40, pp. 693–694).

Nysius vinitor (Bergroth). Rutherglen bug. Wheat, cotton, vegetables, and fruit. Australia and Tasmania (Anonymous, 1957, No. 40, pp. 799–800).

MIRIDAE

Dionconotus cruentatus (Brullé). Orange blossom bug. Grasses, citrus blossoms, legumes, and other plants. Eggs laid in grasses and remain in the diapause state for long periods. Southern Europe, Near East, Tunisia (Anonymous, 1968, No. 18, pp. 377–378).

Plesiocoris rugicollis (Fallén). Apple capsid. Fruit and ornamentals. Palearctic regions of Eurasia (Anonymous, 1958, No. 24, pp. 523–524).

PENTATOMIDAE

Aelia rostrata (Boheman) and related species of *Aelia*. Wheat stink bugs. Graminaceous plants, cultivated and wild. Throughout Western Europe, Near and Middle East, and parts of North Africa (Anonymous, 1962, No. 27, pp. 749–750; Paulian and Popov, 1980, pp. 70–71).

Coridius janus (Fabricius). Red pumpkin bug. Principally cucurbits. Burma, India, Sri Lanka, Pakistan (Anonymous, 1973, No. 13, pp. 197–198).

Eurydema oleraceum (Linnaeus) and related species of *Eurydema*. Cabbage bugs. Generally on crucifers; also numerous other vegetable crops. Throughout Europe, Turkey, Turkestan, and sections of Siberia (Anonymous, 1959, No. 6, pp. 81–82).

Scotinophara lurida (Burmeister). Rice stink bug. Rice and related grasses. Japan, China, and Sri Lanka. Adults aestivate between rice crops (Hill, 1975, p. 216, as *S. coarctata*).

SCUTELLERIDAE

Eurygaster integriceps (Puton) and two related species of *Eurygaster*. Senn (or Sunn) pest. Grain crops and numerous other plants. Present in essentially all Middle East countries, Eastern Europe, and some parts of North Africa (Anonymous, 1957, No. 5, p. 88; Paulian and Popov, 1980, pp. 70–76).

PIESMIDAE

Piesma quadratum (Fieber). Beet bug. Completes development only on Chenopodiaceae, but feeds on many plants. Transmits leaf crinkle virus of sugar beets in Europe. Throughout Europe, including British Isles, the USSR, and Tunisia (Anonymous, 1957, No. 47, pp. 891–892).

TESSARATOMIDAE

Rhoecocoris sulciventris (Stål). Bronze orange bug. Citrus. Overwinter as second instars in protected places underneath leaves on trees. Coastal areas of New South Wales and Queensland, Australia (Anonymous, 1961, No. 5, pp. 51–52, under Pentatomidae).

TINGIDAE

Monosteira unicostata (Mulsant and Rey). Almond bug. Apple and pear; also stone fruits. Mediterranean region and Middle East (Anonymous, 1960, No. 12, pp. 207–208).

Stephanitis pyri (Fabricius). Pear lace bug. Foliage of a wide variety of deciduous fruits. Generally distributed in Europe; also in the Near and Middle East, Afghanistan, and the USSR (Turkestan Caucasus, Siberia) (Anonymous, 1960, No. 18, pp. 345–346).

Hemiptera/Homoptera

Among the Homoptera, those belonging to the series Sternorrhyncha (psyllids, aphids, coccids, and aleyrodids) are well known for their capability to emigrate. The Homoptera in general have contributed more species to our immigrant fauna than any other order, and most of them are Sternorrhyncha (Sailer, 1978). This group also includes many species that transmit plant pathogens, although their chief damage may result simply from the enormous numbers that develop. In contrast to the Sternorrhyncha, the Homoptera Auchenorrhyncha are primarily of concern as vectors of plant pathogens, with leaf hoppers (Cicadellidae) and plant hoppers (Delphacidae) being the more important culprits. The role of the Homoptera as vectors of plant pathogens is discussed elsewhere (Section IV).

Cicadellidae

Cicadulina, Nephotettix, Orosius (Ghauri, 1966, 1971a, 1971b; Nielson, 1968, 1979; Ruppel, 1965; Shikata, 1979).

Delphacidae

Laodelphax, Nilaparvata, Sogatella, Sogatodes (Harpaz, 1972; Mochida *et al.*, 1978).

Aleyrodidae

Aleurocanthus spiniferus (Quaintance). Orange spiny whitefly. Citrus primarily, also several other trees or shrubs. Asia, including Japan, and Guam (Anonymous, 1959, No. 17, pp. 321–322).

Aphididae

Aphis citricidus (Kirkaldy). Oriental black citrus aphid. Citrus, the primary vector of tristezia disease. Most of Asia, Africa (generally south of the Sahara), Australia, New Zealand, citrus-growing regions of South America, Pacific islands (Samoa, Fiji, Hawaii) (Anonymous, 1957, No. 38, pp. 767–768).

Cuernavaca noxius (Mordvilko). Barley aphid. Small grains. Southern USSR, Mediterranean basin, eastern Africa, Zimbabwe, Libya, Morrocco, Spain, and Great Britain (Anonymous, 1963, No. 47, pp. 1357–1358).

Pterochlorus persicae (Cholodkovsky). Clouded peach bark aphid. Stone fruits. Middle East to Afghanistan and Pakistan, southern USSR, and Egypt (Anonymous, 1969, No. 43, pp. 809–810).

COCCOIDEA

Ceroplastes rusci (Linnaeus). Fig wax scale. Fig primarily, but also various other hosts. Mediterranean area (Anonymous, 1960, No. 38, p. 886).

Icerya aegyptica (Douglas). Egyptian flued scale. More than 100 recorded hosts; among the important ones are citrus. Australia, China, Japan, South Asia, Somaliland, Tanganyika, Zanzibar, Tahiti, and Wake Island (Anonymous, 1960, No. 31, pp. 727–728).

Parlatoria ziziphi (Lucas). Black *Parlatoria* scale. Citrus. Southern Europe, Mediterranean area, South Africa, China, South Asia, Philippines, Okinawa, Micronesia, northern Australia, West Indies, Guiana, Argentina (Anonymous, 1960, No. 8, pp. 111–112).

Unaspis yanonensis (Kuwana). Arrowhead scale. Citrus. China (citrus areas of the southeast mainland), Japan, France (Cote d'Azur) (Anonymous, 1968, No. 17, pp. 349–350).

PSYLLIDAE

Diaphorina citri (Kuwayama). Citrus psylla. Citrus and other Rutaceae. Widespread in tropical and subtropical Asia (Anonymous, 1959, No. 26, pp. 593–594).

Psylla mali (Schmidberger). Apple sucker. Apple and other Pomaceae. Western Europe, northeastern Australia (Nova Scotia and New Brunswick) (Anonymous, 1957, No. 50, pp. 925–926).

Spanioza erythreae (Del Guercio). Two-spotted citrus psyllid. Citrus and other members of the family. East Africa, Ethiopia, South Africa, Sudan (Anonymous, 1967, No. 34, pp. 801–802).

Thysanoptera

As with the Homoptera Sternorrhyncha, thrips have demonstrated the ability to move about in commerce far out of proportion to the number of species involved, in comparison with other major groups of insects. A considerable number of economically important species have attained essentially cosmopolitan distribution (Ananthakrishnan, 1971). The biology of thrips, and taxonomic problems encountered in the group, have been reviewed by Ananthakrishnan (1979). The following exotic species are of concern:

THRIPIDAE

Kakothrips pisivorus (Westwood). Pea thrips. Legumes, especially peas and beans; also numerous wild and cultivated plants. Throughout Europe, including the Caucasus (Anonymous, 1958, No. 7, pp. 121–122).

Retithrips syriacus (Mayet). Black vine thrips. Numerous hosts, the more favored being grape, cotton, and a variety of fruits. Eastern Mediterranean, eastern Africa, India, and Brazil (Anonymous, 1967, No. 17, pp. 354–355; Ananthakrishnan, 1971).

Scirtothrips aurantii Faure. South African citrus thrips. Citrus, cotton, and numerous other plants. Egypt, Malawi, Sudan, South Africa, Zimbabwe (Anonymous, 1967, No. 43, pp. 965–966).

Taeniothrips laricivorus (Kratochvil and Farsky). Larch thrips. Responsible for die-back disease of larch through injection of a toxin. Occurs extensively in Central Europe and into the USSR; precise range not known (Anonymous, 1960, No. 25, pp. 541–542).

Thrips imaginis Bagnall. Apple thrips. Flowers of deciduous fruit. Indigenous to Australia, occurs throughout southern Australia and Tasmania (Anonymous, 1958, No. 52, pp. 1029–1030).

Thrips angusticeps Uzel and *T. linarius* Uzel. Cabbage thrips. A wide variety of field and vegetable crops, including flax and crucifers. Western Eurasia and North Africa (Anonymous, 1962, No. 16, pp. 400–402).

Coleoptera

Numerous families of beetles include crop pests, but those most likely to join our immigrant fauna, by virtue of their biologies and behavior, are thought to be members of the families Chrysomelidae and Curculionidae and the phytophagous representatives of the Coccinellidae. Several exotic species of Buprestidae and Cerambycidae are important pests, but their long developmental periods and association with living wood probably severely limit their potential for invasion. Turf pests and defoliators (chafers, family Scarabaeidae), although obviously capable of invading, as illustrated by past performances [e.g., *Amphimallon majalis* (Razoumowsky), the European chafer, and *Phyllophaga brunneri* Chapin, a turf pest, *Popillia japonica* Newman (the Japanese beetle), and several others], presumably invade primarily as adults. Under present quarantine policies, their large size should make them rather unlikely to escape notice. The U.S. Department of Agriculture reports contain information regarding Buprestidae (Anonymous, 1960, No. 49, pp. 1129–1130), Carabidae (Anonymous, 1960, No. 10, pp. 151–152), Cerambycidae (Anonymous, 1962, No. 4, pp. 57–58; 1967, No. 18, pp. 373–374; 1968, No. 30, pp. 717–718), Scarabaeidae (Anonymous, 1959, No. 30, pp. 693–694, No. 45, pp. 989–990, No. 48, pp. 1025–1026; 1962, No. 20, pp. 505–506; 1963, No. 6, pp. 98–100; 1969, No. 28, pp. 535–536; 1971, No. 20, pp. 335–336), and Scolytidae (Anonymous, 1959, No. 8, pp. 125–126; 1967, No. 51, pp. 1077–1078). Burke (1976) has provided a review of the bionomics of the anthonomine weevils, many of which are of economic importance.

ALLECULIDAE

Omophlus lepturoides (Fabricius). Damages flowers of many plants; also cole crops, grain, and late-blooming fruit trees. Overwinter as larvae in the soil; adults emerge in spring. Southern Europe, Southeast Asia, and the USSR from the Ukraine to the Caucasus (Anonymous, 1980, No. 22, pp. 424–426).

CHRYSOMELIDAE

Aphthona euphorbiae (Schrank). Large flax flea beetle. A major pest of flax, feeding also on beets, Euphorbiaceae, and other plants; flax required for sexual maturity. Western Europe, Syria, Libya, European USSR, and some Far Eastern areas (Anonymous, 1969, No. 44, pp. 821–822).

Aulacophora abdominalis (Fabricius). A pumpkin beetle. Cucumber, melons, and other cucurbits. Northeastern Australia, Indonesia, New Guinea, Solomon Islands, and islands within this distributional range. Adults overwinter in dead vegetation or under dead bark (Anonymous, 1979, No. 1, pp. 13–14).

Chaetocnema concinna (Marsham) and *C. tibialis* (Illiger) Flea beetles. Sugarbeets; also numerous other plants. *Chaetocnema concinna* is present in most of Europe and eastward into Siberia; *C. tibialis* occurs throughout Mediterranean Europe and middle Eurasia, and in the Ryukyu Islands. Hibernate as adults (Anonymous, 1961, No. 37, pp. 879–882). N.B.: *C. concinna* is recorded in Massachusetts as of 1979 (Anonymous, 1980, No. 20, pp. 374) and is presumed to be established in the United States.

Chaetocnema aridula (Gyllenhall) and *C. hortensis* (Geoffrey). Flea beetles. Grains, primarily oats, wheat, and rye. Throughout Europe and eastward into Asia; *C. hortensis* also occurs in northern Africa. Overwinter as adults (Anonymous, 1961, No. 37, pp. 879–882).

Colaspidema atrum (Olivier). Black alfalfa leaf beetle. Alfalfa, but in the absence of the favored host will feed on other plants of diverse families. Iberian Peninsula, Algeria, Morocco; possibly also in Italy and the Kiev district of the USSR. Overwinter as adults in soil (Anonymous, 1959, No. 42, pp. 943–944).

Diabrotica speciosa Germar. Cucurbit beetle. Foliage, flowers, and fruit of cucurbits, maize, sorghum, beets, crucifers, peas, beans, cotton, potato, tomato, apple, peach, and citrus. Argentina, Brazil, Uruguay, Colombia,

Bolivia, Costa Rica, Peru, Venezuela, Panama (Anonymous, 1957, No. 52, pp. 949–950).

Dicladispa armigera (Olivier). Rice hispid. Paddy rice. South and Southeast Asia, Indonesia, Taiwan, and coastal China mainland (Anonymous, 1958, No. 40, pp. 857–858).

Galerucella tenella (Linnaeus). Strawberry leaf beetle. Strawberry, ladysmantle, European meadowsweet, spirea. Throughout most of Europe and into western Siberia. Overwinter as adults (Anonymous, 1960, No. 51, pp. 1137–1138).

Marseulia dilativentris (Reiche). A leaf beetle. Many cultivated hosts, wheat, and barley preferred. Israel, Jordan, Lebanon, Syria, and Turkey (Anonymous, 1967, No. 13, pp. 237–238).

Medythia suturalis (Motschulsky). A striped leaf beetle. Peas and beans. South China, east India, Philippines, Ryukyu Islands, Sunda Island, Taiwan, Vietnam; also a subspecies in north and central China (including Manchuria), Japan, Korea, and southeastern Siberia in the USSR (Anonymous, 1979, No. 9, pp. 95–96).

Phytodecta fornicata Brüggemann. Lucerne beetle. Alfalfa and black medic. Throughout most of middle Europe and parts of the Near East; possibly also in England and North Africa. Overwinter as adults in the soil (Anonymous, 1962, No. 35, pp. 985–986).

Psylloides attenuata Koch. Hop flea beetle. Hop, hemp, and *Urtica dioica*. Throughout most of Europe except the extreme north and some southern areas, and eastward into Siberia in the USSR. Overwinter as adults in debris, crevices, hop poles, and hop bins (Anonymous, 1961, No. 49, pp. 1101–1102).

Psylloides chrysocephala (Linnaeus). Cabbage stem flea beetle. Primarily cultivated and wild crucifers; also sugarbeet, flax, vetch, soybeans, and *Mathiola incana*. Throughout Europe; also in Newfoundland, Canada. Overwinter as adults (Anonymous, 1958, No. 44, pp. 923–924).

Raphidopalpa spp. [*R. foveicollis* (Lucas) and others]. Red pumpkin beetle. Cucurbits preferred, but also numerous other plants. Many areas in Australia, Asia, Africa, Europe, and islands near those continents. Overwinter as adults under debris and fallen leaves (Anonymous, 1961, No. 4, pp. 39–40).

Coccinellidae

Epilachna chrysomelina (Fabricius). Twelve-spotted melon beetle. Watermelon, muskmelon, cucumber, pumpkin, and other cultivated and wild species of cucurbits; also other crop plants. Widespread in Africa, southern Europe, Middle East, and Near East. Hibernate as adults (Anonymous, 1959, No. 33, pp. 765–766).

Epilachna paenulata (Germar). A leaf-feeding coccinellid. Many hosts, the most important being cucurbits and beans. South America, about 12° to about 40° south latitude, except the west coast area. Overwinter as adults (Anonymous, 1958, No. 42, pp. 885–886).

CURCULIONIDAE

Amnemus quadrituberculatus (Boheman). Clover root weevil. Subterranean, red, crimson, and white clovers primarily. Coastal districts of New South Wales and Queensland, Australia. Adults survive winters (Anonymous, 1959, No. 36, pp. 837–838).

Anthonomus pomorum (Linnaeus). Apple blossom weevil. Pomaceous fruit primarily. All of Europe across the USSR to China, Korea, and Japan. Hibernate as adults from midsummer to the following spring (Anonymous, 1959, No. 41, pp. 925–926).

Anthonomus vestitus Boheman. Peruvian square weevil. Cotton. Peru and Ecuador (Anonymous, 1959, No. 13, pp. 227–228; Burke, 1976, p. 288).

Baris granulipennis (Tournier). Melon weevil. Watermelons, melons, and cucumber. Egypt, Iran, Israel, and adjacent USSR (Anonymous, 1967, No. 20, pp. 431–432).

Ceutorhynchus pleurostigma (Marsham). Turnip gall weevil. Most crucifers. Larval feeding damage causes galls that enlarge as plant growth continues. Throughout most of Europe (Anonymous, 1957, No. 43, pp. 843–844).

Curculio elephas (Gyllenhal). Chestnut weevil. Chestnuts and acorns. Southeast Europe and Algeria. Overwinter as larvae, normally in soil, but pupation may occur in nuts, from which adults emerge (Anonymous, 1958, No. 50, pp. 1003–1004).

Epicaerus cognatus Sharp. Potato weevil. Potato and other *Solanum* species. Mexico (DF, Puebla, Tlaxcala, Veracruz, Hidalgo) (Anonymous, 1959, No. 44, pp. 971–972).

Furcipus rectirostris (Linnaeus). Stone fruit weevil. Stone fruits, primarily cherries. Northern and Central Europe, the USSR, and Japan (probably also China). Overwinter as adults in ground debris (Anonymous, 1973, Nos. 1–4, pp. 23–24).

Hylobius abietis (Linnaeus). Large pine weevil. All conifers and many hardwoods including oak, birch, and sycamore. Scotch pine and Douglas fir are favorite hosts in Britain. Throughout Europe and northern Asia to Korea and Japan (Anonymous, 1961, No. 30, pp. 707–708).

Hyperodes bonariensis (Kuschel). Wheat stem weevil. Wheat, oats, barley, Italian and perennial ryegrasses, orchard grass, and other cultivated and wild grasses. Native to Argentina and adventive to New Zealand (Anonymous, 1961, No. 3, pp. 31–32).

Lixus junci (Boheman). Beet curculionid and related species. Sugarbeets. General in the Mediterranean area. Overwinter as adults (Anonymous, 1959, No. 4, pp. 43–44).

Pissodes notatus (Fabricius). Banded pine weevil. Pine, spruce, larch, and fir. Most of Europe, including Britain, into eastern Siberia: also Uruguay and Algeria. Adults long-lived (Anonymous, 1958, No. 14, pp. 271–272).

Premnotrypes spp. Andean potato weevils. Potato: may also attack *Solanum wittmackii*. Bolivia, Peru, Chile, Colombia, Ecuador (Anonymous, 1960, No. 48, pp. 1107–1108).

Rhabdoscelus obscurus (Boisduval). New Guinea sugarcane weevil. Sugarcane, coconut, several species of palm and banana. Australia, Indonesia, Taiwan, numerous South Pacific islands, Hawaii, Japan (greenhouses) (Anonymous, 1967, No. 32, pp. 749–750).

Rhynchites cupreus (Linnaeus). Plum borer. Stone and pome fruits: also *Sorbus* spp., hazel, birch, hawthorne, and grape. Throughout most of Europe. the USSR, and Japan. Hibernate as adults (Anonymous, 1959, No. 7, pp. 101–102).

Rhynchites heros Roelofs. Fruit weevil, peach weevil. Peach, pear, apple, cherry, apricot, plum, quince, and loquat. Japan, Korea, Taiwan, and some parts of China (Anonymous, 1958, No. 15, pp. 289–290).

Tanymecus dilaticollis Gyllenhal. Gray corn weevil. Maize and sorghum preferred; also on many other crop plants, including sugarbeets, wheat, and vegetable and fruit crops. Southern USSR, Southern Europe, Turkey, and Cyprus. Overwinter as adults (Anonymous, 1973, No. 26, pp. 413–414).

Lepidoptera

There are a great many important agricultural pests among the Lepidoptera, especially in the family Noctuidae, which includes a large number of polyphagous species. However, under present quarantine regulations and practices, the noctuids are probably less likely than some others to invade North America. One species stands out as a potentially high-risk invader because it deposits eggs in batches and overwinters in that stage: the nun moth, *Lymantria monacha* (Linnaeus), family Lymantriidae. Another species with similar overwintering and egg deposit habits is the lackey moth, *Malacasoma nuestria* (Linnaeus), family Lasiocampidae. However, the lackey moth larvae are gregarious and spin conspicuous webs as they develop, so that establishment of a breeding population in a new area should be readily detected. Larvae of the pine processionary moth are also gregarious and web spinners. *Tortrix viridana* Linnaeus, a defoliator of deciduous trees, also overwinters in the egg stage.

Stem borers of the genus *Chilo,* family Pyralidae, are notorious as pests of graminaceous crops. Hill (1975, pp. 258–262) discusses two important species not mentioned in the following list.

CARPOSINIDAE

Carposina niponensis Walsingham. Peach fruit moth. Pome fruit, peach, pear, plum, apricot, quince, nectarines, and others. Damage to peaches easily confused with that caused by the Oriental fruit moth. Japan, Korea, Manchuria, China, and Soviet Far East (Anonymous, 1958, No. 34, pp. 751–752).

GELECHIIDAE

Gnorimoschema heliopa (Lower). Tobacco stem borer. Cultivated and wild tobacco and egg plant; possibly also wild Solanaceae. Australia, Indonesia, Fiji, New Guinea, Samoa, Philippines, South Asia, southern Africa, Greece, Turkey, and Israel (Anonymous, 1957, No. 26, pp. 523–524).

Gnorimoschema ocellatella (Boyd). Sugar beet crown borer. *Beta cicla, B. vulgaris, B. maritima,* and *B. sacchariera.* Europe, North Africa, and Middle East (Anonymous, 1960, No. 7, pp. 91–92).

GEOMETRIDAE

Bupalus piniarius (Linnaeus). Pine looper. Pine. Most of Europe except Iberian Peninsula and small areas in southern Europe (Anonymous, 1959, No. 20, pp. 397–398).

GRACILLARIDAE

Phyllocnistis citrella Stainton. Citrus leafminer. Citrus. Throughout tropical Asia and the western Pacific islands; also Japan and adjacent Asiatic mainland. Overwinter as adults (Anonymous, 1958, No. 45, pp. 935–936).

INCURVARIIDAE

Incurvaria rubiella (Bjerkander). Raspberry moth. A bud borer of raspberry, blackberry, and loganberry. Overwinter in canes as larvae. Western Europe. N.B.: Established in Canada (New Brunswick, Prince Edward Island) (Anonymous, 1958, No. 18, pp. 355–356).

LASIOCAMPIDAE

Dendrolimnus pini Linnaeus. Pine lappet. Pine. Throughout Europe and Asia; also in Morocco (Anonymous, 1957, No. 39, pp. 785–786).

Gastropacha quercifolia (Linnaeus). Lappet moth. Various deciduous shade trees and fruit trees. Widespread in Western and Central Europe and adjacent USSR, eastern areas of mainland China and USSR, Japan (Anonymous, 1971, No. 26, pp. 463–464).

Malacosoma nuestria (Linnaeus). Lackey moth. Almond, apple, pear, plum, peach, cherry; also deciduous shade trees and woody ornamentals: a defoliator. Throughout most of Europe and Asia (Anonymous, 1958, No. 6, pp. 101–102).

LYCAENIDAE

Lampides boeticus (Linnaeus). Bean butterfly. Legumes of many kinds. Throughout tropical and temperate parts of Africa and most of Asia; also Pacific islands, including Hawaii (Anonymous, 1960, No. 30, pp. 695–696).

LYMANTRIIDAE

Dasychira pudibunda (Linnaeus). Red-tail moth. Deciduous forests and shrubs. A defoliator. Throughout the British Isles, Europe, and most of Asia (Anonymous, 1959, No. 47, pp. 1013–1014).

Lymantria monacha (Linnaeus). Nun moth. Major defoliator of conifers and deciduous trees. Throughout Europe and Asia, including Spain in the south, Britain in the west, and Japan, Korea, and China to the east (Anonymous, 1957, No. 12, p. 227).

NOCTUIDAE

Agrotis segetum (Denis and Schiffermuller). Turnip moth. General feeder (cutworm) on grains, crucifers, beets, cotton, potatoes, and many other crops. Throughout Europe, Asia, Africa, and the Azores (Anonymous, 1957, No. 41, pp. 819–820).

Autographa gamma (Linnaeus). Silver-Y moth. A general feeder on potatoes, beets, crucifers, flax, hemp, and legumes: also cereals and grasses at times. Throughout Europe eastward through Asia to India and China: also in North Africa (Anonymous, 1958, No. 23, pp. 497–498). Rivnay (1962, pp. 125–127) provides a detailed account of the species in the Near East.

Busseola fusca Fuller. Maize stalk borer. Maize, sorghum, millet, and numerous other graminaceous hosts. Widespread in maize-growing areas south of the Sahara, Africa (Anonymous, 1957, No. 24, pp. 477–478; Walters, 1979).

Mamestra brassicae (Linnaeus). Cabbage moth. A general feeder, although crucifers are the most frequent hosts. Tomatoes, tobacco, lettuce, maize, beans, etc. are also attacked. Throughout Europe and Asia (Anonymous, 1958, No. 41, pp. 873–874).

Panolis flammea (Denis and Schiffermuller). Pine moth. Pine preferred, but also attacks silver fir, Douglas fir, spruce, juniper, European larch, and some broad-leaved trees. British Isles, continental Europe, and Asia (Anonymous, 1958, No. 51, pp. 1017–1018).

Sesamia calamistis Hampson. Pink stalk borer. Maize, sorghum, millet, rice, and sugarcane. Occurs in most of Africa but is important primarily in the sub-Saharan region (Hill, 1975, p. 298–299; Walters, 1979).

Sesamia cretica Lederer. Durra stalk borer. Maize, broomcorn, and sorghum. Mediterranean region generally: also Sudan (Anonymous, 1957, No. 7, p. 128).

Spodoptera exempta (Walker). African armyworm. Maize, rice, sorghum, and a wide range of grasses and other cereals. Africa south of the Sahara generally, southern Asia, and islands between southern Asia and Australia, also areas on both east and west coasts of Australia (Hill, 1975, pp. 303–304; Anonymous, 1960, No. 47, pp. 1095–1096).

Spodoptera littoralis (Boisduval). Egyptian cottonworm. A general feeder that attacks many crops, most commonly cotton, tobacco, tomatoes, and maize. Mediterranean basin and many areas of Africa south of the Sahara (Hill, 1975, p. 307).

Spodoptera litura (Fabricius). A polyphagous species; cotton, rice, tomato, and tobacco preferred. South and Southeast Asia from Japan and Korea to Indonesia; also coastal areas of eastern, northern, and western Australia, and Fiji and Hawaii (Hill, 1975, p. 308).

Spodoptera mauritia (Boisduval). Paddy armyworm. Rice, primarily; also maize, sugarcane, other graminaceous crops, and Cruciferae. Essentially the same distribution as *S. litura,* but not including Japan and Korea (Hill, 1975, p. 306; Anonymous, 1960, No. 38, pp. 887–888).

OLETHREUTIDAE

Leguminivora glycinivorella (Matsumura). Soybean pod borer. Soybeans; also cowpeas and *Lupinus perennis.* Throughout Japan, Korea, parts of China, and the Soviet Far East (Anonymous, 1958, No. 1, pp. 11–12) [as *Grapholitha glycinivorella* (Matsumura)].

Lobesia botrana (Schiffermuller). Vine moth. Primarily grapes, but also other hosts, including berries. Southern and middle Europe, North Africa, Near East, and Japan (Anonymous, 1957, No. 30, pp. 611–612).

PAPILIONIDAE

Papilio demoleus (Linnaeus). Lemon butterfly. Citrus, especially young nursery plants. Throughout Africa, southern Asia, and northern Australia (Anonymous, 1958, No. 46, pp. 949–950).

PIERIDAE

Aporia crataegi (Linnaeus). Black-veined white butterfly. A general feeding defoliator; prefers rosaceous plants but also attacks shade trees. Throughout Europe and temperate Asia; also northwest Africa adjacent to the Mediterranean (Anonymous, 1962, No. 52, pp. 1280–1282).

Colias lesbia (Fabricius). Lucerne caterpillar. Alfalfa. Argentina only, although other subspecies are present in southern Brazil, Uruguay, Chile, Peru, and Ecuador (Anonymous, 1960, No. 2, pp. 21–22).

Pieris brassicae (Linnaeus). Large white butterfly. Crucifers preferred, and many kinds attacked; also feeds on garden crops and ornamentals. A migratory species that occurs as a pest throughout Europe, North Africa, and Middle East generally; range extends through northern India and southern Siberia to Tibet and China (Anonymous, 1958, No. 28, pp. 621–622).

PYRALIDAE

Chilo partellus (Swinhoe). Spotted stalk borer. Maize, sorghum, bulrush millet, sugarcane, rice, and various wild grasses. Afghanistan, India, Nepal, Bangladesh, Sri Lanka, Sikkim, Thailand, East Africa, Sudan, and Malawi (Hill, 1975, pp. 258–259).

Chilo suppressalis (Walker). Rice stalk borer. Rice, maize; also various millets, wild species of *Oryza,* and many wild grasses. Southeast Asia from Japan, Korea, and adjacent areas of eastern China south to include Pakistan, India, and the islands north of Australia; also northern Australia, Egypt, Nyasaland, Zanzibar, Spain, Portugal, and Hawaii (Anonymous, 1967, No. 44, pp. 855–856; Hill, 1975, p. 262).

Crocidolomia binotalis Zeller. Cabbage caterpillar. A wide range of crucifers, nasturtium, and various flowering plants. Much of Africa south and east of the Sahara, South Asia, northeastern Australia, and Australasian islands (Anonymous, 1968, No. 52, pp. 1153–1154).

Dichocrocis punctiferalis Gueneé. Yellow peach moth. General feeder on foliage and fruit of many plants, although typically a pod borer. Peach in China and cotton in Australia (boll damage) are major crops and areas of concern. Southern and eastern Asia, including Sri Lanka, Burma, Malaya, China, Taiwan, Japan, and Korea: also Australia, Indonesia, and New Guinea (Anonymous, 1957, No. 34, pp. 697–698).

Leucinodes orbonalis Gueneé. Eggplant fruit borer. Eggplant, potato, tomato, and other cultivated and wild solanaceous plants. Zaire, Sierra Leone, Somalia, Uganda, South Africa, South Asia, Indonesia, and probably numerous other African and Asiatic localities (Anonymous, 1960, No. 17, pp. 321–322).

Omphisa anastomasalis Gueneé. Sweet potato stem borer. Sweet potato. China, Taiwan, India, Sri Lanka, Japan, Burma, Indonesia, New Guinea, Thailand, Philippines, Malaya, and Hawaii (Anonymous, 1960, No. 44, pp. 1047–1048).

PSYCHIDAE

Kotochalia junodi (Heylaerts). Wattle bagworm. *Acacia decurrens mollis* and other species of *Acacia*. South Africa (Anonymous, 1961, No. 45, pp. 1039–1040).

SCYTHRIDIDAE

Syringopais temperatella (Lederer). Cereal leafminer. Wheat, barley, and oats preferred, but many additional hosts recorded. Cyprus, Turkey, Lebanon, Syria, Jordan, Iraq, Israel, and Iran (Anonymous, 1959, No. 38, pp. 873–874).

STENOMIDAE

Stenoma catenifer Walsingham. Avocado seed moth. Avocado (*Persea americana*), coyo (*P. schiedeana*), wild species of *Persea,* and species of *Beilschmedia*. Brazil, Colombia, Ecuador, Peru, Mexico, and all Central American countries (Anonymous, 1980, No. 18, pp. 352–355).

THAUMETOPOEIDAE

Thaumetopoea pityocampa (Denis and Schiffermuller). Pine processionary moth. *Pinus* spp. Italy, Spain, Switzerland, Southern Europe, Syria, and Turkey (Anonymous, 1959, No. 50, pp. 1045–1046).

TORTRICIDAE

Acleris comariana (Zeller). Strawberry tortrix. Plants of the family Rosaceae, especially *Fragaria, Rosa, Rubus,* and *Potentilla*. Middle and Southern Europe, the Balkans, eastern USSR, Japan (Hokkaido), and Canada (British Columbia) (Anonymous, 1973, No. 10, pp. 141–142).

Austrotortrix postvittana (Walker). Light brown apple moth. Apple and a great many other hosts of diverse plant groups. Damage is from defoliation and pitting fruit. Indigenous to Australia and now in Tasmania and parts of New Zealand; also in New Caledonia, England, and Hawaii (Anonymous, 1957, No. 10, p. 187).

Cryptophlebia leucotreta Meyrick. False codling moth. Fruit of many plants, especially citrus and cotton; also sorghum, maize, tangerine, walnut, okra, plum, peach, and many others. Tropical and south temperate Africa from Ethiopia, Senegal, Ivory Coast, Togo, and Upper Volta south to Africa, Madagascar, and Mauritius (Anonymous, 1960, No. 5, pp. 67–68; Hill, 1975, pp. 252–253).

Grapholitha funebrana (Treitschke). Plum fruit moth. Plum, peach, cherry, and other stone fruits. Temperate Europe through Siberia (USSR), Asia Minor, and North Africa (Anonymous, 1958, No. 49, pp. 989–990, as *Laspeyrisia funebrana*).

Tortrix viridana Linnaeus. Green oak tortrix. Defoliator of oak forests; may also attack beech, linden, maple, and other deciduous trees. Throughout Europe, including Britain; also Turkey and Morocco (Anonymous, 1958, No. 12, pp. 229–230).

YPONOMEUTIDAE

Acrolepia assectella (Zeller). Leek moth. Onion, leek, garlic, chives. General in Europe, including the British Isles; also reported from Hawaii (Anonymous, 1960, No. 14, pp. 241–242).

Prays citri Millière. Citrus flower moth. Many species of citrus. Algeria, France, Spain, Italy, Greece, Syria, Israel, Morocco, Pakistan, India, Malaysia, Mauritius, Philippines, Sri Lanka, Fiji, Australia (New South Wales) (Anonymous, 1967, No. 50, pp. 1061–1062).

Diptera

Among the Diptera, several species of fruit flies, family Tephritidae, stand out as successful invaders, e.g., the Mediterranean fruit fly, *Ceratitis capitata* (Wiedemann), and species of the genus *Dacus*. Fortunately, through long periods of study of these species in Pacific islands where they did or now occur, very effective suppressive tactics have been developed. As of 1982, all incipient infestations in the United States have been eradicated.

Diptera least likely to be eradicated, should populations become established in our region, are members of the family Cecidomyiidae.

AGROMYZIDAE

Agromyza oryzae Munakata. Japanese rice leafminer. Cultivated and wild rice, reed, and foxtail grasses. North temperate Japan (Anonymous, 1959, No. 10, pp. 163–164).

Ophiomyia phaseoli (Tyron). Bean fly. Primarily beans, including snap and lima; also cowpeas, chickpeas, soybeans, and some nightshades. Africa (Tanganyika, South Africa, Kenya, Uganda, Belgian Congo, Egypt), East Asia, Philippines, Indonesia, Australia, New Guinea, and adjacent islands (Anonymous, 1957, No. 37, pp. 751–752, as *Melanagromyza*) (Hill, 1975, pp. 329–330).

ANTHOMYIIDAE

Hylemya coarctata Fallén. Wheat bulb fly. Winter wheat, primarily; also winter rye and barley. Western Europe, British Isles, and USSR (European and Siberia) (Anonymous, 1957, No. 48, pp. 903–904).

CECIDOMYIIDAE

Contarinia medicaginis (Kieffer). Alfalfa flower midge. Alfalfa and spotted medic. Widely distributed in Europe, its northern limits vague; also in southeastern USSR (Anonymous, 1961, No. 9, pp. 139–140).

Contarinia nasturtii (Kieffer). Swede midge. Cultivated and wild crucifers. Western Europe and the Ukraine (USSR) (Anonymous, 1962, No. 21, pp. 541–542).

Pachydiplosis oryzae (Wood-Mason). Rice stem gall midge. Various parts of Southeast Asia and the islands northwest of Australia; may also be in the Sudan (Anonymous, 1959, No. 49, pp. 1035–1036; Hill, 1975, pp. 320–321).

CHLOROPIDAE

Chlorops pumilionis (Bjerkander). Gout fly. Winter and spring wheat, barley, rye, and *Agropyron repens* (quackgrass), thought to be its only wild host. Throughout Europe and into the Siberian USSR. Presumably also in the Soviet Far East (Anonymous, 1961, No. 48, pp. 1093–1094).

TEPHRITIDAE

Acanthiophilus eluta (Meigen). Safflower fruit fly. Safflower and native Compositae. Occurs from England and Canary Islands across Southern and Central Europe and North Africa to the Mediterranean area, Middle East, northern Himalayas, and India (Anonymous, 1963, No. 49, pp. 1389–1390).

Ceratitis rosa (Karsch). Natal fruit fly. Most kinds of orchard fruit, peaches, and guavas especially favored. Africa, Angola, Kenya, Mozambique, Nigeria, Malawi, South Africa, Zimbabwe, Swaziland, Tanganyika, Uganda, and islands of Mauritius, Reunion, and Zanzibar (Anonymous, 1963, No. 38, pp. 1132–1134; Hill, 1975, pp. 328–329).

Dacus ciliatus (Loew). Lesser melon fly. Cucurbits primarily; also numerous other hosts. Eastern and central Africa, Mauritius, Arabian Peninsula, and India (Anonymous, 1960, No. 52, pp. 1149–1150).

Dacus cucurbitae (Coquillet). Melon fly. Cucurbits preferred, but also attacks tomato, beans, cowpeas, eggplant, papaya, mango, peach, and others, more than 80 in all. Eastern Africa, South Asia, Philippines, Indonesia, Guam, Saipan, Tinian, Hawaii, Ryukyu Islands, northern Australia (Anonymous, 1959, No. 19, pp. 367–368; Hill, 1975, p. 323).

Dacus dorsalis Hendel. Oriental fruit fly. Fruit of many kinds; more than 150 hosts recorded. Southeast Asia, Bonin Islands, Mariana Islands, Hawaii (Anonymous, 1959, No. 24, pp. 529–530).

Dacus tsuneonis Miyake. Japanese orange fly. Citrus. Japan (Kyushu and Amami-O-shima); also reported from Szechwan Province, China (Anonymous, 1961, No. 51, pp. 1122–1124).

Dacus tyroni (Froggatt). Queensland fruit fly. Pome and stone fruits; also some varieties of citrus and a wide range of other fruit and vegetable hosts. Australia (New South Wales, Queensland, South Australia, and Victoria) (Anonymous, 1957, No. 3, pp. 43–44).

Euleia heraclei (Linnaeus). Celery fly. Celery and parsnip; also *Heracleum* and *Angelica* as wild hosts. Generally throughout Europe; also Morocco, North Africa, and Asia Minor (Anonymous, 1958, No. 19, pp. 375–376, as *Acidia heraclei*).

Myiopardalis pardalina (Bigot). Baluchistan melon fly. Watermelon, cucumber, musk melon, snake melon, pumpkin. India, Pakistan, Iran, Iraq, Israel, Lebanon, Syria, Turkey, and the Caucasus of the USSR (Anonymous, 1957, No. 6, p. 108).

Platyparea poeciloptera Schrank. Asparagus fly. Asparagus the only cultivated host. Generally throughout Central and Southern Europe, Sweden, and the Kiev district of the USSR (Anonymous, 1958, No. 38, pp. 823–824).

Rhagoletis cerasi (Linnaeus). Cherry fruit fly. Cherries preferred, but also other species of *Prunus,* with two species of *Lonicera* as alternate hosts. Most of continental Europe, parts of Turkey and Iran, and the USSR from southern Leningrad Province to the Crimean and southeastern Kazakhstan (Anonymous, 1958, No. 30, pp. 663–664).

Hymenoptera

Among the phytophagous Hymenoptera, several species of sawflies have demonstrated their ability as emigrants. The following species should be considered as potential invaders of North America.

Diprionidae

Diprion pini (Linnaeus). Pine sawfly. Periodic defoliator of pines, particularly Scotch pine, but will attack spruce and fir. Throughout most of Europe, Algeria, and into Siberia; exact eastward limits uncertain (Anonymous, 1959, No. 35, pp. 817–818).

Tenthredinidae

Athalia colibri (Christ). Turnip sawfly, beet sawfly. Cultivated and wild crucifers, sugarbeet, carrot, flax, grape. Throughout most of Europe, Asiatic

USSR, China, Japan, Korea, Taiwan, Iran, Turkey, Morocco (Anonymous, 1957, No. 28, pp. 563–564).

Hoplocampa brevis (Klug). Pear sawfly. Pear; also plum and apple. European USSR, Western and Southern Europe, Syria, and Turkestan (Anonymous, 1958, No. 26, pp. 573–574).

Pristophora abietina (Christ). Small spruce sawfly. Engleman, Norway, Sitka, Blue and Siberian spruce. Finland, Sweden, Estonian SSR, Denmark, Poland, Netherlands, Germany, France, Switzerland, Czechoslovakia, Belgium, and Yugoslavia (Anonymous, 1963, No. 4, pp. 59–60).

IV. DOMESTIC PEST THREATS

A. Home-Grown and Cryptic Hazards

Pest situations that arise from home-grown or hidden hazards are not unlike those caused by the invasion of exotic pests except in their geographic points of departure. Such hazards may be of several sorts. They may be recognized (i.e., known) as likely to occur under certain bioclimatic conditions; they may be suspected to be pests that occur occasionally in the absence of information on the circumstances that trigger them; or they may be completely unknown and unsuspected except as experience reminds us that wherever life exists, things are in a state of flux and changes should be anticipated. These hidden hazards become pest situations when some disturbance alters the status of one or more components of a balanced environment, a phenomenon similar to that of invasion by an exotic pest. The combination of circumstances that enabled a cicada, *Mogannia minuta* (Matsumura), to become a pest of sugarcane on Okinawa is an example of such a situation (Ito and Nagamina, 1981).

Organisms evolve in a variable environment (den Boer, 1968, 1971, 1977), a generalization that applies to both plants and the phytophagous arthropods they host. It is important to remember that very few of our crop plants are indigenous to North America (Waterworth, 1981), and that the genetic composition of those few, as well as that of the many brought here from abroad, has been materially altered during the domestication process. The domestication process to which crop plants have been subjected has, generally speaking, resulted in reduction of their genetic diversity. With few exceptions, phytophagous arthropods have not been subjected to domestication, with the result that most species have retained the ability to adjust to environmental variation within reasonable limits. Therein lies the explanation of the potential for home-grown pest situations to be generated by genetic homogeneity in crop plants and/or monocultural agricultural practices. Development of crop plant varieties resistant to insect attack is an attempt to reverse the domestication trend by restoring to plants the characteristics that make them resistant to specific herbivore species. But that is a one-to-one adjustment, and the agroecosystem is an exceedingly complex structure involving many physical and biological variables (Stinner *et al.*, 1982). No methodology has as yet been developed to permit *a priori* recognition of the many kinds of home-grown threats, such as the European corn borer (an adventive species) becoming a serious pest of snap beans in some bean-producing areas (Sanborn *et al.*, 1982).

Soybean is native to Eastern Asia. Although introduced into North America in the early 1800s, its exploitation as a widely grown crop plant in the United States has been relatively recent, and in 80% of the production areas of the world, the crop has been grown for less than 50 years. Survey data from four states (Illinois, Missouri, North Carolina, Ohio) showed a total of 453 phytophagous species recorded on soybean. Of these, 40 are oligophagous, restricted to legumes; 101 are polyphagous, feeding on soybean only as a secondary host; and the remainder are of somewhat uncertain food habits or only incidentally associated with soybean (Kogan, 1981). Of 10 major pest species attacking the crop in the midwestern, southern and eastern United States, all but three are indigenous to the Nearctic Region and only one, the southern green stink bug [*Nezara viridula* (Linnaeus)] is of Asiatic origin. The other nine are either polyphagous or are oliogophagous legume feeders that existed in the area before the introduction of soybean as a crop. However, even this fauna is not specialized to soybean feeding, as is much of the Oriental soybean fauna, certain species of which are highly adapted to the plant, and some of which depend upon perfect synchronization with crop phenology (Kogan, 1981). Thus, in comparison with soybean arthropod communities in the Orient, those in North America seem immature. Although immigrant species already well adapted to soybean must continue to be viewed as the greater potential threat, it is clear that shifts in food preference by native species, and greater adaptation to soybean feeding together with range expansion, still constitute serious threats.

One of the major soybean pests is the velvetbean caterpillar (VBC), *Anticarsia gemmatalis* (Hubner), a tropical species that apparently cannot tolerate winter temperatures in the regions where it causes major damage (Buschman *et al.*, 1981). Like many other tropical and subtropical species that are crop pests, including other species of noctuid moths, VBC moths migrate north during each growing season. The areas where damage occurs often depend upon the vagaries of air currents associated with storm systems at the times when moth populations are dispersing from their more southerly breeding area.

Another moth species, *Homoesoma electellum* (Hulst), the sunflower moth, is a serious pest of the cultivated sunflower in California, Texas, Kansas, and Nebraska, but only occasionally so in the Dakotas, Minnesota, and the Canadian prairie provinces. As with the VBC, the pest status of the species in the more northerly regions depends upon the arrival of female moths when newly opened sunflower blooms are available. This timing depends upon the development of weather systems that cause warm southerly winds during late June and early July (Arthur and Bauer, 1981).

Cropping practices may considerably influence the development of homegrown problems. The western corn rootworm (WCRW), *Diabrotica virgifera* LeConte, was first reported to be damaging corn in Colorado in the early 1900s (1909–11), and was first noted in the southwest corner of Nebraska in 1929–30, although it caused little damage during the drought years of the 1930s. It continued its eastward spread and is now the dominant rootworm species throughout most of Nebraska, having displaced the northern corn rootworm as a result of the change from corn in short rotation, a practice deleterious to the WCRW populations, to continuous corn planting in much of the region. An increase in irrigation farming apparently also favored the WCRW (Hill and Mayo, 1980), although that practice may be helpful in reducing spider mite populations in field corn (Chandler *et al.*, 1979). Clearly, the economic status of the several species of corn rootworms is changing as agronomic practices change (Krysan *et al.*, 1980).

Urbanization may cause changes in a pest situation. Lethal yellowing disease of palms was first discovered on the Florida mainland in 1971. Adults of the fulgorid planthopper, *Haplaxius crudus* (Van Duzee), are commonly associated with palms and are thought to transmit the disease agent. However, the *H. crudus* populations develop on turf grasses that are widely used in southern Florida landscapes, including house lawns, industrial plantings, parks, and golf courses where palms are also planted (Reinert, 1977, 1980).

Transmission of plant diseases by phytophagous sucking arthropods is one of the more insidious and complex problems involving crop plants. The imponderables in problems of this sort may include uncertainty about vector species; variation among populations of vectors in their ability to transmit pathogens (Kisimoto, 1967); environmental sources of disease agents; vector–pathogen compatibility; species or strains of pathogens; plant and vector host range of pathogen(s); and numerous other plant–arthropod interactions, including the role of agronomic practices in the maintenance of reservoirs of disease agents.

Aphids transmit about 60% of all known insect-borne plant viruses and other pathogens. There are 192 known vector species that transmit about 164 viruses. But this statistic gives a much too conservative picture of the situation. Out of some 3742 species of aphids, only about 300 have been tested as vectors of any of about 300 different viruses known to occur in about the same number of plant species (Harris, 1979). Further, a discussion of aphid involvement in plant pathogen transmission should not ignore the poorly understood phenomenon of aphid polymorphism. Alary and sexual polymorphism in aphids are environmentally determined, with temperature and photoperiod as factors of special significance. Other important factors include crowding, nutrition, and host plants. Aphids can exhibit alary and sexual polymorphism in one population at the same time, thus possessing the capacity to respond rapidly to either favorable or unfavorable conditions for survival. This biological diversity contributes to making aphids the most capable of all vectors of plant diseases (Lees, 1966; Hille Ris Lambers, 1966; Kodet and Nielson, 1980; Sylvester, 1980).

Weed species in or adjacent to agricultural areas are often sources of both plant pathogens and vectors. In studies in the Yakima Valley in Washington (Tamaki and Olsen, 1979; Tamaki *et al.*, 1980; Tamaki and Fox, 1982), production of winged migrants of the green peach aphid (GPA) was estimated to be as high as 70 million per hectare. The GPA is a vector of beet western yellows virus (BWYV), and disease readings of indicator plants from 15 weed species upon which the GPA had fed showed production of more than 500,000 BWYV vectors per hectare on weeds in peach orchards.

Peach is not a host of BWYV, but is affected by X disease, the causal agent of which is believed to be a mycoplasma-like organism (MLO) transmitted by several species of leafhoppers, most of which develop on wild hosts (McClure, 1980, and references cited therein). Studies in Connecticut showed that the highest incidence of X disease in orchards occurred where leafhopper vectors were most abundant, and that vector abundance was greatest in orchards with ground cover comprised primarily of wild host plants of vector species (Lacy *et al.*, 1979). The proximity and composition of wild host plants of vector species also affected colonization of an orchard (McClure, 1982).

An understanding of the cause(s) and source(s) of arthropod-borne plant diseases may be obscured by variation among different populations in the ability to transmit viruses to healthy plants or to their progeny. Whitcomb and Davis (1970, p. 442) point out that clones of vectors, derived from different geographic localities (or for selection from laboratory stock), can differ widely in transmitting ability. The inheritance of such characters is usually, but not always, complex.

The fortuitous association of plant pathogens and potential vectors may often bring to light previously unsuspected pest situations. Present evidence suggests that MLOs are likely candidates for transmission by a broad spectrum of vector species. "Flavescence doree" of grape (*Vitis*) in Europe may be caused by an MLO (Whitcomb and Davis, 1970). The disease is not known to occur in North America, yet it is transmitted in Southern Europe by a leafhopper [*Scaphoideus littoralis* (Ball), considered by Barnett (1977, p. 537) to be conspecific with *S. titanus* (Ball)] indigenous to North America but adventive to Europe.

Corn stunt was first observed in the early 1940s in south Texas. Initial studies revealed that the causal agent was transmissible by two species of leafhopper, *Dalbulus maidis* (DeLong and Wolcott) and *D. eliminatus* (Ball). Following the discovery and spread of the disease in the southern and midwestern United States, several additional leafhopper species (*Graminella nigrifrons* (Forbes), *G. sonorus* (Ball), and *Baldulus tripsaci* (Kramer and Whitcomb) were shown to be capable of transmitting the disease agent, a spiroplasma [see Nault and Bradfute (1979) for a summary of the discovery and subsequent investigations of vector species].

Citrus stubborn disease has been observed in California since 1915. It is caused by an MLO, *Spiroplasma citri*. The organism is now known to be transmissible by three species of leafhoppers [*Scaphytopius nitridus* (DeLong), *S. acutus delongi* (Young), and *Circulifer tenellus* (Baker)] and is known to occur in numerous field plants and weeds, both inside and outside the citrus culture area (Kaloostian *et al.*, 1979). *Spiroplasma citri* may have numerous leafhopper vectors and may be the cause of a disease in turnips.

B. Pest Potential of New Crop Development

The development of a new crop from a native plant or from a newly introduced plant creates a new set of conditions for native and introduced arthropod species. It may be difficult to anticipate exactly which group of organisms may utilize the crops. Cultivars of the new seed oil crop meadowfoam were developed from several native species of *Limnanthes* occurring in California and Oregon (Jolliff *et al.*, 1981). Insects occurring on the native species of *Limnanthes* were surveyed for possible pest species (K. J. West and J. D. Lattin, unpublished). Insect species judged most likely to be pests were native *Lygus* (Hemiptera: Miridae) and the 12-spotted cucumber beetle, *Diabrotica undecimpunctata undecimpunctata* (Mannerheim) (Coleoptera: Chrysomelidae), also a native species. Several other native insects were considered potential pests, none of them found on *Limnanthes* exclusively. Several species of crested wheatgrass, *Agropyron* spp., have been introduced into western North America from central Asia for rangeland improvement. *Labops hesperius* (Uhler), a native species of Miridae (Hemiptera) known to feed on several native grasses, moved over to the introduced grass species and caused damage (Todd and Kamm, 1974). At least one species of *Labops* occurring in central Asia is reported to occur on *Agropyron* in that area. One might have anticipated the likelihood of a native North American species becoming a pest on the introduced plant. Conversely, the development of new grass cultivars from native grass species provided suitable host plant material for several species of Miridae (Hemiptera) introduced into the Pacific Northwest from Europe, or secondarily introduced from a primary introduction on the eastern seaboard [e.g., *Leptopterna dolobrata* (Fallén) and *Megaloceraea recticornis* (Geoffrey)] (J. D. Lattin, unpublished).

C. Internal Movement of Pests Already Established

Although entirely new pests are now being introduced into North America, perhaps the most common occurrence is the internal movement within the United

States of pests already established here. Both native pest species and introduced pests may be involved. Perhaps the most notorious example is the gypsy moth, long established on the East Coast, having been introduced from Europe and now being discovered along the West Coast. The easy transport of the egg masses on vehicles provides countless opportunities for relocation, often over great distances. The apple maggot is another example; a native pest on apples in the East, it has now been detected in the Pacific Coast states of Oregon and Washington. That the movement also occurs from west to east is documented in the occurrence of a native western species of the ash bug, genus *Tropidosteptes,* on ornamental ash trees in the eastern state of Pennsylvania (Wheeler and Henry, 1974). The possibility of transporting species introduced into the United States from Europe to other countries such as China should not be overlooked. There has been little faunal exchange between China and Western Europe, but the possibility of accidental introduction into China of European species already established in North America does exist.

V. CONCLUSIONS AND RECOMMENDATIONS

Environmental similarity of geographic regions is an inadequate criterion for determining potential sources of agricultural pests. Although species inhabiting such regions may be intrinsically well suited to survive in North American crop areas, the importance of environmental similarity as a major factor is tempered by the realizations that (1) the more capable immigrants from such regions have already arrived, and (2) the capacity for invasion and establishment involves different biological and behavioral traits than the capacity to cause damage.

The great increase in world commerce, the wide global dissemination of crop varieties, and explosive tourism have greatly increased the number of source points from which pest species may emigrate; hence, there is a need for greater concern about species attributes, as opposed to geographic points of departure. Species attributes that enhance the potential for invasion and establishment include (1) long periods of quiescence (diapause, aestivation, hibernation), especially if the quiescent stage has a small size, secretive habits, or noncrop-associated traits that tend to make detection difficult; (2) development in plant parts that are customarily transported outside commercial channels and consumed without cooking (e.g., fruits, nuts) by individuals unfamiliar with exotic pest risks; (3) high levels of fecundity, especially if followed by short developmental cycles; high fecundity is naturally enhanced by efficient mate finding of bisexual species or by parthenogenesis.

Species complexes (sibling species, semispecies, biotypes, etc.) with wide geographic ranges are indicative of genetic plasticity, and in consequence, the ability to adjust to a wide range of environmental conditions. Such species complexes are potential sources of hidden hazards and deserve particular scrutiny as potential pests.

Given the complexities and volume of modern commercial shipping, the increasing levels of tourist traffic, and the limited personnel engaged in quarantine enforcement, our first line of defense against exotic pests is inadequate and likely to remain so. The logical alternatives include (1) a greatly expanded, cooperative effort to improve knowledge of the world's phytophagous arthropod fauna and (2) more effective surveillance and detection methods, and implementation of vigorous eradication and/or containment programs against invading pest species.

REFERENCES

Ananthakrishnan, T. N. (1971). Thrips (Thysanoptera) in agriculture, horticulture, and forestry—Diagnosis, bionomics and control. *J. Sci. Ind. Res.* **30**(3), 113–146.

Ananthakrishnan, T. N. (1979). Biosystematics of Thysanoptera. *Annu. Rev. Entomol.* **24,** 159–183.

Anonymous (1951–1980). "Distribution Maps of Pests," Series A (Agricultural) Maps Nos. 1–416. Commonwealth Institute of Entomology, London.

Anonymous (1957). Cooperative economic insect report. *U.S. Dept. Agric., Agric. Res. Serv., Plant Pest Control Div.* **7,** 1–950.

Anonymous (1958). Cooperative economic insect report. *U.S. Dept. Agric., Agric. Res. Serv., Plant Pest Control Div.* **8,** 1–1030.

Anonymous (1959). Cooperative economic insect report. *U.S. Dept. Agric., Agric. Res. Serv., Plant Pest Control Div.* **9,** 1–1064.

Anonymous (1960). Cooperative economic insect report. *U.S. Dept. Agric., Agric. Res. Serv., Plant Pest Control Div.* **10,** 1–1165.

Anonymous (1961). Cooperative economic insect report. *U.S. Dept. Agric., Agric. Res. Serv., Plant Pest Control Div.* **11,** 1–1135.

Anonymous (1962). Cooperative economic insect report. *U.S. Dept. Agric., Agric. Res. Serv., Plant Pest Control Div.* **12,** 1–1202.

Anonymous (1963). Cooperative economic insect report. *U.S. Dept. Agric., Agric. Res. Serv., Plant Pest Control Div.* **13,** 1–1426.

Anonymous (1967). Cooperative economic insect report. *U.S. Dept. Agric., Agric. Res. Serv., Plant Pest Control Div.* **17,** 1–1094.

Anonymous (1968). Cooperative economic insect report. *U.S. Dept. Agric., Agric. Res. Serv., Plant Pest Control Div.* **18,** 1–1154.

Anonymous (1969). Cooperative economic insect report. *U.S. Dept. Agric., Agric. Res. Serv., Plant Pest Control Div.* **19,** 1–908.

Anonymous (1971). Cooperative economic insect report. *U.S. Dept. Agric., Agric. Res. Serv., Plant Pest Control Div.* **21,** 1–784.

Anonymous (1973). Cooperative economic insect report. *U.S. Dept. Agric., Agric. Res. Serv., Plant Pest Control Div.* **23,** 1–800.

Anonymous (1974). Cooperative economic insect report. *U.S. Dept. Agric., Agric. Res. Serv., Plant Pest Control Div.* **24,** 1–904.

Anonymous (1976). Cooperative plant pest report. *U.S. Dept. Agric., Anim. Plant Health Inspec. Serv.* **1,** 1–902.

Anonymous (1979). Cooperative plant pest report. *U.S. Dept. Agric., Anim. Plant Health Inspect. Serv.* **4,** 1–849.

Anonymous (1980). Cooperative plant pest report. *U.S. Dept. Agric., Anim. Plant Health Inspect. Serv.* **5,** 1–704.

Arthur, A. P., and Bauer, D. J. (1981). Evidence of northerly dispersal of the sunflower moth by warm winds. *Environ. Entomol.* **10**(4), 528–533.

Ashley, T. R. (1981). FAMULUS: A reprint classification system for the research scientist. *Bull. Entomol. Soc. Am.* **27**(3), 161–165.

Baker, H. G., and Stebbins, G. L., eds. (1965). "The Genetics of Colonizing Species." Academic Press, New York.

Barnett, D. E. (1977). A revision of the Nearctic species of the genus *Scaphoideus* (Homoptera: Cicadellidae). *Trans. Am. Entomol. Soc.* **102,** 485–593.

Batra, L. R., Whitehead, D. R., Terrell, E. E., Golden, A. M., and Lichtenfels, J. R. (1977). Overview of predictiveness of agricultural biosystematics. *Beltsville Symp. Agric. Res.* **2,** 275–301.

Berlocher, S. H. (1979). Biochemical approaches to strain, race and species discrimination. *In* "Genetics in Relation to Insect Management" (M. A. Hoy and J. J. McLelvey, Jr., eds.), pp. 135–144. Rockefeller Found., New York.

Berlocher, S. H. (1980). An electrophoretic key for distinguishing species of the genus *Rhagoletis* (Diptera: Tephritidae) as larvae, pupae, or adults. *Ann. Entomol. Soc. Am.* **73**(2), 131–137.

Bierne, B. P. (1954). The *Prunus*- and *Rubus*-feeding species of *Macropsis* (Homoptera: Cicadellidae). *Can. Entomol.* **86,** 86–90.

Brader, L. (1979). Integrated pest control in the developing world. *Annu. Rev. Entomol.* **24,** 225–254.

Burke, H. R. (1976). Bionomics of the anthonomine weevils. *Annu. Rev. Entomol.* **21,** 283–303.

Buschman, L. L., Pitrie, H. N., Hovermale, C. H., and Edwards, N. C., Jr. (1981). Occurrence of the velvetbean caterpillar in Mississippi: Winter survival or immigration. *Environ. Entomol.* **10**(1), 45–52.

Chandler, L. D., Archer, T. L., Ward, C. R., and Lyle, W. M. (1979). Influences of irrigation practices on spider mite densities on field corn. *Environ. Entomol.* **8**(2), 196–201.

Chiang, H. C. (1978). Pest management in corn. *Annu. Rev. Entomol.* **23,** 101–123.

Cronquist, A. (1977). Once again, what is a species? *Beltsville Symp. Agric. Res.* **2,** 3–20.

den Boer, P. J. (1968). Spreading risk and stabilization of annual numbers. *Acta Biotheor.* **18,** 165–194.

den Boer, P. J., ed. (1971). Dispersal and dispersal power of Carabid beetles. *Misc. Pap.—Landbouwhogesch. Wageningen* **8,** 1–151.

den Boer, P. J. (1977). Dispersal power and survival: Carabids in a cultivated countryside. *Misc. Pap.—Landbouwhogesch. Wageningen* **14,** 1–190.

Elton, C. S. (1958). "The Ecology of Invasions by Animals and Plants." Methuen, London.

Embree, D. G. (1965). The population dynamics of the winter moth in Nova Scotia. 1954–1962. *Mem. Entomol. Soc. Can.* **46,** 1–57.

Espenshade, E. B., ed. (1979). "Goode's World Atlas," 15th ed. Rand McNally, Chicago, Illinois.

Everett, T. R., and Lamey, H. A. (1969). Hoja blanca. *In* "Viruses, Vectors and Vegetation" (K. Maramorosch, ed.), Wiley (Interscience), New York.

Ferguson, D. C. (1978). Pests not known to occur in the United States or of limited distribution. Winter moth, *Operophtera brumata* (L.) (Lepidoptera: Geometridae). *U.S. Dep. Agric. Coop. Plant Pest Rep.* **3** (48–52), 687–694.

Gallun, R. L., Starks, K. J., and Guthrie, W. D. (1975). Plant resistance to insects attacking cereals. *Annu. Rev. Entomol.* **20,** 337–357.

Ghani, M. A., and Muzzaffar, N. (1974). "Relations between Parasite-Predator Complex and the Host-Plants of Scale Insects in Pakistan," Misc. Publ. No. 5. Commonw. Inst. Biol. Control, Commonw. Agric. Bur. Slough, England.

Ghauri, M. S. K. (1966). Revision of the genus *Orosius* Distånt (Homoptera: Cicadelloidea). *Bull. Br. Mus. (Nat. Hist.), Entomol.* **18**(7), 231–252.

Ghauri, M. S. K. (1971a). Revision of genus *Nephotettix* Matsumura (Homoptera: Cicadelloidea: Euscelidae) based on the type material. *Bull. Entomol. Res.* **60,** 481–512.

Ghauri, M. S. K. (1971b). A remarkable new species of *Cicadulina* China (Homoptera: Cicadelloidea) from East Africa. *Bull. Entomol. Res.* **60,** 631–633.

Gillespie, D. R., Finlayson, T., Tonks, N. V., and Ross, D. A. (1978). Occurrence of the winter moth, *Operophtera brumata* (Lepidoptera: Geometridae) on Southern Vancouver Island, British Columbia. *Can. Entomol.* **110,** 223–224.

Gordon, R. D. (1975). A revision of the Epilachninae of the Western Hemisphere (Coleoptera: Coccinellidae). *U.S. Dep. Agric., Tech. Bull.* **1493,** 1–409.

Grover, P., and Prasad, S. N. (1976). "Bionomics of "Silver-shoot" Gall Fly, *Pachydiplosis oryzae* (Wood-Mason), a Pest of Rice in India," Final Tech. Rep. Department of Zoology, University of Allahabad, Allahabad, India.

Hamid, S. (1976). "Natural Enemies of Graminaceous Aphids," Final Tech. Rep. Pakistan Station (Rawalpinidi), Commonw. Inst. Biol. Control (mimeo).

Harpaz, I. (1972). "Maize Rough Dwarf, A Plant Hopper Virus Disease Affecting Maize, Rice, Small Grain and Grasses." Israel Univ. Press, Jerusalem.

Harris, K. F. (1979). Leafhoppers and aphids as biological vectors: Vector-virus relationships. *In* "Leafhopper Vectors and Plant Disease Agents" (K. Maramorosch and K. F. Harris, eds.), pp. 217–308. Academic Press, New York.

Hendrick, P. W., Ginevan, M. E., and Ewing, E. P. (1976). Genetic polymorphism in heterogeneous environments. *Annu. Rev. Ecol. Syst.* **7,** 1–32.

Hill, D. (1975). "Agricultural Insect Pests of the Tropics and Their Control." Cambridge Univ. Press, London and New York.

Hill, R. E., and Mayo, Z. B. (1980). Distribution and abundance of corn rootworm species as influenced by topography and crop rotation in eastern Nebraska. *Environ. Entomol.* **9**(1), 122–127.

Hille Ris Lambers, D. (1966). Polymorphism in Aphidadae. *Annu. Rev. Entomol.* **11,** 47–78.

Ibrahim, Amira Abd El-Hamid (1979). "Survey, Biological and Ecological Studies of Parasites and Predators of Certain Aphids Infesting Crops in Egypt," Final Tech. Rep. Inst. Plant Prot., Agric. Res. Cent., Dokki, Cairo, Egypt.

Isa, A. L. (1979). "Studies on Sugarcane Borers in Egypt," Final Tech. Rep. Plant Prot. Res. Inst., Minist. Agric., Egypt.

Israel, P., and Padmanabhan, S. Y. (1976). "Biological Control of Stem Borers of Rice in India," Final Tech. Rep. Cent. Rice Res. Inst., Indian Counc. Agric. Res. (mimeo).

Ito, Y., and Nagamina, M. (1981). Why a cicada, *Mogannia minuta* Matsumura, became a pest of sugarcane: An hypothesis based on the theory of "escape." *Ecol. Entomol.* **6**(3), 273–283.

Johnson, C. G. (1969). "Migration and Dispersal of Insects by Flight." Methuen, London.

Jolliff, G. D., Tinsley, I. J., Calhoun, W., and Crane, J. M. (1981). Meadowfoam (*Limanthes alba*) its research and development as a potential new oils seed crop for the Willamette Valley of Oregon. *Oreg., Agric. Exp. Stn., Tech. Bull.* **648,** 1–17.

Jones, D. P. (1973). Agricultural entomology. *In* "History of Entomology," pp. 307–332. Annu. Rev. Entomol., Palto Alto, California.

Kaloostian, G. H., Oldfield, G. N., Pierce, H. D., and Calavan, E. C. (1979). *Spiroplasma citri* and its transmission to citrus and other plants by leafhoppers. *In* "Leafhopper Vectors and Plant Disease Agents" (K. Maramorosch and K. F. Harris, eds.), Academic Press, New York.

Kiritani, K. (1979). Pest management in rice. *Annu. Rev. Entomol.* **24,** 279–312.

Kisimoto, R. (1967). Genetic variation in the ability of a plant-hopper vector, *Laodelphax striatellus* (Fallen) to acquire the rice strip virus. *Virology* **32,** 144–152.

Klassen, W., Shay, J. R., Day, B., and Brown, A. W. A. (1975). *In* "Crop Productivity—Research Imperatives" (A. W. A. Brown, T. C. Byerly, M. Gibbs, and A. San Pietro, eds.), pp. 275–308. Mich. Agric. Exp. Stn., East Lansing.

Knipling, E. F. (1979). The basic principles of insect population suppression and management. *U.S., Dep. Agric., Agric. Handb.* **512,** 1–659.

Kodet, R. T., and Nielson, M. W. (1980). Effect of temperature and photoperiod on polymorphism of the blue alfalfa aphid, *Acyrtosiphon kondoi. Environ. Entomol.* **9**(1), 94–96.

Kogan, M. (1981). Dynamics of insect adaptations to soybean: Impact of integrated pest management. *Environ. Entomol.* **10**(3), 363–371.

Krantz, G. W., and Lindquist, E. E. (1979). Evolution of phytophagous mites (Acari). *Annu. Rev. Entomol.* **24,** 121–158.

Krysan, J. L., Smith, R. F., Branson, T. F., and Guss, P. L. (1980). A new subspecies of *Diabrotica virgifera* (Coleoptera: Chrysomelidae): Description, distribution, and sexual compatibility. *Ann. Entomol. Soc. Am.* **73**(2), 123–130.

Lacy, G. H., McClure, M. S., and Andreadis, T. G. (1979). Reducing populations of vector leafhoppers is a new approach to X-disease control. *Front. Plant Sci.* **32,** 2–4.

Lattin, J. D. (1980). Scientific and technological needs for identification of arthropods of significance to human welfare. *FAO Plant Prot. Bull.* **29**(1), 25–28.

Lattin, J. D., and Knutson, L. (1982). Taxonomic information and services on arthropods of importance to human welfare in Central and South America. *FAO Plant Prot. Bull.* **30**(1), 9–12.

Lees, A. D. (1966). The control of polymorphism in aphids. *Adv. Insect Physiol.* **3,** 207–277.

Lipa, J. J. (1975). "Interaction of Spore-forming Bacteria and Viruses of Noctuids," Final Tech. Rep. Inst. Plant Prot., Miczurina, Poznan, Poland (mimeo).

Linnavuori, R., and DeLong, D. M. (1977). The leafhoppers (Homoptera: Cicadellidae) known from Chile. *Bernesia* **12/13,** 163–267.

McClure, M. S. (1980). Role of wild host plants in the feeding, oviposition, and dispersal of *Scaphytopius acutus* (Homoptera: Cicadelliade), a vector of peach X-disease. *Environ. Entomol.* **9**(2), 265–274.

McClure, M. S. (1982). Factors affecting colonization of an orchard by leafhopper (Homoptera: Cicadellidae) vectors of peach X-disease. *Environ. Entomol.* **11**(3), 695–700.

Metcalf, Z. P. (1962–1968). "General Catalog of the Homoptera," Fasc. 1–17. Agric. Res. Serv., U.S. Department of Agriculture, Washington, D.C.

Miller, J. C., and Cronhardt, J. E. (1982). Life history and seasonal development of the western winter moth, *Operophtera occidentalis* (Lepidoptera: Geometridae), in western Oregon. *Can. Entomol.* **114,** 629–636.

Mochida, O., Suryana, T., Hendarsih, and Wahyu, A. (1978). "Identification, Biology, Occurrence and Appearance of the Brown Plant Hopper. The Brown Planthopper (*Nilaparvata lugens* Stal), pp. 1–39. Indonesia Institute of Science, Jakarta.

Mound, L. A., and Waloff, N. (1978). Diversity of insect faunas. *Symp. R. Entomol. Soc. London* **9,** 1–204.

Nault, L. R., and Bradfute, O. E. (1979). Corn stunt: Involvement of a complex of leafhopper-borne pathogens. *In* "Leafhopper Vectors and Plant Disease Agents" (K. Maramorosch and K. F. Harris, eds.), pp. 561–586. Academic Press, New York.

Nielson, M. W. (1968). The leafhopper vectors of phytopathogenic viruses (Homoptera: Cicadellidae), taxonomy, biology and virus transmission. *U.S. Dep. Agric., Tech. Bull.* **1382,** 1–386.

Nielson, M. W. (1979). Taxonomic relationships of leafhopper vectors of plant pathogens. *In* "Leafhopper Vectors and Plant Disease Agents" (K. Maramorosch and K. F. Harris, eds.), pp. 3–27. Academic Press, New York.

Niemczyk, E., Miszczak, M., and Olszak, R. (1976). "The Effectiveness of Some Predaceous Insects in the Control of Phytophagous Mites and Aphids on Apple Trees," Final Tech. Rep. Res. Inst. Pomol., Skierniewice, Poland.

Nuttonson, M. Y. (1965). "Global Agroclimatic Analogues for the Northern Great Plains Region of the United States and an Outline of its Physiography, Climate and Farmcrops." Am. Inst. Crop Ecol., Washington, D.C.

Paulian, F., and Popov, C. (1980). Sunn pest or cereal bugs. *In* "Wheat" (E. Hafliger, ed.), pp. 69–74. Documenta Ciba-Geigy, Ciba-Geigy Ltd., Basel, Switzerland.

Radcliffe, E. B. (1982). Insect pests of potato. *Annu. Rev. Entomol.* **27,** 173–204.

Reinert, J. A. (1977). Field biology and control of *Haplaxius crudus* on St. Augustine grass and Christmas palm. *J. Econ. Entomol.* **70**(1), 54–56.

Reinert, J. A. (1980). Phenology and density of *Haplaxius crudus* (Homptera: Cixiidae) on three southern turfgrasses. *Environ. Entomol.* **9**(1), 13–15.

Rivnay, E. (1962). Field crop pest in the Near East. *Monogr. Biol.* **10,** 1–450.

Romberger, J. A., ed. (1978). "Beltsville Symposia on Agricultural Research," Vol. 2. Allanheld, Osmun, Montclair, New Jersey.

Ruppel, R. F. (1965). A review of the genus *Cicadulina* (Hemiptera: Cicadellidae). *Publ. Mus., Mich. State Univ., Biol. Ser.* **2**(2), 385–428.

Sailer, R. I. (1978). Our immigrant insect fauna. *Bull. Entomol. Soc. Am.* **24**(1), 3–11.

Sanborn, S. M., Wyman, J. A., and Chapman, R. K. (1982). Studies on the European corn borer in relation to its management on snap beans. *J. Econ. Entomol.* **75**(3), 551–555.

Shaikh, M. R. (1978). "Studies on Crystaliferous Bacteria Infecting Lepidopterous Pests of Pakistan," Final Tech. Rep. Department of Physiology, University of Karachi, Karachi-32, Pakistan.

Shikata, E. (1979). Rice viruses and MLO's and leafhopper vectors. *In* "Leafhopper Vectors and Plant Disease Agents" (K. Maramorosch and K. F. Harris, eds.), pp. 515–527. Academic Press, New York.

Singh, S. R., and van Emden, H. F. (1979). Insect pests of grain legumes. *Annu. Rev. Entomol.* **24,** 255–278.

Smith, H. D., Maltby, H. L., and Jimenez, J. E. (1964). Biological control of the citrus blackfly in Mexico. *U.S. Dep. Agric., Tech. Bull.* **1311,** 1–30.

Stinner, R. E., Regniere, J., and Wilson, K. (1982). Differential effects of agroecosystem structure on dynamics of three soybean herbivores. *Environ. Entomol.* **11**(3), 538–543.

Sylvester, E. D. (1980). Circulative and propagative virus transmission by aphids. *Annu. Rev. Entomol.* **25,** 257–286.

Tamaki, G., and Fox, L. (1982). Weed species hosting viruliferous green peach aphids, vector of beet western yellows virus. *Environ. Entomol.* **11**(1), 115–117.

Tamaki, G., and Olsen, D. (1979). Evaluation of orchard weed hosts of green peach aphid and the production of winged migrants. *Environ. Entomol.* **8**(2), 314–317.

Tamaki, G., Fox, L., and Chauvin, R. L. (1980). Green peach aphid: Orchard weeds are host to fundatrix. *Environ. Entomol.* **9**(1), 62–66.

Tao, C.-C., and Chiu, S.-C. (1971). "Biological Control of Citrus, Vegetables and Tobacco Aphids," Spec. Publ. No. 10. Taiwan Agric. Res. Inst., Taipei, Taiwan, China.

Todd, J. G., and Kamm, J. A. (1974). Biology and impact of a grass bug *Labops hesperius* Uhler in Oregon Rangeland. *J. Range Manage.* **27**(6), 453–458.

U.S. Department of Agriculture (1978). "Biological Agents for Pest Control. Status and Prospects." USDA in cooperation with the Land-Grant Universities and the Agricultural Research Institute, Washington, D.C.

Varley, G. C., and Gradwell, G. R. (1968). Population models for the winter moth. *Symp. R. Entomol. Soc. London* **4,** 132–142.

Wagner, E., and Weber, H. H. (1964). "Faune de France," No. 67. Miridae, Paris.

Wallwork, J. A. (1976). "The Distribution and Diversity of Soil Fauna." Academic Press, New York.

Walters, M. C. (1979). Maize pests of sub-Saharan Africa. *In* "Maize" (E. Hafliger, ed.), pp. 66–71. Documenta Ciba-Geigy, Ciba-Geigy Ltd., Basel, Switzerland.

Waterworth, H. E. (1981). Our plants' ancestors immigrated too. *BioScience* **31**(9), 698.

Wheeler, A. G., Jr., and Henry, T. J. (1974). *Tropidosteptes pacificus,* a western ash plant bug introduced into Pennsylvania with nursery stock (Hemiptera: Miridae). *U.S. Dep. Agric., Coop. Econ. Insect Rep.* **24**(30), 588–589.

Whitcomb, R. F., and Davis, R. E. (1970). Mycoplasma and phytaboviruses as plant pathogens persistently transmitted by insects. *Annu. Rev. Entomol.* **15,** 405–464.

Wiackowski, S., Chlodny, J., Tomków, M., Witrylak, M., and Kolk, A. (1976). "Studies on the Entomofauna of Larch, Alder and Birch in Different Environmental Conditions and its Ecological Relationships with Insect Pests of More Important Forest Tree Species," Final Tech. Rep. For. Res. Inst. & Educ. Univ., Kielce, Warsaw, Poland.

Yen, D. (1973). "A Natural Enemy List of Insects of Taiwan." College of Agriculture, National Taiwan University.

CHAPTER 6

Where Are the Exotic Disease Threats?

C. L. SCHOULTIES, C. P. SEYMOUR, and J. W. MILLER

Division of Plant Industry
Florida Department of Agriculture and Consumer Services
Gainesville, Florida

I. INTRODUCTION

No matter what destruction introduced pathogens may wreak upon our plants, it is we who perceive the threat and who can act to mitigate the introduction and establishment of exotic pathogens within our borders. Suffice it to say that exotic disease threats are possibilities in our minds, whereas exotic diseases are possibilities for our agriculture. There is an inherent danger in writing about exotic disease threats and exotic diseases as possibilities, because some people will

EXOTIC PLANT PESTS AND
NORTH AMERICAN AGRICULTURE

ISBN 0-12-757880-3

invariably take possibilities to mean predictions. We cannot, however, predict which diseases will exist in the future. This is beyond our present comprehension.

In this chapter, we turn our attention to plant diseases that occur outside the mainland of the United States and that could damage our agriculture if the causative agent were introduced and established. We shall approach the subject of exotic disease threats from the following perspectives: (1) awareness of exotic diseases and exotic disease threats; (2) realization or actualization of exotic diseases in our agriculture; and (3) discussion of foreign places, in particular their climate, agriculture, relevant trade, citizens' travel to the United States, and some of their diseases.

II. AWARENESS OF EXOTIC DISEASES AND EXOTIC DISEASE THREATS

Despite the catastrophic losses attributable to diseases that were once exotic to us, the awareness of most plant pathologists and associated scientists of diseases that are now exotic, and of the risks that these diseases pose to our agriculture, is slight. Awareness of threatening exotic diseases can aid in preventing the inadvertent introduction of a plant pathogen into a new area from which it might have been excluded and can give the people of a country or region time to prepare for the exotic disease and thereby minimize losses (Thurston, 1973).

Because we cannot possibly discuss all the threatening exotic diseases of all our agricultural plants, answers to the questions which follow about the awareness of exotic disease threats seem in order. How does one learn about exotic diseases? How does one perceive exotic diseases to be threats? While revealing these sources and insights, we shall review some of the pertinent literature on exotic disease threats.

A. Indexes of Exotic Diseases

To prepare an index of diseases that are exotic to the United States, it is first necessary to establish what diseases occur here. Stevenson (1926) prepared the first U.S. index of exotic diseases in the United States. This 56-year-old publication lists several thousand exotic diseases on nearly 1000 different hosts. Subsequently, Spaulding (1961) prepared an index of foreign diseases of forest trees, and Watson (1971) listed 1492 exotic fungal and bacterial diseases of agronomic and ornamental plants. The more recent compilers of indexes had the advantage of having access to the "Index of Plant Diseases in the United States" (Anonymous, 1960).

B. Publications on Exotic Disease Threats

Spaulding (1914) addressed the Massachusetts Horticultural Society just 2 years after the enactment of the Plant Quarantine Act of 1912 and warned that we as a nation were in constant and increasing danger from serious foreign plant diseases. He cited several foreign pathogen introductions, noted the economic losses that they caused, and named some threatening foreign diseases such as black wart of potato, powdery scab of potato, and alfalfa wart. He lauded the passage of the Quarantine Act as a heretofore unused method of safeguarding our forestry, horticultural, and agricultural interests and urged international cooperation by plant pathologists in the development of their newly emerging science.

Hunt (1946) reviewed the subject of destructive plant diseases not established in North America by indicating potential sources of exotic pathogens, by citing numerous examples of exotic diseases on agronomic, silvicultural, and ornamental plants, and by exploring various factors that affected the destructiveness of pathogens. He pointed out that destructiveness varies from place to place. Two of our most notorious diseases, citrus canker and chestnut blight, were unknown until their pathogens began destroying their respective hosts here. Yet, two other diseases, white pine blister rust and Dutch elm disease, were known to be destructive before their respective pathogens reached the United States. On the other hand, the powdery scab of potato was considered serious elsewhere and was feared in the United States, until it was found to be well established here and usually unimportant. Thus, Hunt (1946) concluded that based upon present knowledge, it was not possible to predict with certainty what would be the future impact on our agriculture of specific foreign pathogens that might be introduced into the United States.

Spaulding (1961), in his index of foreign diseases of forest trees of the world, selected 30 exotic diseases from a compilation of more than 3000 diseases as being potentially the most dangerous to North American forests.

McGregor (1973, 1978) and his collaborators selected and ranked 551 threatening exotic plant diseases from an initial grouping of approximately 2000 diseases. The selection of diseases was based mainly on the economic value of the host, and the ranking of the 551 diseases was based on a mathematical model that took into account the probability of the pathogen becoming established in the United States and the economic impact if it should become established. Besides selecting and ranking exotic diseases, McGregor discussed pathways of pathogen entry, colonization and establishment of introduced pathogens, the effectiveness of quarantine programs, and various international quarantine programs.

Stakman and Harrar (1957), a working group of forest disease researchers (Anonymous, 1963), and Thurston (1973) addressed the subject of threatening exotic diseases from an international perspective. Stakman and Harrar (1957) stressed the threat which plant pathogens pose to the production of economic crops of international importance and emphasized the plant pathologist's role in seeking ways to prevent the spread and development of plant disease epiphytotics. The forestry working group (Anonymous, 1963) called for international cooperation to recognize and characterize those forest tree pathogens on each continent considered to be potentially dangerous if introduced to other continents and to promote measures to restrict their spread. Thurston (1973) reviewed tropical disease threats and devised a relative scale to rank various exotic plant diseases. Highly threatening plant diseases were so designated because they had a high potential for spread to other countries, a demonstrated ability to cause serious crop losses, and a poor history of control. Diseases of intermediate threat were those in which pathogens did not clearly possess the potential for rapid spread to other countries or for which efficient, economic control measures were not clearly available. Diseases of limited threat potential were those serious diseases which had spread very slowly or which had efficient, economic controls available in case the pathogen spread.

C. Periodicals and Bulletins

For a given area, new pathogens sometimes occur, old pathogens often take on new forms, and exotic pathogens are introduced. Such events usually first appear in periodicals or in government-issued bulletins in the country where the resulting disease occurs. Occasionally, the disease can become serious and widespread before it is noted in the literature. This delay in reporting often reflects a lack of trained people and/or economic and political considerations (i.e., lack of monetary resources and fear of foreign quarantines). One can search the literature using abstract journals, current titles, and computers and eventually find reported diseases—both exotic and domestic. The problem is, however, to select readily only those titles (articles) that deal with new diseases that do not occur in the United States. This, in our experience, is an arduous task that involves plant-by-plant sorting of exotic from domestic diseases.

Having found a title or abstract about an exotic disease, obtaining and, if need be, translating the article can sometimes be time-consuming, expensive, and difficult. Occasionally, we had to settle for being fortunate enough to have the abstract from *Review of Plant Pathology* (*Review of Applied Mycology*).

Evaluating the threat potential of the disease from the articles is most often speculative. The assessment of disease loss (hence, much of the potential threat to us) is often couched in subjective terms such as "highly destructive," "severe," or "catastrophic," and not in objective, quantitative terms.

We lack a periodical that carries information and articles exclusively on exotic pests that threaten U.S. agriculture. The 35 member nations of the European and Mediterranean Plant Protection Organization (EPPO) do publish a regular scientific journal, the *EPPO Bulletin,* on exotic pests that threaten their countries (Van

Tiel, 1976). We do have the journal *Plant Disease,* which is international in scope and which contains regularly appearing sections entitled "New Diseases and Epidemics" and "Focus." The former section often contains new diseases that are exotic to us, and the latter carries snippets about many relevant topics, including reports on exotic diseases that appear in other journals.

D. Distribution Maps and Descriptions of Plant Pathogens

It was convenient for us to find an exotic disease that we had been tracking and see its points of establishment on a map (CMI Distribution Maps of Plant Diseases) and/or read a brief summary of the pathogen ("CMI Descriptions of Pathogenic Fungi and Bacteria," "CMI/AAB Descriptions of Plant Viruses," and "CIH Descriptions of Plant-parasitic Nematodes").

E. Compendia

The various disease compendia which have been published by the American Phytopathological Society within the last 9 years on nine different plants provide significant aid in determining the nature and distribution of exotic pathogens. For example, the "Compendium of Wheat Diseases" (Wiese, 1977) discusses 73 important worldwide pathogens of wheat, 20 of which are exotic to the United States mainland. In addition, second editions of the corn and soybean compendia are available. Having successive editions provides a "window in time" to view what is new and what has changed relative to diseases of these respective crops. For example, the first edition of the "Compendium of Soybean Diseases" (Sinclair and Shurtleff, 1975) includes *Pyrenochaeta glycines,* which had been reported in Ethiopia and which caused leaf spotting, leaf necrosis, and premature leaf drop on seedlings and mature plants of soybean. No data on yield losses were included in the first edition. In the second edition of the "Compendium of Soybean Diseases" (Sinclair, 1982), *Pyrenochaeta* leaf spot had been detected in Cameroon and Zambia, as well as Ethiopia. Yield losses of 50% were reported in some early-planted fields under wet conditions in Zambia.

F. Colleagues

Colleagues who are knowledgeable about crop pathogens and widely traveled can often provide sound advice on exotic disease threats and can help to sift and sort the literature on this difficult subject. For example, a colleague on an agricultural exchange to Zimbabwe was told that *Pyrenochaeta* leaf spot of soybean occurred there (Rothwell, 1980) and that the disease was inflicting heavy losses (P. S. Lehman, personal communication). This personal communication corroborates the "uprising" of this disease as discussed in Section II,E.

In our view, *Pyrenochaeta* leaf spot of soybean should qualify as a potential exotic disease threat.

III. REALIZATION OR ACTUALIZATION OF EXOTIC DISEASES IN OUR AGRICULTURE

An exotic plant disease is threatening to us not only because we know that the disease exists and can be destructive elsewhere, but also because we know that the cause of the destruction is capable of movement and introduction via certain operative pathways and that our plants may be vulnerable to the introduced exotic plant pathogen.

A. Movement and Introduction of Exotic Plant Pathogens

Most of us were never in fear of extraterrestrial pathogens until our astronauts took a round trip to the moon in the late 1960s. A pathway for exotic moon pathogens was opened, a potential threat was perceived, and our government accordingly imposed quarantine restrictions on the returning space travelers to minimize the threat. Fortunately, neither we nor our fellow creatures nor our plants were affected by anything from the moon—except the threat.

Through the nineteenth century, little or no attention was paid to the movement of plant pathogens by most countries (Mathys and Baker, 1980). This was understandable, because scientists were just beginning to appreciate the microbial etiology of infectious plant diseases. Knowledge about the causality of plant diseases paved the way to understanding how exotic pathogens could be moved and introduced. In the twentieth century, the U.S. federal government enacted the Plant Quarantine Act of 1912, the Organic Act of 1944, and the Federal Pest Act of 1957 to limit the movement of pathogens by authorizing governmental activities to regulate the movement and introduction of exotic plant pathogens. This has been accomplished by specific embargoes, by inspections conducted at various places (i.e., fields, ports of entry, and ports of origin), by controlled entry (i.e., disinfestation treatments), and by postentry quarantine requirements (Mathys and Baker, 1980).

1. *Trade and Travelers*

The movement of plants with any accompanying pathogens began early in our history, long before we were aware of microbial plant pathogens. For example, citrus came to Haiti in 1493 with the second voyage of Columbus (Ray and Walheim, 1980). It prospered in the Caribbean area and was transported from Cuba by Cortez and his men, who took seeds and planted them in Vera Cruz, Mexico, in 1518. By the 1700s, Spanish missionaries, using this Mexican source, were planting citrus in California and Arizona. In 1565, the first citrus crop was started in Florida at St. Augustine (Ray and Walheim, 1980). Fortunately, the early movement of citrus appears to have been by seed, and pathogen movement tends to be less with seed than with other forms of plant germplasm (McGregor, 1973). Eventually, other types of citrus germplasm (fruit, whole plants, cuttings, bud sticks, etc.) began arriving, and with these arrivals came the increased threat of exotic pathogens. One such exotic pathogen was *Xanthomonas campestris* pv. *citri,* which arrived in the United States on infected citrus rootstock from Japan and became rampant on citrus in the southeastern Gulf Coast states in the early years of this century, until it was eradicated (Stall and Seymour, 1983).

Most nations restrict or prohibit the movement of persons or materials likely to carry pathogens that could affect domestic plants (McGregor, 1973). Only 11 of 171 countries reviewed by McGregor had no form of regulation on import restrictions or inspection of either entering travelers or cargo. Further, most governments that regulate such movements give a higher priority to cargo than to travelers. That is, 82% of the countries regulate cargo, whereas only 22% regulate travelers. The success of regulatory efforts to preclude the entry of foreign pests can be demonstrated by citing statistics on the interception of serious exotic pathogens at our ports of entry. In a recent 1-year period, citrus canker (*X. campestris* pv. *citri*) was intercepted 379 times, sweet orange scab (*Elsinoe australis*) 594 times, citrus black spot (*Guignardia citricarpa*) 508 times, and chrysanthemum rust (*Puccinia horiana*) six times (Anonymous, 1981c). These interception statistics indicate a form of success; however, there is no way to know what or how much went by undetected (Mathys and Baker, 1980). Moreover, it is apparent from reading the lists that only those plants or plant parts showing signs of the pathogen or obvious symptoms were intercepted. For example, no *Phytophthora* species or virus interceptions were posted in the above listing. These absences probably reflect the fact that a thorough, time-consuming diagnosis was not possible.

Not to be overlooked is the importation of various vectors of domestic and exotic plant pathogens. Foster (1982), using government-compiled statistics, determined that more than 30,000 known vectors (mostly arthropods) were intercepted at U.S. ports of entry from July 1973 to September 1977 and that 431 known arthropod vectors of plant pathogens were brought into the United States for scientific investigations by the appropriate regulatory agencies from September 1977 to August 1979. Another problem with introduced vectors is that some can retain their pathogens for life, and some can transovarially pass the pathogen to the next generation. Thus, some exotic vectors could arrive in our country harboring exotic pathogens. Nematodes, fungi, and insects in the families Aleyrodidae, Cicadellidae, Cixiidae, Delphacidae, Miridae, Piesmatidae, Psyllidae, and Thripidae, which transmit viruses, mycoplasmalike organisms, rickettsialike organisms, and spiroplasmas, may retain their pathogens for periods approaching their life span (Foster, 1982).

The form in which commodity imports arrive is important relative to the risk of introducing exotic pathogens. Many of our crop imports arrive as fruit, produce, seeds, or refined products (Anonymous, 1981d). On the other hand, ornamental plants and their propagative parts are shipped to us in abundance from various places and in numbers far exceeding those of other agricultural plants (Baker and Linderman, 1979). Most of our forestry imports arrive as seed or as sawn wood (E. L. Barnard, personal communication). McGregor (1973) ranks the level of risk of pathogen introduction in descending order as follows: (1) living plants, (2) soil adhering to roots of living plants, (3) fruits, (4) seeds, and (5) soil not adhering to roots of living plants. Thus, with the exception of ornamentals, we are importing plant materials at the lower risk levels. However, offsetting these rankings of McGregor, at least for seeds, is the growing list of known seed-transmitted pathogens (Neergaard, 1977; Mandahar, 1981), the possibility of seeds being carriers of pathogens, and the sheer if not astronomical number of seeds imported into the United States (Anonymous, 1981d).

A relatively infrequent occurrence that carries a high risk for the introduction of exotic pathogens is the emergency need for plants or plant parts. If a significant portion of a crop is devastated, particularly in the seedling stage, market and political demands combine to press for the emergency importation of plants. For example, in 1979, most of south Florida's tomato seedlings were lost due to cold injury. To fill the void and to ensure the first crop of tomatoes of the new year, seedlings were rapidly brought into Florida from Mexico. Fortunately, no known disease or pest introduction resulted from the massive shipments. When the need for emergency plant imports arises, it is important for regulatory officials to know the exotic diseases and pests of a particular crop at its source and to use appropriate measures to minimize the risks of importing exotic pathogens.

Travelers arriving at our ports of entry are increasing in number every year. In 1980, 32 million people came to the United States, in contrast to 23 million only 3 years earlier (Ford, 1981). Further, nearly 1 million of those arriving in 1980 carried some sort of agricultural contraband, and approximately half of them failed to declare the material. Travel to various easily accessible places outside the country by private plane or boat presents additional opportunities to introduce exotic pests (Dowling *et al.*, 1982). In Florida particularly, legal and illegal immigrants may knowingly or unknowingly carry prohibited materials that harbor pathogens.

More rapid means of transportation historically have increased the chances of introducing exotic pathogens. For example, Yarwood (Chapter 3, this volume) suggests that in the initial introduction of potatoes from South America to Europe, *Phytophthora infestans* could not survive the long voyage through the tropics but could survive in later voyages when faster steamships were available.

Since the early 1940s, aircraft have assumed increasing importance as carriers of goods and people accompanied by plant pathogens and their insect vectors (McGregor, 1973). Today, with the availability of subsonic and supersonic transport, the threat of exotic diseases looms ever larger. Moreover, the number of rapidly moving aircraft is becoming greater and the number of locations that may be visited by a single flight is increasing. In 1980, more than 270,000 aircraft arrived in the United States from all parts of the world, a 10% increase over the previous year (Ford, 1981).

Within the past 25 years, containerization has revolutionized ocean shipping and dramatically increased the threat of exotic pests moving into the United States by making cargo less accessible for inspection and by carrying cargo into the interior of the country. However, containerization does reduce the threat of pest introduction between the port of entry and the destination because the likelihood of a pest exiting a container is very low (McGregor, 1973).

2. *Scientific Pathways*

Plant scientists, hobbyists, and growers worldwide have contributed to our horticultural and silvicultural well-being. Sunflower is our only major crop to have originated and been domesticated in the United States (Chapter 3, this volume). It is obvious, therefore, that there has been a vast number of exotic plant introductions into the United States for our country to be so preeminent in agriculture. This plant movement was somewhat haphazard and uncontrolled until the end of the nineteenth century (Waterworth, 1981). In 1898, the federal government began conducting plant explorations to introduce new and unusual plants and began maintaining systematic and detailed records on plant introductions. As of early January 1981, nearly 450,000 plant introductions had been made through this system (Waterworth, 1981). Most germplasm enters the United States as a result of government-sponsored explorations or as a result of requests by nurserymen, botanical gardens, arboreta, hobbyists, and various scientists. Germplasm from original or secondary gene centers is now in great demand. With the high priority awarded to resistance (Granhall, 1981), introduction often comes from developing countries with very limited quarantine resources and limited information on endemic diseases. Further, endemic diseases are common in the gene centers for plants, because these same centers are also the likely gene centers for the pathogens as well.

Only in recent years has consideration been given to the importation of certain plant pathogens that may accompany plant introduction. For example, citrus researchers in Riverside, California, from 1910 to 1940 freely introduced budwood from various foreign sources into their vast citrus variety collection (Roistacher, 1981). According to Roistacher, introductions of budwood at earlier times were undoubtedly responsible for the introduction of tristeza virus and its seedling yellow forms into California. Because of our own experience and that of others with uncontrolled plant entry and because of scientific and technological innovations, the federal government, acting under the authority of the three quarantine acts, now places most plant introductions into three quarantine categories (Waterworth and White, 1982): (1) restricted, subject to inspection and possible disinfestation treatments; (2) postentry, subject to containment under close supervision after inspection and/or treatment; and (3) prohibited, subject to stringent and specific requirements.

Just as exotic plants are introduced in a specific manner, so are exotic plant pests destined for scientific study in the United States. The Federal Plant Pest Act of 1957 made provisions for the movement of exotic pathogens into this country, as well as movement of domestic pathogens within the country (Waterworth and White, 1982). There has been and will continue to be attempts by some to bring plant material and/or pathogens into the United States by knowingly ignoring or being unaware of regulatory laws. Often, those who knowingly violate regulations may do so because they are zealous in their own pursuits, may think that their efforts are beyond bureaucratic scrutiny or may think that their requests will be delayed or even rejected. There are no proven instances in which researchers have intentionally introduced restricted pests into the United States, but circumstantial evidence tends to suggest that such practices occur (Dowling *et al.*, 1982). Two examples from overseas illustrate the risks with scientific studies. Tobacco blue mold (*Peronospora tabacina*) was first introduced into Europe for the purpose of laboratory studies (Zadoks and Schein, 1979) and subsequently became rampant on that continent. A recent Associated Press release from Caracas, Venezuela, indicates that the country is having problems with marauding killer bees which may sting human or animal victims up to 200 times (Anonymous, 1982a). These bees were reportedly imported from Africa to Brazil by entomologists who were studying ways to breed out the bee's aggressive tendencies.

McGregor and his task force, in their extensive report on emigrant pests (McGregor, 1973), speculate that the scientific pathway of entry of exotic pests (pathogens) in this time of rapid travel will become more important as the need for genetic control of pests increases. According to McGregor, the danger lies not only in introducing new pests but also in increasing the gene pool of established pests. For example, *Pyricularia oryzae* in rice has several hundred pathogenic races making up 32 racial groups (Neergaard, 1981). Pathogenic races new to rice-growing areas in the Upper Volta, Nigeria, and South Korea were introduced with imported rice cultures. As a consequence, according to Neergaard, the resistance of domestic cultivars broke down in the sense that they were not resistant to the exotic pathogenic races uncritically introduced with seeds. Yet, another consideration in this modern scientific age is the emergence of recombinant DNA science and technology. This 7-year-old methodology holds promise for another green revolution in agriculture by the end of the twentieth century

(Boddé, 1982). On the other hand, there is the possibility that exotic plant pathogens (i.e., DNA-containing viruses or bacteria) might be imported for recombination studies and may escape laboratory-greenhouse confinement.

The benefits to agriculture of importing exotic germplasm or plant pathogens often outweigh the inherent risks. To minimize the risks, the following requirements are suggested for the importation of germplasm and pathogens: (1) approval by state and federal regulatory officials, and (2) strict compliance with all precautions stipulated by regulatory officials, including the use of quarantine facilities.

3. *Wind*

Among the natural forces, it is usually the wind that transports exotic pathogens from distant places within a reasonable time and for the necessary distance. Some historical examples of wind dissemination are given by Yarwood (Chapter 3, this volume). Although we cannot control what the wind brings us, we can be aware of which diseases (i.e., rusts, smuts, powdery mildews, and some incited by higher phycomycetes) are menacing in other countries and can thus be prudently prepared. For example, soybean rust (*Phakopsora pachyrhizi*) has been reported on soybeans in Puerto Rico (Vakili and Bromfield, 1976) and is a potential threat to at least some of our vast soybean acreage, particularly in the warmer and more humid states. This rust could be blown into the southeastern United States by hurricane winds. By analogy, two recently introduced pathogens, sugarcane rust (*Puccinia melanocephala*) and sugarcane smut (*Ustilago scitaminea*), prior to coming to Florida were rapidly and extensively established in the Caribbean area, and there is no reason to assume that the soybean rust fungus might not follow the same scenario as the sugarcane pathogens. With the sugarcane pathogens, however, it is difficult to determine whether the wind carried spores into the sugarcane fields of south Florida or whether migrant cane workers from the Caribbean somehow moved the pathogens into the United States, to be subsequently dispersed by the wind after their arrival. In either case, aerial dissemination of these sugarcane pathogens meant rapid and uncontrolled movement of the pathogen. Thus, the capacity of pathogens to be wind-disseminated increases the threat potential of any exotic plant pathogen.

B. Vulnerability of Agriculture to Introduced Exotic Plant Pathogens

Our agricultural plants are potentially vulnerable to the harm that introduced exotic plant pathogens may cause. This vulnerability is part of our perceived threat of exotic pathogens.

1. *Disease Complex Considerations*

The major task for the inadvertently introduced exotic pathogen in its new location is to encounter a susceptible host under proper environmental conditions to cause disease. This fact alone must thwart significantly the potential of most introduced pathogens. From a logical and historical point of view, introduced plant pathogens from a given climatic area are more likely to find similar and susceptible hosts and the proper environmental conditions in a place with a climate similar to that of their area of origin. Most of our agriculture is located in temperate and subtropical climates, with little occurring in tropical climates. It follows that most exotic disease threats will come from other temperate and subtropical areas of the world. Tropical exotic diseases do sometimes pose a threat to the same crop in subtropical areas but tend to be of minimal threat to the same crop in temperate areas, except where higher elevations override latitudinal considerations.

The ultimate success of a pathogen is not entirely left to chance. With many pathogens, a saprophytic mode of existence or a dormant phase can perpetuate the pathogen until such time as a susceptible host and a suitable environment occur. With most viruses and other obligate pathogens, a host is soon required for reproduction and is frequently required for preservation. Compensating for this host dependency in viruses is the polyphagous nature of many viruses, which are capable of infecting both wild and cultivated hosts (Bos, 1981).

2. *Vector Considerations*

With some diseases, vectors such as insects, mites, nematodes, or fungi are required for transmission of the pathogen. Neither man nor the wind and/or the rain can effectively transmit these pathogens. Thus, the vector can, in such cases, be the principal determinant of a pathogen's importance.

Occasionally, the vector is native, but the pathogen is exotic. Plum pox (Sharka) is a serious viral disease of plum, peach, and apricot in Europe (Sutic and Pine, 1968). The virus is transmitted by the aphid *Myzus persicae,* which occurs in the United States, as well as by grafting and mechanical means. Further, in Europe, it has been demonstrated that many perennial weeds growing near affected hosts can be infected and can serve as reservoirs of this virus (Bos, 1981). If the plum pox virus were introduced into the United States, it would seem that efficient insect vectoring could occur and that a diverse array of susceptible hosts could make plum pox a very serious disease. Additionally, a disease of apple, apple proliferation or apple witches' broom, poses a threat to our apple production and is an interesting variation on the situational theme—pathogen exotic, vector domestic. Apple proliferation is caused by a mycoplasmalike organism (MLO) which occurs throughout much of Europe and inflicts heavy yield and size losses on apple (Anonymous, 1978b). Transmission of the MLO is known to occur only by root fusion or by grafting. Insect vectors are not known but are suspected (Marenaud *et al.*, 1978). The apple leafhopper, *Empoascas maligna,* is distributed throughout the apple-growing areas of North America, but it does not occur in Europe, where the apple proliferation disease occurs (Metcalf, 1968). Two species of *Empoascas* are known to transmit the disease agent of papaya bunchy top (Nielson, 1979). Therefore, it is within the realm of possibility that the apple leafhopper that occurs here could be a vector for the apple proliferation MLO, should the pathogen be introduced.

Frequently, both vector and pathogen will be exotic. Unless the pathogen has effective means of transmission other than vectoring, the threat offered by the pathogen alone is minimal until a means of vectoring either arrives and survives or develops. Vectors of various kinds arrive by much the same pathways as pathogens, and survival, like that of plants and pathogens, depends upon abiotic as well as biotic factors. Climatic considerations and zoogeographic information are useful when examining the distribution of various vectors and the threat of exotic vectored pathogens. For example, the following exotic diseases affecting various grain crops have pathogens that are transmitted by leafhoppers of the genus *Cicadulina:* maize streak, which occurs in Africa, south of the Sahara Desert, India, and various islands in the Indian Ocean (Shurtleff, 1980); enanismo (cereal dwarf), which occurs in Colombia and Ecuador (Wiese, 1977); and Eastern wheat striate, which occurs in India. The leafhopper genus *Cicadulina* is not known to occur in the United States and occurs in other distinct zoogeographical realms as evidenced by the areas where the various viral diseases occur. Accordingly, these aforementioned diseases seem to be of minimal threat to our various grain crops. If exotic pathogens transmitted by the exotic planthopper, *Laodelphas striatellus,* are considered, the distribution of the various diseases could cause alarm. For example, *L. striatellus* transmits viruses of the following exotic diseases which affect various grain crops: barley yellow striate mosaic, which occurs in Italy (Wiese, 1977); cereal tillering, which occurs in Italy and Sweden (Wiese, 1977); maize rough dwarf, which occurs in southern Europe, Israel, and India (Shurtleff, 1980); and northern cereal mosaic, which occurs in Japan (Wiese, 1977). These diseases occur in climates similar to ours and thus pose a threat. Nielson (1979) introduces an optimistic note by stating that successful introductions of leafhopper vectors from one zoogeographic realm to another are rare. He cites five introductions from the Palearctic to the Neartic Region, one from the Neartic to the Paleartic Region, and one from the Oriental to the Australian Region.

3. *Agricultural Considerations*

Our agronomic plants are particularly vulnerable to exotic pathogens because nearly all of them originated and were domesticated elsewhere (Chapter 3, this volume). Because pathogens tend to coevolve with their respective hosts

(Granhall, 1981), introduced exotic pathogens frequently are reunited with their long-departed hosts within our mainland. Not only were pathogens left behind, but also the microbial associates that may have effectively suppressed the disease in its former environment. Presumably, more reunions of hosts and their pathogens would occur if we imported more agricultural commodities (a major pathway of exotic pathogen introduction). Fortunately, as a nation, we are well beyond self-sufficiency with most agronomic crops and are a major world producer of many crops such as soybean, maize, and cereal grains (Anonymous, 1979a). What we do import in large quantities are commodities that we apparently are unable to grow (i.e., bananas, coffee, and cocoa) or probably need to fill seasonal market demands (i.e., potatoes and tomatoes) (Anonymous, 1979b). However, offsetting this favorable situation of few agricultural imports is the sometimes enormous crop monoculture that we have established to achieve high levels of productivity. Moreover, we have, in many cases, made the monocultural situation more explosive for pathogens by narrowing the genetic base of many of our crops (Zadoks and Schein, 1979). Yield, marketing considerations, or labor cost reductions are sometimes valued more than disease or pest resistance by crop breeders, seed producers, and growers. One only has to be reminded of the southern corn leaf blight (*Helminthosporium maydis*) epiphytotic of 1970 in the United States to realize the hazard. At that time, nearly all of our cultivated genetic lines of corn contained the Texas male-sterile cytoplasmic line, which was susceptible to race T of the pathogen. Within a single season, the pathogens spread throughout the corn-growing areas of the eastern United States (Zadoks and Schein, 1979).

Whereas with many agronomic plants the genetic base is quite narrow, the opposite can be said for most ornamental plants. Genera of ornamentals include many species and often many cultivars within the species (Baker and Linderman, 1979). For example, as Baker and Linderman cite Bailey and Bailey (1976), approximately 20,000 varieties of rose, 7000 of gladiolus, and 300 of sweet pea have been developed and named. However, ornamentals are shipped more often as whole plants (without soil) and as propagative plant parts than as seed, as are other kinds of plants (Baker and Linderman, 1979), and the risk of introduction of exotic pathogens tends to be high with the shipment of whole plants and their vegetative propagative parts (McGregor, 1973). The United States receives significant amounts of bedding and nursery stock from Western European countries, which are temperate in climate, and many of our flowers and foliage plants from countries in tropical Central and South America (Section IV). Following an earlier stated assumption, our ornamentals would tend to be more vulnerable to pathogens introduced from temperate rather than tropical climates. However, much of our ornamental industry takes place in subtropical climates and would therefore be vulnerable to many tropical pathogens.

Many of our worst exotic disease disasters have involved forest trees (i.e., Dutch elm disease, chestnut blight, and white pine blister rust). It is not surprising that the first five exotic plant disease threats listed by McGregor (1973, 1978) as being the most dangerous to the United States are diseases of trees. Why is the threat of exotic diseases so much more intense with forest trees than with other plants? Most agronomic crops have shorter life spans than trees, which have life expectancies measured in the decades and sometimes centuries. Trees are, in effect, long-standing targets for plant pathogens. Agronomic and ornamental plants are often more limited in space than are trees, which may form transcontinental ranges. The course of disease epiphytotics can often be very swift with our food crops. The 1970 epiphytotic of southern corn leaf blight (*H. maydis*) was essentially completed in a single season because breeders and pathologists were able to restore resistance by utilizing resistant genetic sources (Zadoks and Schein, 1979). However, consider the current status of the American elm and American chestnut and their respective epiphytotics. Despite substantive efforts to provide hope in the form of resistant varieties or hybrids of these species, we still find ourselves decades later showing no progress on these problems (Hart, 1980; Hepting, 1974). Another important consideration that heightens the threat of exotic diseases to our forest resources is the difficulty of detecting a localized, newly introduced exotic pathogen within the sheer vastness of our forest resources. Although early detection, containment, and/or control have eradicated some important pathogens of agronomic and ornamental crops, such procedures have not eradicated important pathogens of forest trees.

The introduction of an exotic plant can widen the host range or serve as an alternate host of an otherwise innocuous domestic pathogen, and the resulting disease from the exotic plant introduction can be as severe as one that would result if a new, destructive exotic pathogen were introduced. Klinkowski (1970), citing the work of E. Gäumann, exemplifies this point with the worldwide epiphytotic of white pine blister rust. A susceptible alternate host, five-needled white pine, was brought into the natural *Ribes* (currants and gooseberries) area in Russia, where a rust fungus, *Cronartium ribicola,* was epidemic. With this introduction of pine, a new infection chain was established by man, resulting in a bridge from Russia around the earth, eventually reaching the East Coast and later the West Coast of the United States.

4. *Preparatory Considerations*

Ideally, to avoid or minimize the consequences of exotic pathogen introductions, the agricultural community should assemble reliable strategies to anticipate these events and to prepare adequately for them (Chiarappa, 1981). Contingency plans for regulatory action (i.e., exclusion, containment, and control, including eradication) have been formulated by the USDA's Emergency

Program Unit for Exotic Pests for citrus canker [(*X. campestris* pv. *citri*), potato black wart (*Synchytrium endobioticum*), and chrysanthemum rust (*P. horiana*)] (S. Wilson, personal communication). Contingency plans for other exotic diseases will be forthcoming from this group. Early detection and rapid deployment of plans for control are central, even vital, to each of these strategies, and their existence reduces vulnerability to exotic pathogens and costs of control.

Another important aspect of preparation for exotic diseases is communication and cooperation between American plant pathologists and regulatory officials and their counterparts in areas where threatening diseases occur. Not only can cooperation prepare us much better for an exotic disease, but also our contributions may help to control it. By controlling a major disease threat elsewhere, our vulnerability at home is reduced accordingly.

C. Exotic Diseases: Probability Considerations

There are two instances, perhaps three, in which we feel that if the exotic pathogen were introduced under the right circumstances, it would have a high probability of establishment and possible spread. The two surer instances occur (1) when the pathogen and disease have been reported here, but the pathogen was subsequently eradicated and is now once again exotic, and (2) when the pathogen and disease are here, but exotic variants occur. The third and more speculative instance occurs when the pathogen has already been reported here but the disease has not yet been reported.

1. *Disease Once Here But Now Exotic*

With at least three diseases, citrus canker (*X. campestris* pv. *citri*), chrysanthemum rust (*P. horiana*), and potato black wart (*S. endobioticum*), concerted eradicative efforts by state and federal regulatory agencies succeeded in eliminating these pathogens from the U.S. mainland. Reports indicate that *X. campestris* pv. *citri* was found in the citrus-growing regions of the Gulf Coast states from 1910 to 1941 (Dopson, 1964); *P. horiana* was found in pockets in chrysanthemum-growing areas of Pennsylvania, New York, and New Jersey from 1977 to 1979 (S. Wilson, personal communication); and *S. endobioticum* was found in isolated potato-growing areas of Maryland, Pennsylvania, and West Virginia from about 1915 (Walker, 1957) to the early 1960s, when the pathogen was declared eradicated (S. Wilson, personal communication). Chrysanthemum rust and potato black wart were contained after their initial outbreak by strict domestic quarantine measures. Such measures were nonexistent in the early 1900s, when the citrus canker pathogen first found its way to the United States. The threat of these three exotic pathogens is unquestionable. Indeed, it was precisely these diseases which were first selected by the recently formed USDA Emergency Programs Unit for Exotic Pests to draft contingency plans for regulatory action (Section III,B,4). The threat posed by citrus canker and chrysanthemum rust is enhanced even more by the fact that these diseases are regularly intercepted at U.S. ports of entry (Section III,A,1). Also noteworthy with chrysanthemum rust is an unusually long period (up to 1 month) from the time of infection to the appearance of rust pustules (Zadoks and Schein, 1979). This rust disease seemingly exploded in Western Europe during the period 1964–1967, despite strict quarantine measures. Unrooted and rust-infected cuttings produced in South Africa were air-freighted to Europe, rooted, and resold throughout Europe to commercial producers before rust pustules appeared from latent infections. In the United States we were fortunate that the rust pathogen was detected early and immediately prevented from spreading throughout the country. We could, however, easily experience the same situation that occurred in Western Europe. If so, chrysanthemum rust would probably be a permanent, serious problem in the United States. Despite our previous success with eradicating the widely distributed citrus canker pathogen and our general state of preparedness for the pathogen's return, the threat of canker still looms large. In Florida, citrus production has increased from 6.4 million boxes in 1910 to 283.6 million boxes in 1980, with more than 850,000 acres presently planted in citrus (Anonymous, 1981b). Thus, the intensity of cultivation has vastly increased since the canker pathogen last invaded Florida, and the bacterium would now find a vast canker-susceptible monoculture instead of the scattered citrus plantings that existed earlier. Further, societal constraints upon eradication efforts are certain to be greater today and more so in the future than they were previously. One only has to reflect upon the California Mediterranean fruit fly crisis of 1981 to come to this realization. The key to eradication of the canker bacterium, as well as many other exotic pathogens, then, is early detection coupled with swift and precise regulatory action.

2. *Pathogen Here But Exotic Strain(s) Elsewhere*

The existence of exotic variants (strains, substrains, races, biotypes, pathovars, etc.) of pathogens that are already in this country sometimes involves predictable consequences should we import the exotic relative. In some cases, the very existence of variants will not be known until the variant finds its way to a new place. Dutch elm disease (*Ceratocystis ulmi*) painfully illustrates this point. In Europe, a new series of epidemics of this disease began in the late 1960s (Stipes and Campana, 1981). In Great Britain, the new epidemic was particularly destructive, whereas the old epidemic of 1927 was relatively mild there. As it turns out, a new, much more virulent strain of the pathogen was involved, and this strain may have entered Great Britain on imported elm logs (Stipes and Campana, 1981). As a result of this new and virulent strain, between 1971 and 1978 more than 70% of southern Great Britain's 22 million elms were killed. Subsequently, the new strain was reported in the Netherlands, France,

Germany, and Russia. Further, recent evidence indicates yet another strain of the virulent type. What is to prevent the pathogen from producing even newer strains that could attack heretofore resistant species of American elm? In this regard, man's technology can do little, if anything. In other instances, we often know of the existence of exotic variants and can prevent their entry. For example, entry of foreign strains of *Urocystis agropyri* (flag smut of wheat), as well as other exotic pathogens, is prohibited by restricting the following imports from 34 named countries (Anonymous, 1981a): grain, most straw, hulls, chaff, and most milling products of *Triticum* spp., *Aegilops* spp. (goat grass), and any intergeneric cross of *Triticum aegilops,* seeds of *Melilotus indica* (annual yellow sweetclover), and seeds of any other field crops that have been separated from wheat during the screening process. Another threat posed by an exotic variant is that of *Pseudomonas solanacearum* race 3 (Buddenhagen *et al.,* 1962) or biotype 2 (Hayward, 1964) on potatoes. This exotic race or biotype occurs on four continents and is considered a cool temperature form. The race of *P. solanacearum* that causes brown rot in potato-growing regions of the southeastern United States is a warm temperature form. The introduction of the exotic cool temperature form would then possibly threaten potato production in our more northern climates, where the intensity of potato cultivation is great.

A certain amount of confusion with strain formation in viruses can sometimes arise. This seems to stem from inherent weakness in viral "taxonomy." For example, sugarcane mosaic virus (SCMV), which affects sugarcane and maize (temperate as well as tropical), produced innumerable strains and substrains which are variously distributed either worldwide or in specific exotic locations (Shurtleff, 1980). These variants of SCMV are maize dwarf mosaic, maize mosaic, abaca mosaic, sorghum concentric ring blotch, sorghum red stripe, and Transvaal grass mosaic, and the variants can be distinguished only by extensive host range studies (Shurtleff, 1980).

Finally, and probably the most overlooked issue relating to exotic strains, is that if it occurs in an exotic location, it can also occur here (i.e., be formed from existing domestic strains). Literature from the Philippines in the early 1960s indicates that lines of corn that carried Texas male-sterile cytoplasm were very susceptible to *Helminthosporium maydis* (Mercado and Lantican, 1961). By 1970, race T of *H. maydis* was somehow present in the United States, nearly all of our corn carried the Texas male-sterile cytoplasm, and we experienced an epiphytotic (Zadoks and Schein, 1979). Another and more recent example involves tomato and fusarium wilt (*Fusarium oxysporum* f. sp. *lycopersici*). A race 3 of the pathogen had been reported from Australia (Grattidge, 1982). During the winter of 1982, tomato plants in the Manatee-Ruskin tomato-growing area of Florida exhibited wilt symptoms, and *F. oxysporum* f. sp. *lycopersici* was isolated from vascular tissue of tomato varieties that were heretofore wilt-resistant (Jones *et al.,* 1982). The race was "typed" and found to be race 3 (J. P. Jones, personal communication). There are two possibilities that could have occurred. Race 3 had either evolved here from preexisting races or was introduced from Australia.

3. *Exotic Pathogen Here But Disease Not Established*

The third instance in which some speculation is warranted occurs when the pathogen is already here but the disease is not clearly established. This occurrence is probably not all that rare and could represent the transition state between pathogen establishment and obvious spread to a host. Beet necrotic yellow vein virus (BNYVV) has been associated with the "rhizomania" disease of sugarbeets in Japan and Europe (Tamada, 1975). BNYVV is vectored by the plasmodiophoromycetous fungus, *Polymyxa betae*. Both virus and vector have been detected in the United States. BNYVV was "baited" from soils beneath cherry trees in Washington state using *Gomphrena globosa* as a host (Musa and Mink, 1981). *Polymyxa betae* has been detected in sugarbeet soils in the Salinas Valley of California (Falk and Duffus, 1977) and, according to these authors, is probably more widespread than in the Salinas Valley. It could be that the pathogen, vector, and host(s) have never met here and/or that the abiotic and biotic environments were not conducive to an observable expression of the disease in plant hosts.

IV. WHERE ARE THE EXOTIC DISEASES?

Plant disease occurs wherever plants grow, but fortunately for agriculture, not all plant diseases occur everywhere. Hence, every country has diseases that are exotic to it, and nearly every country has regulations that attempt to exclude exotic pathogens. By examining exotic diseases where they occur, we can gain some perspective on our exotic disease threats.

A. Scientific Strategies

Every pathogen has a geographic distribution. Some distributions are small or narrow, whereas others are wide or cosmopolitan. This distribution, in turn, reflects the pathogen's genetic potential, the relationship of the individual pathogen with its environment, and the interrelationships of the various organisms [i.e., pathogen, host(s), vectors(s), microbial associates, man, and animals] with their environment. If the geographic distributions of a pathogen and its host are known and are sufficiently defined by appropriate climatological data, and if the ecological requirements of both host and pathogen are sufficiently understood, it becomes possible to give a prognosis for the likelihood of disease occurrence in previously noninfested areas or in places where the host is introduced as a new crop (Weltzien, 1972, 1978). The limitations of the scientific approach of Weltzien are couched in the words "if" and "sufficiently." Much of the information required is at best inadequate and would come only with costly expenditures for research. Mapping becomes confusing when the number of disease factors per map increases beyond three. However, it would seem that Weltzien's approach is the most logical strategy to employ in evaluating exotic disease threats.

Soybean rust has been reported in this hemisphere in Puerto Rico (Vakili and Bromfield, 1976), Brazil, and Colombia (Sinclair, 1982). In addition to the rust in our hemisphere, we have highly susceptible genotypes of soybean in production in the United States (Bromfield, 1976), as well as other susceptible leguminous hosts (Sinclair, 1982). The CMI distribution map of soybean rust (Anonymous, 1975) indicates that the disease is largely contained within latitude 34° and that some disease occurs outside of latitude 34° (extending northward to northeastern China and southeastern Russia, and southward to mideastern Australia). The northernmost extremity of rust distribution is approximately 43°–45° latitude, corresponding to Madison, Wisconsin, in the United States. However, according to Sinclair (1982), the rust pathogen is especially destructive on soybeans growing in subtropical and tropical areas of the Orient. Nonetheless, a recent U.S. delegation of scientists to China observed soybean rust in the temperate northeastern provinces of Shandong (latitude 35°–40° N) and Liaoning (latitude 40°–45° N) (Ford *et al.*, 1981). Climate then becomes a critical determinant in evaluating the threat that soybean rust poses. Soybean rust would probably find conditions more favorable in areas of the United States where summers are warm, humid, and rainy, and where soybeans are intensely cropped. The Gulf Coast states would most likely approximate these conditions. If rust were to be established there, inoculum could spread northward to the main soybean-growing regions of the United States, and the fungus could possibly overwinter in the South on its many leguminous hosts. Moreover, strain differences do occur with the soybean rust pathogen, and certain strains may have more temperate climatic requirements than others. Strain differences relative to climatic requirements are suspected with isolates in Puerto Rico and Taiwan (Vakili, 1981). Further, differences in virulence have been observed with various isolates of the soybean rust pathogen from three continents (Bromfield *et al.*, 1980).

Maize production is largely in the temperate zone, but, rather recently, its culture in subtropical and tropical climates has increased markedly as a result of the development of high-yielding, adapted hybrids and open-pollinated varieties (Renfro and Ullstrup, 1976). The genetic diversity of maize in the tropics is much greater than in temperate areas, and this difference of diversity vs. uniformity can be an important factor in disease development, as previously mentioned in our discussion of vulnerability (Section III,B,3). Cultural practices for maize differ in temperate and tropical areas (Renfro and Ullstrup, 1976). Much of the

TABLE I

The Relative Prevalence[a] and Importance[b] of Some Exotic Maize Diseases of the World[c]

Exotic disease (pathogen)	Temperate (outside of latitude 34°)	Subtropical (within latitude 34°)	Tropical (within latitude 23°15′)		
			Highland >1000 m	Lowland (<1000 m)	
				Winter	Summer
Rust					
Tropical rust (*Physopella zeae*)	−0	+1	−0	+1	+1
Downy mildews					
Java (*Peronosclerospora maydis*)	−0	−0	−0	++3	+++3
Philippines (*P. philippinensis*)	−0	++2	−0	+++3	+++3
Sugarcane (*P. sacchari*)	−0	++2	−0	+++3	+++3
Spontanea (*P. spontanea*)	−0	−0	−0	+1	++2
Brown stripe (*Sclerospora macrospora* = *Sclerophthora macrospora*)	−0	+++3	−0	−0	++2
Virus					
Maize streak	−0	++2	+1	++2	+++3
Maize rough dwarf	++2	++2	−0	−0	−0
Late wilt (*Cephalosporium maydis*)	−0	+++3	−0	+++3	+++3
Ear and kernel disease					
Ergot (*Claviceps gigantea*)	−0	−0	+++1	−0	−0

[a] Prevalence: +++, abundant; ++, common; +, occasional; −, absent.
[b] Importance: 3, major; 2, moderate; 1, minor; 0, rare or absent.
[c] Data from Renfro and Ullstrup (1976).

maize in the tropics is produced with hand labor, whereas in temperate areas maize production is, for the most part, mechanized. Generally, row spacing is narrower in temperate areas, and this, coupled with greater applications of fertilizers, has influenced certain maize diseases in temperate areas. Continuous cropping in warmer climates permits the maintenance of a high inoculum potential of the pathogen and of vectors (Renfro and Ullstrup, 1976; Niblett *et al.*, 1981). Where there is an intervening cold season in temperate areas or a dry period in the tropics, the inoculum tends to be depleted and then builds up again with the onset of the maize planting season (Renfro and Ullstrup, 1976). Renfro and Ullstrup have segregated various world maize diseases on a temperate–subtropical–tropical basis and have rated the various diseases in relation to their prevalence and occurrence. We, in turn, have determined which of these diseases are exotic to us. The results are presented in Table I. Of all the exotic diseases, only one is in the temperate zone. Maize rough dwarf virus (MRDV) disease occurs in Southern and Western Europe, Israel, India (Shurtleff, 1980), and China (J. H. Tsai, personal communication). In nature, occurrence of MRDV disease is dependent on the planthopper vector, *Laodelphax striatellus,* which also happens to be exotic. All other presented exotic maize diseases are subtropical or tropical. Ergot of maize is found only in the tropical highlands of Mexico, where conditions conducive (mean temperature of 13°–15° C and rainfall > 1000 mm) to disease occur (Shurtleff, 1980). Ergot is of minimal threat to our agriculture because these conditions are not commonly found in the continental United States. Two downy mildews of maize are subtropical to tropical. Of the downy mildew pathogens, *Peronosclerospora maydis* and *P. spontanea* appear to be more tropical than *P. philippinensis, P. sacchari,* or *Sclerophthora macrospora.* To us, *P. sacchari* (sugarcane downy mildew) appears threatening because of its ability to infect maize and sugarcane and because of south Florida's vast, nearly year-round production of sugarcane and maize. Tropical rust of corn appears minimally threatening. Maize streak virus is transmitted by five leafhopper species of the genus *Cicadulina* which are not present in the United States but are widespread in tropical Africa, India, and some islands in the Indian Ocean (Shurtleff, 1980).

B. Geographic Places: How We Relate

The discussion which follows reviews 14 foreign geographic areas relative to previous pathogen introductions into the United States, climates, crop domestications and origins, agricultural imports of selected commodities (Table II), numbers of people traveling to the United States from foreign places (Table III), and some exotic diseases within these areas (Table IV). Regarding Table IV, it should be borne in mind that some of the data under the heading "Transmission" is based upon our own knowledge of the particular pathogen or related patho-

TABLE II

Percentage of Agricultural Imports from World Areas during Calendar Year 1980[a]

World area	Imported agricultural products							
	Agronomic				Forest products		Ornamentals	
	Orange	Potato	Tomato	Tobacco	Sawn hardwood	Sawn softwood	Flowers and foliage	Nursery and bedding stock
Africa	<1	0	0	4	1	<1	1	1
Australia and New Zealand	<1	0	0	0	<1	<1	1	1
Canada	<1	99	<1	1	23	99	10	14
Central America	0	0	<1	3	<1	<1	1	14
China	<1	0	0	4	<1	0	<1	<1
Near East	12	0	0	5	0	<1	2	1
Eastern Europe	0	0	0	6	0	<1	<1	<1
South Asia	0	0	0	1	<1	<1	<1	<1
East Asia and South Pacific	<1	<1	0	4	36	<1	1	<1
Japan	9	0	0	0	<1	<1	1	1
Mexico	75	0	99	7	<1	<1	<1	<1
South America	<1	0	<1	9	37	<1	72	2
West Indies and the Antilles	<1	0	<1	2	<1	<1	1	1
Western Europe	3	0	0	60	1	<1	9	61

[a] Anonymous (1981d).

TABLE III

Travel from World Areas during the First 6 Months of 1979 to the United States[a]

World area	No. of travelers	World area	No. of travelers
Africa	42,662	Far East	
Australia and New Zealand	127,777	Hong Kong	18,213
		Indonesia	4,358
Canada	NA	Korea	14,187
Central America	107,605	Malaysia	3,735
China	1,047	Philippines	28,455
Near East		Singapore	5,447
Iran	31,114	Taiwan	17,811
Israel	36,630	Thailand	6,126
Lebanon	7,021	Japan	458,101
Eastern Europe		Mexico	984,926
USSR	3,045	South America	376,265
South Asia		West Indies	181,862
India	22,572	Western Europe	1,245,027

[a] Anonymous (1980b).

gens, and that citations appearing under the heading "References" are often reports that justify distribution records and not necessarily individual, first reports from various world locations. Finally, it should be realized that tabulating exotic diseases is an open-ended undertaking and subject to varying points of view. What is included in our tabulation are examples of exotic diseases that have been consequential to people and their plants. Some important plants, as well as some important diseases, have probably been omitted. Moreover, some exotic diseases listed in Table IV may never find a suitable home in the United States because the proper conditions may simply not occur here. Thus, we acknowledge our acts of omission and commission but make no apologies for our efforts to elicit meaningful facts from a difficult, complex array of diverse thought and information.

1. *Africa*

Africa is divided between a northern area (Paleartic) which is similar in flora and fauna to Europe, a broad tropical belt south of the Sahara desert, and the southern temperate region (McGregor, 1973). The indigenous flora of tropical Africa and temperate South Africa are very different from those of Europe and the United States. However, South Africa historically has traded heavily with Europe, and we have interacted significantly with Europe (McGregor, 1973). Consequently, many diseases are common to the three areas, whereas others are found only in Africa.

TABLE IV

Some Potential Exotic Disease Threats

Host(s)	Disease (pathogen)	Distribution	Transmission	Reference(s)
Alfalfa	Alfalfa enation (bacilliform-shaped virus)	France, Spain, Switzerland, Bulgaria, Romania, and Morocco	Aphid, *Aphis craccivora*	Graham *et al.* (1979b)
Alfalfa	Transient streak (spherical-shaped virus)	Australia	Natural transmission unknown	Graham *et al.* (1979b)
Apple	Apple proliferation, or apple witches' broom (mycoplasmalike organism)	Eastern Europe (scattered) and Turkey	Grafting, root fusion, possible leafhopper vector, *Empoasca maligna*	Anonymous (1978b); Marenaud *et al.* (1978)
Apple	Root-knot nematode (*Meloidogyne mali*)	Japan	Soil-borne and infected plants	Itoh *et al.* (1969)
Bean (common)	Bean summer death (double-virus particles)	Australia	Leafhopper, *Orosius argentatus*	Bock (1982)
Bean (*Vicia faba*): bean (common) pea (English)	Broad-bean stain (spherical-shaped virus)	England and Germany	Seed-borne and mechanically; natural transmission unknown	Gibbs *et al.* (1968); Smith (1972)
Carrot	"Carrot sickness" (*Heterodera carotae*)	India, Poland, USSR, and Western Europe (scattered)	Soil-borne and infected plants	Matthews (1975)
Chrysanthemum	White rust (*Puccinia horiana*)	Africa (scattered), Australia, New Zealand, China, Japan, and Western Europe (scattered)	Infected stock	Punithalingam (1968); Seymour (1977)
Citrus	Citrus black spot (*Guignardia citricarpa*)	Present on all continents except North America	Nursery stock and fruit	Knorr (1973)

(*continued*)

TABLE IV *Continued*

Host(s)	Disease (pathogen)	Distribution	Transmission	Reference(s)
Citrus	Citrus canker (*Xanthomonas campestris* pv. *citri* = *X. citri*)	Africa, China, South Asia (scattered), Far East (scattered), Japan, Brazil, Argentina, Uruguay, Paraguay, Mexico, Guam, and Hawaii	Fruit, plant parts, and nursery stock	Knorr (1973); Miller *et al.* (1977); Stall and Seymour (1983); J. M. Pender (personal communication)
Citrus	Greening, yellow shoot, and likubin (mycoplasmalike organism)	South Africa, India, China, Taiwan, Philippines, and Indonesia	Psyllids, *Trioza erytreae* and *Diaphorina citri*	Kaloostian (1980)
Conifers and hardwoods	Violet root rot (*Helicobasidium mompa*)	Taiwan, Japan, and Korea	Infested soil and infected stock	Anonymous (1963)
Cotton and hibiscus	Cotton leaf curl (virus)	Nigeria and Sudan	Whitefly, *Bemisia gossypiperda*	Anonymous (1980c); Muniyappa (1980)
Cotton	Escobilla, or witches' broom (*Colletotrichum gossyppi* var. *cephalosporioides*)	Venezuela and Brazil	Infected and infested seeds	Watkins (1981)
Cotton	Phyllody (mycoplasmalike organism)	Upper Volta, Mali, and the Ivory Coast	Leafhopper, *Orosius cellulosus*	Chiykowski (1981); Halliwell (1981)
Cucumber	Phytophthora blight (*Phytophthora sinensis*)	China	Soil-borne and infected plants	Yu and Zhuang (1982)
Cucumber, melons, and squash	Phytophthora blight (*Phytophthora melonis*)	Japan, China, and Taiwan	Soil-borne and infected plants	Katsura (1976)
Maize, wheat, barley, oats, and some grasses	Maize rough dwarf (spherical-shaped virus)	Southern Europe, Israel, and India	Planthopper, *Laodelphax striatellus*	Shurtleff (1980)
Maize, wheat, barley, oats, rye, rice, sugarcane, teosinte, and many wild grasses	Maize streak (double-virus particles)	Africa (south of the Sahara Desert), Mauritius, Reunion, Madagascar, and India	Leafhoppers, *Cicadulina* spp.	Shurtleff (1980); Bock (1982)

Oak and elm(?)	Vascular mycosis(es) (*Ophiostoma* spp.)	USSR, Rumania, and Yugoslavia	Infected raw wood products	Wilson (1967, 1970)
Peanut	Peanut clump (virus)	Upper Volta and Senegal	Soil-borne and mechanical	Thourenel *et al.* (1976)
Peanut	Peanut rosette (virus)	Australia, Central and South Africa, Fiji, India, and Argentina	Aphid, *Aphis craccivora,* mechanically and seed-borne	Anonymous (1976); Benigo and Favali-Hedaynt (1977)
Pepper and tomato	Pepper veinal mottle (virus)	Nigeria and Ghana	Aphid, *Myzus persicae*	Brunt and Kenten, (1972); Lapida and Roberts (1977)
Pine (*Eucalyptus saligna* and *Araucaria angustifolia*)	*Cylindrocladium* root rot (*Cylindrocladium clavatum*)	Brazil		Hodges and May (1972)
Pine	Needle blight of pine, *Cercospora* pine blight (*Cercospora pini-densiflorae*)	Malawi, Zimbabwe, Tanzania, Zambia, Hong Kong, India, Japan, Korea, Taiwan, and West Malaysia	Infected stock (latent period 6 weeks)	Anonymous (1980a)
Pine (especially *Pinus sylvestris* and related species)	Scots pine blister rust (*Cronartium flaccidium* = *Peridermium pini*)	Eastern and Western Europe (scattered), northern Asia, Japan, and Korea	Infected stock and alternate hosts	Anonymous (1963)
Plum, apricot, peach, and perennial weeds	Plum pox or Sharka (potyvirus group)	Eastern and Western Europe (scattered)	Grafting, mechanical means, and the leafhopper, *Empoascus flavescens*	Anonymous (1970b); Smith (1972); Bos (1981)
Poplar	Bacterial canker (*Xanthomonas populi* = *Aplanobacter populi*)	Eastern and Western Europe (scattered)	Infected planting material and cankered wood	Anonymous (1963, 1978a, 1980d)
Potato	Andean potato mottle (virus)	Andean region of South America and in Andean germplasm collection in Europe	Mechanically and possibly beetles	Fribourg *et al.* (1979)
Potato	Andean potato latent (virus)	Andean region of South America and Andean germplasm collection in Europe	Beetle-potato flea beetle, mechanically and seed-borne	Jones and Fribourg (1981)

(*continued*)

TABLE IV *Continued*

Host(s)	Disease (pathogen)	Distribution	Transmission	Reference(s)
Potato	Brown rot (*Pseudomonas solanacearum* race 3 or biotype 2) (cool temperature form)	Egypt, Kenya, Sri Lanka, Australia, Costa Rica, Colombia, Peru, and Sweden	Infected potatoes and soil-borne	Buddenhagen *et al.* (1962); Ciampi and Sequeira (1980); El-Goorani (1976); Graham *et al.* (1979a); Harris (1976); Hayward (1976); Lloyd (1976); Olsson (1976)
Potato	Gangrene (*Phoma exigua* var. *foveata*)	Australia, Romania, and Western Europe (scattered)	Infected potatoes and soil-borne	Anonymous (1982b)
Potato	Phoma leaf spot (*Phoma andina*)	Peru and Bolivia	Infected plants	Turkensteen (1981)
Potato	Potato black ringspot (virus)	Peru	Infected potatoes and mechanically	Salazar and Harrison (1978)
Potato, tomato, eggplant, and solanaceous weeds	Potato cyst nematode (*Globodera pallida*)	Colombia, Ecuador, Peru, Bolivia, Argentina, and Great Britain	Soil-borne, cyst nematode contaminated tubers, and infected plants	Mai *et al.* (1981)
Potato	Potato mop-top (virus)	Andean region of South America, Northern and Central Europe	Fungus, *Spongospora subterranea*, infested seed potatoes, and mechanically	Harrison (1974); Jones (1981)
Potato, tomato and *Solanum dermium*	Potato rust (*Puccinia pittieriana*)	Colombia, Costa Rica, Ecuador, Peru, Venezuela, Mexico, and possibly Bolivia and Brazil	Wind-borne or infected or infested foliage	French (1981)
Potato	Potato wart (*Synchytrium endobioticum*)	South Africa, China, India, Nepal, New Zealand, Canada, Mexico, and South America (scattered)	Infected potatoes and soil-borne	Anonymous (1982c)

Potato	Thecaphora smut (*Angiosorus solani* = *Thecaphora solani*)	Mexico, Bolivia, Chile, Colombia, Ecuador, Peru, and Venezuela	Infected potatoes	Page (1981)
Rice	Bacterial leaf blight (*Xanthomonas oryzae*)	Africa, Asia, Australia, Central America, and Mexico	Infested rice seed, infected (?) rice	Anonymous (1974); Bradbury (1969); Lozand
Rice	Nematode (*Heterodera oryzae*)	Ivory Coast	Infested soil and infected plants	Luc and Taylor (1977)
Rice	Transitory yellowing disease, Huang-yet-ping (bullet-shaped virus)	Taiwan	Leafhoppers, *Nephotettix nigropictus, N. cineticeps,* and *N. virescens*	Ling (1972); Nielson (1979); Foster (1982)
Rice	Tungro disease, penyakit merah, yellow-orange leaf, and leaf yellowing (spherical-shaped virus)	Philippines, Malaysia, Thailand, India, Indonesia, East Pakistan	Leafhoppers, *Nephotettix, N. virescens*	Ling (1972); Nielson (1979); Foster (1982)
Rice	''Ufra'' disease (*Ditylenchus angustus*)	Bangladesh, Burma, Maylaya, Philippines, Madagascar, Thailand, and Egypt	Infected soil, infected plants, dry glumes and stubble	Seshadri and Dasgupta (1975)
Rice	Yellow dwarf (virus or mycoplasmalike organism)	Japan, Taiwan, Malaysia, Ceylon, India, East Pakistan, Philippines, China, and Thailand	Leafhoppers, *Nephotettix cincticeps, N. nigropictus, N. virescens*	Ling (1972); Nielson (1979); Foster (1982)
Soybean, cowpea, cucumber, and tobacco	Choanephora leaf blight (*Choanephora infundibulifera*)	Thailand	Infected plants	Sinclair (1982)
Soybean	Pyrenochaeta leaf spot (*Pyrenochaeta glycines*)	Cameroon, Ethiopia, Zambia, Zimbabwe	Infected plants	Rothwell (1980); Sinclair (1982)
Soybean	Soybean dwarf (virus)	Japan	Infected plants and by the aphid *Aulacorthum solani*	Sinclair (1982)
Soybean	Pod rot or blight (*Macrophoma mame*)	China	Infected plants seed-borne	Ford *et al.* (1981); Liu (1948)

(*continued*)

TABLE IV *Continued*

Host(s)	Disease (pathogen)	Distribution	Transmission	Reference(s)
Soybean	Soybean blast, sleeping leaf or brown stripe (*Septogloeum sojae*)	China	Infected plants	Ford *et al.* (1981); Nishizawa *et al.* (1955)
Soybean and numerous other legumes	Soybean rust (*Phakopsora pachyrhizi*)	Australia, China, Japan, India, Taiwan, Thailand, India, Puerto Rico, Brazil, Colombia, and much of the South Pacific	Wind-borne, infected stock, and infested seed	Anonymous (1975); Sinclair (1982)
Soybean	Soybean scab	Japan	Infected stock	Jenkins (1951); Sinclair (1982)
Sugar beet	Yellow wilt (mycoplasmalike or rickettsialike organism)	Argentina and Chile	Leafhopper, *Atanus exitiosus* (= *Paratanus exitiosus*), and infected plants and plant parts	Beamer (1943); Urbina-Vidal (1974)
Sugarcane and rice	Nematode (*Heterodera sacchari*)	Africa, India, and Jamaica	Infected plants and infested soil	Luc (1974)
Sugarcane, maize, grain, sorghum, teosinte, gamagrass, and broomcorn	Sugarcane downy mildew (*Peronosclerospora sacchari*)	Australia, Fiji Islands, New Guinea, Inola, Philippines, and Taiwan	Infected stock	Shurtleff (1980)
Sunflower and tobacco	Witchweed (*Orobanche cernua*)	Hungary, USSR, and India	Seeds are soil-borne or contaminate various natural commodities	Musselman (1980)
Tobacco	Tobacco leaf curl (double-virus particle)	Japan	Whitefly, *Bemisia tabaci*	Bock (1982)
Tobacco and 32 species in seven families	Tobacco yellow dwarf (double-virus particles)	Australia	Leafhopper, *Orosius argentatus*	Bock (1982)
Tomato	Yellow leaf curl (virus)	Near East and South Asia	Whitefly, *Bemisia tabaci*	Makkouk (1978); Mazyad *et al.* (1979)

Wheat	Alternaria leaf blight (*Alternaria triticina*)	India	Seed-borne	Wiese (1977)
Wheat, barley, rye, oats, and other grasses	Cereal tillering (spherical-shaped virus)	Italy and Sweden	Planthoppers, *Laodelphax striatellus* and *Dicranotropis hamata*	Wiese (1977)
Wheat, oats, barley, rye, maize, and other grasses	(European) wheat striate mosaic (causal agent not identified)	Western Europe (scattered)	Planthoppers, *Javesella pellucida* and *J. dubia*	Wiese (1977)
Wheat	Karnal bunt (*Neovossia indica*)	India, Pakistan, and Mexico	Seed- and soil-borne	Wiese (1977)
Wheat, barley, oats, rice, rye, and 15 grasses	(Northern) cereal mosaic (bullet-shaped virus)	Japan	Planthoppers, *Laodelphax striatellus, Unkanodes sapporonus,* and *Delphacodes* spp.	Wiese (1977)
Wheat, oats, barley, rye, and noncultivated grasses	(Russian) winter wheat mosaic (flexunous rod-shaped virus)	China, USSR, and other Eastern European countries	Leafhoppers, *Psammotettix striatus* and *Macrosteles laevis*	Kelman and Cook (1977); Wiese (1977)
Wheat and barley	Wheat yellow leaf (flexuous rod-shaped virus)	Japan	Aphid, *Rhopalosiphum maidis*	Wiese (1977)
Wheat	Wheat dwarf or blue dwarf (suspected virus)	USSR, Czechoslovakia, Sweden, and China	Leafhoppers, *Psammotettix striatus* and *Macrosteles laevis*	Kelman and Cook (1977); Wiese (1977)

Africa seems to have a disproportionate share of the world's known parasitic weeds (Musselman, 1980). As a result, at least seeds of maize from Africa (and Asia) are prohibited from entering the United States (Waterworth and White, 1982). *Striga asiatica,* which became important in South Africa in the early part of this century on grain crops, was discovered in the late 1950s in the Carolinas and is believed by some to have been introduced from South Africa into the United States by the transport of wool (Musselman, 1980).

Agricultural imports from Africa are relatively low (Table II), as is travel from African nations (Table III). Of the 67 exotic diseases listed in Table IV, 15 (22%) are found in Africa. Of the 49 most threatening exotic diseases reported by McGregor (1973), 13 (27%) are in Africa.

2. *Australia and New Zealand*

Australia has both tropical and temperate climates and a central and westward desert area. New Zealand has a temperate climate. The two countries, like the United States, are not considered centers of origin or development of agronomic crops (Harlan, 1976). Thus, they too have suffered the consequences of exotic disease introductions, and their pest problems have been similar to ours, having resulted from foreign introductions (McGregor, 1973).

Trade in the selected agricultural imports from Australia and New Zealand is relatively low (Table II), as is tourism from these countries (Table III). Of the 67 exotic diseases listed in Table IV, 11 (16%) are found in this area. Of the 49 diseases cited by McGregor (1973), six (12%) were located in Australia and New Zealand.

3. *Canada*

The life zones of the northern United States continue on into Canada, as evidenced by the adaptation of indigenous insects to life in both countries within the same zone (McGregor, 1973). Rust pathogens of cereals first became established in the United States and then moved to Canada.

The volume of agricultural imports is relatively high from Canada (Table II). Especially high are potatoes and forestry products. Data on Canadian travel to the United States are not maintained by the U.S. Travel Service, because visas or various forms are not required of Canadians upon entry. Of the 67 exotic diseases listed in Table IV, only one (2%) is found in Canada. This finding is not surprising, given the common geographic boundary, the similarity of flora and fauna, and the heavy traffic of commerce and people between the United States and Canada. Of the 49 diseases cited by McGregor (1973), four (8%) are found in Canada.

4. *Central America*

Many of our agroeconomic crops, such as maize, tomato, the common bean, and cotton, had their origin and development in the subtropical and tropical areas of Central America (Harlan, 1976). Undoubtedly, we have not only gathered germplasm from this area but also have brought pathogens of these crops into our country.

Other than nursery and budding stock, imports of selected agricultural commodities from Central American countries are low (Table II). Tourism from this area is also relatively low (Table III). Of the 67 exotic diseases listed in Table IV, nine (13%) are found in this area. McGregor (1973) cited no threatening exotic diseases from Central America in his 49 selections.

5. *People's Republic of China*

The normalization of relations with the People's Republic of China beginning in the early 1970s provides another location from which to introduce exotic plant pathogens into the United States. Because of China's large area, its 1 billion people (one-fourth of the world's population), and its isolation from the West during the last 3 decades, China should be examined closely from the standpoint of exotic disease threats.

China is not yet a major exporter of agricultural plants and plant products. By 1976, China was nearly self-sufficient in its agriculture and exported some rice to offset imports of wheat (Kelman and Cook, 1977). Our current imports from China of the various agricultural commodities presented in Table II are very low. Tourism from China is also very low (Table III).

Crop for crop, most plant diseases that predominate in the United States are also common in China (Ford *et al.*, 1981). This is not all that surprising because historically there has been much exchange of plants and plant propagative parts with China. Soybean, citrus, and peach are crops that had their origin in and around China. Some examples of exotic diseases reported from China by Ford *et al.* (1981) include late wilt of corn caused by *Cephalosporium maydis,* rust of soybean caused by *Phakopsora pachyrhizi,* scab of soybean caused by *Sphaceloma glycines,* sleeping leaf or brown stripe of soybean caused by *Septogloeum sojae,* and pod rot or blight of soybean caused by *Macrophoma mame*. Some examples of exotic diseases reported from China by Kelman and Cook (1977) include northern cereal mosaic on wheat, Russian winter wheat mosaic on wheat, and citrus yellow shoot on citrus. Only one of these diseases, pod rot or blight of soybean, is reported exclusively in China (Table IV). Of the 67 exotic diseases listed in Table IV, 15 (22%) are found in China. Of the 49 diseases reported by McGregor (1973), five (10%) were found in the same area. Perhaps much of this difference between the two exotic disease compilations reflects the recent normalization of relations between China and the United States.

Perhaps our greatest agricultural opportunity that may arise in regard to China is the potential for obtaining germplasm of its native plants and using these plants in breeding for disease resistance. Some of these plants would include soybean, cabbage, onion, and peach (Harlan, 1976). This benefit may far outweigh any risk that we may receive from exotic diseases in China. Germplasm of some agroeconomic crops is already in quarantine facilities in the United States and is being indexed for various pathogens.

6. *Near (or Middle) East*

The area of Near East Asia is the center of origin of many of our crops (Harlan, 1976), including wheat, barley, onion, pea, lentil, chickpea, pear, date, and possibly apple. Many of the diseases of these and other crops probably spread to Europe at a very early time, and we, in turn, received some of these diseases from Europe. The climate varies in this area, ranging from warm-dry Mediterranean, to warm-dry like much of the southwestern United States, to hot desert.

We receive significant imports of oranges and tobacco from the Near Eastern countries (Table II). Tourism from this area is low, with most tourists coming from Israel (Table III). Of the 67 exotic diseases listed in Table IV, three (4%) are found in this area. Of the 49 diseases reported by McGregor (1973), three (6%) were in Near East Asia.

7. *Eastern Europe*

All of Europe shares with North America a common climate and many species of plants which are indigenous and common to both areas (McGregor, 1973). McGregor feels that Western Europe has frequently acted as a way station for pests in Eastern Europe which eventually came to North America. For example, white pine blister rust was first prominent in the USSR, then moved into Western Europe, and finally arrived in the United States (Chapter 3, this volume). On the other hand, the Russian naturalist I. V. Michurin pointed out in 1911 that many plant pests and diseases, such as phylloxera (*Viteus vitifolii*), American gooseberry mildew (*Sphaerotheca morsuvae*), and dry corn rot (*Diplodia zeae*), were brought from abroad on plants and seeds, and that quarantine measures should be implemented to prevent pest and pathogen invasion, especially from America and Japan (Anonymous, 1970a).

Trade in the selected agricultural commodities (Table II) and tourism (Table III) are very low from Eastern European countries. Of the 67 exotic diseases reported in Table IV, 11 (16%) are found in this area. Of the 49 diseases cited by McGregor (1973), six (12%) were found in Eastern Europe.

8. *South Asia*

South Asia, the Indian subcontinent and Pakistan, was the area of development of the eggplant, cucumber, and pigeon pea (Harlan, 1976). The climate here varies from tropical monsoon, to desert hot, to temperate in the mountainous areas.

Trade in the selected agricultural commodities (Table II) and tourism are very low from countries in this area (Table III). Of the 67 exotic diseases in Table IV,

20 (30%) are found in this area. Of the 49 diseases reported by McGregor, 17 (35%) were from South Asian countries.

9. *Far East and the South Pacific*

Various crops, such as rice, yams, sugarcane, and citrus, came from this subtropical and tropical area (Harlan, 1976).

Other than sawn hardwood and tobacco, trade in the selected commodities is relatively low with countries in this area (Table II), as is tourism (Table III). Of the 67 exotic diseases listed in Table IV, 14 (21%) are found in this area. Of the 49 diseases reported by McGregor (1973), 12 (24%) were found here.

10. *Japan*

A tropical monsoonal climate prevails in Japan but is moderated by latitude and the sea. Winters are cold, and summers are hot and humid with high rainfall.

Other than oranges, trade in the selected commodities is relatively low with Japan (Table II). However, tourism is high (Table III). Of the 67 exotic diseases listed in Table IV, 15 (22%) are found in this country. Of the 49 diseases reported by McGregor (1973), 12 (24%) were found in Japan.

11. *Mexico*

Life and agricultural zones are contiguous between Mexico and southern parts of the United States (McGregor, 1973). McGregor believes that most pests that could adapt to the more northern climate of the United States are here, and most of the others have had an opportunity but could not survive. Of the 49 diseases reported by McGregor (1973), one (2%) was from Mexico. Of the 67 diseases in Table IV, five (8%) are from this area.

Trade with Mexico in the selected agricultural commodities is relatively high (Table II). Nearly all of our imported tomatoes come from Mexico, as well as three-fourths of our fresh oranges. This is alarming in light of a recent report that as of August 1982, citrus canker was found in Mexico (J. M. Pender, personal communication). Tourism is very high, with nearly 1 million people arriving in the first 6 months of 1979 (Table III).

12. *South America*

Like Africa, South America is divided into tropical and temperate areas, with the temperate zone extending into tropical areas along the Andes Mountains (McGregor, 1973). It is in this highland region of South America that peanut, potato, and lima bean originated, and the common bean and cotton were domesticated (Harlan, 1976). Many of the pathogens of these crops, such as *P. infestans* on potato, have found their way to the United States. Citrus canker is currently found in Argentina, Brazil, Uruguay, and Paraguay (Table IV).

Flowers and foliage imports from South America are very high and account for

72% of our imports in this category (Table II). Tourism from South America is high (Table III). Of the 67 exotic diseases in Table IV, 15 (22%) are found in South America. Of the 49 diseases reported by McGregor (1973), five (10%) were found on this continent.

13. *West Indies and the Antilles*

Within the past 5 years, sugarcane smut and sugarcane rust have jumped from the Caribbean area to Florida. Soybean rust is currently in Puerto Rico (Table IV). Most of the diseases that threaten the United States from this subtropical to tropical area would affect mainly the Gulf Coast states.

Agricultural imports of the selected commodities are relatively low from this area (Table II). Tourism was moderate during the first 6 months of 1979 (Table III). Of the 67 exotic diseases listed in Table IV, two (3%) are found in this area. Of the 49 diseases reported by McGregor (1973), four (8%) were found in the Caribbean area.

14. *Western Europe*

Many of our most infamous diseases came to us via Europe. Stem rust of wheat, Dutch elm disease, and white pine blister rust are among such diseases (Chapter 3, this volume). Some of these and other diseases did not originate in Western Europe but were transplanted from the Near and Far East, South America, and the heartlands of Russia. Western Europe has much in common with us. There is a similarity in climates and crops, as well as close cultural ties. This has traditionally resulted in much commerce of goods and people between the two areas (McGregor, 1973).

Agricultural imports in the selected commodities were relatively high from the Western Europe countries (Table II). Especially high among the imports were tobacco and nursery bedding stock. Tourism from European countries is high, with more than one-third of our tourists coming from this area in the first half of 1979 (Table III). Of the 67 exotic diseases listed in Table IV, 19 (28%) are found in this area. Of the 49 diseases reported by McGregor (1973), 23 (47%) were found in Western European countries.

V. CONCLUSIONS

The threat to our agriculture from exotic diseases impels us to take action by seeking ways to minimize our vulnerability to exotic pathogens. We have emphasized the importance of an informed awareness as a necessary first step in countering the possibility of experiencing exotic diseases. One reason is that scientific pursuits provide opportunities to introduce exotic pathogens into this country. By having scientists understand and appreciate the inherent risks of their ventures, it is hoped that they will be inclined to determine the proper procedures, accompanied by legal approval, to import exotic plants, plant propagules, and pathogens, and to comply with the precautions and prescriptions of these agreements.

In regard to the central question, where are the exotic disease threats?, the so-called scientific pathway of pathogen introduction potentially opens our country to all geographic areas and their respective pathogens. Because resistance has been awarded a high priority in scientific research, it follows that areas that were centers of crop origin–domestications (hence, pathogen origin) will be the starting points for much research. In this regard, China is alluring, especially in light of its recently ended three-decade isolation from the Western World of nations. One obvious way to enhance our awareness of exotic diseases from all geographic areas would be to have a scientific journal, similar to the *EPPO Bulletin,* for our country or hemisphere on threatening exotic pests.

When other more traditional avenues of pathogen introduction, such as trade and travel, are considered relative to the question, where are the exotic disease threats?, it is the Western European countries that have many exotic pathogens and provide us with many agricultural imports and tourists. We may tend to view this area with more alarm than is necessary because we have received some of our most devastating pathogens via Western Europe. However, these nations, like the United States, were not primary centers of crop origin–domestication and were subjected to the ravages of introduced pathogens. Countries of Central and South America may be "rising stars" as sources of exotic pathogens. These assessments are based upon the numbers and kinds of exotic pathogens, the level of trade and tourism, the center of crop origin–domestication, and the expanding agriculture. Japan, although not a large exporter of agricultural goods to the United States, does have many exotic pathogens and vectors, has a high level of tourism with the United States, and is geographically close to China and countries of Far East Asia which have many threatening, highly destructive diseases.

Lastly, we would remind the reader that exotic disease threat assessment (i.e., the fear) and exotic disease actualization in our agriculture (i.e., the reality) are based upon an indeterminate number of possibilities and probabilities, respectively. It is usually only after the fact that the relevant bits and pieces fit together. However, such hindsight can contribute to our "informed awareness," and this foresight can counter the possibility of experiencing future exotic plant diseases.

ACKNOWLEDGMENTS

The authors wish to express their utmost appreciation to J. C. Temple for typing the manuscript; our colleagues, N. E. El-Gholl, J. J. McRitchie, T. S. Schubert, and G. C. Wisler for their advice and support while we were writing; D. W. Dickson, R. N. Huettel, and D. W. Stokes for suggestions on exotic nematode threats; E. L. Barnard for suggestions on exotic disease threats to our forest resources; J. H. Tsai for suggestions relative to China and for his review; S. A. Alfieri, Jr., for his encouragement, suggestions, and reviews; and our Division Director, H. L. Jones, for graciously allowing us to do this study.

REFERENCES

Anonymous (1960). Index of plant diseases in the United States. *U.S. Dep. Agric., Agric. Handb.* **165.**

Anonymous (1963). Internationally dangerous forest tree diseases. *Misc. Publ.—U.S. Dep. Agric.* **939.**

Anonymous (1970a). ''A Handbook of Pests, Diseases, and Weeds of Quarantine Significance,'' 2nd ed. Kolos Publishers, Moscow (translated and published by Amerind Publ. Co. Pvt.-Ltd., New Delhi, 1978).

Anonymous (1970b). Plum pox virus. *CMI Distribution Maps Plant Dis.* No. 392.

Anonymous (1974). *Xanthomonas oryzae* (Uyeda & Ishiyama) Dowson. *CMI Distribution Maps Plant Dis.* Map No. 304, ed. 3.

Anonymous (1975). *Phakopsora pachyrhizi. CMI Distribution Maps Plant Dis.* No. 49.

Anonymous (1976). Groundnut rosette disease. *CMI Distribution Maps Plant Dis.* No. 49.

Anonymous (1978a). *Aplanobacter populi Ridé. Data sheets on quarantine organisms. Bull. OEPP* **8,** Set 1.

Anonymous (1978b). Apple Proliferation (Mycoplasm). Data sheets on quarantine organisms. *Bull. OEPP* **8,** Set 1.

Anonymous (1979a). ''1978 FAO Production Yearbook,'' FAO Stat. Ser. No. 22. FAO, Rome.

Anonymous (1979b). ''1978 FAO Trade Yearbook,'' FAO Stat. Ser. No. 24. FAO, Rome.

Anonymous (1980a). *Cercospora pini-densiflorae* Hori & Nambu. Data sheets on quarantine organisms. *Bull. OEPP* **10,** Set 3.

Anonymous (1980b). ''Foreign Tourist Arrivals by Selected States and Ports-of-entry January-June 1979.'' U.S. Dep. Commer, U.S. Travel Serv., Washington, D.C.

Anonymous (1980c). ''Prohibited Articles,'' Federal Quarantine Manual, 319.37-2C. U.S. Dep. Agric., Washington, D.C.

Anonymous (1980d). *Xanthomonas populi* (Ridé) M. Ridé & S. Ridé. *CMI Distribution Maps Plant Dis.* Map. No. 422, ed. 2.

Anonymous (1981a). ''Flag Smut,'' Federal Certification Manual, 319.59. U.S. Dep. Agric., Washington, D.C.

Anonymous (1981b). ''Florida Agricultural Statistics. Citrus Summary 1980.'' Fla. Crop & Livestock Reporting Serv., Orlando.

Anonymous (1981c). ''List of Intercepted Plant Pests (Pests reported from October 1, 1978 through September 30, 1979). ''U.S. Dep. Agric., Washington, D.C.

Anonymous (1981d). ''U.S. and General Imports. Ft 150/Annual 1980,'' Schedule 1 A commodity grouping by world area. U.S. Dep. Comm. Bureau of the Census, Washington, D.C.

Anonymous (1982a). Associated Press, Caracas, Venezuela, appearing in the *Gainesville Sun* on Feb. 17, 1982, p. 8A.

Anonymous (1982b). *Phoma exigua* Desm. var. *foveata* (Foister) Boerema. Data sheets on quarantine organisms. *Bull. OEPP* **12,** Set. 5.

Anonymous (1982c). *Synchytrium endobioticum* Schilb.) Percival. Data sheets on quarantine organisms. *Bull. OEPP* **12,** Set 5.

Bailey, L. H., and Bailey, L. Z. (1976). ''Hortus Third.'' Macmillan, New York.

Baker, K. F., and Linderman, R. G. (1979). Unique features of the pathology of ornamental plants. *Annu. Rev. Phytopathol.* **17,** 253-277.

Beamer, R. H. (1943). A new *Atanus* from Argentina. *Proc. Entomol. Soc. Wash.* **45,** 178–179.

Benigo, D. A., and Favali-Hedaynt, M. A. (1977). Investigation on previously unreported or noteworthy plant viruses and virus diseases in the Philippines. *FAO Plant Prot. Bull.* **25,** 78–84.

Bock, K. R. (1982). Geminivirus diseases in tropical crops. *Plant Dis.* **66,** 266–270.

Boddé, T. (1982). Genetic engineering in agriculture: Another green revolution? *BioScience* **32,** 572–575.

Bos, L. (1981). Wild plants in the ecology of virus diseases. *In* "Plant Diseases and Vectors: Ecology and Epidemiology" (K. Maramorosch and K. F. Harris, eds.), pp. 1–33. Academic Press, New York.

Bradbury, J. F. (1969). *Xanthomonas oryzae. CMI Description Pathog. Fungi Bact.* No. 239.

Bromfield, K. R., Melching, J. S., and Kingsolver, C. H. (1980). Virulence and aggressiveness of *Phakopsora pachyrhizi* isolates causing soybean rust. *Phytopathology* **70,** 17–21.

Brunt, A. A., and Kenten, R. H. (1972). Pepper veinal mottle virus. *CMI/AAB Descriptions Plant Viruses* No. 104.

Buddenhagen, J. W., Sequeira, L., and Kelman, A. (1962). Designation of races of *Pseudomonas solanacearum. Phytopathology* **52,** 726 (abstr.).

Chiarappa, L. (1981). Man-made epidemiological hazards in major crops of developing countries. *In* "Plant Diseases and Vectors: Ecology and Epidemiology" (K. Maramorosch and K. F. Harris, eds.), pp. 319–339. Academic Press, New York.

Chiykowski, L. N. (1981). Epidemiology of diseases caused by leafhopper-borne pathogens. *In* "Plant Diseases and Vectors, Ecology and Epidemiology" (K. Maramorosch and K. F. Harris, eds.), pp. 105–159. Academic Press, New York.

Ciampi, L., and Sequeira, L. (1980). Influence of temperature on virulence of race 3 strains of *Pseudomonas solanacearum. Am. Potato J.* **57,** 307–317.

Dopson, R. N., Jr. (1964). The eradication of citrus canker. *Plant Dis. Rep.* **48,** 30–31.

Dowling, C. F., Jr., Graham, A. E., and Alfieri, S. A., Jr. (1982). Plant Inspections and Certifications. *Plant Dis.* **66,** 345–351.

El-Goorani, M. A. (1976). Current status of the potato brown rot research in Egypt. *In* "Proceedings of the lst International Planning Conference and Workshop on the Ecology and Control of Bacterial Wilt Caused by *Pseudomonas solanacearum*" (L. W. Sequerira and A. Kelman, eds.), pp. 68–72. North Carolina State University, Raleigh.

Falk, B. W., and Duffus, J. E. (1977). The first report of *Polymyxa betae* in the western hemisphere. *Plant Dis. Rep.* **61,** 492–494.

Ford, H. L. (1981). "Minutes of Port of Entry Passenger Baggage Inspection Meeting with Industry." U.S. Dep. Agric., Washington, D.C.

Ford, R. E., Bissonette, H. L., Horsfall, J. G., Millar, R. L., Schlegel, D. E., Tweedy, B. G., and Weathers, L. G. (1981). Plant pathology in China. *Plant Dis.* **65,** 706–716.

Foster, J. A. (1982). Plant quarantine problems in preventing the entry into the United States of vector-borne plant pathogens. *In* "Pathogens, Vectors, and Plant Diseases: Approaches to Control" (K. F. Harris and K. Maramarosch, eds.), pp. 151–185. Academic Press, New York.

French, E. R. (1981). Common rust. *In* "Compendium of Potato Diseases" (W. J. Hooker, ed.), p. 78. Am. Phytopathol. Soc., St. Paul, Minnesota.

Fribourg, C. E., Jones, R. A. C., and Koenig, R. (1979). Andean Potato Mottle Virus. *CMI/AAB Descriptions Plant Viruses* No. 203.

Gibbs, A. J., Giussani-Belli, G., and Smith, H. G. (1968). Broad-bean stain and true broad-bean mosaic viruses. *Ann. Appl. Biol.* **61,** 99–107.

Graham, J. H., Jones, D. A., and Lloyd, A. B. (1979a). Survival of *Pseudomonas solanacearum* race 3 in plant debris and in latently infected potato tubers. *Phytopathology* **69,** 1100–1103.

Graham, J. H., Frosheiser, F. I., Stuteville, D. L., and Erwin, D. C. (1979b). "A Compendium of Alfalfa Diseases." Am. Phytopathol. Soc., St. Paul, Minnesota.

Granhall, I. (1981). Old and new problems in the field of plant quarantine. *Bull. OEPP* **11,** 139–144.

Grattidge, R. (1982). Occurrence of a third race of fusarium wilt of tomatoes in Queensland. *Plant Dis.* **66,** 165–166.

Halliwell, R. S. (1981). Viruses and mycoplasma-like organisms. *In* "Compendium of Cotton Diseases" (G. M. Watkins, ed.), pp. 56–59. Am. Phytopathol. Soc., St. Paul.

Harlan, J. R. (1976). The plants and animals that nourish man. *Sci. Am.* **235,** 89–97.

Harris, D. C. (1976). Bacterial wilt in Kenya with particular reference to potatoes. *In* "Proceedings of the 1st International Planning Conference and Workshop on the Ecology and Control of Bacterial Wilt Caused by *Pseudomonas solanacearum*" (L. W. Sequerira and A. Kelman, eds.), pp. 84–88. North Carolina State University, Raleigh.

Harrison, B. D. (1974). Potato Mop-Top Virus. *CMI/AAB Descriptions Plant Viruses* No. 38.

Hart, J. L. (1980). Tragedy of dutch elm disease bears hope for modern control. *Weeds, Trees, Turf* **19,** 16–20, 23–24.

Hayward, A. C. (1964). Characteristics of *Pseudomonas solanacearum. J. Appl. Bacteriol.* **27,** 265–277.

Hayward, A. C. (1976). Systematics and relationships of *Pseudomonas solanacearum. In* "Proceedings of the 1st International Planning Conference and Workshop on the Ecology and Control of Bacterial Wilt Caused by *Pseudomonas solanacearum*" (L. W. Sequerira and A. Kelman, eds.), pp. 6–21. North Carolina State University, Raleigh.

Hepting, G. H. (1974). Death of the American chestnut. *J. For. Hist.* **18,** 61–67.

Hodges, C. S., and May, L. C. (1972). A root disease of pine, *Araucaria,* and *Eucalyptus* in Brazil caused by a new species of *Cylindrocladium. Phytopathology* **62,** 898–901.

Hunt, R. N. (1946). Destructive plant diseases not yet established in North America. *Bot. Rev.* **12,** 593–627.

Itoh, Y., Oshima, Y., and Ichniohe, M. (1969). A root-knot nematode, *Meloidogyne mali* n. sp. on apple-tree from Japan (Tylenchida: Heteroderidae). *Appl. Entomol. Zool.* **4,** 194–202.

Jenkins, A. E. (1951). Sphaceloma scab, a new disease of soybeans discovered by plant pathologists in Japan. *Plant Dis. Rep.* **35,** 110–111.

Jones, J. P., Jones, J. B., and Miller, J. W. (1982). Fusarium wilt of tomato. *Fla. Dep. Agric. & Consumer Serv., Div. Plant Ind., Plant Pathol. Circ.* No. 237.

Jones, R. A. C. (1981). Potato mop-top virus. *In* "Compendium of Potato Diseases" (W. J. Hooker, ed.), pp. 79–80. Am. Phytopathol. Soc., St. Paul, Minnesota.

Jones, R. A. C., and Fribourg, C. E. (1981). Andean potato latent virus. *In* "Compendium of Potato Diseases" (W. J. Hooker, ed.), p. 78. Am. Phytopathol. Soc., St. Paul, Minnesota.

Kaloostian, G. H. (1980). Psyllids. *In* "Vectors of Plant Pathogens" (K. F. Harris and K. Maramorosch, eds.), pp. 87–91. Academic Press, New York.

Katsura, K. (1976). Two new species of *Phytophthora* causing damping-off of cucumber and trunk rot of chestnut. *Trans. Mycol. Soc. J.* **17,** 238–242.

Kelman, A., and Cook, R. J. (1977). Plant pathology in the People's Republic of China. *Annu. Rev. Phytopathol.* **17,** 409–429.

Klinkowski, M. (1970). Catastrophic plant diseases. *Annu. Rev. Phytopathol.* **8,** 37–60.

Knorr, L. C. (1973). Citrus diseases—a bibliography. *PANS* **19,** 441–463.

Lapida, J. L., and Roberts, J. M. (1977). Pepper veinal mottle virus associated with a streak disease of tomato in Nigeria. *Ann. Appl. Biol.* **87,** 133–138.

Ling, K. C. (1972). "Rice Virus Diseases." Int. Rice Res. Inst., Los Banos, Philippines.

Liu, S. T. (1948). Seed-borne disease of soybean. *Bot. Bull. Acad. Sin.* **2,** 69–80 (not seen); *Rev. Appl. Mycol.* **20,** 156 (1949).

Lloyd, A. B. (1976). Bacterial wilt in a cold-temperate climate of Australia. *In* ''Proceedings of the 1st International Planning Conference and Workshop on the Ecology and Control of Bacterial Wilt Caused by *Pseudomonas solanacearum*'' (L. W. Sequerira and A. Kelman, eds.), pp. 134–135. North Carolina State University, Raleigh.

Lozand, J. C. (1977). Identification of bacterial leaf blight in rice, caused by *Xanthomonas oryzae,* in America. *Plant Dis. Rep.* **61,** 644–648.

Luc, M. (1974). *Heterodera sacchari. CIH Descriptions Plant-parasitic Nematodes* No. 48, Set 4.

Luc, M., and Taylor, D. P. (1977). *Heterodera oryzae. CIH Descriptions Plant-parasitic Nematodes* No. 91, Set 7.

McGregor, R. C. (1973). ''The Emigrant Pests'' (mimeo report to the Administrator, Animal and Plant Health Inspection Service). U.S. Dep. Agric., Washington, D.C.

McGregor, R. C. (1978). People-placed pathogens: The emigrant pests. *In* ''Plant Pathology: An Advanced Treatise'' (J. G. Horsfall and E. B. Cowling, eds.), Vol. 2, pp. 383–396. Academic Press, New York.

Mai, W. F., Brodie, B. B., Harrison, M. B., and Jatula, P. (1981). Potato cyst nematodes. *In* ''Compendium of Potato Diseases'' (W. J. Hooker, ed.), pp. 94–97. Am. Phytopathol. Soc., St. Paul, Minnesota.

Makkouk, K. M. (1978). A study on tomato viruses in the Jordan Valley with special emphasis on tomato yellow leaf curl. *Plant Dis. Rep.* **62,** 259–262.

Mandahar, C. L. (1981). Virus transmission through seed and pollen. *In* ''Plant Diseases and Vectors: Ecology and Epidemiology'' (K. Maramorosch and K. F. Harris, eds.), pp. 241–292. Academic Press, New York.

Marenaud, C., Mazy, K., and Lansac, M. (1978). Apple proliferation: A strange and dangerous disease. *Rev. Hortic.* **188,** 41–50 (not seen); *Rev. Plant Pathol.* **57,** 500(1978) (abstr.).

Mathys, G., and Baker, E. A. (1980). An appraisal of the effectiveness of quarantines. *Annu. Rev. Phytopathol.* **18,** 85–101.

Matthews, H. J. P. (1975). *Heterodera carotae. CIH Descriptions Plant-parasitic Nematodes* No. 61, Set 5.

Mazyad, H. M., Omar, F., Al-Taher, K., and Salha, M. (1979). Observations on the epidemiology of tomato yellow leaf curl disease on tomato plants. *Plant Dis. Rep.* **63,** 695–698.

Metcalf, Z. P. (1968). ''General Catalogue of the Homoptora,'' Fasc. VI, Part 17. U.S. Dep. Agric., Washington, D.C.

Mercado, A. C., Jr., and Lantican, R. M. (1961). The susceptibility of cytoplasmic male sterile lines of corn to *Helminthosporium maydis* Nish & Miy. *Philipp. Agric.* **45,** 235243.

Miller, J. W., Seymour, C. P., and Burnett, H. C. (1977). Citrus canker. *Fla. Dep. Agric. & Consumer Serv., Div. Plant Ind., Plant Pathol. Circ.* No. 180.

Muniyappa, V. (1980). Whiteflies. *In* ''Vectors of Plant Pathogens'' (K. F. Harris and K. Maramorosch, eds.), pp. 39–85. Academic Press, New York.

Musa, A., and Mink, G. I. (1981). Beet necrotic yellow vein virus in North America. *Phytopathology* **71,** 773–776.

Musselman, L. J. (1980). The biology of *Striga, Orobanche,* and other root-parasitic weeds. *Annu. Rev. Phytopathol.* **18,** 463–489.

Neergaard, P. (1977). ''Seed Pathology,'' Vol. I. Wiley, New York.

Neergaard, P. (1981). Risks for the EPPO region from seed-borne pathogens. *Bull OEPP* **11,** 207–212.

Niblett, C. L., Tsai, J. H., and Falk, B. W. (1981). Virus and mycoplasma diseases of corn in Florida. *In* ''Proceedings of the 36th Annual Corn and Sorghum Industry Research Conference'' (H. D. Loden and D. Wilkinson, eds.), pp. 78–88. Am. Seed Trade Assoc., Washington, D.C.

Nielson, M. W. (1979). Taxonomic relationships of leafhopper vectors of plant pathogens *In* "Leafhopper Vectors and Plant Disease Agents" (K. Maramorosch and K. F. Harris, eds.), pp. 3–27. Academic Press, New York.

Nishizawa, T., Kinoshita, S., and Yoshii, H. (1955). On the soybean blast and its causal fungus *Septogloeum sojae* n. sp. *Ann. Phytopathol. Soc. J.* **20,** 11–15 (not seen); *Rev. Appl. Mycol.* **35,** 862 (1956).

Olsson, K. (1976). Overwintering of *Pseudomonas solanacearum* in Sweden. *In* "Proceedings of the 1st International Planning Conference and Workshop on the Ecology and Control of Bacterial Wilt Caused by *Pseudomonas solanacearum*" (L. W. Sequerira and A. Kelman, eds.), pp. 105–109. North Carolina State University, Raleigh.

Page, O. T. (1981). Thecaphora smut. *In* "Compendium of Potato Diseases" (W. J. Hooker, ed.), pp. 63–65. Am. Phytopathol. Soc., St. Paul, Minnesota.

Punithalingam, E. (1968). *Puccinia horiana. CMI Descriptions Pathog. Fungi Bact.* No. 176.

Ray, R., and Walheim, L. (1980). "Citrus: How to Select, Grow, and Enjoy." Horticultural Publ., Tucson, Arizona.

Renfro, B. L., and Ullstrup, A. J. (1976). A comparison of maize diseases in temperate and tropical environments. *PANS* **22,** 491–498.

Roistacher, C. N. (1981). A blueprint for disease. Part one. The history of seedling yellows disease. *Citrograph* **67,** 4, 5, 24.

Rothwell, A. (1980). A revised list of plant diseases in Zimbabwe. Additions: 1973–78. *Kirkia* **12,** 183–190.

Salazar, L. F., and Harrison, B. D. (1978). Host range and properties of potato black ringspot virus. *Ann. Appl. Biol.* **90,** 375–386.

Seshadri, A. R., and Dasgupta, D. R. (1975). *Ditylenchus angustus. CIH Descriptions Plant-parasitic Nematodes* No. 64, Set 5.

Seymour, C. P. (1977). White rust of chrysanthemum. *Fla. Dep. Agric. & Consumer Serv., Div. Plant Ind., Plant Pathol. Circ.* No. 180.

Shurtleff, M. C., ed. (1980). "Compendium of Corn Diseases," 2nd ed. Am. Phytopathol. Soc., St. Paul, Minnesota.

Sinclair, J. B., ed. (1982). "Compendium of Soybean Diseases," 2nd ed. Am. Phytopathol. Soc., St. Paul, Minnesota.

Sinclair, J. B., and Shurtleff, M. C., eds. (1975). "Compendium of Soybean Diseases," Am. Phytopathol. Soc., St. Paul, Minnesota.

Smith, K. M. (1972). "A Textbook of Plant Virus Diseases." Academic Press, New York.

Spaulding, P. (1914). Undesirable foreign plant disease. *Mass. Hortic. Soc. Trans.* Part 1, pp. 153–179.

Spaulding, P. (1961). Foreign diseases of forest trees of the world. *U.S. Dep. Agric., Agric. Handb.* **197.**

Stakman, E. C., and Harrar, J. G. (1957). "Principles of Plant Pathology." Ronald Press, New York.

Stall, R. E., and Seymour, C. P. (1983). Canker, a threat to citrus in the Gulf-Coast States. *Plant Dis.* **67,** 581–585.

Stevenson, J. A. (1926). "Foreign Plant Disease, a Manual of Economic Plant Disease which are New to or Widely Distributed in the United States" (unnumbered publication). U.S. Dep. Agric., Washington, D.C.

Stipes, R. J., and Campana, R. J., eds. (1981). "Compendium of Elm Diseases." Am. Phytopathol. Soc., St. Paul, Minnesota.

Sutic, D., and Pine, T. S. (1968). Sharka (plum pox) disease. *Plant Dis. Rep.* **52,** 253–255.

Tamada, T. (1975). Beet necrotic yellow vein virus. *CMI/AAB Descriptions Plant Viruses* No. 144, pp. 1–4.

Thourenel, J. C., Dallet, M., and Fanquet, C. (1976). Some properties of peanut clump, a newly discovered virus. *Ann. Appl. Biol.* **84,** 311–320.

Thurston, H. D. (1973). Threatening plant diseases. *Annu. Rev. Phytopathol.* **11,** 27–52.

Turkensteen, L. J. (1981). Phoma leaf spot. *In* "Compendium of Potato Diseases" (W. J. Hooker, ed.), pp. 47–48. Am. Phytopathol. Soc., St. Paul, Minnesota.

Urbina-Vidal, C. (1974). Non-viral agents associated with sugar beet yellow wilt in Chile. *Phytopathol. Z.* **81,** 114–123.

Vakili, N. G. (1981). Distribution of *Phakopsora pachyrhizi* on *Lablab purpureus* in Puerto Rico. *Plant Dis.* **65,** 817–819.

Vakili, N. G., and Bromfield, K. R. (1976). Phakopsora rust on soybeans and other legumes in Puerto Rico. *Plant Dis. Rep.* **60,** 995–999.

Van Tiel, N. (1976). "Reflections on 25 Years of EPPO," Special issue of the European and Mediterranean Plant Protection Organization (EPPO), *25th Anniversary of EPPO,* pp. 17–24. OEPP, Paris.

Walker, J. C. (1957). "Plant Pathology," 2nd ed. McGraw-Hill, New York.

Waterworth, H. E. (1981). Our plants' ancestors immigrated too. *BioScience* **31,** 698.

Waterworth, H. E., and White, G. A. (1982). Plant introductions and quarantine: The need for both. *Plant Dis.* **66,** 87–90.

Watkins, G. M. (1981). Escobilla (witches' broom). *In* "Compendium of Cotton Diseases" (G. M. Watkins, ed.), pp. 31–32. Am. Phytopathol. Soc., St. Paul, Minnesota.

Watson, A. J. (1971). Foreign bacterial and fungus diseases of food, forage, and fiber crops. An annotated list. *U.S. Dep. Agric., Agric. Handb.* **418.**

Weltzien, H. C. (1972). Geophytopathology. *Annu. Rev. Phytopathol.* **10,** 277–298.

Weltzien, H. C. (1978). Geophytopathology. *In* "Plant Pathology: An Advanced Treatise" (J. G. Horsfall and E. B. Cowling, eds.), Vol. 2, pp. 339–360. Academic Press, New York.

Wiese, M. V. (1977). "Compendium of Wheat Diseases." Am. Phytopathol. Soc., St. Paul, Minnesota.

Wilson, C. L. (1967). Vascular mycosis of oak in Russia. *Plant Dis. Rep.* **51,** 739–741.

Wilson, C. L. (1970). Vascular mycosis of oak in eastern Europe. *Plant Dis. Rep.* **54,** 905–906.

Yu, Y., and Zhuang, W. (1982). *Phytophthora sinensis,* a new species causing blight on *Cucumis sativus. Mycotaxon* **14,** 181–188.

Zadoks, J. C., and Schein, R. C. (1979). "Epidemiology and Plant Disease Management." Oxford Univ. Press, London and New York.

CHAPTER 7

Where Are the Principal Exotic Weed Pests?

ROBERT L. ZIMDAHL

Weed Research Laboratory
Department of Botany and Plant Pathology
Colorado State University
Fort Collins, Colorado

EXOTIC PLANT PESTS AND
NORTH AMERICAN AGRICULTURE

ISBN 0-12-757880-3

Our ability to move about in an ever-shrinking global community has given us enormous benefits and offers great potential for the development of the human community. It also offers great potential for agricultural disaster, as all kinds of new and different weeds easily move thousands of miles to new and potentially hospitable environments.

To illustrate, beside my desk, as I write, is a box of soil. Two days ago it was in Bologna, Italy. U.S. federal regulations required that my laboratory be inspected prior to receipt of the parcel to determine whether or not the soil would be handled and disposed of properly. What is in the box, besides soil? I do not know. The potential for weed importation is there. I will follow the federal rules because I believe they are right. Will others follow the rules even when they do not know them and, moreover, probably would not recognize their importance if they did?

Last fall, a student brought a strange-looking, presumably weedy plant to my office for identification. It was large (about 4 feet tall), bushy, green and purple, and looked like a member of the *Amaranthaceae* family. Further investigation revealed that it was an amaranth, but one heretofore unknown in Colorado. It was collected in a garden area where it had been planted, and subsequently abandoned, by some African students who used the plant for greens and roasted its abundant seed supply. It was imported by the students because they ate it at home and wanted it here. They were doing what we do when we take our favorite sweet corn variety to a new location. Apparently, this member of the *Amaranthaceae* is not weedy, but many others are, and they could arrive in the same way. Will they?

I. DEFINITION OF A WEED

Weed scientists need to be aware of the possibility of weed importation and the actual threat of exotic plants. To evaluate exotic weed threats, we need criteria to define "weediness" or the potential to become weedy. The Weed Science Society of America (WSSA) has defined a weed as "a plant growing where it is not desired" (Buchholtz, 1967, p. 389). This is an adequate definition of what a weed is, but it does not tell us why a particular plant is weedy. To define exotic weed pests, one must rely, in part, on appraisals of weed scientists in other countries. However, one must also determine what characteristics any plant could possess which may make it weedy and if certain combinations of these are potentially more weedy than others.

Baker (1974, p. 1) defined a plant as weedy if "in any specified geographical area, its populations grow entirely or predominantly in situations markedly disturbed by man (without, of course, being deliberately cultivated plants)." He noted that the extent or limitation of a species in its present environment, with or without human influence, is not a certain indication of its potential success or failure as a weed in other environments (Baker, 1974). This is why appraisals by weed scientists are necessary and useful, but not sufficient as determinants of a plant's potential weediness in a new environment.

Baker (1965) has listed the characteristics which an ideal weed should possess (Table I). Fortunately, although there are serious contenders for the title (Holm *et al.*, 1977), there are no claimants.

In another discussion of weeds, Baker (1974, p. 4) defined evolutionary success as something that "is to be measured in terms of numbers of individuals in existence, the extent of their reproductive output, the area of the world's surface they occupy, the range of habitats they can enter and their potential for putting their descendants in a position to continue the genetic line through time." Invading weeds succeed for one of three reasons (Baker, 1972):

1. They change genetically.
2. They find unoccupied (perhaps because they are recently created) ecological niches.
3. They successfully escape from the inhibitory attention of other pests.

The composition of a particular plant community also depends on the structure of local soils, their nutritional content, water capacity, and aeration, as well as the area's climate (King, 1966). The area occupied by a new weed is almost always limited, at first, to a few isolated plants. Subsequently, a number of seemingly (or actually) unrelated populations appear, and no further spread may ensue. The area eventually occupied is most often limited by ecological interactions of the

TABLE I

Characteristics of the Ideal Weed[a]

1.	No special environmental requirements for germination
2.	Discontinuous, self-controlled germination and seed longevity in soil
3.	Rapid seedling growth
4.	Short vegetative period prior to flowering
5.	Long period of seed production with favorable growing conditions
6.	High seed production in a favorable environment
7.	Produces some seed in a wide range of environmental conditions
8.	Has special adaptations for short- and long-distance dispersal of seeds
9.	Is self-compatible but is not an obligate self-pollinator or apomictic
10.	Cross-pollination achieved by nonspecialized flower visitors or wind
11.	Has a high tolerance (and often plasticity) in the face of climatic and edaphic variation
12.	If a perennial, has vigorous vegetative reproduction
13.	If a perennial, is brittle at the lower nodes of rhizomes or rootstocks
14.	If a perennial, shows ability to regenerate from several portions of rootstock
15.	Has ability to compete by special means, e.g., rosette formation, allelopathy

[a] From Baker (1965).

weed in its new environment (Polunin, 1960). Environmental adaptation and, thus, evolutionary success is ultimately genetically based in traits such as self-fertilization, morphological differences, and phenotypic plasticity (Baker, 1974). These traits are expressed through physiological mechanisms such as photosynthetic pathways, seed dormancy, and growth rate (Baker, 1974; Polunin, 1960).

The most successful weeds are colonizing opportunists with what could best be described as a general-purpose genotype (Young and Evans, 1976). These general-purpose or cosmopolitan and evolutionarily successful weeds tend to tolerate a wide range of environmental conditions and are thus able to invade available niches within some climatic limit (Polunin, 1960). Other weeds are quite exacting in their requirements and are thus restricted to small areas or regions.

We can conclude that we do know, in a general sense, what criteria to use to define weediness. But we must also conclude that we are still struggling to apply the criteria in the specific sense, and the difficulty is compounded when geographic translocation is a factor. Using some of the traits of the ideal weed and the observed characteristics of two species of *Ageratum,* Baker (1972) has shown that one cannot assume weediness by association or genetic relation (Table II). *Ageratum conyzoides* is a highly successful pantropical weed in more than 20 countries (Holm *et al.*, 1979). Its close relative, *A. microcarpum,* is only slightly weedy in the same areas (Baker, 1972). Therefore, we must also conclude that

TABLE II

Comparative Features of *Ageratum microcarpum* and *Ageratum conyzoides* Revealed by Controlled Environmental Experiments[a]

Ageratum microcarpum (scarcely weedy)	*Ageratum conyzoides* (widespread weed)
Light requirement for germination	No light requirement for germination
Perennial	Life span for 1 year
Flowers in second season of growth	Germination to flowering in 6–8 wk
Flowering inhibited by high nighttime temperatures	Flowering at low (10°C) or high (27°C) nighttime temperatures
Flowering better with long (12-hour) nights	No photoperiodic control of flowering
Mesophyte	Tolerates waterlogging, drought
Self-incompatible	Self-compatible (largely self-fertilized)
Not very phenotypically plastic (n = 10)	Phenotypically plastic (n = 20)

[a] From Baker (1972).

evolutionary success and weediness cannot be predicted based on genotypic similarity.

II. DISTRIBUTION OF WEEDS

The most comprehensive report on world plant distribution is that by Ridley (1930). The distribution of weeds has not been examined as carefully, but it has not been ignored. The generalization that weed flora generated in areas with long agricultural histories tend to provide successful invaders has been attributed (Baker, 1974) to Gray (1879). This is a perfectly logical assumption because these plants have already been subjected to several generations of selection for weedy traits. It is also true that weeds have moved as the shadow of history (Renfrew, 1973; Young and Evans, 1976). For example, many species mentioned as prehistoric food plants are names found in the current weed literature. However, primitive agriculturalists cannot be blamed for developing our weeds, because they simply did not have the energy to modify the environment and create new ecological niches in which extant plants could express or develop their weediness (Young and Evans, 1976).

A. The Role of Man

We know that the primary foreign sources of weeds are such things as ship ballast, wool, straw, and soil. These sources all function most effectively when they are linked to man. Perhaps the best way for a plant to express its colonizing ability is to become associated with man (Heiser, 1965). King (1966) has distinguished anthropophytes (man-encouraged plants) from apophytes (native species venturing into largely man-created habitats). He asserts that it is the anthropophytes that have been successful and that many of our weeds are properly regarded as cast-off plants, remnants, or discards of our usage and later abandonment. Young and Evans (1976) suggested that the introduction of weedy species to new environments, with or without their associated pests and parasites, may be man's greatest environmental manipulation. Although others might regard this suggestion as hyperbole, it seems logical to assume that with the ever quickening pace of change and speed of transport, such introductions and their evolutionary results will dominate the attention of succeeding generations of weed scientists.

Man's role as an agent of dispersal is well documented. Of 89 common weeds in New York State, 40% were common in England (Salisbury, 1961). Salisbury (1961) also cited the unique case of New Zealand. A study of 500 weedy species showed that 80% were also common in Europe, 44 were from the United States, 16 from South Africa, but only 31 from Australia. If natural dispersal was the primary mode of transport, one would expect more weeds from the nearest large land mass. Man is the only logical agent that could have dispersed the majority of weeds between Australia and other centers and New Zealand.

Another example of man's role in international transfer is the recent report that of 40 weeds imported to Israel, 30% were from the United States, which has had a continued interaction with Israel (Dafni and Heller, 1980). Johnsongrass (*Sorghum halepense* L.) was probably purposefully introduced into the United States in the early nineteenth century from Turkey (McWhorter, 1971) as a forage plant (Williams, 1980) and probably entered as seed (McWhorter, 1971).

B. Plants That Have Become Weeds

Muenscher (1955) listed 42 plants (Appendix 1) intentionally introduced by man for cultural purposes that escaped to become important weeds. It is interesting to note that 31 of the species were perennials, five biennials, and six annuals. If interspecific competition is primarily responsible for success, it is during seed germination and seedling establishment that success or failure will be determined (Harper, 1965). However, a perennial that reproduces by a transportable vegetative organ (e.g., tuber, rhizome, stolon, aerial bulblet) may effectively escape the rigors of early seedling competition that an annual faces.

Of the plants on Muenscher's list, 21 were imported as ornamentals, four as vegetables, seven as drug plants, six as herbs, and four as ornamental shrubs. One might conclude that entry of agricultural commodities is not the primary avenue to watch for new weeds. On the other hand, one might also assume that escaped ornamentals are more readily observed.

In a later study, Williams (1980) reported 36 purposefully introduced plants that have become noxious or poisonous weeds in the United States (Appendix 2). There were only six duplicates with Muenscher's (1955) earlier list, and only in the case of tansy (*Tanacetum vulgare* L.) did they disagree on its source; one said that it was imported as a herb (Williams, 1980) and the other as a drug (Muenscher, 1955).

III. EVALUATING EXOTIC PLANTS

Williams (1980) cited four factors that should be evaluated when considering the threat posed by a particular exotic plant. Like Baker (1965) (Table 1) he considered (1) methods of dissemination and (2) longevity of seed and vegetative reproductive organs to be important. He added the more nebulous but no less important factor, (3) capacity to invade. This could be viewed as a general heading for characteristics 1–4 and 11 in Table 1. Young *et al.* (1972) illustrated the capacity to invade in their report of the man-aided invasion of native sagebrush (*Artemisia* spp.) associations by alien annuals such as downy brome (*Bromus tectorum* L.). Annuals are inherently vigorous competitors that close vacant niches in sagebrush communities. In undisturbed communities, native perennials do not need to reproduce every year to maintain their stand. Alien annuals have the apparent disadvantage of annual reproduction, but they accomplish this via phenotypic and genotypic plasticity that allows them to survive in years with below-normal precipitation and to preempt environmental potential in years with an abundance of precipitation (Young *et al.*, 1972). Seeds are the linchpin in this survival strategy and expression of the capacity to invade. Downy brome seed germinates over a long period of time and has highly developed dispersal mechanisms. Williams (1980) also added the presence of other undesirable characteristics. Here he revealed an appropriate bias as a weed scientist which Baker's ecological thinking did not include. These characteristics include lack of palatability, potential for mechanical damage to animals, allergenic effects, ability to harbor insects or diseases, and poisonous nature (Williams' fourth factor).

Purposeful introductions by man are important but are not the only source of weed introductions. One should not neglect crop mimics such as the weedy species of *Oryza* and *Avena*. Because of similarity of seed size, germination time, and cultural requirements, these make good weeds and are especially well adapted for transport with crop seeds.

There are not many examples, but weeds of hybrid crops can be important.

Within the last decade, growers in the western United States have become concerned about shattercane [*Sorghum bicolor* (L.) Moench]. Many hybrids are developed via the male sterile technique, which includes pollination by an inbred (male-fertile) line. Sorghums often become pollinated by the possibly weedy sudangrass [*S. sudanense* (Piper) Stapf.] or the actually weedy johnsongrass [*S. halepense* (L.) Pers.] or sorghum almum (*S. almum* Parodi—a variable species) (Baker, 1974). One may get a seed-bearing annual such as shattercane or a sterile line with vigorous vegetative reproduction. Such an event should not necessarily be regarded as the introduction of an exotic species, but it should be considered in evaluating the weedy potential of exotic and domestic species.

Weed scientists should recognize that an exotic colonizer is not a plant out of place but rather a plant very much in its place (Mulligan, 1965). Man has been the foremost agency of transport for colonizers, and then, in turn, has defined them as out of place when they expressed their weediness. We have aided all natural methods of dispersal. We have also diminished the pressure of competition between species by checking the establishment of a continuous plant cover and creating crop environments in which some species can survive and others dominate. These same species could not (or would be less likely to) dominate under the more severe competition of the natural environment.

Cohabitation is the rule among species, rather than a life-and-death competitive struggle within and between all trophic levels (Harper, 1965). It is slight differences that make one plant succeed in competition. Harper (1965) suggested that where light is limiting, long hypocotyls may provide the necessary competitive edge. When water is limited, rapid root growth can be very important. In most cases, early seed germination and rapid seedling growth are advantageous, but, as in many realms, acquisition of advantage involves risk. Early growth makes a plant more susceptible to drought and frost.

We are concerned about weeds because they compete, but competitive ability alone has low heritability and may not be associated with fitness or responses to increasing plant density (Sakai, 1965). Competitional variance in a mixed population results in the enlargement of nonhereditary variation of traits related to fitness, resulting in less effective selection for fitness.

A review of the characteristics of an ideal weed (Table I), consideration of the preceding discussion, and the apparent fact (Baker, 1974) that the extent or limitation of a species in one environment is not a certain indication of its success in another leaves one with the insecure feeling that the threat posed by a particular exotic plant cannot be predicted. The usual plea is for more research on each species, which is a logical but not always reasonable request. When science and the legislative realm interact, as they must if we are to restrict importation (a legislative decision) of potentially dangerous plants (a scientific determination), then some decisions will inevitably be made on the basis of incomplete data. Neither the scientist nor the legislator will know with certainty.

Sources of Information and Criteria

Presently, there are three general sources of information on exotic weed threats. These are the two volumes by Holm *et al.* (1977, 1979) and the still unpublished work of the U.S. Department of Agriculture Technical Committee to Evaluate Noxious Weeds (TCENW).[1] These were used extensively in preparing this chapter, which is limited by available space and therefore includes only 24 species. These were selected from several hundred that could have been included. The criteria in the preceding discussion were used, as were those in the TCENW report. Among the more important criteria used to select the weeds were:

1. World distribution—wide distribution offers greater potential for invasion.
2. Countries where the species is listed in the weed law. Prior recognition of weediness by several countries was interpreted as evidence of their threat. (Little information was found).
3. Edaphic and climatic requirements.
4. Life cycle.
5. Phenotype and phenotypic plasticity.
6. Competitive ability or aggressiveness.
7. Adequacy of dispersal mechanisms and presence of seed dormancy.
8. Tolerance of control methods.
9. Harmful aspects such as effects on crop yield, toxicity, allergenic and poisonous qualities, and reservoir for other pests.

Three additional criteria were used:

1. A species presently in the continental United States was excluded even if the infestation was minor, *except* when it had only been reported as inhabiting ore piles or ballast dump sites and was presently not an agricultural weed.
2. Sufficient information had to be available to permit an intelligent decision, especially regarding several species in one genus.
3. The species had to be identified as an important weed somewhere. This, of course, precluded consideration of plants that may not be weedy where they occur but have great potential for weediness elsewhere.

No attempt was made to make these criteria quantitative. In the final analysis, informed judgment and shared opinions concerning the application of the criteria were often decisive.

[1]For information contact: The Secretariat TCENW, Room 238, Building 001, BARC-West, Beltsville, Maryland 20705.

Application of these criteria resulted in exclusion of the perennial tropical and subtropical grass weeds *Pennisetum clandestinum* and *P. purpureum* (although one other *Pennisetum* species is included) and the annual grass *Rottboellia exaltata* because each is present in the United States as an important agricultural weed. This decision is not intended to minimize the importance of reducing movement or controlling further importation of these species. The decision was based on the fact that if the entire country is considered, these species cannot be regarded as exotic (i.e., foreign). Reluctantly, the many *Orobanche* species were also excluded. At least 20 species of *Orobanchaceae* are root parasites of dicotyledonous plants, including some of the world's important crops. At least two of the important species are already in the United States (i.e., *Orobanche minor* and *O. ramosa*) (Gunn *et al.*, 1981).

The many parasitic herbaceous annual species of *Striga* were also excluded. Two species, *Striga lutea* and *S. gesnerioides,* are already present in the United States. There are at least 40 other species (Gunn *et al.,* 1981), and all are potential exotic threats. Not all could be included, and it was not possible to identify one that was more important than all others; therefore, all were excluded. The same line of reasoning was used with the more than 20 species of the perennial woody species of the genus *Prosopis*. None are present in the continental United States, and all may be important threats. Plants in this genus form impenetrable spiny thickets that preclude the use of rangeland for grazing (Gunn *et al.,* 1981). However, it was again impossible to decide which species were most important; so, reluctantly, all were omitted.

Alopecurus myosuroides Huds. was listed by Holm *et al.* (1979) as a serious weed in the United Kingdom. It is present as a weed in the United States (Holm *et al.,* 1979), but has not become an important problem and is not included in this chapter. The situation may be similar to that of common ragweed (*Ambrosia elatior* L.), which competes well in many crops in large portions of the United States, but has not been a successful weed in the United Kingdom. The reason may be climatic and related to the relatively late maturation of fruits in the United Kingdom (Salisbury, 1961).

A third illustration of the problem of selection is offered by the pantropical submerged aquatic perennial *Hydrilla verticillata* (L.) Royle, which was probably introduced from South America for use in aquaria (Haller, 1978). The female plants are widespread in the southern United States, and current efforts are focused on excluding importation of male plants. The species reproduces vegetatively by four separate means (Haller, 1978). If male plants enter the United States, sexual reproduction will follow, genetic variability and adaptive potential will increase, and one assumes that control will become more difficult.

Fortunately, the problem of identifying important exotic weeds has been worked on by Holm *et al.* (1977, 1979) and the TCENW. The problem of selecting the most important weed species remains. The 24 species selected for inclusion herein are shown in Table III. Following Table III there is a discussion of each species, which includes a botanical description, its distribution, and, if available, evidence of spread, weedy characteristics, and agricultural importance.

TABLE III

Exotic Weeds That Are Potential Threats to U.S. Agriculture

Weed species	Life history[a]	Plant family[b]	Habitat[c]	Location	FNWA status[d]
Alternanthera sessilis	A	D	A	Africa, Asia	2
Asphodelus fistulosus	A	D	T	Asia	3
Azolla pinnata	A–P	Fern	A (ff)	Asia	2
Borreria alata	A	D	T	Asia, South America	2
Chromolaena odorata	P	D	T	Asia, Africa	3
Commelina benghalensis	P	D	T	Africa, Asia	1
Digitaria scalarum	P	M	T	Africa, India	1
Digitaria velutina	A	M	T	Africa	2
Drymaria arenarioides	P	D	T	Mexico	1
Euphorbia prunifolia	A	D	T	Asia	2
Imperata brasiliensis	P	M	T	South America	1
Ipomoea aquatica	P	D	A (fl)	Asia, Africa	2
Ischaemum rugosum	A	M	T	Asia, South America	1
Mikania cordata	P	M	T	Asia, Africa	2
Mikania micrantha	P	M	T	Asia, Africa	1
Mimosa invisa	B	D	T	Asia	2
Monochoria hastata	A–P	M	A (E)	Asia	1
Monochoria vaginalis	P	M	A (E)	Asia	1
Pennisetum polystachion	A–P	M	T	Africa, Asia	2
Saccharum spontaneum	P	M	T	Asia, Africa	2
Sagittaria sagittifolia	A–P	M	A (E)	Europe, Asia, South America	2
Salvinia molesta	A–P	Fern	A	Asia, Africa, South America	1
Setaria pallide-fusca	A	M	T	Africa	3
Tridax procumbens	A–P	D	T	Africa, Asia, South America	2

[a] A = annual; B = biennial; P = perennial.
[b] D = dicotyledon; M = monocotyledon.
[c] A = Aquatic (ff, free floating; fl, floating; E, emergent); T = terrestrial.
[d] FNWA = Federal Noxious Weed Act; 1 = included; 2 = recommended for inclusion; 3 = not recommended.

IV. EXOTIC WEEDS THAT MAY THREATEN U.S. AGRICULTURE

A. *Alternanthera sessilis* (L.) DC. (Amaranthaceae: Sessile Joyweed)

This spreading, branched, prostrate annual herbaceous weed propagates by seed. Synonyms include *Gomphrena sessilis* (L.), *Illecebrum sessilis* (L.) (Reed and Hughes, 1977), *A. triandra* Lam., and *A. repens* (L.) Link. (Holm *et al.*, 1979). It is a weed of cultivated areas throughout tropical lands including Africa, South Asia, the West Indies, the Middle East, Southeast Asia, China, East Asia, Australia, and the Pacific basin. It is regarded as a serious weed in Puerto Rico, Taiwan, and Hawaii, and it has been reported as present in the continental United States, but only on ore piles in Maryland (Reed, 1961). The plant exists as a pantropical weed commonly growing in damp pastures, on pond margins, and in other damp areas. It grows in areas that are inundated with water for long periods, where it roots in the substratum of mud or clay. It can grow in association with other amphibious marsh species. Its prostrate or decumbent habit allows it to root from the lower nodes.

Holm *et al.* (1979) list the weed as serious in the Ivory Coast, Mozambique, Nigeria, the Philippines, and Thailand. It is a principal[2] weed in four countries in Southeast Asia and is common in eight additional countries (Holm *et al.*, 1979). Most of these countries are in Africa or Asia, and as yet, it is not an important weed outside of those areas. It is a weed of rice and grows in association with other rice weeds such as *Echinochloa colonum, Monochoria vaginalis, Aponogeton* spp., and *Marsilea erosa.* Its present habitat is aquatic, but it is a problem in upland rice and thus has the potential to invade areas that are not exclusively aquatic. It therefore should be regarded as a potential problem in rice and in shallow ponds, irrigation ditches, and other waterways. *Alternanthera sessilis* is related to and smaller than *A. philoxeroides* (floating alligator weed), which is a serious aquatic weed in the United States.

[2]Holm *et al.* (1979) used the terms ''serious,'' ''principal,'' and ''common'' to describe the relative importance of weeds in particular countries. These terms are used to describe the distribution and importance of several of the weed species in this chapter. Holm *et al.* defined ''serious'' as a word which has a commonality of understanding. They assert that when a worker states that a certain species ranks as one of the two or three most ''serious'' weeds in a crop, there is no problem with meaning. Similarly, ''principal'' indicated one of the five most troublesome species in a particular crop. The designation ''common'' referred to a weed that was widespread, required constant attention, but never seriously threatened crops. They also designated some weeds as ''present,'' i.e., the species was present and behaved as a weed in the country, but its importance was unknown. The final category referred to a plant present in the flora but showing no evidence of weedy behavior.

B. *Asphodelus fistulosus* L. (Liliaceae: Onion Weed)

This weed is often referred to by its common name, "onion weed," although it has no onion odor. It may be referred to as *A. aestivus* Reichb. (Gunn *et al.*, 1981; King, 1966), or *A. tenuifolius* Cav. (Gunn *et al.*, 1981). It is native to the Mediterranean region, northern India, and much of the Middle East (Reed and Hughes, 1977). It is regarded as a serious weed in India and Pakistan and a principal weed in Australia. It is common in Italy, Lebanon, and Nepal, and has been reported as a sparsely occurring weed in gardens in Mexico, Florida, and California (Gunn *et al.*, 1981).

This weed is capable of growing in temperate and subtropical climates in several cultivated crops. Indian studies have found that it favors soils with neutral pH (Gunn *et al.*, 1981). It is a prime candidate for infestation of small grains in the United States because the seed is similar in size and shape to wheat, barley, and oats, and it probably matures at the same time.

C. *Azolla pinnata* R. Br. (Salviniaceae: Azolla)

Synonymous names for *Azolla* include *A. africana* Desv. (Gunn *et al.*, 1981; Holm *et al.*, 1979), *A. guineensis* Schumann, *A. imbricata* (Roxb.) Nakai, and *Salvinia imbricata* Roxb. ex. Griff. (Gunn *et al.*, 1981).

Azolla is generally regarded as an annual, but multiplies rapidly both sexually and asexually by fragmentation. The main rhizome bears several alternating branches with attached lateral branches. At the point of attachment, each branch has an abscission layer which is important in vegetative reproduction (Lumpkin and Plucknett, 1980). Under favorable conditions, *Azolla* rapidly develops dense mats. It is reported to double its biomass in 35 days in the laboratory and in as little as 5–10 days in the field (Gunn *et al.*, 1981; Lumpkin and Plucknett, 1980). Beneath the dense mat, light penetration is inhibited and often nearly prevented. *Azolla* can survive in a pH range of 3.5–10.0, but optimum growth is observed in the range 4.5–7.0 (Lumpkin and Plucknett, 1980). The most favorable temperature for growth is between 20° and 30°C. Outside of this range, growth decreases until the plant begins to die at temperatures below 5°C and above 45°C (Lumpkin and Plucknett, 1980). Thus, the plant is a weed of tropical, subtropical, and warm temperate climates but not of the northern temperate zones. It is found in ponds, ditches, and rice paddy fields, where there is minimum aquatic disturbance, gradual water level changes, and a substrate of loose soil.

It is usually found growing in association with other aquatic weeds, such as *Eichhornia pistia, Stratiotes lemna, Spirodela polyrhiza,* and *Salvinia* spp. (Gunn *et al.*, 1981). Wherever these species occur could be considered potential habitats for *Azolla*.

Azolla can extract nitrogen from the aquatic environment and thus survive. However, its most unique characteristic is its symbiotic relationship with the nitrogen-fixing blue-green alga *Anabaena azollae*. The rate of nitrogen fixation in the *Azolla–Anabaena* symbiosis is equal to or greater than that for the *Rhizobium*–legume symbiosis. Thus, we are caught in a controversy over *Azolla*. Is it a weed, or is it a very useful plant that can be exploited in agriculture? *Azolla* has been reported to disrupt fishing and livestock watering (Florida Department of Natural Resources), clog pumps (Chomchalow and Ponepangan, 1973), impede water flow in ditches, clog pipes and floodgates (Blackburn and Weldon, 1965; Eady, 1974; Edwards, 1975; Kleinschmidt, 1969; Matthews, 1963; Oosthuizen and Walters, 1961), and interfere with watercress cultivation in Maryland (Lumpkin and Plucknett, 1980; Reed, 1951–1953). However, there is an equal amount of contrary and quite forceful evidence attesting to *Azolla's* beneficial effects. A thick mat of *Azolla* will suppress weed development in rice paddys (Lumpkin and Plucknett, 1980) and not affect rice. In some rice fields, the benefit from *Azolla* weed suppression may even surpass that from nitrogen fixation (Lumpkin and Plucknett, 1980). Some weeds can, of course, push through the *Azolla* mat, but that does not lessen its suppression of others. It is also useful as a fodder crop because of its high growth rate and protein content (Lumpkin and Plucknett, 1980). Lumpkin and Plucknett (1980) reported that it is useful to man because of its ability to fix atmospheric nitrogen in what are commonly nitrogen-deficient water communities, i.e., rice paddies. It also will grow in a flooded paddy condition where traditional nitrogen-fixing green manure crops will not grow. Its secondary potentials include use as a fodder crop for carp and pigs, as a compost for upland crops, and possibly as a photosynthetically driven nitrogen, hydrogen, or protein factory.

There seems to be no question that *Azolla* has weedy characteristics, and if allowed to spread in an uncontrolled fashion, it would become an important aquatic weed in the southern United States. Holm *et al.* (1979) do not report it as a serious weed in any country, but it is a principal weed in India and Thailand. However, it also has important beneficial characteristics, and these may outweigh its weedy potential.

D. *Borreria alata* (Aubl.) DC. (Rubiaceae)

Synonymous names for this annual weed include *B. latifolia* (Aubl.) Schum. and *Spermacoce latifolia* Aubl. It is found in Asia and South America.

The weed is native to the West Indies and South America and is now spreading in Southeast Asia.[3] It is ranked as a serious weed in Borneo and Indonesia, is common in Senegal, and is unranked in 17 other countries including Mexico. It

[3]L. G. Holm (personal communication, 1980).

becomes established easily in open sandy places and is a weed in cultivated fields, rubber plantations, tea, coconut, and pineapple. Fourteen other species are known in this genus, and none are important weed problems in any of the countries where *B. alata* has been identified as a weed.

E. *Chromolaena odorata* (L.) R. M. King and H. Robinson (Asteraceae: Jack-in-the-Bush)

This is an aggressive, perennial, spreading shrub which branches profusely to form tangled thickets (Holm *et al.*, 1977). It grows from 3 to 7 m high and has a deep, massive taproot. The taproot is the vegetative reproductive organ and regenerates a new plant following cutting or burning. It is not a creeping perennial.

There is a profuse production of airborne seeds. There is also evidence of production of plant growth inhibitors and insect repellent oils by the plant, which reduces competition and attack by insects (Gunn *et al.*, 1981). The weed is an aggressive competitor for nutrients in cotton, corn, rice, sugarcane, tobacco, rubber, tea, teak, coconuts, and pasture. It is able to replace grasses in open grasslands. It may also be responsible for livestock poisoning at certain times of the year. As its name implies, the leaves have a pungent odor when crushed. It grows on many soil types but seems to prefer well-drained sites with adequate sunlight or partial shade. It does not survive vigorous competition for light (Holm *et al.*, 1977).

Chromolaena odorata has been referred to as *Eupatorium odoratum,* and Holm *et al.* (1977) list it as a serious weed in India and a principal weed in Australia and Thailand. It is present as a weed in five other countries and Hawaii. It has been found in the continental United States and is regarded as a widely adventive species along the coast from Florida to Louisiana and in southeast Texas (Gunn *et al.*, 1981; King and Robinson, 1970). It is not regarded as an important weed in the United States at the present time. Its aggressive habit and perennial nature, its ability to assume dominance over many grassy species, and its already established reputation for weediness in several important crops lead one to believe that it could be an important problem in the United States.

F. *Commelina benghalensis* (Commelinaceae: Wandering Jew/Tropical Spiderwort)

This is a somewhat fleshy or succulent creeping annual or perennial herb often referred to as *C. africana* L.

Commelina benghalensis is present as a weed in Hawaii but not in the continental United States. It is a weed in Africa, except northern Africa, and also in southern Asia, Southeast Asia, East Asia, Australia, and the Pacific Basin. Thus, it is really a weed found throughout the tropics from sea level to an elevation of 1300 m.

Because the plant roots at nodes, cultivation can be detrimental to control because it cuts up the stems, spreads them, and produces more plants. The plant has a great ability to grow readily from vegetative cuttings. King (1966) reports that several species of *Commelina* are classed as wetland hydrophytes, or plants which are rooted in soil that is water saturated. It grows best under conditions of high soil moisture and fertility but will persist in sandy, rocky, or dry soils even under very dry conditions and then grow rapidly with the onset of rain (Holm *et al.*, 1977).

Its importance as a weed is due to its great persistence in cultivated land and the difficulty of control. These characteristics are aggravated by its adaptation to several different soils and soil moisture conditions and its ability to compete vigorously under these variable conditions. The stoloniferous growth forms dense stands which smother out several crop plants, especially low-growing ones. According to Holm *et al.* (1977), it is one of the three most important weeds in coffee in Tanzania, and it is one of the principal weeds of corn in Kenya. It is important as a weed in sugarcane in South Africa, coffee in Kenya and Tanzania, upland rice in India and the Philippines, cotton in Kenya, tea in India, wheat in Angola, and soybeans in the Philippines (Holm *et al.*, 1977). It is ranked as a serious weed in Borneo, India, Mozambique, the Philippines, Tanzania, and Zambia. It is a principal weed in Thailand, Indonesia, and six African countries (Holm *et al.*, 1979).

Commelina diffusa Burman F. is already present as a weed in the southern United States up to Virginia and in Ohio, Indiana, Illinois, Missouri, and eastern Kansas. *Commelina benghalensis* is reported as a weed in 25 crops in 28 countries (Holm *et al.*, 1979), and the presence in this country of *C. diffusa* is certainly evidence that if it spreads to this country, it could become as important a weed here as it is in other parts of the world.

G. *Digitaria scalarum* (Schweinf.) Chiov. (Poaceae: African Couchgrass/Fingergrass)

This creeping, rhizomatous perennial is widely distributed in relatively moist, shady places and roadsides from sea level up to 3300 m in East Africa (Reed and Hughes, 1977). According to Holm *et al.* (1977), it may be the worst weed of the important crops of that area, including coffee, cotton, sisal, sugarcane, tea, pineapple, flax, and wheat. Ivens (1967) regarded it as the most troublesome of all East African weeds. It is palatable for cattle when young, but apparently it is not productive enough to be used for grazing. Once established, it is nearly impossible to control by cultivation because of its extensive underground rhizome system.

Holm *et al.* (1977) reported that this weed is distinctly set apart among the world's important weeds because of its narrow regional distribution in East Africa. It has not yet spread beyond East and South Africa but is present in India (Holm *et al.*, 1979). The leaves are softer than those of *Cynodon dactylon* and are much darker green. The presence of the membranous ligule also differentiates this plant from *Cynodon* (Ivens, 1967). According to Holm *et al.* (1977), it is closely related to *Digitaria decumbens* (Pangola grass), which is an important grass in South Africa and is becoming one of the important pasture grasses of the tropical and subtropical regions throughout the world. Holm *et al.* (1977) further reported that the plant appears to become dominant when soil fertility declines. There have been reports that *D. scalarum* replaces *Pennisetum clandestinum,* which is also an aggressive, creeping perennial grass that spreads by stolons and rhizomes and is an important weed in its own right. It is a vigorous competitor and, like most rhizomatous grasses, is extremely difficult to control by hand or with tillage equipment.

H. *Digitaria velutina* (Forsk.) Beauv. (Poaceae: Velvet Fingergrass/Annual Couchgrass)

Digitaria velutina is a short-lived annual weed of arable land and frequently invades pastures and cultivated fields in tropical Africa. It is present across the western, eastern, and southern parts of Africa.

The weed grows in soil types ranging from sand to volcanic ash to clay loam. It grows in areas where rainfall averages as much as 197 cm/year and in areas where rainfall is very seasonal. It is a weed in coffee, pineapple, rice, tea, wheat, cotton, peanuts, sorghum, corn, barley, and beans, and is troublesome in young rubber plantations. It is reported to be a serious weed in the Ivory Coast, Kenya, Tanzania, and Uganda and a principal weed in Senegal (Holm *et al.*, 1979). In areas where herbicides have been used to reduce broadleaf competition in crops, *D. velutina* is one of the first annual grasses to move in. Because of the ubiquity of niches for weedy annual grasses in the U.S. cropping system, the escape of this weed from Africa to the United States could once again complicate the already difficult problem of the control of annual weedy grasses in our crops.

I. *Drymaria arenarioides* Humboldt and Bonpland ex Romer and Schultes (Caryophyllaceae: Lightning Weed)

This is a short-lived, perennial, prostrate, herbaceous plant. The primary reason for concern about this plant is not its widespread distribution but its proximity to the United States. It is well established in North Sonora and Chihuahua in Mexico and extends south to central Mexico. It inhabits hill and plains areas. Very small amounts are lethal to cattle, sheep, and goats (Gunn *et al.*, 1981). Because of the importance of the cattle industry in the areas of the United States bordering Mexico, this weed presents a very real threat if imported.

J. *Euphorbia prunifolia* Jacq. (Euphorbiaceae: Painted Euphorbia)

In their 1979 "Geographical Atlas of World Weeds," Holm *et al.* (1979) listed 80 different species of *Euphorbia* as weeds of the world. At least 18 of those species are already present as weeds in the United States. A presentation of this kind would be remiss if the Euphorbiaceae were not at least examined for their potential threat. Unfortunately, little information is available on the many species which are not present in the United States.

Euphorbia prunifolia is also known as *E. geniculata* and perhaps incorrectly as *E. heterophylla,* which may now be considered a separate and far more important species (Holm *et al.*, 1979; Reed and Hughes, 1977). *Euphorbia prunifolia* is a serious weed in rice, cotton, peanuts, and corn. It has been known to invade areas where *Imperata cylindrica* has been controlled (Gunn *et al.*, 1981). Holm *et al.* (1979) list it as a serious weed in Indonesia, the Philippines, and Thailand and a principal weed in Brazil, India, Melanesia, and Nigeria.

Given the ubiquity of Euphorbiaceae throughout the world, there is every reason to suspect that this weed and many other Euphorbiaceae could become important weeds if transferred to new places.

K. *Imperata brasiliensis* Trin. (Poaceae: Brazilian Satintail)

This aggressive rhizomatous perennial grass is similar to and often confused with *I. cylindrica,* which is now widespread in the United States, having been introduced into southern Alabama (Tabor, 1952). The two are similar, but there is no guarantee that they will respond in the same manner to control measures.

One of the characteristics of this plant which makes it a good weed is the spikelet, which is long (4.5 mm) and extremely fragile, thus scattering seeds easily on the soil. The caryopsis is 1.5 mm long, lanceolate, and striate, with pale yellow-amber color grading to a dark red-brown at the two apices. The plant has been reported as present in Florida and Alabama but apparently is not yet recognized as an important weed (Gunn *et al.*, 1981). It infests perennial crops but is itself of little forage value. It is a principal weed in Trinidad and is found in Argentina, Brazil, and Bolivia.

When choosing the weeds to include in this chapter, it was most difficult to exclude *I. cylindrica,* which Holm *et al.* (1977) reported as a serious weed in 18 countries and a principal weed in six others. It has spread since its introduction to the southern United States but is still mainly a weed of noncultivated areas (Dickens, 1974). Holm *et al.* (1977) classified it as one of the world's worst weeds and found that 73 countries reported it as a weed in 35 different crops. Why hasn't it spread more in the United States? Will it spread? We don't know. All experiential evidence and common sense lead us to conclude that we should limit its spread and, if possible, preclude the introduction of close relatives (such as *I. brasiliensis*) that may become serious weeds and offer opportunities for interspecific hybridization. This process could lead to dead ends or it could create new, better adapted, phenotypically similar, and more vigorous weed ecotypes.

L. *Ipomoea aquatica* Forsk. (Convolvulaceae: Swamp Morning-Glory/Water Spinach)

Synonymous names for this tropical and subtropical species include *I. reptans* (L.) Poir., *I. repens* Roth, and *I. subdentata* Miq. There is good evidence that this plant, like many others, is not always a weed (Edie and Ho, 1969). It is one of the few aquatics grown as a green vegetable (Boyd, 1974) and supplies up to 15% of the local vegetable output per season in Hong Kong (Edie and Ho, 1969) and presumably elsewhere in Asia. There is also no doubt that it possesses the appropriate attributes of a good weed and will be competitive in many aquatic and some terrestrial environments in the United States. It is regarded as a serious weed in India, Mozambique, and Thailand, and a principal weed in three other countries. It is apparently present in Florida but is not regarded as a weed problem (Gunn *et al.*, 1981).

M. *Ischaemum rugosum* Salisb. (Poaceae: Murainograss/Saromaccagrass)

Ischaemum rugosum is a serious weed in rice because the seeds mature at the same time as rice and are harvested with the crop. It has been difficult to control and eradicate (Parham, 1958). This annual grass is a native of tropical Asia and a weed of warm, humid tropical regions (Bor, 1960). The plant is universally categorized as an aggressive and variable species found in wet and dry habitats (Bor, 1960).

Ischaemum rugosum is a serious weed in rice in Sri Lanka, India, Madagascar, and Thailand, and is also important in Fiji, Surinam, and four other countries (Holm *et al.*, 1979). It is also found in South America, parts of western Africa and the Philippines (Holm *et al.*, 1977). Holm *et al.* (1977) reported that it is a special problem in many parts of the world because its vegetative habit is similar to that of rice; it is not recognized as a weed and removed at the time of hand weeding. The plant has been used as a fodder grass when young, but is not a good pasture grass because it is an annual (Parham, 1958). The seed has been eaten as grain in India during times of famine (Bor, 1960; Reed and Hughes, 1977).

There is no question that this weed could invade rice crops in the United States. Whether it could move into other crops in the warmer regions of the United States is at this point unknown, but there is every reason to suspect that it could.

N. *Mikania cordata* (Burm. f.) B. L. Robins (Asteraceae: Mile-a-Minute) and *Mikania micrantha* HBK (Asteraceae)

These species of the large genus *Mikania* are related to *M. scandens,* which already exists as a weed problem in the United States and is strictly a North American species (Robinson, 1934). Most of the species of this genus are herbaceous or slightly woody vines, but they are quite variable. The two species mentioned are both rapidly growing, creeping and twining perennial vine-type plants. They each reproduce vegetatively by stem fragments and seeds.

Because *M. scandens* is present, the United States already has a problem, which will undoubtedly be complicated if the other two species invade. *Mikania cordata* is probably native to Southeast Asia and East Africa and is widespread in those areas and in the Pacific Islands (Holm *et al.,* 1977). The species is presently confined to the tropical regions of the world, where it is found in many different places. *Mikania micrantha,* on the other hand, is widespread throughout South America and Central America but is weedy in only a few places (Holm *et al.,* 1977). It has also been reported as a serious weed in Fiji, India, and Polynesia. Parker (1972) reported that *M. micrantha* has been identified in several countries of Asia and suggested that it may be the most aggressive of the three species.

Holm *et al.* (1977) considered *M. cordata* to be the most important weed of the genus *Mikania.* They identified it as a weed of plantation crops and perhaps most serious in rubber and tea. It is also reported as a weed problem in coffee, coconuts, cacao, oil palms, and bananas. It may have the ability to grow more rapidly than other species of this genus and can grow from very small stem fragments. The ability to control it by hand or with tillage is limited. The combination of perennial characteristics, ability to reproduce from small stem fragments (Holm *et al.,* 1977), a creeping, twining, climbing habit, great competitive ability, wind dispersal of seed, and the apparent adaptability for competition in several crops make this weed a serious threat to most agricultural environments. Although the evidence shows that it grows especially well in tropical environments, we have no direct evidence that it will not grow in the warmer temperate zones of the world.

O. *Mimosa invisa* Mart. (Leguminosae: Giant Sensitive Plant)

Relatives of this plant are present in the United States. Most students of botany have observed that when it is touched, the bipinnate leaves immediately fold shut. Botanists have puzzled over this action for many years. Weed scientists have not spent equal time worrying about its weedy potential. It is a native of Brazil and is present in most of South America. It can behave as a biennial but perhaps more commonly as a perennial. The seeds are borne in spiny pods and are adapted for external transport. It vigorously competes with other vegetation, its masses of spiny stems make it nearly impossible to control by hand, and it forms impenetrable thickets in which neither animals nor man care to move about.

Cutting is one method often recommended for control of this semiwoody species, but it fails because rapid stump regeneration occurs from buds on the stem at the base of the plant. Seed germination is enhanced by burning. *Mimosa invisa* is called ''giant sensitive plant'' to differentiate it from its more common and familiar relative *Mimosa pudica* (''sensitive plant''). It is less sensitive to touch than *M. pudica.* Holm *et al.* (1977) reported it as a weed in 13 crops in 18 countries. It is a serious weed in Borneo, Fiji, Malaysia, Melanesia, Nigeria, the Philippines, Polynesia, and Taiwan (Holm *et al.,* 1979). Apparently, it has not yet been reported in most of Asia, Africa, Europe, or North America, whereas *M. pudica* has been (Holm *et al.,* 1979). It is an important weed of plantation crops such as rubber, coconuts, sugarcane, and tea. It is also a weed in cultivated crops, such as upland rice, cassava, soybeans, and corn. The vigor of its growth indicates that it will be able to compete effectively against many of our presently serious weeds once it becomes established in areas such as the warm pasturelands of the southeastern United States.

Its habit of growth and the obvious inability to control it by hand or mechanical methods suggest reliance on synthetic, organic herbicides which are presently not available for the selective control of this weed in important agricultural environments in the United States. This is not to say that these herbicides do not exist, but only that the research to determine the sensitivity of the weed to them and their selectivity in the crops has not been done.

P. *Monochoria hastata* (L.) Solms (Pontederiaceae: Pickerel Weed) and *Monochoria vaginalis* (Burm. f.) Presl. (Pontederiaceae: Monochoria)

A monochord is a musical instrument composed of a sounding board with a single string. It was used in the eleventh century in singing schools to teach the intervals of plain song. Later, it was used as an instrument for the mathematical determination of musical intervals. Thus, the generic name *Monochoria* means "single chord" or "single string," and indeed, plants of this genus have only one stamen. They are natives of tropical Asia and Africa and grow in subaquatic or marshy sites.

According to Holm *et al.* (1977) *M. hastata* is a weed of paddy rice occurring in Korea and Japan through the Pacific Islands to the mainland to Southeast Asia, and across to India. It is a member of the same plant family as the water hyacinth, *Eichhornia crassipes,* and its weedy potential is not to be doubted. In Taiwan, *M. vaginalis* produced higher fresh weight yields in fields of paddy rice than any other weed found in those fields (Lin, 1968). It produced more dry matter than *Echinochloa crus-galli.* In other studies in the Philippines, it was not as competitive as *E. crus-galli* (Lubigan and Vega, 1971). It apparently competes more vigorously for nutrients than for light because it is low-growing and roots in the mud but nevertheless is a vigorous competitor in rice culture. *Monochoria vaginalis* was reported to be a serious weed in Borneo, Indonesia, Japan, Korea, the Philippines, and Taiwan (Holm *et al.*, 1979). It is a principal weed in three other countries and Hawaii. *Monochoria hastata* is a serious weed in Borneo and Fiji and a principal weed in three additional countries (Holm *et al.*, 1979).

Q. *Pennisetum polystachion* (L.) Schult. (Poaceae: Thin Napier Grass)

Holm *et al.* (1977) cite 14 species of *Pennisetum* which are important as weeds in the world. Of these, *P. clandestinum* and *P. purpureum* are already present as weeds in the United States, but the extent and importance of the infestation is unknown. *Pennisetum clandestinum* and *P. polystachion* resemble *P. purpureum* in growth habits. Each can behave as a perennial, although *P. polystachion* often behaves as an annual. They are large, upright bunch grasses that have been spread widely throughout the world as fodder and pasture crops (Holm *et al.*, 1977). These grasses are natives of tropical Africa, where they spread and colonize wastelands and infest perennial tropical crops. They do best in tropical wet conditions but will survive drought and fires (Holm *et al.*, 1977).

Infestations of any one of these species can render land nearly unsuitable for cultivation and crop production. They are weeds of the tropics and, according to Holm *et al.* (1977), rarely extend beyond latitude 23° north and 23° south. However, a recent report indicates that *P. pedicellatum* may be found in Florida (Gunn *et al.*, 1981), and there is no reason to suspect that it, like many other grasses typically regarded as tropical, cannot and will not spread throughout the agricultural area of the southern United States. One or more of the species has often been introduced as a pasture grass, from which it spreads as a weed. Holm *et al.* (1977) reported that *P. pedicellatum* is a serious weed in Nigeria and Thailand and a principal weed in Australia. *Pennisetum polystachion* is a serious weed in Thailand, a principal weed in India, and present in eight other countries (Holm *et al.*, 1979).

R. *Saccharum spontaneum* L. (Poaceae: Wild Sugarcane)

This weed is widely distributed in the warmer regions of the Old World and is often cultivated as a good soil-binding crop. It is a tall, dense, tufted, rhizomatous perennial grass with deep-spreading roots. Bor (1960) reported that it flowers and fruits at the end of the rainy season and therefore is capable of colonizing many areas left bare after this season has ended. It can be controlled by burning during the dry season, but it is replaced by other weedy species. Bor (1960) also suggested that the species is quite variable, and some races may merit specific rank.

Holm *et al,* (1979) rank *S. spontaneum* as a serious weed in Indonesia, India, and Thailand and as a principal weed in the Philippines and Puerto Rico. It is present as a weed in 20 additional countries ranging from Asia to Africa to the USSR and the Arabian Peninsula. It should not be considered as a weed problem only in sugarcane, even though it is a member of the same genus and is referred to by the common name "wild sugarcane." Its vigor, great height, and perennial rhizomatous nature lead one to suspect that it could be a vigorous competitor in many crops. Its present distribution in the warmer regions of the world is no reason to suspect that it could not compete in the more temperate zones. Its allelopathic potential in interfering with the growth of three wheat varieties has been suggested (Amritphale and Mall, 1978).

S. *Sagittaria sagittifolia* L. (Alismataceae: Arrowhead)

This annual or perennial is an emergent aquatic monocotyledonous weed. There are about 12 other weedy species, but the other important weedy member of the family is *S. guayanensis* HBK.

This is a weed of swamps, reservoirs, rice paddies, canals, and lakes. It is acknowledged as a principal weed in Italy, Sweden, Germany, Hawaii, and Taiwan, and is found in other places in Europe (Holm *et al.*, 1979). It is present as a weed in such diverse places as India, China, Australia, Argentina, England, Norway, and 16 other countries. Its presence in diverse regions of the world enhances the opportunity for spread. At the present time, neither of the two important species is in the United States, but at least six other species are present as aquatic weeds (Holm *et al.*, 1979; Tarver *et al.*, 1978).

T. *Salvinia molesta* D. S. Mitchell (Salviniaceae: Salvinia)

This weed is part of the *S. auriculata* complex, which includes *S. auriculata* Aublet, *S. bilobia* Raddi, *S. herzogii* de Lasota and *S. molesta* D. S. Mitchell.

This complex was first recorded as a weed in East Africa when it was accidentally introduced in a lake near Kitale in Kenya. Within months, it had completely covered the lake surface (Ivens, 1967).

According to Mitchell (1974), *S. molesta* was first seen on Lake Kariba, which forms part of the border between Rhodesia and Zimbabwe, in May 1959, 5 months after closure of the dam. By 1962, Mitchell reported that it covered an area of more than 1000 km^2 with a mat up to 10 in thick (Ivens, 1967). By 1963 the area of Lake Kariba suitable for colonization by *S. molesta* was limited mainly by wave action, which served as a partial control. Mean relative growth rates range from 8.61% to 4.85% per day in terms of increase in leaf number and 6.8% to 4.0% per day in terms of dry weight. Its doubling time in the field is reported to be as low as 4.6 days (Mitchell, 1974). It is regarded as a serious weed in Sri Lanka, Guyana, Indonesia, and Zambia. It is a principal weed in India, Brazil, and three African countries.

This weed is one of the free-floating hydrophyte fern allies which has become very specialized through evolution and no longer resembles a fern. The free-floating habit leads to dispersal by wind and water currents or by the action of man and animals. *Salvinia molesta* has successfully escaped the necessity for sexual reproduction and, as a result, is a more serious weed threat. The vegetative propagules are more capable of surviving adverse environmental conditions than is the plant itself and recommence growth when conditions are favorable. It behaves as an annual or a perennial. The slender, horizontal branched stems are usually 5–10 cm long but are capable of growing up to 30 cm long. Each stem has numerous leaves (fronds) produced in groups of three, each with two broad-leaved, ovate, entire, undivided green aerial leaves with a distinct midrib from the base to the apex. The upper surface is covered with many close parallel rows of numerous hairs that terminate in a cagelike, club-shaped tip and prevent the leaves from getting wet. It has been described as a bird cage arrangement with groups of four hairs on the leaf surface. The lower leaf surface is smooth except for some simple hairs, which Holm *et al.* (1977) suggested may be rootlets. The third submerged rootlike water leaf is situated ventrally, is much divided, feathery, brown, up to 25 cm long, and resembles and functions as a true root. This leaf bears the sporocarp which is the best way to distinguish the species in the *S. auriculata* complex. The sporocarps are globose, 2–3 mm in diameter on a short 1-mm stalk. They are densely hairy, indehiscent, and monoecious, and occur in clusters on the stem at the base of the submerged leaf (Holm *et al.,* 1979; Ivens, 1967; Reed and Hughes, 1977). The numerous leaves form mats up to 2.5 cm thick, but under favorable conditions much thicker mats form (Ivens, 1967).

Although the normal habitat of this weed as a free-floating plant is aquatic, it is capable of surviving on moist soil in a reduced form, particularly in humid conditions (Mitchell, 1974). Mitchell (1974) reported that *S. molesta* has grown on the edge of a rain forest opposite the Victoria Falls above Lake Kariba, where it is kept continually moist by spray from the falls. Therefore, one might conclude that it is ideally adapted to infesting and surviving in irrigation reservoirs and ditches. It apparently will not survive cold winters, but it certainly should be regarded as a threat to agriculture in tropical and temperate zones. It is unlikely that man-made irrigation canals and the system for which they are designed are capable of incorporating fluctuations in water level sufficient to control an aquatic weed such as *S. molesta*.

Mitchell (1974) reported that aquatic plants have provided some of the most spectacular examples of the adverse consequences of invasions and subsequent population explosions by undesirable weedy species. The aquatic plant also has an avenue for spread that is not as commonly available to terrestrial species: the worldwide network of people who deal in aquaria and aquarium supplies. It is also true that pisciculturists have spread aquatic weeds as they transfer fish from place to place. No one would advocate the complete restriction of trade in either of these areas. Advocacy of greater awareness of the possibilities for transport of undesirable weedy species should be accepted by all. We do know that *S. molesta* does not survive in salt water; therefore, it will not be of concern in estuaries or coastal areas. It is capable of reproducing by spores, but does so rarely and relies instead on vegetative reproduction, which facilitates its transport. Holm *et al.* (1977) reported that it has never been a cause of serious concern in waterways in South America, but they provide no explanation for this lack of concern. One might assume that given its characteristics, it would be as serious a problem in South America as it has been in Asia and Africa. Particularly in Africa, *S. molesta* has been capable of disrupting food production and man's mobility, work, and health (Holm *et al.*, 1977). It also disrupts power production, fishing, and water transport for irrigation. Furthermore, when ditches and waterways become clogged, an environment is created in which vectors of human and animal disease thrive. *Salvinia molesta* is, of course, an important problem in rice and should be regarded as a potential rice weed in the United States. We probably should be more concerned about it because of its potential as a weed in waterways as well as in crops.

U. *Setaria pallide-fusca* (Schumach.) Stapf & Hubb. (Poaceae: Cat-Tail Grass)

It is apparent that the phenotype plasticity within the *Setaria* genus is causing problems in specific identification of this plant. It is phenotypically similar to *S. lutescens* and *S. viridis*. There is also some problem with spelling of the specific name, which is often listed as *S. pallidifusca, S. pallide-fuscum,* or *S. pallidafusca*.

Setaria pallide-fusca is an important weed in corn fields and pastures and is widely distributed in the tropics of the world. Holm *et al.* (1979) list it as an important weed in Senegal, Sudan, Uganda, and Zambia. It is a principal weed in Fiji, India, Kenya, Mauritius, and Zimbabwe. Animals will graze on it when the plants are young, and it may be used as fodder, although its annual habit precludes its use as a permanent pasture grass. The United Nations Food and Agriculture Organization recommends it as a grass cover crop in Africa (Gunn *et al.*, 1981). It has been identified in the United States in Louisiana but is not recognized as a weed in this country. Given the problems caused by the *Setaria* species that are already in the United States and the potential for hybridization, every effort should be made to keep this species from entering.

V. *Tridax procumbens* L. (Poaceae: Tridax Daisy/Coat Buttons)

The genus *Tridax* contains about 26 species (Powell, 1965), which are restricted to Central America and adjacent territories. However, *T. procumbens* has become a prominent weed throughout the tropics, and Baker *et al.* (1972) are studying it to find out why it has been able to spread when its congeners are so restricted. One might assume that it is similar to *Ageratum conyzoides* (Table II).

Tridax procumbens is another example of an exception to the rule set forth in the introduction of this chapter. This species has been found on all continents and is present in southern Florida (Gunn *et al.*, 1981; Long and Lakela, 1971). However, Holm *et al.* (1979) do not cite it, nor has anyone reported it, as a particularly serious weed in the United States. Holm *et al.* (1979) reported it as a serious weed in five countries (Ghana, Ivory Coast, Mozambique, Nigeria, and Thailand) and as a principal weed in six others. It is present in 48 countries. It may also be serious in India, Taiwan, and several areas of East Africa and Asia. It grows over a wide range of soils and elevations. It is commonly found in fallow and waste areas and neglected pastures. It most commonly grows in dry situations (Parham, 1958). Cattle will eat *T. procumbens* but the numerous hairs on the plant deter grazing. They also serve as an aid to dissemination of the seed. If an animal bumps the plant or pulls on it, or if it is pulled by hand, it breaks off at a basal node and leaves the root system intact. It is competitive in upland rice, corn, sugarcane, pineapple, cotton, peanuts, soybeans, cowpeas, sweet potatoes, tea, bananas, sorghum, cassava, and sisal.

Some aspects of intraspecific competition have been studied and show that increasing population density leads to reduced growth and less flowering (Oladokun, 1978). The range of crops infested by *T. procumbens* and its competitive ability give it great potential as a weed in the United States.

V. CONCLUDING COMMENTS

A brief review of Table III will show that the primary sources of weeds selected as potential threats to U.S. agriculture are Asia and Africa. Even the unknowing might assume this intuitively. One suspects that some equilibrium has been reached among Europe, other developed regions of the world, and the United States, because of the extensive commerce and human interchange that have occurred for so many years. Although there has been an increasing level of commercial and human interchange among Asia, Africa, and the United States, it has not been on the scale of that with the European continent. We expect such interchanges to increase and also recognize a great diversity of flora in those areas, many species of which are largely unknown in the United States.

The diversity of the selected weeds is also obvious. Among those selected are 12 monocots, 10 dicots, and two ferns. It is interesting to note that seven are aquatic species. Most weed scientists in the world work with terrestrial species which infest agronomic or horticultural crops. Weed scientists have given too little attention to aquatic species and their control. As water becomes more limiting in arid regions because of urban and industrial demands, we may find that we are less able to allow aquatic species to utilize water intended for agricultural, industrial, or urban consumption.

Of the 24 weeds selected for inclusion in this chapter, nine are already listed in the Federal Noxious Weed Act and 12 are recommended for inclusion (Gunn *et al.*, 1981). The three other weeds (*Asphodelus fistulosus, Chromolaena odorata,* and *Setaria pallide-fusca*) may be considered for inclusion in the future.

It will inevitably be asked on what authority the weeds were chosen. Some of that authority is evident in the references selected for inclusion, but, as is stated at several points in the chapter, much of the authority is based on personal decisions and information from colleagues. Any of the weeds included in this chapter could fail to be important to the United States or, if imported, could prove to be of little or no consequence. The evidence suggests that for the included species this would not be true, but many of the guesses, even educated guesses, that man makes about the behavior of the natural world have proven to be incorrect. For example, 10 of the species selected are perennials and 6 sometimes behave as perennials. One must ask if the definition of perennial, as we know it in the temperate zone, is applicable to weeds that originate in the tropical areas of the world. There is no doubt that many tropical plants live for a long time, but are they simply persistent annuals rather than perennials, as we know them, which escape the rigors of changing seasons and frozen soils? If these plants are introduced to the United States, will they become only troublesome annuals or will their perennial nature persist? We don't know in all cases, but the rigors of some environments may reduce the presumed perennial threat.

The example of interspecific hybrids was cited earlier in the chapter with reference to the weed shattercane [*Sorghum bicolor* (L.) Moench.]. This may, in fact, be a greater threat to U.S. agriculture than the mere presence of a new weed. Evolution and genetic plasticity may lead to the development of species uniquely adapted to the new environment in which they have been placed, probably through the agency of man.

However, one must not assume that everything that is important is bad or has the potential to become bad. A recent article (Musselman and Parker, 1981) on *Striga gesnerioides* shows that the host range of this plant is very limited and that no major crop species (including tobacco, pigeon pea, sweet potato, cotton, green pepper, celery, chickpea, rape, broad bean, and a few others) were seriously affected. The authors concluded that *S. gesnerioides* (indigo witchweed) poses little danger to American agriculture. The weed is common throughout much of Africa, the Arabian Peninsula, and the Indian subcontinent. Holm *et al.* (1979) listed it as a principal weed in Nigeria and an important weed in Rhodesia and South Africa. It is an important weed in tobacco in South Africa. Genetic interchange is unlikely to produce altered host specificity (Musselman and Parker, 1981). But this possibility is particularly remote in the case of the American strains because of their complete isolation from other strains. Therefore, one must not conclude that ''exotic'' equals ''dangerous'' or that importation of an exotic plant automatically leads to increased weed losses in the United States. On the other hand, one cannot conclude that all exotics are innocuous. The recent example of *Crupina vulgaris* Cass. in Idaho is illustrative. It was introduced from the Mediterranean area and first reported in Idaho in 1968, and now infests 23,000 acres of rangeland.[4]

Thus, decisions about exotic threats are difficult and controversial. Science is not a prophetic enterprise, and attempts to predict the future are inevitably prone to error.

This brief exposition of potential threats intentionally excludes any mention of control of these weeds. Part of the reason is that many control measures are largely unknown, and we do not know for sure whether the control would be needed if the weeds invaded.

[4]D. Thill, University of Idaho, *Weed Topics,* March 1981.

APPENDIX 1

Some Introduced Cultivated Plants That Have Escaped and Become Weeds in Some Sections of the United States[a]

	Annuals	
O[b]	*Centaurea cyanus*	Bachelor's button
O	*Chenopodium botrys*	Jerusalem oak
O	*Ipomoea purpurea*	Morning glory
O	*Kochia scoparia*	Summer cypress
O	*Nicandra physalodes*	Apple of Peru
V[c]	*Raphanus sativus*	Wild radish

APPENDIX 1 *Continued*

	Biennials	
H[d]	*Carum carvi*	Caraway
D[e]	*Chelidonium majus*	Celandine
D	*Conium maculatum*	Poison hemlock
O, D	*Digitalis purpurea*	Foxglove
V	*Pastinaca sativa*	Wild parsnip
	Perennials	
H	*Artemisia absinthium*	Wormwood
O	*Bellis perennis*	English daisy
OS[f]	*Berberis vulgaris*	Barberry
O	*Campanula rapunculoides*	Bellflower
V	*Chenopodium bonus-henricus*	Good King Henry
V	*Cichorium intybus*	Chicory
OS	*Cytisus scoparius*	Scotch broom
O	*Euphorbia cyparissias*	Cypress spurge
OS	*Genista tinctoria*	Dyers broom
O	*Hibiscus trionum*	Flower-of-an-hour
D	*Inula helenium*	Elecampane
O	*Knautia arvensis*	Field scabious
D	*Leonurus cardiaca*	Motherwort
O	*Lonicera japonica*	Japanese honeysuckle
O	*Lysimachia nummularia*	Moneywort
O	*Malva moschata*	Musk mallow
H	*Marrubium vulgare*	Horehound
H	*Mentha spicata*	Spearmint
H	*Nepeta cataria*	Catnip
O	*Glechoma hederacea*	Ground ivy
O	*Ornithogalum umbellatum*	Star-of-Bethlehem
O	*Physalis alkekengi*	Chinese lantern plant
O	*Polygonum cuspidatum*	Japanese knotweed
OS	*Rosa eglanteria*	Sweet brier
H	*Rumex acetosa*	Sour dock
O	*Saponaria officinalis*	Bouncing Bet
O	*Sedum acre*	Stonecrop
O	*Sedum purpureum*	Live-for-ever
D	*Symphytum officinale*	Comfrey
D	*Tanacetum vulgare*	Tansy
O	*Veronica filiformis*	Veronica

[a] From Muenscher (1955).
[b] O = ornamental.
[c] V = vegetable.
[d] H = herb.
[e] D = drug plant.
[f] OS = ornamental shrub.

APPENDIX 2

Purposeful Plant Introductions That Have Become Weeds in the United States[a]

Species	Purpose of introduction	Weed type	Problem areas	Toxic principle
Belladonna *Atropa bella-donna* L.	Herb	Annual	Roadsides, waste areas	Atropine
Bermudagrass *Cynodon dactylon* (L.) Pers.	Forage	Perennial	Pastures	None
Bouncingbet *Saponaria officinalis* L.	Ornamental	Perennial	Pastures	Saponins
Brazilian peppertree *Schinus terebinthifolius* Raddi.	Ornamental	Perennial	Parks, forests, yards	Irritants
Buckthorn *Rhamnus* spp.	Ornamental	Perennial	Grazing areas	Anthraquinone
Chinaberry *Melia azedarach* L.	Ornamental	Perennial	Grazing areas	Unknown
Cogongrass *Imperata cylindrica* (L.) Beauv.	Forage	Perennial	Southern farms	None
Corn cockle *Agrostemma githago* L.	Ornamental	Annual	Wheat, grasslands	Saponins
Crotalaria *Crotalaria spectabilis* Roth *Crotalaria retusa* L.	Green manure, hay	Annual to perennial	Ranges, waste areas, soybeans	Monocrotaline
Dalmation toadflax *Linaria genistifolia* (L.) Mill subspp. *dalmatica* (L.) Maire & Petitmengin	Ornamental	Perennial	Rangeland	None
Dyers woad *Isatis tinctoria* L.	Dyes	Biennial	Rangeland, crops	None

Foxglove *Digitalis purpurea* L.	Ornamental	Biennial	Pastures, waste areas	Aglycones
French tamarisk *Tamarix gallica* L.	Ornamental	Perennial	Pastures, floodplains, waterways	None
Goatsrue *Galega officinalis* L.	Forage	Perennial	Pastures, canal banks	Galegin
Hemp *Cannabis sativa* L.	Fiber, medicine	Annual	Pastures, waste areas	Tetrahydrocannabinol
Henbane *Hyoscyamus niger* L.	Medicine	Annual or biennial	Roadsides, waste areas	Atropine, scopolamine, hyoscyamine
Hydrilla *Hydrilla verticillata* (L.f.) Royle	Aquarium trade	Perennial	Lakes, reservoirs, waterways	None
Japanese honeysuckle *Lonicera japonica* Thunb.	Ornamental	Perennial	Wooded areas, pastures	None
Japanese knotweed *Polygonum cuspidatum* Sieb. & Zucc.	Ornamental	Perennial	Lowlands, homesites	None
Jimsonweed *Datura stramonium* L.	Ornamental	Annual	Pastures, cropland	Solanaceous alkaloids
Johnsongrass *Sorghum halepense* (L.) Pers.	Forage	Perennial	Cropland, pastures	Cyanogenetic glycosides
Kochia *Kochia scoparia* (L.) Schrad.	Forage	Annual	Widespread	None
Kudzu *Pueraria lobata* (Willd.) Ohwi	Ornamental, erosion control, forage	Perennial	Forests, rights of way, field borders	None
Lantana *Lantana camara* L.	Ornamental	Perennial	Fence rows, ditchbanks, fields	Lantadene
Macartney rose *Rosa bracteata* Wendl.	Ornamental	Perennial	Pastures	None

(*continued*)

APPENDIX 2 *Continued*

Species	Purpose of introduction	Weed type	Problem areas	Toxic principle
Melaleuca *Melaleuca leucadendron* (L.) L.	Ornamental	Perennial	Swamps, cities	Respiratory and skin irritants
Multiflora rose *Rosa multiflora* Thunb. ex Murr.	Windbreaks, cover plantings	Perennial	Pastures	None
Precatory bean *Abrus precatorius* L.	Ornamental	Perennial	Fence rows, roadsides	Abrin
Reed canarygrass *Phalaris arundinacea* L.	Forage	Perennial	Canals, ditchbanks	None
Sicklepod milkvetch *Astragalus falcatus* Lam.	Forage	Perennial	Rangeland	Nitro compounds
Strangler vine *Morrenia odorata* (Hook. & Arn.) Lindl.	Ornamental	Perennial	Citrus	None
Tansy *Tanacetum vulgare* L.	Herb	Perennial	Old gardens, roadsides	Unknown
Water fern *Salvinia auriculata* Aubl.	Ornamental	Perennial	Canals, waterways	None
Water hyacinth *Eichhornia crassipes* (Mart.) Solms	Ornamental	Perennial	Canals, lakes, waterways	None
Wild melon *Cucumis melo* L.	For observation	Annual	Imperial Valley, cropland	None
Yellow toadflax *Linaria vulgaris* Mill.	Ornamental	Perennial	Rangelands	None

[a] From Williams (1980).

ACKNOWLEDGMENTS

Preparation of this chapter would have been far more difficult if the work by Holm *et al.* 1977, 1979 and the work of the Technical Committee to Evaluate Noxious Weeds (Gunn *et al.*, 1981) had not been available. The personal assistance of L. G. Holm is acknowledged with gratitude.

REFERENCES

Amritphale, D., and Mall, L. P. (1978). Allelopathic influence of *Saccharum spontaneum* L. on the growth of three varieties of wheat. *Sci. Cult.* **44,** 28–30.

Baker, H. G. (1962). Weeds–Native and introduced. *J. Calif. Hortic. Soc.* **23,** 97–104.

Baker, H. G. (1965). Characteristics and modes of origin of weeds. *In* "The Genetics of Colonizing Species " (H. G. Baker and G. L. Stebbins, eds.), pp. 147–168. Academic Press, New York.

Baker, H. G. (1972). Migrations of weeds. *In* "Taxonomy, Phytogeography and Evolution" (D. H. Valentine, ed.), pp. 327–347. Academic Press, New York.

Baker, H. G. (1974). The evolution of weeds. *Annu. Rev. Ecol. Syst.* **5,** 1–24.

Blackburn, D. D., and Weldon, L. W. (1965). The sensitivity of duckweed (Lemnaceae) and Azolla to diquat and paraquat. *Weeds* **13,** 147–149.

Bor, N. L. (1960). "The Grasses of Burma, Ceylon and Pakistan." Pergamon, Oxford.

Boyd, C. E. (1974). Utilization of aquatic plants. *In* "Aquatic Vegetation and Its Use and Control" (D. S. Mitchell, ed.), pp. 107–115. UNESCO, Paris.

Buchholtz, K. P. (1967). Report of the Terminology Committee. *Weeds* **15,** 388–389.

Chomchalow, N., and Ponepangan, S. (1973). Types of aquatic weeds. *In* "Aquatic Weeds in S. E. Asia" (C. K. Varshney and J. Rzoska, eds.), pp. 43–50. D. W. Junk, The Hague.

Dafni, A., and Heller, D. (1980). The threat posed by alien weeds in Israel. *Weed Res.* **20,** 277–283.

Dickens, R. (1974). Cogongrass in Alabama after sixty years. *Weed Sci.* **22,** 177–179.

Eady, F. (1974). The aquatic weed problem. 1. Identification. *N. Z. J. Agric.* **8,** 40–45.

Edie, H. H., and Ho, B. N.C. *Ipomoea aquatica* as a vegetable crop in Hong Kong. *Econ. Bot.* **23,** 32–36.

Edwards, D. J. (1975). Taking a bite at the waterweed problem. *N. Z. J. Agric.* **130,** 33–36.

Florida Department of Natural Resources (1973). "Aquatic Weed Identification and Control Manual" (staff, eds.), p. 37. Bur. Aquatic Plant Res. Control, Tallahassee, Florida.

Gray, A. (1879). The predominance and pertinacity of weeds. *Am. J. Sci.* **118,** 161–167.

Gunn, C. R., Ritchie, C. A., and Poole, L. (1981). "Exotic Weed Alert" (unpublished alphabetical printout prepared for Technical Committee to Evaluate Noxious Weeds). TCENW, BARC-West, Beltsville, Maryland.

Haller, W. T. (1978). Hydrilla: A new and rapidly spreading aquatic weed problem. *Circ.—Univ. Fl., Inst. Food Agric. Sci.,* **S-245,** 1–13 (first printing, 1976).

Harper, J. R. (1965). Establishment, aggression and cohabitation in weedy species. *In* "Genetics of Colonizing Species" (H. G. Baker and G. L. Stebbins, eds.), pp. 243–265. Academic Press, New York.

Heiser, C. B., Jr. (1965). Sunflowers, weeds and cultivated plants. *In* "The Genetics of Colonizing Species" (H. G. Baker, and G. L. Stebbins, eds.), pp. 391–398. Academic Press, New York.

Holm, L. G., Plucknett, D. L., Pancho, J. V., and Herberger, J. P. (1977). "The World's Worst Weeds: Distribution and Biology." The East-West Center, Univ. of Hawaii Press, Honolulu.

Holm, L. G., Pancho, J. V., Herberger, J. P., and Plucknett, D. L. (1979). "A Geographical Atlas of World Weeds." Wiley (Interscience), New York.

Ivens, G. (1967). "East African Weeds and Their Control." Oxford Univ. Press, Nairobi, Kenya.

King, L. J. (1966). "Weeds of the World: Biology and Control." Hill, London.

King, R. M., and Robinson, H. (1970). Studies in the Eupatorieae (Compositae). XXIX. The Genus Chromolaena. *Phytologia* **20,** 196–209.

Kleinschmidt, H. E. (1969). Effect of granular 2,4-D on some waterweeds and its persistence. *Queensl. J. Agric. Anim. Sci.* **26,** 587–592.

Lin, C. (1968). "Weeds Found on Cultivated Land in Taiwan," Vols. 1 and 2. College of Agriculture, National Taiwan University, Taipei.

Long, R. W., and Lakela, O. (1971). "A Flora of Tropical Florida." Univ. of Miami Press, Coral Gables, Florida.

Lubigan, R., and Vega, M. (1971). The effect of different densities and duration of competition of *Echniochloa crusgalli* (L.) Beauv. and *Monochoria vaginalis* Burm. f. Presl. on the yield of lowland rice. *In* "Weed Science Report, 1970–71," pp. 19–23. Dep. Agric. Bot., University of Philippines, College of Agriculture, Los Banos.

Lumpkin, T. A., and Plucknett, D. L. (1980). Azolla: Botany, physiology, and use as a green manure. *Econ. Bot.* **34,** 111–153.

McWhorter, C. G. (1971). Introduction and spread of johnsongrass in the United States. *Weed Sci.* **19,** 496–500.

Matthews, L. J. (1963). Weed identification and control: *Azolla rubra. N. Z. J. Agric.* **106,** 297.

Mitchell, D. S. (1974). Chap. 3. The development of excessive populations of aquatic plants. Chap. 5. Environmental management in relation to aquatic weed problems. *In* "Aquatic Vegetation and Its Use and Control" (D. S. Mitchell, ed.), pp. 38–49 and 57–71. UNESCO, Paris.

Muenscher, W. G. (1955). "Weeds," 2nd ed. Macmillan, New York.

Mulligan, G. A. (1965). Recent colonization by herbaceous plants in Canada. *In* "The Genetics of Colonizing Species" (H. G. Baker and G. L. Stebbin, eds.), pp. 127–143. Academic Press, New York.

Musselman, L. J., and Parker, C. (1981). Studies on Indigo Witchweed, the American strain of *Striga gesnerioides* (Scrophulariaceae). *Weed Sci.* **29,** 594-596.

Oladokun, M. A. O. (1978). Nigerian weed species: Intraspecific competition. *Weed Sci.* **26,** 713–718.

Oosthuizen, G. J., and Walters, M. M. (1961). Control water fern with diesoline. *Farming S. Afr.* **37,** 35–37.

Parham, J. W. (1958). The weeds of Fiji. *Dep. Agric. Fiji Bull.* **35,** 1–200.

Parker, C. (1972). The *Mikania* problem. PANS **18,** 312–315.

Polunin, N. (1960). "Introduction to Plant Geography." McGraw-Hill, New York.

Powell, A. M. (1965). Taxonomy of Tridax (Compositae). *Brittonia* **17,** 47–96.

Reed, C. F. (1951–1953). *Azolla caroliniana* in Maryland. *Castanea* **16–18,** 143–144.

Reed, C. F. (1961). Amaranthaceae new to eastern United States. *Castanea* **26,** 123–128.

Reed, C. F., and Hughes, R. O. (1977). Economically important foreign weeds—Potential problems in the United States. *U.S., Dep. Agric., Agric. Handb.* **498,** 1–261.

Renfrew, J. M. (1973). "Paleoethnobotany: The Prehistoric Food Plants of the Near East and Europe." Columbia Univ. Press, New York.

Ridley, H. N. (1930). "The Dispersal of Plants Throughout the World." L. Reeve & Co, Ltd. Ashford, England.

Robinson, B. L. (1934). *Mikania scandens* and its near relatives. *Contrib. Gray Herb. Harv. Univ.* **104,** 55–70.

Sakai, K.-I. (1965). Contributions to the problem of species colonization from the viewpoint of competition and migration. *In* "The Genetics of Colonizing Species" (H. G. Baker and G. L. Stebbins, eds.), pp. 215–239. Academic Press, New York.

Salisbury, E. J. (1961). "Weeds and Aliens." Collins, London.

Tabor, P. (1952). Comments on cogon and torpedo grasses: A challenge to weed workers. *Weeds* **1,** 374–375.

Tarver, D. P., Rodgers, J. A., Mahler, N. J., and Lazar, R. L. (1978). "Aquatic and Wetland Plants of Florida," Bur. Aquatic Plant Res. Control, Fla. Dep. Nat. Resour., Tallahassee.

Williams, M. C. (1980). Purposefully introduced plants that have become noxious or poisonous weeds. *Weed Sci.* **28,** 300–305.

Young, J. A., and Evans, R. A. (1976). Responses of weed populations to human manipulations of the natural environment. *Weed Sci.* **24,** 186–190.

Young, J. A., Evans, R. A., and Major, J. (1972). Alien plants in the Great Basin. *J. Range Manage.* **25,** 194–201.

CHAPTER 8

Ecology and Genetics of Exotics

RALPH SCORZA

Appalachian Fruit Research Station
Agricultural Research Service
United States Department of Agriculture
Kearneysville, West Virginia

I. INTRODUCTION

Land bridges during the Paleocene, Miocene, and Pliocene periods allowed terrestrial organisms to migrate over much of the earth's surface. It is believed, for example, that although the origin of the Sclerotiniaceae is North America, most of the species migrated from North America to Europe during the Tertiary era (Reichert, 1958). Camels and tapirs, now exotic beasts in North American zoological parks, are believed to have evolved on that continent, only to migrate during the Pliocene or Pleistocene to Eurasia, where they flourished (Elton, 1958). Gressit (1974) gives an account of the prehistoric migrations of insects throughout the world following their appearance during the Devonian. The term

ISBN 0-12-757880-3

"exotic," therefore, denotes an organism only at a specific point in time during our planet's history.

To be exotic is not an intrinsic property of an organism. Regardless of its habitat, the organism will retain the same basic physiological and genetic systems. The survival of an organism in its native habitat, or the successful establishment of an exotic organism, depends upon three sets of factors: (1) climatic and other physical parameters, including temperature, humidity, precipitation, and soil pH; (2) biological parameters, including the presence of parasites, competing organisms, and susceptible hosts; and (3) the genetic plasticity to adapt to new and varied sets of physical and biological parameters. The odds are overwhelmingly against successful establishment of an exotic organism; even a slight alteration in the environment can spell doom. Indeed, it has been estimated that 99.9% of the species that have lived on the earth since life began are now extinct due to a failure to adapt to changes in the physical and biological environments (Smith, 1971).

The difficulty of establishing beneficial insect populations and new species of exotic game mammals provides historic examples of which we are well aware (DeBach, 1965; de Vos *et al.*, 1956). It is therefore appropriate to briefly summarize those factors, both physical and biological, that affect the success or failure of an exotic organism and to comment upon the role of genetic plasticity, keeping in mind that these are general principles that affect all organisms. The discussion will center upon pest species and the particular phenomena associated with host–pathogen interactions.

II. DENSITY DEPENDENCE AND DENSITY INDEPENDENCE

Survival and growth of populations may be regulated by density-dependent and/or density-independent factors. The relative importance of these factors is a subject of debate (Boughey, 1973; Huffaker *et al.*, 1971; Andrewartha and Birch, 1954) but may depend upon the particular organism of interest. The exotic pest is affected by both sets of factors. Initial survival and reproduction in a new area may be most heavily dependent upon density-independent factors—climate, soil, pH, nutrients, day length, topography, inherent genetic defects, accidents, and agricultural practices. These factors will be critical for initial survival of the pest and success in location of the host. Thus, density-independent factors determine the ecological area where a species can or cannot exist (Knipling, 1979).

Following this brief initial introduction period, density-dependent factors such as host availability and density, buildup of natural enemies, and competition from other organisms for host, light, moisture, other nutritional sources, and reproductive niches will assume an increasingly important role. Under certain circumstances, an exotic may find an area climatically suitable and meet with little competition from native organisms. In this case, density-independent and density-dependent factors should play prominent intraspecific roles in the growth of the exotic population. The initial period of population buildup will be density independent and will follow an exponential pattern, since the exotic population is small. As population density increases, density-dependent factors will play a more important role and ultimately act as the chief population-regulating force as long as climate and other density-independent factors remain favorable. The population level at which these density-dependent factors hold the exotic organism determines whether it becomes a pest or simply a relatively innocuous addition to the native flora and fauna.

III. TEMPERATURE

Most biological processes are temperature dependent (Larcher *et al.*, 1973). It is to be expected, then, that temperature would exert a major effect upon the survival, reproduction, and distribution of an exotic pest. Colhoun (1973) has reviewed the effects of temperature on plant disease, including host predisposition, spore production and discharge, spore germination, host susceptibility, incubation period, symptom expression, pathogen survival, and fungal mutation rate. Unfavorable temperature at any one of these eight points may prevent successful establishment of an exotic pest. Temperature is an easily measured parameter. Thus, research has been skewed toward investigation of this factor, possibly to the exclusion of other equally important factors and interactions. Nonetheless, the importance of temperature limits for the dispersal of many organisms cannot be denied.

Phymatotrichum omnivorum root rot survives freezing temperatures for only a day or two (Ezekiel, 1945), and as such, the fungus is restricted in the United States to the area south of approximately 35° north latitude. *Pellicularia rolfsii* and *Pseudomonas solanacearum,* both very important pathogens, are likewise restricted to subtropical and tropical areas (Walker, 1969). Low temperatures necessary for infection restrict onion smut, *Urocystis cepulae,* to the northern United States and Europe, although it is repeatedly distributed in southern areas on infected onion sets (Walker and Wellman, 1926). It is believed that chrysanthemum rust, *Puccinia chrysanthemi,* has not become established in the inland United States despite repeated introductions due to high summer temperatures and low winter temperatures. High temperatures reduce spore germination, eradicate infection, and destroy spores in existing uredia. Low temperatures prevent spore germination (Campbell and Dimock, 1955).

The rate of spread of the chestnut blight, *Endothia parasitica,* has been related to the influence of temperature on its growth rate (Stevens, 1917). Several highly vagile insects, including *Junonia coenia, Vanessa cardui, Celerio lineata,* and

Colias eurytheme, are known to immigrate regularly to North America, only to be annually destroyed by low winter temperatures (Remington, 1968).

IV. MOISTURE

Moisture is essential to growth and development. Plant pathogens require moisture for spore germination and infection. For most, a water film on the plant's surface is necessary, but for the powdery mildews, high air humidity is essential—a water film being inhibitory. Many plant pathogens rely upon a water film, rain splashing, or water droplets for dissemination (Brook, 1969; Zentmyer and Bald, 1977). The duration of wetness is a critical factor for pathogen survival. *Phytophthora infestans,* for example, requires at least 3 hours of moist conditions for infection (Lapwood, 1968), whereas *Venturia inaequalis* may require 17–26 hours (Sys and Soenen, 1970). Soil moisture is critical to organisms such as *Plasmodiophora brassicae* (club root of crucifers) with a motile stage in the life cycle. Dry soils favor some of the *Fusarium* species of cereals and *Urocystis agropyri* on wheat (Colhoun, 1953). In the case of *F. culmorum,* germ tubes are better able to resist lysis by bacteria in dry soils (Cook and Papendick, 1970), and, more indirectly, the growth rate of wheat seedlings is such that they remain in a susceptible condition for a longer period (Malalasekera and Colhoun, 1968). Moisture as a limiting factor for disease spread may be cited for several diseases of crucifers on the Pacific Coast of the United States. Cabbage ring spot (*Mycosphaerella brassicola*) is restricted to the Pacific Coast due to the high humidity in this area. The disease is not important east of the hundredth meridian (Nelson and Pound, 1959). Black rot and black leg are rare in the Pacific Coast area. Seed beds are established in the dry season, and the fact that establishment of these organisms cannot occur during this period precludes development of these diseases during the favorable moisture conditions that subsequently occur (Walker, 1922). Worldwide distribution of citrus canker and citrus scab is a function of the amount, frequency, and seasonal distribution of precipitation (Peltier, 1926).

The two major limiting factors, temperature and moisture, have been used to characterize an area ecologically and, further, to explain the success or lack of success of introduced organisms. The climatic classification system devised by Koppen (Trewartha, 1954), based upon temperature, precipitation, and seasonal characteristics, or Thornthwaite's system (Thornthwaite, 1931), based on a precipitation–evaporation index, are useful for gross mapping of vegetation types and therefore would give some indication of the potential spread of pests on typical plant species. Climographs of mean temperature and precipitation are also useful in charting the ecological range of organisms. Reichert (1958) has incorporated basic temperature and moisture data with data on recorded occurrence, geographic origin, and migration of the organism, formulating a pathogeographical approach for the prediction of plant disease (Reichert and Palti, 1967). The approaches to the study of population distribution and invasion potential are only as useful as our ability to deduce accurately the effects of the climatic parameters on the organisms of interest.

V. OTHER FACTORS

Light intensity and duration, soil pH, soil nutrients, wind, and air pollution have also been cited as limiting factors influencing the survival, development, and dispersal of plant pathogens (Zentmyer and Bald, 1977; Bowden *et al.*, 1971; Colhoun, 1973). Although it has been demonstrated that a particular environmental parameter such as temperature or moisture may have a profound effect upon the growth and reproduction of a pathogen, interactions among factors such as temperature, moisture, and light have been noted (Colhoun, 1973; Zentmyer and Bald, 1977). It is likely that most organisms are affected by a number of potentially limiting factors subtly interacting in patterns, yet undeciphered. Based upon our present knowledge of those strict environmental constraints that have been found to govern certain organisms, it is reasonable to assume that myriad interactions, although less easily measured, would exert upon potential invaders the same strict limitations. Indeed, this has probably been the case for many exotic organisms that have entered the United States, unrecorded victims of a hostile environment.

VI. BIOLOGICAL COMPETITION

Assuming that an organism can survive and reproduce under the climatic regimes of a new environment, it is also thrown into a milieu of other organisms. The exotic organism, in this case a predator–parasite, may react in one of three ways: (1) it may coexist with other predator–parasite species; (2) it may drive out other species; or (3) it may be driven out by other species. If coexistence is the mode of interaction, an exotic may act independently of native organisms, may compete with natives, or may be preyed upon or parasitized by natives. The ability to coexist with or dominate potential rivals is a major factor determining successful colonization by an exotic. These interspecific interactions are probably the least understood facets of exotic pest establishment. Observations of successful colonization are numerous (Chapters 2, 3, and 4, this volume; Sailer, 1978), but failure to colonize is rarely treated. The failure of exotics to colonize has been little studied due to the difficulty in documenting nonsurvival and the absence of an adverse economic or ecological impact arising from unsuccessful establishment. Introduced plant species may successfully become established in disturbed habitats such as annual crop lands, roadsides, or lumbered woods, but rarely do they invade undisturbed communities (Dansereau, 1957). Ridley (1930) noted that few exotic species except weeds of corn fields have been successful in establishing themselves in Great Britain. Elton (1958) observed that the dead nettle, *Laminum album,* although a roadside weed, has never been able to penetrate the woodlands of the British Isles, although in its original home in the Caucasus, it is a successful woodland plant. The experience with intentionally introduced mammals has been that the majority of introductions failed due to the intense competition from native fauna (de Vos *et al.,* 1956). Further, most of the successful introductions have become pest species, presumably due to the absence of competition or the ability to overwhelm competitors (de Vos *et al.,* 1956).

The inability to compete with other species is dramatically illustrated in the domesticated plants. Dansereau (1957) cites a few examples, including Norway maple, European spruce, maize, wheat, oleander, Boston fern, and impatiens. These plants prosper under cultivation, yet when freed of human intervention, fail to survive. Competition from other organisms, including predation, parasitization, and competition for light, moisture, and nutrients, is too great. If the establishment of exotic mammals and introduced plants can be compared to the establishment of exotic plant pests, the biological barrier that these exotics face is considerable. Research on control by natural enemies has elucidated some general principles of natural enemy–host interaction. The most effective control generally results from a density-dependent interaction—the host population regulated by its enemy and the enemy regulated by the number of hosts (Huffaker *et al.,* 1971). Interactions of this nature will not eradicate a pest but have the potential of keeping the pest population at a stable low and uneconomic level, and at this level, vulnerable as a population to eradication by other environmental factors or human control efforts. Control by natural enemies is most effective when density-independent mortality is high and consistent, since only a small amount of density-related mortality would be required for effective control (Huffaker *et al.,* 1971). Control of an exotic pest by natural enemies would depend upon concurrent introduction of host-specific enemies and the ability of these enemies to survive and reproduce or the presence of non-host-specific enemies which could attack the exotic. Thus, although the effect of natural enemies on an introduced pest may be great, this control depends upon the host–enemy population interaction characteristics.

The introduction of natural enemies following the establishment of an exotic pest has met with both outstanding success and utter failure (see Chapter 14, this volume). Such introductions carry the danger of attacking not only the intended pest but also valued species. Yet, unlike climate, once the exotic pest is established, biological control agents represent a parameter that may be manipulated in a program of control (Huffaker *et al.*, 1971).

VII. NATURAL VERSUS AGRICULTURAL ECOSYSTEMS

The precise reasons for the seemingly greater susceptibility of agricultural and disturbed areas to invasion are unclear but may relate to the following factors: (1) the host species that dominate these areas are generally themselves exotic introductions with few native pests, and as such are subject to damage by introduction of pests from their native habitats; (2) these areas are frequented by humans, domesticated animals, farm machinery, and other human traffic which can serve as transport agents for introduced pests; (3) due to chronic disturbances, ecological niches are constantly created and destroyed, providing colonization opportunities for new species; and (4) pesticide use and pollution of these areas can destroy potential predators, parasites, and species which could compete with invaders. Complex ecosystems such as forests may resist the entrance of exotic species due to the intense interspecific competition that can develop among the numerous species of organisms in such an ecosystem. However, when this barrier is overcome, the results can be devastating.

Before the 1905 report of chestnut blight, *E. parasitica,* in New York (Merkel, 1905), the American chestnut comprised 25% of the eastern U.S. hardwood forests, with a range of more than 200 million acres. By 1950 the blight had infected more than 80% of the trees throughout this range. The American chestnut is now a minor species in the forests of the eastern United States (Kuhlman, 1978).

Upon introduction into eastern North America between 1900 and 1915, pine blister rust, *Cronartium rubicola,* spread rapidly and established itself throughout most of this area's white pine region (van Arsdel *et al.* 1956). In the 1970s, 70% of the elms in southern England were destroyed by Dutch elm disease, *Ceratocystis ulmi* (Stipes and Campana, 1981). The European spruce saw fly, *Archips hercyniae,* was introduced into northeastern North America in the 1920s. By 1937 nearly two-thirds of the white spruce and one-fourth of the black spruce on the Gaspé Peninsula of southern Quebec had been destroyed, and *A. hercyniae* was determined to be the most abundant spruce-feeding insect in northeastern North America. The lack of parasites of this pest was noted as a striking feature of the outbreak (Elton, 1958).

The gypsy moth, *Porthetria dispar,* was accidentally introduced into Massachusetts in 1869 and rapidly spread throughout New England. It is presently found in southeastern Canada, the north central and southern United States, and in 1976 was discovered in San Jose, California. It is an important forest pest and causes extensive hardwood losses (Corliss, 1952; Davidson and Lyon, 1979).

The rapid spread of this insect was related to the absence of natural enemies. Introduction of natural enemies from Europe was of limited value due to the unavailability of alternate hosts (Corliss, 1952).

The dramatic spread and disease loss caused by exotics in the complex forest ecosystem may be the result of (1) the difficulty of applying control measures in such an environment; (2) the difficulty of establishing an enemy of the exotic in such a highly competitive ecosystem; (3) the perennial nature of forest tree species and the long generation cycles, which do not allow for rapid replacement of resistant or tolerant genotypes for susceptible genotypes, although there is some suggestion that individual trees are composed of genetically distinct subunits (Whitham and Slobodchikoff, 1981); and (4) the fundamental potential instability of such a complex ecosystem.

VIII. STABILITY–COMPLEXITY OF NATURAL ECOSYSTEMS

The potential instability of the complex forest ecosystems may be surprising, since the concepts of stability arising from complexity (forests) and instability resulting from ecological simplification (croplands) have often been presented. Elton (1958) offered six lines of evidence that formed the foundation of this hypothesis: (1) mathematical models of simple one predator–one prey systems illustrate profound instability; (2) experimental two-species systems are violently unstable, exhibiting extreme population oscillations (Gause, 1934; Gause *et al.*, 1936; Park, 1954), as predicted mathematically; (3) natural habitats on small islands are more vulnerable to invading populations than are those of the continents; (4) successful invasions most often occur on cultivated land (with artificially simplified habitats); (5) observations of tropical rain forests rich in species diversity show these areas to be ecologically quite stable, whereas the oscillations observed in arctic and boreal populations may be due to a lack of species diversity (Hutchinson, 1959); and (6) orchards, although resembling woodlands, are more intensely managed and suffer a greater number of foreign invasions. These six factors form the basis of what some have considered to be the theory of trophic web complexity begetting increased stability. In fact, there may be special cases in which increased trophic web complexity leads to increased ecosystem stability. For example, the vegetation–herbivore system is mathematically unstable, but the addition of a predator to the model produces a stable point. Thus, the more complex vegetation–herbivore–predator system is mathematically more stable than the vegetation–herbivore system (May, 1974). Yet in general, the stability-complexity concept does not appear entirely valid. Insightful criticisms of the six cornerstones of the stability–complexity hypothesis are as follows (May, 1974; van Emden and Williams, 1974; Goodman, 1975):

1. Whereas mathematical models of simple one predator–one prey systems predict instability, comparison of one predator–one prey and more complex models predicts greater instability with greater complexity.
2. Experimental two-species systems are unstable, but demonstration of greater stability in more complex systems is necessary for support of the stability–complexity hypothesis. This has not been demonstrated. Empirical studies of diversity-stability relationships through descriptive correlative analysis and short-term perturbation analysis suggest that more diverse communities are less stable than simpler ones (Hurd *et al.*, 1971; Murdoch *et al.*, 1972).
3. The rapid colonization of islands by invading species is not necessarily a function of instability of these systems but rather a result of vulnerability. May (1974) observed that the complexity of linkages in the trophic web of the North American continental forest system was not effective in stabilizing the system when the Japanese beetle, European gypsy moth, and chestnut blight were introduced. He further suggested that the relationships among organisms in a community are nonlinear, and therefore, stability cannot be simply related to the sum of linkages.
4. Although successful invasions are often recorded in agricultural lands, this is by no means necessarily a function of reduced complexity of the ecosystem. Goodman (1975) points out that agricultural systems are in a constant state of upheaval through tillage, successional replacement of species, and other human-directed inputs and have not been allowed to stabilize. A stable level of interaction, though, may not be favorable to survival of cultivated species at an economic level (van Emden and Williams, 1974).
5. The stability of tropical rain forests may simply be an illusion attributable to insufficient study of these systems. Observations that do appear in the literature suggest that population fluctuations can be quite violent (Smith, 1970; Goodman, 1975). The extreme susceptibility of rain forest ecosystems to human intervention has become increasingly clear, especially as colonization efforts have extended into tropical rain forest regions (Gomez-Pompa *et al.*, 1972; Bretsky *et al.*, 1973).
6. The apparently greater susceptibility of orchards over woodlands is a special case of (4) in which human inputs have prevented the establishment of an equilibrium state, since such an equilibrium may not favor economic production of the intended crop.

Although the concept of ecological stability resulting from trophic web complexity is ill supported, May (1974) suggests an alternate hypothesis. Complexity of an ecosystem may be important in, for example, full exploitation of a habitat. In those situations in which complexity is advantageous, it may be allowed to arise only under stable environmental conditions. Complex systems are composed of a number of interconnecting interactions, with each interaction having its characteristic equilibrium point. Each interaction may exist at equilibrium or disequilibrium. The greater the number of interactions, the greater will be the chance for unstable interactions and instability of the system as a whole. An unstable environment can easily drive complex systems to disequilibrium. Less complex systems may be better able to withstand unstable environments and recover from disequilibrium states. Trophic web complexity, then, may result from environmental stability and not stability from complexity. The association of these two phenomena is validated, although Elton's (1958) interpretation is reversed.

The question of stability–complexity is an important one if we are to predict the possibilities for establishment of an exotic invader in a particular ecosystem. Initiation and continuation of empirical studies are necessary.

IX. GENETIC INTERACTIONS

Although environment regulates the distribution and establishment of exotic organisms, the response to environment is under genetic control as are intra- and interspecific competition (the predator–prey and host–parasite relationships). Thus, a discussion of environmental and interorganismal parameters is incomplete without a discussion of the genetic bases of these interactions.

A. The Ecogenetic Model

Invasion of a new habitat by an exotic pest is generally followed by a flurry of research concerning the host range, historical spread, natural enemies, and physiology of the pest. Rarely is the source population from which the pest emigrated studied. Yet, here may be a key to the prediction of the success of establishment, the spread of the pest, and the choice of control strategies. The genetic structure of the source population can affect the initial establishment of an exotic organism and initial growth of the introduced population. Remington (1968), in his ecogenetic model of populations, characterizes marginal and core populations. Subpopulations near the geographic core of the species (generally the most favorable environment) tend to be large. Individuals are highly heterozygous, highly adaptive to a range of microenvironments, and vigorous, but show a low specific adaptability and a large genetic load due to the great number of alleles at each locus. It has also been suggested that in place of highly adaptive individuals are highly adaptive core populations as a whole, due to a great number of individual genotypes specifically adapted to different areas (Baker, 1965; Waddington, 1965). Populations in ecologically marginal areas are under adaptive stress and under intense selection for adaptation. These populations are characteristically small and highly homozygous. Further, as illustrated with *Drosophila* species, individuals carry fewer lethals, semilethals, and subvitals when compared with individuals from the highly heterozygous core populations (Townsend, 1952; Dobzhansky, 1956; Dobzhansky *et al.*, 1963; Spassky *et al.*, 1960). Remington (1968) suggests several important implications of this ecogenetic population model. In the areas of accidental introduction without control measures, he suggests the following:

1. The most successful establishment will come from many founders from a core population.
2. The least establishment will come from few founders from a core population.
3. The best potential for early survival, plus development of a new, well-adapted polytypic population, is to be expected from many founders from a marginal population.

In the area of controlling unwanted introduced insects, he concludes:

1. Few founders from a core population are likely to become extinct, even without a control program.
2. Few or many founders from a marginal population are the easiest to extirpate soon after arrival but are likely (a) to become the most dangerous later on because of their high potential for evolving a new superior type and (b) to become the least vulnerable to introduction of enemies from their former home.
3. Many founders from a core population are the hardest to extirpate if the new environment is similar to that of the source, but they will be the most vulnerable to the introduction of enemies from the former home.

Although aspects of the ecogenetic model await validation, the concepts appear sound. The interaction of ecosystem and genetics is implicit in this model. Genetic variation of the colonizing population, whether a result of colony size or of genetic variability of individuals, is less important as the ecosystem more closely approaches that ecosystem from whence the colony emigrated. Conversely, the greater the genetic variability in the founder stock, the less important are the ecosystem similarities—to a point, naturally. If we assume that the ecosystems of the areas of emigration and immigration are similar enough to allow the founder colony to survive, we then become concerned with an invasion of exotic genes. At this level, we are concerned with the number of genes, their arrangement on chromosomes, recombination (linkages), mutation rates, and viability (lethality).

B. Sources of Genetic Variability

Exotic invaders owe their ability to adapt to a new ecosystem to the basic genetic mechanisms of recombination and mutation common to all organisms. These two processes may occur in both sexual and somatic cells, depending upon the organism.

1. *Recombination*

Recombination is the key to adaptation, and adaptation is the key to successful invasion and colonization. Through sexual reproduction, independent assortment and crossing over of chromosome sections occur. It is through these processes that new variants of an organism may arise with unique combinations of genes for fitness and/or virulence.

2. *Mating Systems*

Recombination occurs during the sexual cycle of most organisms. The potential for recombination events is a function of the amount of outbreeding. It may then be presumed that a characteristic of a successful colonizing plant species is the propensity for outbreeding. As Allard (1965) and Stebbins (1965) illustrate, this is not the case; instead, a number of successful exotic weed species are self-pollinating. Through comparisons of outbreeding and inbreeding populations, Allard (1965) observed that whereas outbreeders are more variable within populations, inbreeders are more variable between populations. Within an area, no two populations of inbreeders are genetically alike. The chance outcrossing that does occur provides for the formation of individuals with gene combinations imparting greater adaptation to the existing, changing, or a completely new ecosystem. Most individuals resulting from outcrossing will not be better adapted, but those few that are, through inbreeding, will be able to develop an adapted population relatively free from further recombination that results from outcrossing. The inbreeding system as envisioned by Allard (1965) offers an invading species the advantages of both inbreeding and outbreeding. The small percentage of outbreeding between populations provides for the formation of uniquely adapted individuals, and the high rate of inbreeding allows for the buildup of a population of the adapted genotype to fill a new ecological niche.

The process repeats itself as a species advances or is carried into new areas, and it colonizes as it progresses. Allard (1963, 1965) suggests another aspect in the success of colonization by inbreeding plant species. These genetic systems may possess a plasticity which allows for flexibility in the amount of outcrossing and the rate of crossing over. In populations of *Avena fatua,* the amount of outcrossing may range from less than 1% to greater than 10%. The successful plant invader may be envisioned as spreading and colonizing not only by means of a system of recombination through limited outcrossing and fixation of adapted types through inbreeding but also by regulating populations through selection of types with higher levels of outcrossing, higher crossover rates, and other variability-increasing mechanisms. From this pool of variability, adapted types arise. Through selection of genotypes with low levels of outcrossing, low crossover rates, and mechanisms which limit variability, these adapted types are fixed and able to colonize successfully the ecosystems for which they are adapted. Self-compatibility and inbreeding are also important to the migration and establishment of plant species in the sense that theoretically only one propagule is necessary to establish a population in a new area (Baker, 1955, 1965; Mulligan, 1965).

3. *Somatic Recombination*

A number of important fungal pathogens only rarely undergo a sexual cycle, and recombination occurs in somatic cells. In these fungi, recombination may occur between two genetically different nuclei in the heterokaryotic cells. The basidiomycetes, which include the rust and smut fungi, characteristically possess binucleate cells. Conidia that develop from heterokaryotic cells may contain the different nuclei and develop a better-adapted race. Fungi with multinucleate cells may alter the proportions of different nuclei in response to selection (Parmeter *et al.*, 1963). Anastomosis (hyphal) fusion is another mechanism through which different nuclei may be brought together in one organism to create a third potentially fitter race. Anastomosis between two different races of *Puccinia striiformis* has been shown to give rise to a third race (Little and Manners, 1969; Goddard, 1976).

4. *The Parasexual Cycle*

The variation in some fungi arises from the fusion of two genetically distinct nuclei in heterokaryons. When somatic crossing over of these chromosomes occurs in the diploid, haploid individuals with new gene combinations may subsequently be produced. This parasexual cycle may be the major source of variation in some plant pathogens (Russell, 1978).

5. *Mutation*

Mutations may involve alterations of single genes, changes in chromosome number (ploidy), or alterations of chromosome structure including translocations, inversions, deletions, and duplications. Mutation is considered the basic mechanism for evolutionary change and adaptation. Rates vary between 10^{-4} and 10^{-9} mutational events per gene per replication (Levine, 1968). Although these rates are low and most mutations are considered deleterious, mutation may have a dramatic impact on an organism when the effects enhance its adaptability. Recombination may further modify the effects by bringing the mutant gene or gene combination together with other fitness-imparting genes, further enhancing the general fitness of the genotype. Gene mutations for virulence commonly enable fungal pathogens to overcome host resistance. Although the mutation may reduce general fitness (van der Plank, 1968), the ability to parasitize an otherwise resistant host is of such great advantage that the mutation may readily become established in a population. When the virulence gene is no longer necessary due to a change in host genotype, it may be carried as a recessive in the heterozygous condition to await the possibility of future need. The appearance of virulent races of plant pathogenic fungi is quite common, causing great difficulty in maintaining disease-resistant crop cultivars (Dekker, 1976). The development of mutant races of fungi resistant to fungicides causes similar problems with respect to chemical disease control (Dekker, 1976; Georgopoulos and Zaracovitis, 1967).

Bacteria are especially well suited to the development of variation through mutation. With mutation rates on the order of one in every 2000 cells (Russell, 1978) and populations that can double in as little as 15–20 minutes (Pelczar and Reid, 1972), the buildup of mutant populations can proceed very rapidly. Although mutant races of plant pathogenic bacteria have overcome resistant cultivars (Russell, 1978) and races resistant to chemical control agents have been identified (Moller *et al.*, 1981), it is surprising that resistance-breaking races have not been more of a problem to agriculture than has been recorded (Crosse, 1975).

6. *Polyploidy*

In general, it is apparent that polyploidy is not necessary for successful colonization by plant species (Stebbins, 1965; Mulligan, 1960, 1965), although it may be quite important for particular species or specialized habitats. Mulligan (1965) noted a relatively high percentage of polyploidy in row crop weeds and suggested that greater variability of polyploids over diploids, especially in regard to seed germination, made them better adapted to frequent disturbance resulting from cultivation. Polyploid *Aegilops* species exhibit greater variability than their diploid counterparts and therefore are more successful weeds (Zohary, 1965). Stebbins (1965) found that in the species he studied, polyploids had a greater tendency toward weediness than did corresponding diploids. He also noted that polyploidy in *Claytonie perfoliata* imparts an advantage in terms of larger seed size. Polyploid weeds are characterized by their genetic variability. The enhanced variability results from the ability to draw upon genetic diversity from ecologically distinct parental diploids and to recombine this genetic material at the polyploid level. Recombination between genomes is thought to be facilitated by the buffering provided by a common diploid genome (Stebbins, 1956, 1965; Ehrendorfer, 1965).

7. *Balanced Lethals*

Chromosome translocations can be important for the maintenance of adapted genotypes, the classic example being the balanced lethal system of *Oenothera*

(Cleland, 1962). In this system, races possess nonallelic recessive lethal genes each on different homologous chromosomes. When interbred, races appear to breed true, but in fact, one-half of the progeny are lethals and fail to germinate. The translocations thus provide for the maintenance of heterozygosity and its associated vigor. The balanced lethal system may, at least in part, account for the weediness of certain *Euoenothera* forms (Zohary, 1965). Such balanced lethal systems have been identified in *Drosophila* (Muller, 1918) and *Rhoeo discolor* (Strickberger, 1968).

8. *Heterosis*

Heterosis is the increase in vigor, survival, and reproductive ability of a genotype. Heterosis can impart to a genotype the ability to establish itself in a new environment or to outcompete native biota and successfully colonize the habitat and extend its range. Heterosis is of considerable theoretical significance in the evolution and colonization by organisms and is of great practical importance in plant and animal breeding; yet, the genetic and physiological bases for heterosis are not understood (Allard, 1960).

Heterosis is generally a function of heterozygosity; thus, mechanisms which increase or maintain heterozygosity will, in general, impart an adaptive advantage to a population. Chromosome translocations and the development of balanced lethal systems are important heterozygosity-maintaining mechanisms for a few species, and other systems have been identified. An obvious system for genotype fixation is the ability to bypass the sexual cycle. This may be accomplished through vegetative reproduction or apomixis.

Vegetative reproduction may be achieved through the production of rhizomes, stolens, tubers, bulbs, corms, or runners. The spread of a number of weeds, including water hyacinth (*Eichhornia crassipes*) and *Oxalis* spp., is achieved through vegetative reproduction (Baker, 1965). Some weedy species, including *Hemerocallis fulva, Mentha spicata,* and *M. piperita,* are not known to produce seed in the United States. *Allium vineale* and *Euphorbia lucida* are known only rarely to produce seed in the northern states (Muenscher, 1952).

9. *Apomixis*

The formation of seed without fertilization is thought to be an important factor in maintaining some adaptive weed genotypes (Baker, 1965). Plants with stable apomictic reproduction are frequently found in cross-pollinated perennial species that have the capacity for vegetative reproduction. Powers (1945) suggests that a precise combination of mutations affecting the sexual cycle is necessary for the appearance of apomixis in a population. Cross-pollination, perennial habit, and vegetative reproduction are thought to facilitate the preservation of individual mutants that eventually lead to apomixis. Remote hybridization and polyploidy induce reduced fertility and heterosis. The combination of these factors may greatly favor apomictic individuals and lead to strongly apomictic genotypes.

A weedy plant species may be represented as one of two genetic types. The general-purpose genotype is highly heterozygous, tolerant of wide environmental variation, and able to maintain this plastic genotype through self-pollination, a low chromosome number, vegetative propagation, apomixis, balanced lethals, or other genetic systems. Such species would be most likely to invade disturbed habitats exposed to continually fluctuating conditions, such as those characteristic of agricultural areas and roadsides. Other weedy species may be more specifically adapted, owing their competitive strength to a high level of adaptation to a particular niche. These weeds have higher levels of homozygosity that are maintained through extensive self-pollination. A species of this group would be suited to invasion and colonization of relatively undisturbed, stable habitats in which the benefits of precise adaptation would be of greatest advantage. Colonization of new areas by these species may be achieved through the development of new genotypes that occur through limited outcrossing and recombination.

X. CONCLUSIONS

Climate is the first potential barrier to an exotic organism. If climate allows survival, other barriers may be considered. Beyond the absolute climatic effects, the effects of a marginal climate in which the exotic survives but is at a competitive disadvantage may be significant. Biological interaction between the exotic and other life forms, including humans, are little understood but are of major importance in the survival and colonization by exotic organisms. The invaded area may supply the exotic with a readily available host, or it may present the invader with strong competitors for the host, or species which prey upon the exotic. The ultimate capacity for survival and colonization by an exotic species in the face of climatic and biological stresses lies in its genetic makeup. A number of systems are known which provide the genetic plasticity required for an organism to become established and prosper. These include adaptive mating systems, recombination, mutation, balanced lethals, heterosis, and apomixis. A greater understanding of the interactions between organisms and their environment will naturally lead to a better understanding of exotic invaders.

REFERENCES

Allard, R. W. (1960). "Principles of Plant Breeding." Wiley, New York.

Allard, R. W. (1963). Evidence for genetic restriction of recombination in the lima bean. *Genetics* **48,** 1389–1395.

Allard, R. W. (1965). Genetic systems associated with colonizing ability in predominantly self-pollinated species. *In* "The Genetics of Colonizing Species" (H. G. Baker and G. L. Stebbins, eds.), pp. 49–76. Academic Press, New York.

Andrewartha, H. G., and Birch, L. C. (1954). "The Distribution and Abundance of Animals." Univ. of Chicago Press, Chicago, Illinois.

Baker, H. G. (1955). Self-compatibility and establishment after "long distance" dispersal. *Evolution* **9**, 347–349.
Baker, H. G. (1965). Discussion of paper by H. L. Carson. *In* "The Genetics of Colonizing Species" (H. G. Baker and G. L. Stebbins, eds.), pp. 548–549. Academic Press, New York.
Boughey, A. S. (1973). "Ecology of Populations." Macmillan, New York.
Bowden, J., Gregory, P. H., and Johnson, C. G. (1971). Possible wind transport of coffee leaf rust across the Atlantic Ocean. *Nature (London)* **229**, 500–501.
Bretsky, P. W., Bretsky, S. S., Levinton, J., and Loretz, D. M. (1973). Fragile ecosystems. *Science* **179**, 1147.
Brook, D. J. (1969). Effects of light, temperature, and moisture on release of ascospores by *Venturia inaequalis* (Cke.) Wint. *N. Z. J. Agric. Res.* **12**, 214–227.
Campbell, C. E., and Dimock, A. W. (1955). Temperature and the geographical distribution of chrysanthemum rust. *Phytopathology* **45**, 644–648.
Cleland, R. E. (1962). The cytogenetics of Oenothera. *Adv. Genet.* **11**, 147–237.
Colhoun, J. (1953). A study of the epidemiology of clubroot disease of Brassicae. *Ann. Appl. Biol.* **40**, 262–283.
Colhoun, J. (1973). Effects of environmental factors on plant disease. *Annu. Rev. Phytopathol.* **11**, 343–364.
Cook, R. J., and Papendick, R. I. (1970). Soil water potential as a factor in the ecology of *Fusarium roseum* F. sp. *cerealis* 'Culmorum.' *Plant Soil* **32**, 131–145.
Corliss, J. M. (1952). The gypsy moth. *In* "Insects—The Yearbook of Agriculture, 1952," pp. 694–698. U.S. Govt. Printing Office, Washington, D.C.
Crosse, J. E. (1975). Variation amongst plant pathogenic bacteria. *Ann. Appl. Biol.* **81**, 438 (abstr.).
Dansereau, P. (1957). "Biogeography—An Ecological Perspective." Ronald Press, New York.
Davidson, R. H., and Lyon, W. F. (1979). "Insect Pests of Farm Garden and Orchard," 7th ed. Wiley, New York.
DeBach, P. (1965). Some biological and ecological phenomena associated with colonizing entomophagous insects. *In* "The Genetics of Colonizing Species" (H. G. Baker and G. L. Stebbins, eds.), pp. 287–306. Academic Press, New York.
Dekker, J. (1976). Acquired resistance to fungicides. *Annu. Rev. Phytopathol.* **14**, 405–428.
de Vos, A., Manville, R. H., and van Gelder, R. G. (1956). Introduced mammals and their influence on native biota. *Zoologica (N. Y.)* **41**, 163–194.
Dobzhansky, T. (1956). Genetics of natural populations. XXV. Genetic changes in populations of *Drosophila pseudoobscura* and *Drosophila persimilis* in some locations in California. *Evolution* **10**, 82–92.
Dobzhansky, T., Hunter, A. S., Paulovsky, O., Spassky, B., and Wallace, B. (1963). Genetics of natural populations. XXXI. Genetics of an isolated marginal population of *Drosophila pseudoobscura*. *Genetics* **48**, 91–103.
Ehrendorfer, F. (1965). Dispersal mechanisms, genetic systems, and colonizing abilities in some flowering plant families. *In* "The Genetics of Colonizing Species" (H. G. Baker and G. L. Stebbins, eds.), pp. 331–352. Academic Press, New York.
Elton, C. S. (1958). "The Ecology of Invasions by Animals and Plants." Wiley, New York.
Ezekiel, W. N. (1945). Effects of low temperatures on survival of *Phymatotrichum omnivorum*. *Phytopathology* **35**, 296–301.
Gause, G. F. (1934). "The Struggle for Existence." Williams & Wilkins, Baltimore, Maryland.
Gause, G. F., Smaragdova, N. P., and Witt, A. A. (1936). Further studies of interaction between predators and prey. *J. Anim. Ecol.* **5**, 1–18.
Georgopoulos, S. G., and Zaracovitis, C. (1967). Tolerance of fungi to organic fungicides. *Annu. Rev. Phytopathol.* **5**, 109–130.
Goddard, M. V. (1976). The production of a new race, 105 E 137 of *Puccinia striiformis* in glasshouse experiments. *Trans. Br. Mycol. Soc.* **67**, 395–398.

Gomez-Pompa, A., Vazquez-Yanes, C., and Guevara, S. (1972). The tropical rain forest: A non-renewable resource. *Science* **177,** 762–765.

Goodman, D. (1975). The theory of diversity-stability relationships in ecology. *Q. Rev. Biol.* **50,** 237–266.

Gressitt, J. L. (1974). Insect biogeography. *Annu. Rev. Entomol.* **19,** 293–321.

Huffaker, C. B., Messenger, P. S., and DeBach, P. (1971). The natural enemy components in natural control and the theory of biological control. *In* "Biological Control" (C. B. Huffaker, ed.), pp. 16–67. Plenum, New York.

Hurd, L. E., Mellinger, M. V., Wolf, L. L., and McNaughton, S. J. (1971). Stability and diversity at three trophic levels in terrestrial ecosystems. *Science (Washington, D.C.)* **173,** 1134–1136.

Hutchinson, G. E. (1959). Homage to Santa Rosalia, or why are there so many kinds of animals? *Am. Nat.* **93,** 145–159.

Knipling, E. F. (1979). The basic principles of insect population suppression and management. *U.S., Dep. Agric., Agric. Handb.* **512.**

Kuhlman, E. G.(1978). The devastation of American chestnut by blight. *In* "Proceedings of the American Chestnut Symposium" (W. L. MacDonald, F. C. Cech, J. Luchok, and C. Smith, eds.), pp. 1–3. West Virginia University, Morgantown.

Lapwood, D. H. (1968). Observations on the infection of potato leaves by *Phytophthora infestans*. *Trans. Br. Mycol. Soc.* **51,** 233–240.

Larcher, W., Heber, U., and Santarius, K. A. (1973). Limiting temperatures for life functions. *In* "Temperature and Life" (H. Precht, J. Christophersen, H. Hensel, and W. Larcher, eds.), pp. 195–231. Springer-Verlag, Berlin and New York.

Levine, R. P. (1968). "Genetics," 2nd ed. Holt, Inc., New York.

Little, R., and Manners, J. G. (1969). Somatic recombination in yellow rust of wheat, *Puccinia striiformis*. *Trans. Br. Mycol. Soc.* **53,** 251–258.

Malalasekera, R. A. P., and Colhoun, J. (1968). *Fusarium* diseases of cereals. III. Water relations and infection of wheat seedlings by *Fusarium culmorum*. *Trans. Br. Mycol. Soc.* **51,** 711–720.

May, R. M. (1974). "Stability and Complexity in Model Ecosystems." Princeton Univ. Press, Princeton, New Jersey.

Merkel, H. W. (1905). A deadly fungus on the American chestnut. *10th Annu. Rep., N.Y. Zool. Soc.* pp. 97–103.

Moller, W. J., Schroth, M. N., and Thompson, S. V. (1981). The scenario of fireblight and streptomycin resistance. *Plant Dis.* **65,** 563–568.

Muenscher, W. C. (1952). "Weeds." Macmillan, New York.

Muller, H. J. (1918). Genetic variability, twin hybrids and constant hybrids in a case of balanced lethal factors. *Genetics* **3,** 422–499.

Mulligan, G. A. (1960). Polyploidy in Canadian weeds. *Can. J. Genet. Cytol.* **2,** 150–161.

Mulligan, G. A. (1965). Recent colonization by herbaceous plants in Canada. *In* "The Genetics of Colonizing Species" (H. G. Baker and G. L. Stebbins, eds.), pp. 127–146. Academic Press, New York.

Murdoch, W. W., Evans, F. C., and Peterson, C. H. (1972). Diversity and pattern in plants and insects. *Ecology* **53,** 819–829.

Nelson, M. R., and Pound, G. S. (1959). The relation of environment to the ringspot (*Mycosphaerella brassicola*) disease of crucifers. *Phytopathology* **49,** 633–640.

Park, T. (1954). Experimental studies of interspecific competition. II. Temperature, humidity, and competition in two species of *Tribolium*. *Physiol. Zool.* **27,** 177–238.

Parmeter, J. R., Jr., Snyder, W. C., and Reichle, R. E. (1963). Hetero-kaiyosis and variability in plant pathogenic fungi. *Annu. Rev. Phytopathol.* **1,** 51–76.

Pelczar, M. J., Jr., and Reid, R. D. (1972). "Microbiology," p. 127. McGraw-Hill, New York.

Peltier, G. L. (1926). Effects of weather on the world distribution and prevalence of citrus canker and citrus scab. *J. Agric. Res. (Washington, D.C.)* **32,** 147–164.

Powers, L. (1945). Fertilization without reduction in guayule (*Parthenium argentatum* Gray) and hypothesis as to the evolution of apomixis and polyploidy. *Genetics* **30,** 323–346.

Reichert, I. (1958). Fungi and plant diseases in relation to biogeography. *Trans. N. Y. Acad. Sci.* [2] **20,** 333–339.

Reichert, I., and Palti, J. (1967). Prediction of plant diseases occurrence: Pathogeographical approach. *Mycopathol. Mycol. Appl.* **32,** 337.

Remington, C. L. (1968). The population genetics of insect introduction. *Annu. Rev. Entomol.* **13,** 415–426.

Ridley, H. N. (1930). "The Dispersal of Plants throughout the World." Ashford, Kent.

Russell, G. E. (1978). "Plant Breeding for Pest and Disease Resistance." Butterworth, London.

Sailer, R. I. (1978). Our immigrant insect fauna. *ESA Bull.* **24,** 3–11.

Smith, H. H. (1971). Broadening the base of genetic variability in plants. *J. Hered.* **62,** 265–276.

Smith, N. G. (1970). On change in biological communities. *Science* **170,** 312–313.

Spassky, B., Spassky, N., Pavlovsky, O., Krimbas, M. G., and Dobzhansky, T. (1960). Genetics of natural populations. XXIX. The magnitude of the genetic load in populations of *Drosophila pseudoobscura*. *Genetics* **45,** 723–740.

Stebbins, G. L. (1956). Artificial polyploidy as a tool in plant breeding. *Brookhaven Symp. Biol.* **9,** 37–52.

Stebbins, G. L. (1965). Colonizing species of the native California flora. *In* "The Genetics of Colonizing Species" (H. G. Baker and G. L. Stebbins, eds.), pp. 173–191. Academic Press, New York.

Stevens, N. E. (1917). The influence of temperature on the growth of *Endothia parasitica*. *Am. J. Bot.* **4,** 112–118.

Stipes, R. J., and Campana, R. J. (1981). "Compendium of Elm Diseases." Am. Phytopathol. Soc., St. Paul, Minnesota.

Strickberger, M. W. (1968). "Genetics," p. 508. Macmillan, New York.

Sys, S., and Soenen, A. (1970). Investigations on the infection criteria of scab (*Venturia inaequalis* (Cke.) Wint.) on apples with respect to the table of Mills and Laplante. *Agricultura (Heverlee, Belg.)* **18,** 3–8; *Rev. Plant Pathol.* **50,** 702 (abstr.).

Thornthwaite, C. W. (1931). The climates of North America according to a new classification. *Geogr. Rev.* **21,** 633–655.

Townsend, J. I., Jr. (1952). Genetics of marginal populations of *Drosophila willistoni*. *Evolution* **6,** 428–442.

Trewartha, G. T. (1954). "An Introduction to Climate." McGraw-Hill, New York.

van Arsdel, E. P., Riker, A. J., and Patton, R. F. (1956). The effects of temperature and moisture on the spread of white pine blister rust. *Phytopathology* **46,** 307–318.

van der Plank, J. E. (1968). "Disease Resistance in Plants." Academic Press, New York.

van Emden, H. F., and Williams, G. F. (1974). Insect stability and diversity in agroecosystems. *Annu. Rev. Entomol.* **19,** 455–475.

Waddington, C. H. (1965). Discussion of paper by H. L. Carson. *In* "The Genetics of Colonizing Species" (H. G. Baker and G. L. Stebbins, eds.), pp. 547–549. Academic Press, New York.

Walker, J. C. (1922). Seed treatment and rainfall in relation to the control of cabbage black leg. *U.S. Dep. Agric., Bull.* **1029.**

Walker, J. C. (1969). "Plant Pathology." McGraw-Hill, New York.

Walker, J. C., and Wellman, F. L. (1926). Relation of temperature to spore germination and growth of *Urocystis cepulae*. *J. Agric. Res. (Washington, D.C.)* **32,** 133–146.

Whitman, T. G., and Slobodchikoff, C. N. (1981). Evolution by individuals, plant-herbivore in-

teractions, and mosaics of genetic variability: The adaptive significance of somatic mutations in plants. *Oecologia* **49,** 287–292.
Zentmyer, G. A., and Bald, J. G. (1977). Management of the environment. *In* ''Plant Disease: An Advanced Treatise'' (J. G. Horsfall and E. B. Cowling, eds.), Vol. 1, pp. 122–140. Academic Press, New York.
Zohary, D. (1965). Colonizer species in the wheat group. *In* ''The Genetics of Colonizing Species'' (H. G. Baker and G. L. Stebbins, eds.), pp. 403–423. Academic Press, New York.

CHAPTER 9

Stopping Pest Introductions

E. CROOKS, K. HAVEL, M. SHANNON, G. SNYDER, and T. WALLENMAIER

Plant Protection and Quarantine
Animal and Plant Health Inspection Service
United States Department of Agriculture
Washington, D.C.

EXOTIC PLANT PESTS AND
NORTH AMERICAN AGRICULTURE

ISBN 0-12-757880-3

I. LEGAL BASIS FOR STOPPING PEST INTRODUCTIONS

Many of the dangerous plant pests located throughout the world are hitchhikers and can be artificially spread long distances from their habitats. Plant pests are often disseminated throughout the world because of inadequate controls applied in the country of origin. Since controls are lacking in many countries, destination countries must depend on laws and regulations to protect against outside pests and to retard pest spread within their borders.

The agency of the U.S. Department of Agriculture (USDA) that is responsible for keeping these foreign pests and diseases from entering the United States is the Animal and Plant Health Inspection Service (APHIS), through its Plant Protection and Quarantine (PPQ) program. Legal authority to prevent pest entry, retard pest spread within the country, and certify plants and plant products for export is provided by statutes and delegated to the Secretary of Agriculture.

The statutes or acts give the Secretary of Agriculture the authority to (1) take emergency action against plant pests that are new or not widespread and against weeds determined to be noxious; (2) establish restrictive and prohibitory regulations against imports that may introduce exotic plant pests and designated noxious weeds; (3) establish regulations for cooperative federal–state programs to prevent the spread of introduced plant pests and designated noxious weeds; (4) carry out cooperative federal–state or international suppression, control, or eradication measures against designated plant pests; and (5) provide export certification for domestic plants and plant products upon request.

The restrictive powers on importations under the Plant Quarantine Act are broad. It seems clear that Congress intended to give the Secretary authority to restrict and control the entry of plants and plant products to the extent necessary to prevent the entry of plant pests. The Plant Quarantine Act was enacted August 20, 1912, and since that time, approximately 90 quarantines and regulations have been promulgated.

The Federal Plant Pest Act, enacted May 23, 1957, gave authority to the inspectors to take emergency action to seize, treat, or destroy articles or products with respect to plant pests new or not widely prevalent in the United States. This act also gave the USDA authority to regulate the movement of plant pests per se into or through the United States. The definition of a plant pest in this act was greatly expanded in comparison to the definition provided in the Plant Quarantine Act.

The Organic Act of 1944 gives the Secretary of Agriculture authority to cooperate with farmers' associations and individuals, as well as states or political subdivisions, to control or eradicate certain designated pests.

The original Organic Act also gives the USDA authority to cooperate with such efforts in Mexico. Amendments to the act in 1976 broadened that authority to include all countries in the Western Hemisphere. The Organic Act also gives the Secretary authority to inspect and certify domestic plants and plant products as needed in order to meet the phytosanitary requirements of foreign countries.

The Federal Noxious Weed Act of 1974 gives authority to the Secretary to restrict the entry of certain designated noxious weeds and any weed determined to be noxious even though not so designated.

As part of its international cooperation, the United States is a party to the International Plant Protection Convention of 1951 and is a member of two regional plant protection organizations: the North American Plant Protection Organization (NAPPO) and the Caribbean Plant Protection Commission (CPPC). It is also actively cooperating with the Inter-American Institute for Cooperation on Agriculture (IICA), a Western Hemisphere organization. This topic is discussed further in Chapter 17, this volume.

II. PEST RISK REDUCTION SYSTEM

The geographic cropping areas of the United States, which the USDA's PPQ strives to protect from pest invasions, encompass a wide range of ecological extremes and crop systems. These vary from tropical areas of Puerto Rico, Hawaii, and Guam to temperate North America, including Alaska. The types of exotic organisms which are of concern are equally broad, ranging from organisms affecting the production of pineapple and sugarcane to northern hardwood forests. PPQ also has a role in facilitating agricultural export trade by assisting exporters to meet import requirements of foreign countries and to exclude exotic animal diseases.

A. Exclusion Strategy

The strategy necessary to exclude an exotic pest organism is determined by evaluating both the pest and the situation in which it occurs. A successful strategy is dependent on information drawn from the disciplines of pest biology, pest distribution, crop culture methods in the United States and foreign areas, treatments and residue tolerances, commercial practices, and past quarantine experiences with imported pest–host complexes. In the absence of this information, the USDA generally adopts a conservative strategy and uses maximum safeguards to prevent the entrance of a pest-carrying product.

B. Technical Information

Sailer (1978, p. 11) summarized the role of information in pest exclusion quite well when he said, "It is my opinion that our best hedge against the dangers posed by foreign species is intelligence. The more information we have about the world insect fauna, the more likely we are to be able to devise means of excluding dangerous species and to take appropriate counteraction against any that do invade the United States.'' Technical information is important both for long-range planning and for support of everyday activities.

C. Long-Range Planning

Planning for the future involves pest risk analyses based on sound biology. The entry status of a commodity should be correlated with the level of pest risk (Kahn, 1979). The need for biological data occurs when determining the risk involved in important commodities; when there is a need to set policy on certain pathways, such as wood-packing materials, or on certain pest groups, such as insect vectors. Important exotic pests must be monitored, and their biology and distribution used in such planning. If an importer has applied for a permit to import a commodity, this permit will not be issued until the pest risk has been determined for the commodity and adequate safeguards recommended as conditions for the permit issuance. When live insects or pathogens are proposed for importation, technical information concerning the biology of these organisms plays an important part in the decision to allow or not to allow their movement. Technical information about pests is also needed for planning and devising treatments for pests.

D. Everyday Activities

Technical information must be available immediately to allow for a rapid response to an urgent quarantine situation. When an unusual pest is found infesting a shipment at a port of entry, a rapid decision must be made as to the course of action to be taken. This decision must be based upon the identification, biology, and economic threat of the pest involved.

E. Contents of Technical Information

Information used in stopping pest introductions is found in the agricultural and biological sciences, such as entomology, nematology, acarology, malacology, plant pathology, botany, agronomy, and horticulture. The data needed can be arranged into three groups: identification, biology, and distribution. Thus, taxonomy becomes a very important area. In a given genus, for example, if species A cannot reliably be separated from species B, then the quarantine actions taken against species of the genus must reflect this broadened grouping. The biologic data consist of life history information and food preferences. Host preferences are very important in determining whether a particular imported commodity is a host for an exotic pest. The description of the life cycle of a pest is often essential. Where an insect spends its pupal stage often determines the pest risk of its host commodity. A midge, for example, that feeds only on mango flowers is not likely to travel in commerce with mango fruits. The distribution of a pest is also necessary to determine where the pest risk may occur.

F. Sources of Technical Information

Information is needed on the hundreds of thousands of species of plant-feeding insects and the thousands of bacteria, fungi, and virus species. The amount of data needed to be familiar with the pest threat is insurmountable. Compounding the problem is an ''almost total absence of basic data on the biology, physiology, and autecology of all but a few species'' (Baker and Bailey, 1979, p. 43). Technical information, then, is crucial for sound decision making.

Throughout the world, scientific journals, monographs, handbooks, and other publications provide valuable sources of information for plant quarantine. The USDA's National Agriculture Library collects this information and is a valuable source of scientific data.

Scientists, both in the United States and in other countries, provide data on pests or pest groups and often contribute advice and recommendations. Records of pest interceptions at ports of entry provide extremely valuable data on pest distribution and arrival frequency. Some countries publish lists of intercepted pests. Condition or status reports on a country's major pest furnish useful information about potential problems.

G. Pest Risk Analysis

Pest risk analysis is the process of defining safeguards for imports by evaluating the pest of concern and factors that impact on its potential for introduction and establishment. Pest risk analysis has two components, a pest component, and a safeguard component.

The pest component considers the organism's potential agricultural economic impact, including social and economic value of the crops attacked, cost of necessary controls, and the feasibility of eradication.

Safeguard components are situational factors that reduce the probability of pest, pathogen, or weed establishment. Obviously, any imported agricultural material involves a risk of pest introduction. However, there are many factors that may act to prevent pest colonization in a new environment.

Pest risk analysis evaluates the number and types of safeguards necessary to make the possibility of exotic pest establishment remote. For example, the Mediterranean fruit fly is a significant exotic pest, and host products with an infestation potential are treated upon entry or prohibited if treatments are not available.

Exclusion strategies for pathogens of crops and certain ornamental plants also illustrate the concept of risk analysis. The threat posed by this type of exotic organism is high because of the great colonizing potential of plant pathogens, the economic value of the crops attacked, and the difficulty and cost of either living with or eradicating the organism, if established. A zero tolerance for *introductions* of many exotic disease agents is necessary because of their high probability of *establishment* in the presence of a host and a favorable environment.

As an example, the most stringent safeguard, total prohibition, is applied to importation of citrus propagules, since safeguards against pest and pathogen introduction and establishment are minimal when citrus is imported for growing. This is because inspection is not an adequate safeguard against pathogen entry. Further, practical eradicative treatments for imported shipments are not available, and there is sufficient risk of pest escape and establishment.

Although the importation of citrus, corn, wheat, or sugarcane may be prohibited, because of the presence of exotic plant diseases in the country of exportation, provisions are made for the importation of germplasm for research purposes under special permit procedures and safeguards. In addition to the presence of adequate safeguards, the intended use of the material for the betterment of U.S. agriculture receives consideration in determining whether to allow the normally prohibited products to be imported.

When a quarantine-important pathogen is not known to occur in a particular country but is present in nearby areas, imports of propagules of susceptible host materials are placed under post entry quarantine for a period adequate to detect the target pathogen should it be present. The plants are directly monitored by federal officers or state cooperators at the importer's growing site throughout a growing period sufficient to detect latent infections.

Most other propagative materials, with the exception of certain low-risk herbaceous seeds, specified bulbs, and high-risk materials imported for research purposes, are allowed entry at specified Inspection Stations. At these locations, special expertise, equipment, and facilities constitute the necessary safeguards. The act of inspection is deemed an adequate safeguard for any insects and/or diseases moving with these materials.

Examples of plants in this entry category are tropical foliage plants, orchids, and bromeliads. The economic impact of pests and diseases entering on these types of commodities is significantly less than that of organisms entering on food, forage, or fiber crops. The level of risk inherent in these importations is such that mandatory treatment or prohibition would not be required to exclude the organism.

Tolerance for introduction of insect pests is, in general, greater than that for plant pathogens. This is reflected in the exclusion strategy applied to most exotic insects. However, some insects are subject to conservative exclusion strategies, including prohibition or mandatory treatment of host materials. Insects which are not readily detectable by inspection are generally named in agricultural quarantines. Examples include the Mediterranean fruit fly (*Ceratitis capitata*), the mango weevil (*Sternochetus mangiferae*), and the sweet potato weevil (*Eusepes postfaciatus*). These economically important agricultural pests are known to be effective colonizers in new environments. Inspection alone is not adequate to detect these internal feeders. In the absence of eradicative treatments that do not damage the product, these host materials are prohibited from areas where the pests occur.

The majority of exotic insect pests are excluded through inspection of cargoes and carriers at ports of entry. Likewise, noxious weed seeds, which may move with cargoes, are excluded by inspection, with action taken based on findings. As with prohibited plant pests which may be imported for scientific research, importations of weeds are controlled by permits specifying stringent safeguards for containment and limitations on use of the material.

Nematode exclusion places emphasis on the cyst-forming types, and their possible introduction is countered by a total prohibition of soil. However, under special conditions and permits, soil may be imported for research purposes. Means of conveyance, plants, and plant products are inspected for soil contamination prior to their entry into commerce. The entry of bagging materials previously used for root crops is regulated, and fumigation is required.

Exotic phytophagous mollusks are also of quarantine concern but are excluded by inspection. Specific products are not regulated because of them.

No one appears to know specifically what factors explain why one organism successfully colonizes and another fails, or how the organism will behave in a new environment. On this point, Sailer (1978) states that of the 212 immigrant species which became important pests in the United States, only 73 were expected to be important, based on current knowledge of their economic significance in their country of origin.

These unpredictable phenomena also apply to control of pests by means of natural enemies. Here, organisms are purposely introduced, with the intent of colonization leading to successful control of their host (Chapter 14, this volume). Such work provides much of our current information on the nature of colonizing organisms.

Huffaker *et al.* (1971) describe a successful natural enemy as being well adapted and usually highly specific to its host, effective in host-searching ability, and having high fecundity, good host discrimination ability, and synchrony with the host life cycle.

The same biological capabilities that determine a successful introduced natural enemy would also make for a successful introduced exotic plant pest (Chapter 14, this volume). Such factors are useful to consider in the pest risk analysis process. They make it clear why certain types of pests, such as crop pathogens, tephritid fruit flies, and certain situations such as imports of propagative materials, generally require most strict safeguards to be imposed by quarantine enforcement agencies.

The quarantine regulations and the administration instruction manuals, developed from the regulations help PPQ officers (college-trained managers and officers) to identify pest–host complexes of high risk and guide their decision making by specifying safeguards to be applied. These guidelines have certain features in common: They identify the pest, pest type, or pest group and the avenue for entry, and specify the safeguards to be applied to counter introduction and establishment.

H. Pest Identification

PPQ officers have the authority to stop and inspect any incoming cargo to determine whether or not living stages of quarantine-significant plant pests or weeds are present, and if found, to take whatever action may be necessary to deal with that risk. Before such action is taken, however, the pest must be authoritatively identified to determine if it is of quarantine significance.

In the early 1960s, to expedite these identifications, the PPQ program created a cadre of identifiers and located them at major ports of entry. Today, there are 20 such positions in the field of entomology, 14 in plant pathology, and 6 in botany. Each position provides a broad range of coverage. Entomologists are responsible for taxonomic coverage for all insect groups, as well as mites and snails. Similarly, plant pathologists cover the entire field, including nematology. Botanists have a comparably broad area of identification responsibility, including both noxious weeds and endangered plant species.

Each identifier must provide decumented proof of his proficiency in identifying a particular organism before action can be taken on cargoes or carriers on the basis of his identification. Where such documented proficiency is lacking, the intercepted specimen(s) must be sent to a more proficient taxonomic authority for identification.

In addition to the more common taxa, the program maintains a high level of taxonomic expertise in the following disciplines: mycology, nematology, malacology, and the insect groups Coccoidea, Thysanoptera, and Aleyrodidae. Intercepted organisms in these groups will generally receive final identification without referral to outside experts. Intercepted organisms requiring definitive identification by experts in other than the above groups are referred to more competent taxonomic authorities within the USDA.

I. Responsibility for Quarantine Treatments

When quarantine-significant pests, pathogens, or weeds are found contaminating cargoes or carriers, PPQ requires that actions be taken to eradicate infestations or infections prior to further movement in commerce. Such actions can include an approved processing procedure, application of an eradicative treatment, reexportation, or destruction. With one exception, these actions are taken at the expense of the owner of the commodity or carrier. Enterable propagative plant materials, processed through plant inspection stations and found to be infected or infested with quarantine-significant pests for which treatments are available, are treated by PPQ personnel without charge to the importer. This policy recognizes the sensitivity of propagative materials to phytotoxic effects of treatments and allows for increased PPQ control to minimize the potential for introduction of pests and pathogens through this high-risk entry pathway.

J. Operational Strategies for Exclusion

The international commerce which is responsible for the artificial dispersal of exotic agricultural pests is concentrated at the many ports of entry. This entry threat posed by exotic pests, diseases, and weeds is countered by enforcement actions at these entry portals.

Field operations in the United States are managed through four national regional officers. More than 90 entry points into the United States for incoming passengers, carriers, cargoes, and mail are staffed by nearly 1000 officers with college-level training in the biological sciences. The Hawaiian Islands, Puerto Rico, and the U.S. Virgin Islands areas are also staffed. These offshore operations involve inspection of passengers, cargoes, and carriers before their departure to U.S. mainland locations. Similar activities take place in two foreign locations, Bermuda and the Bahamas, and in many overseas military bases, where military cooperators are used.

All cargo, carriers, mail, and passenger baggage entering this country from foreign areas and offshore locations are subject to inspection at the first port of entry to determine the presence of exotic plant pests and weeds or prohibited/restricted plant or animal materials, and to ensure that entry conditions specified by regulation are met prior to entry.

Port of entry activities may be categorized into a number of different operational activities: inspection stations, foreign air operations, domestic preclearance activities, maritime operations, land border operations, and foreign mail inspection. A given port may include a combination of two or more of these operations.

1. *Inspection Stations*

Pests entering on imported plant materials intended for propagation have a high potential for colonizing new environments. Recognizing this threat, PPQ maintains 14 plant inspection stations for the entry of such materials. Imports move through these locations under the control of permits which include special shipping labels, directing the shipment to the inspection station nearest the port of entry.

Specialized equipment at the stations includes large-capacity fumigation chambers, many with vacuum capability, autoclaves, growth chambers, x-ray units for examination of seed, tanks for applying dip treatments, and microscopes for examination of plant material and identification of minuscule pest structures.

The station staff usually consists of one entomologist and one plant pathologist, assisted by plant quarantine officers as the volume of imported material may require. Because most propagative plant materials are perishable, the taxonomic expertise at the station expedites the identification required to initiate quarantine action. These locations processed nearly 152 million plant units during FY 1982.

2. *Maritime Operations*

Activities at maritime ports of entry encompass clearance of cargo, carriers, and passengers. Clearance of vessels emphasizes compliance with USDA regulations on the use and handling of vessels, food provisions, and resulting garbage. Ships must maintain all provisions, garbage, pet birds, or any other agricultural materials on board the vessel while it remains within U.S. territorial limits. Ships' garbage must be maintained in closed, leakproof containers on board the vessel, and may be removed only under direct supervision of a PPQ officer for destruction in a biologically safe manner. Many ports have incinerators or cookers approved for this purpose, although environmental concerns have reduced the number of locations where this service is available.

In the case of noncompliance with these requirements, permission to use restricted/prohibited provisions is denied by placing materials under seal. The vessel master may also be prosecuted and fined.

All vessels arriving directly from foreign countries are subject to boarding by a PPQ officer upon arrival at a U.S. port, although certain ships with low-risk histories may be deferred or subject only to occasional monitoring. Inspecting officers inform vessel masters of U.S. agricultural requirements, inspect the vessel and its contents for risk situations, and clear passengers. Appropriate vessel food stores may be sealed if there are infestations posing imminent risk to U.S. agriculture. In the case of infestations of the khapra beetle (*Trogoderma granarium*) in vessel stores or holds, fumigation of such compartments is required.

All passengers and disembarking crew arriving by sea vessels are subject to inspection to meet U.S. Customs, Agriculture, and Immigration requirements. Nonenterable materials in baggage are seized and destroyed by PPQ officers.

Maritime ports are the entry point for the bulk of cargo imported into the United States from foreign countries. Any cargo entering a U.S. port of entry is subject to inspection to ensure that it meets entry requirements and is not contaminated with prohibited materials or infested with exotic pests.

Most commercial agricultural products approved for importation are controlled through a system of permits that tends to eliminate or minimize the risk at the source, and to impose restrictions and safeguards as entry conditions. As discussed earlier, the safeguards applied vary with the pests and diseases that are expected to be associated with the product in the country of origin. These range from eradicative treatment by fumigation or cold treatment to inspection, with quarantine actions being taken if exotic plant pests are found.

Inspection of agricultural products, nonagricultural products, and container vans includes vigilance for soil contamination or prohibited packing materials. Contaminated materials must be removed and disposed of in an approved manner, under PPQ supervision, or the contaminated cargo is refused entry.

Inspections and treatments for khapra beetle (*T. granarium*) are important activities at maritime ports. Certain cargo from specified countries requires fumigation upon arrival as a condition of entry based upon a history of infestation a large proportion of the time. Certain exports from countries where khapra beetle is endemic are inspected prior to release to importers. In general, these are dry plant products such as grains, spices and gums, foodstuffs, cotton piece goods, and most materials contained in burlap bags.

Because of the importance of forestry to the United States, imported logs and lumber receive intensive inspection for the presence of wood-infesting pests. In addition, incoming materials accompanied by wooden crating, pallets, or dunnage are inspected and fumigated with methyl bromide if quarantine-significant pests are found.

The advent of containerized shipping has complicated this inspection process. Unitized cargo is less accessible for inspection, moves from foreign sources to interior U.S. locations in relatively short periods of time, and has increased the use of low-quality wooden dunnage to stabilize cargo during transportation.

3. *Airport Operations*

The international movement of carriers, passengers, and cargo by air into the United States provides a pest introduction pathway of great concern to U.S.

quarantine programs. Accordingly, a great deal of program emphasis is placed on this activity, and all such entry into the United States is subject to inspection at the point of arrival.

All meal provisions and garbage on commercial, military, and private aircraft generally must be removed at the first port of arrival and disposed of by approved cooking or incineration methods. For commercial carriers, this is often accomplished under regularly monitored compliance agreement arrangements. Officers inspect carriers for contraband materials left behind by deplaning passengers. In addition, aircraft holds, particularly those of cargo and military aircraft, are inspected for the presence of hitchhiking adult insects. If these are found, aircraft spaces are treated with an aerosol insecticide.

Passenger baggage poses a high-risk entry pathway for numerous plant and animal pests and diseases. A large volume of pest-carrying materials is moved by passengers quickly from foreign countries to favorable pest-establishment area in the United States.

Air cargo poses a similar threat. Most propagative plant materials are imported by air. These are diverted to inspection stations for final clearance. Other cargo clearance procedures parallel those used at maritime ports.

The introduction threat at international airports is highlighted by recent program data. In FY 1982, 245,863 aircraft were cleared and millions of pieces of baggage opened. More than 850,000 seizures of potential pest- and disease-bearing plant and animal products were seized by PPQ officers. This material was found to be contaminated with exotic plant pests nearly 23,000 times. Seizures of potential animal disease-carrying meat and animal products totaled 72,000 units.

4. *Predeparture Clearance Operations*

U.S. offshore areas, insular states, territories, and possessions pose a unique quarantine problem in that they may act as staging grounds for pest or weed introductions to mainland areas. Special domestic or territorial regulations have been developed to deal with this problem.

Hawaii is an example that illustrates the threat offshore locations pose. Although Hawaii has only 8% of the land area of the continental United States, an average of 16 insect species are detected each year. In contrast, the continental U.S. increment is 12 per year. Fully 38% of the Hawaiian fauna are composed of immigrant species.

Locations in this category, in addition to Hawaii, are Puerto Rico and the U.S. Virgin Islands. Besides staffing of airports at these locations for clearance of arriving international flights, predeparture agriculture inspection is provided for flights to the mainland. Such clearance consists of inspection of all mainland-bound passengers before they board their departing aircraft.

In FY 1982, 193,000 seizures of prohibited and restricted materials were made from such inspections. Unlike clearance of foreign carriers, which is accomplished in cooperation with the U.S. Customs and Immigration Service, this inspection is done solely by PPQ personnel.

5. *Land Border Operations*

PPQ inspectors are also on duty at border crossings from Mexico and Canada into the United States. The vast majority of PPQ resources are allocated to the major border crossing points on the Mexican border. Activities at these locations consist of inspection of cargoes, vehicles, passengers, and rail traffic from Mexico. Three of the PPQ-staffed Mexican border locations include plant inspection station facilities.

Because of the geographic proximity of the United States and Mexico, there are many similarities between their pest fauna. As a result, the number of pests and diseases of concern is more restricted than that from other foreign areas. However, many economically important pests are excluded in the operation.

Canada and Mexico administer entry requirements which are less restrictive in some respects than those of the United States. As a result, materials which enter those countries often are offered for entry at U.S. border crossings. Another consideration is that air passengers enter these foreign countries with less restrictive inspection requirements and then continue on to the United States.

The United States imports more produce from Mexico than any other foreign country. Such cargo is inspected and cleared at border crossing points. In addition, field inspection surveys are conducted in crop production areas in Mexico.

Inspection of vehicles is also a major border activity. A total of 53,000,000 vehicles cleared Mexican border crossings in 1980. There is also a great deal of railway traffic between the United States and Mexico. Railroad cars returning to the United States are inspected and cleaned as required when contamination or agricultural residue materials are present.

6. *Mail Inspection*

International mail facilities operate at many air and maritime ports. Inspections of packages are a cooperative effort with the U.S. Customs Service, generally in the primary role of referring suspect parcels for agricultural inspection. In the process of clearing 1,750,000 packages in 1982, 10,270 lots of prohibited quarantine material were intercepted, including 22,856 pounds of meat and animal products.

K. Quarantine Treatments

In the early days of plant quarantine enforcement, few treatments were available to control undesirable pests which accompanied imported plants and various agricultural commodities. Today, many commodities are permitted entry into the United States subject to some type of treatment.

PPQ policy is that treatments will be required only in situations of actual infestation or infection, or in any situation of imminent dissemination of an agricultural pest or disease, or when prescribed as a condition of entry in the U.S. Code of Federal Regulations (CFR).

The treatment program not only eliminates pest and disease hazards that may accompany international trade but also allows and encourages international trade in commodities that would otherwise be prohibited entry or be unsuitable for export.

Available treatments include a wide variety of nonchemical procedures that are integrated or compatible with trade practices. Where such procedures can be utilized, they are preferred to those using chemical agents.

To safeguard our agricultural and horticultural industries, admissible nursery stock items are imported through special inspection stations equipped to fumigate or otherwise treat plant importations. Fruit or vegetable imports may be restricted to certain ports of entry because of the lack of treatment facilities.

U.S.-grown agricultural commodities are also treated, mostly by fumigation before export, to meet quarantine regulations of other countries. The treatment of agricultural commodities for export appears to be increasing due to additional quarantine restrictions by other countries.

1. *Fumigation*

Fumigation is the type of chemical treatment used most often by quarantine organizations. Fumigants are generally preferred because they form lethal concentrations within enclosures and commodities, whereas other types of insecticides penetrate with difficulty or not at all.

Methyl Bromide. Methyl bromide is the most common fumigant. The main advantages are its high toxicity to many insects, mites, and other pests in all stages of development, including eggs; its generally good tolerance by many plants and commodities; and its stability and high penetrability.

Methyl bromide is the basic treatment for imported nursery stock, plants, and seeds, since some 95% of such plant materials moved in commerce are tolerant to the dosage schedules that will kill the pests concerned.

2. *Other Chemical Treatments*

Aerosol treatments, including the pesticide *d*-phenothrin, are prescribed for certain hitchhiking pests, including fruit flies inside aircraft, and are not applied in the presence of passengers.

A malathion-carbaryl dip may be used for imported plant material which is known to be sensitive to methyl bromide fumigation. It is not as effective as fumigation and is thus limited to use only with certain pests. The entire plant is submerged in the chemical dip for 30 sec, and agitated to eliminate air pockets and provide complete coverage.

3. *Nonchemical Treatments*

Cold treatment (refrigeration), vapor heat, quick freeze (frozen pack), hot water dips, and heat treatments are nonchemical quarantine treatments.

a. Cold Treatment and Quick Freeze. Since the early 1900s, sustained cold treatment has been employed as an effective method for the control of the Mediterranean and certain other tropical fruit flies. Exposing infested fruit to temperatures of 36°F or below for definite periods results in the complete mortality of the various life stages. Under prescribed conditions, this treatment can be used on vessels while the cargo is in transit. The procedures provide for temperature-recording equipment control by requiring calibration tests prior to loading and for air circulation control by a prescribed method of stowage under supervision. Other aspects of the operation and certification are designed to provide practical overall control at the point of origin. The entire treatment record is reviewed for accuracy and completeness when the vessel arrives at a U.S. port of entry.

Imported fresh fruits commonly cold-treated in transit include apples, apricots, citrus, cherries, grapes, nectarines, peaches, pears, and plums. The cold treatment consists of precooling the fruit completely to 31°–36°F and holding it at or below that temperature for 10–22 days, depending upon the fruit fly species and the treatment temperature maintained.

The quick-freeze or frozen pack treatment consists of initial freezing at subzero temperatures (F) and subsequent storage at temperatures not higher than 20°F. It is an effective treatment against fruit flies and other insects.

A combination of methyl bromide fumigation plus refrigeration treatment has been recently developed for the treatment of fruit for certain pests. This combination treatment provides for effective quarantine treatment which could not be accomplished with methyl bromide alone because of injury to the fruit. After fumigation, the fruit is refrigerated for 3–10 days depending upon the fumigation schedule used and the refrigeration temperature. Fresh fruits treated under this schedule include apple, avocado, apricot, cherry, nectarine, peach, pear, and plum.

b. Vapor Heat. This was first used on a large scale on Florida citrus during the 1929 Mediterranean fruit fly outbreak. An 8-hr warmup and an 8-hr holding period was used. Later, alternative schedules were developed of 4-hr and 8¾-hr holding periods. The commodity is heated by air saturated with water vapor at 110°–112°F until the approximate center reaches that temperature. After the exposure period, the fruit is immediately cooled.

c. Heat Treatments. Hot water dip is approved for various bulbs and other plants mainly for nematode control. Schedules range from 110°F for 4 hr to 118°F for 30 min depending upon the pest and the host involved. Other heat treatments include both steam and dry heat, which are used to decontaminate soil-infested equipment to prevent the introduction of serious plant pests.

4. *Treatment Locations*

Although many quarantine treatments are conducted at U.S. ports of entry, a high percentage of fruit shipments are treated in the country of origin under PPQ fruit preshipment clearance programs. The current list includes various fruits from Chile (methyl bromide fumigation), apples from France (cold treatment), oranges from Israel [ethylene dibromide (EDB) fumigation], mangoes from Haiti (EDB), and citrus and mangoes from Mexico (EDB). Large quantities of papaya (Hawaii) and mango (Puerto Rico) are fumigated with EDB before shipment to the U.S. mainland.

5. *Irradiation*

In the future, irradiation may provide a potential quarantine treatment for fruits and vegetables. Research has been conducted during the past 25 years on the effects of ionizing radiation on immature fruit flies and other pests of fruit in Hawaii and elsewhere. In general, these studies support the view that irradiation will prevent the emergence of fruit flies at a dose level between 10 and 20 krad. Further research is necessary on the tolerance of commodities to the treatments and the methodology for their application. Tolerance must also be established by the U.S. Food and Drug Administration for this form of treatment.

III. SUPPRESSION AND ERADICATION PROGRAMS FOR INTRODUCED EXOTICS

In addition to the exclusion and inspection activities of APHIS, the agency conducts programs to combat exotic insects, weeds, and diseases that have gained a foothold in our country. The programs are administered through APHIS's four regional offices. There are basically two types of activity, suppression and eradication. Suppression programs deal with well-established pests while eradication programs deal with those that have become newly established or have a history of reintroduction.

APHIS programs are designed as cooperative programs in those areas where heavy pest infestations occur, and, as federal programs, APHIS activities are not restricted by state boundaries as long as the states recognize the need for pest control action. Covered in this section are examples of some of the APHIS programs in the various areas of responsibility.

A. Insect Suppression

Our primary programs dealing with well-established exotic insect pests include boll weevil (*Anthonomus grandis*), gypsy moth (*Lymantria dispar*), imported fire ants (*Solenopsis richteri* and *invicta*), Japanese beetle (*Popillia japonica*), pink bollworm (*Pectinophora gossypiella*), and the West Indian sugarcane root borer (*Diaprepes abbreviatus*). In addition to the above, APHIS also has some responsibility in the area of unusual outbreaks of endemic pests. Three such programs involve grasshoppers, Mormon crickets, and the rangeland caterpillar.

B. Insect Eradication

Certain pest species are high-risk pests for introduction or reintroduction. APHIS program operations include intensive detection surveys concentrated in high-hazard areas of introduction. Many of these kinds of pests, because they are detected early, can be eradicated before they become well established. Some of the insect pests against which successful eradication programs have been carried out include the Mediterranean fruit fly (*Ceratitis capitata*), the Mexican fruit fly (*Anastrepha ludens*), the Oriental fruit fly (*Dacus dorsalis*), and the khapra beetle (*Trogoderma granarium*). Eradication is usually the result of the use of several control measures, such as quarantine, toxic food and lure baits, fumigation, sterile male releases, and chemical spray programs. All or a limited combination of the above control measures have been used successfully in each eradication program.

C. Noxious Weeds

Weeds cost American farmers and consumers approximately $18 billion annually and many of our most serious weeds are of foreign origin. Congress passed the Federal Noxious Weed Act in 1974, and it was signed by the president on January 3, 1975. The act was first funded in FY 1979. There are approximately 95 weed species on the Federal Noxious Weed List or recommended for the list. One of the first programs initiated was hydrilla control in Florida and California. Efforts to free clogged waterways from hydrilla have continued with APHIS cooperative programs currently in effect in both of these states.

About one-third of the noxious weed species in the United States occur on limited acreage, lending their ecological position to eradication feasibility studies. Three such studies were conducted under cooperative agreements with Louisiana State University (itchgrass, *Rottboellia exaltata*), Utah State University (goat's-rue, *Galega officinalis*), and the Idaho State Department of Agriculture (common crupina, *Crupina vulgaris*). Currently APHIS is also carrying out population suppression and eradication measures against witchweed (*Striga asiatica*). It is an annual parasitic seed-bearing plant that attacks plants in the grass family. This pest is found in North and South Carolina, and since APHIS's involvement witchweed has been eradicated from 10 of the original 38 counties where it was found.

D. Nematode Suppression and Eradication

Cooperative programs with the state of New York seek to contain the golden nematode (*Globodera rostochiensis*), a plant pest that attacks potatoes and tomatoes. APHIS is hoping to reduce populations to an undetectable level on cropland in the United States by the end of calendar year 1988.

E. Plant Disease Quarantine and Eradication

Citrus canker (*Xanthomonas citri*) was eradicated from the United States during the 1940s. Currently, citrus from Mexico is under quarantine because of this disease. Also of concern to APHIS is *Lachnellula willrammii,* a disease of larch that was originally from Europe but is now found in Canada and some of the northeastern states. Quarantines will be proposed to impede its spread.

IV. NEW TRENDS FOR NEW PROBLEMS

Processing travelers' baggage is one of the major problems facing the PPQ Agriculture Quarantine Inspection program at international gateway airports. As greater number of foreigners visit the United States and U.S. citizens travel abroad, passenger processing capabilities of the Federal Inspection Services (FIS) (Customs, Immigration, and Agriculture) decrease. Airports designated as official airports of entry are suffering generally from overcrowded conditions. Many of the international terminals are also crowded because facilities, space, and equipment are outdated; many were built before the advent of jet aircraft.

When new, widebody aircraft unload a full passenger complement, 450 persons or more disembark and clog outdated facilities. At many of the larger airports, up to three or four of these large aircraft arrive simultaneously, discharging 1000–1500 passengers. No wonder many travelers dread their arrival at a U.S. airport, which almost surely involves up to several hours of waiting and entry processing time.

Inundated as the FIS agencies are becoming, they have worked miracles in trying to keep pace. New inspection systems, computerized passport controls, cooperative and cross-designated inspection efforts, and other initiatives have helped, but have barely maintained an acceptable level of dealing with the increases in travelers. Slowly, though, the FIS agencies are losing ground and have realized that drastic actions are called for in order to deal with the problem in the future.

PPQ officers consider baggage inspection to be one of the most critical functions in protecting American agriculture. Many of the food and plant products carried either hidden or unintentionally by travelers are prohibited entry, including fresh fruits and vegetables, plant parts (cuttings, seeds, bulbs, etc.), and fresh, dried, or canned meat products.

As more travelers arrive for processing through the baggage inspection system, fewer bags are inspected to avoid delay. The significant drawback to this expediency is that increased amounts of contraband are missed, thereby increasing the chance for pest or animal disease introductions.

While pressure against the FIS agencies to carry on an effective baggage inspection program has increased, PPQ has developed contingency plans to offset reduced inspection, including innovative approaches to detecting contraband in travelers' baggage while allowing the traveler to move through the clearance process more rapidly. Of the many proposals discussed, several interesting approaches evolved.

A. Detector Dogs

Detector dogs have been considered in searching for prohibited contraband in travelers' baggage. PPQ has tested aggressively trained dogs, an approach that proved effective when working on large pieces of checked baggage. Tests have proved conclusively that dogs can be trained to identify nearly any odor, in addition to their ability to detect the scents of narcotics and explosives. However, since much of the prohibited agricultural material confiscated from travelers is found in hand-carried baggage, an aggressive canine response in the vicinity of travelers could be dangerous and therefore unacceptable.

Nonaggressive or passively trained dogs can be used effectively. These dogs can be used among travelers in the baggage pickup area, sniffing out both the larger pieces of checked baggage and smaller hand-carried items.

Dogs can be trained to intercept several exotic (high-risk) fruits and meat products, including fresh, dried, and even canned meat and meat products. Since dogs adapt well to many situations and training techniques, their future in helping PPQ detect undeclared, prohibited products in travelers' baggage is nearly assured.

B. Soft X-Rays

Soft x-ray imagery, which has been successful for other detection purposes, is being tested to detect agricultural contraband. Adapting this technique to search for plant and animal products required considerable innovative transition research but was accomplished by the USDA's Agriculture Research Service. PPQ's Methods Development Staff worked out imagery training techniques. When expansion of x-ray use by PPQ at selected airports is fully operational, the effectiveness of the system should be immediately beneficial.

C. Mechanical Sniffers

The Agriculture Research Service is also developing carbon dioxide "sniffers" and other electronic/mechanical probe devices which may well allow PPQ to eventually "see into," sample the air of, or otherwise manipulate travelers' baggage to ensure that no prohibited materials are present. All of this is possible without opening the baggage or even slowing down the traveler when moving through the clearance process—unless the presence of contraband is detected or strongly suspected.

D. Profiles

Human judgment plays a vital role for the U.S. Customs and Agriculture officers who separate high-risk from low-risk travelers. To aid their judgment, the Agriculture Profile System (APS) has been developed by PPQ. It utilizes input from many field employees with many years of passenger inspection experience.

The APS, in simple terms, is a detailed listing of the personal features, characteristics, behavioral patterns, and other details regarding individuals, couples, or family units presenting themselves and their baggage for port of entry clearance. For example, a traveler very likely to have prohibited agricultural material is an elderly, plainly dressed woman from a Mediterranean country traveling alone with numerous modest pieces of luggage (some often tied with rope). She is likely to state in broken English that she is coming to visit (or live with) relatives who moved to the United States some years earlier.

The officer who encounters this woman and her baggage would judge her to be a highly probable risk because she is of modest means and is visiting or immigrating to the United States from a foreign country and has been traveling for some time. Many elderly persons from the Old World travel with a supply of food to ensure that eating will not be a problem during their trip. Also, this woman is from a country whose cultural standards demand gifts from relatives. Home-processed foods or home-grown fruits and vegetables are valued gift items. The woman's tied and modest baggage also suggests that it contains more than clothing and personal effects.

Interception records of the many years show that with the above profile individuals carry an 80% probability of having one or more prohibited agricultural items of concern.

By drawing together much of the knowledge gained by experienced officers over the years about travelers from various parts of the world, a fairly accurate set of profile examples has been accumulated. They are taught to new PPQ officers and to Customs and Immigration officers who work where travelers' baggage is inspected for agricultural contraband.

The eventual goal of PPQ is to be able to pick out only those bags of the 3–4% of the travelers most likely to be carrying agricultural materials so that the officers can make an entry decision on the products. These scientific and technological approaches can help the agency fulfill its obligation to protect American agriculture, and do it rapidly and with fewer inspection officers.

V. CONCLUSIONS

Inspection of passengers and animals goes on night and day, year after year. Few people realize the importance of the job agricultural inspectors perform in guarding our country from unwelcome pests. Yet the future of our nation's largest industry, agriculture, with its assets totaling more than $1 trillion, can be profoundly affected by the introduction of an unwanted pest. About 23 million people work in some phase of agriculture. Their jobs are protected just as surely as are the 2.4 million farms in America. And, of course, the budget of every food consumer is protected from the appetites of foreign pests and the waste of diseased food animals.

This protection cost American taxpayers approximately $45.4 million in FY 1982. Based on the total United States population estimated at about 232,114,000 in FY 1982 census, that works out to about 20 cents per person.

REFERENCES

Baker, C. R. B., and Bailey, A. G. (1979). Assessing the threat to British crops from alien diseases and pests. *In* "Plant Health" (D. L. Ebbels and J. E. King, eds.), pp. 43–54. Blackwell, Oxford.

Huffaker, C. B., Messenger, P. S. and DeBach, P. (1971). The natural enemy component in natural control and the theory of biological control. *In* "Biological Control" (C. B. Huffaker, ed.), pp. 16–62. Plenum, New York.

Kahn, R. P. (1979). A concept of pest risk analysis. *Bull. OEPP* **9,** 119–130.

Rapoport, E., Ezcurra, E., and Drausal, B. (1976). The distribution of plant diseases: A look into the biogeography of the future. *J. Biogeogr.* **3,** 365–372.

Sailer, R. I. (1978). Our immigrant insect fauna. *Bull. Entomol. Soc.* **24,** 3–11.

CHAPTER 10

How to Detect and Combat Exotic Pests*

KE CHUNG KIM

The Frost Entomological Museum
Department of Entomology
The Pennsylvania State University
University Park, Pennsylvania

*Authorized for publication on September 29, 1982, as paper no. 6524 in the Journal Series of the Pennsylvania Agricultural Experiment Station, University Park, Pennsylvania 16802.

ISBN 0-12-757880-3

I. INTRODUCTION

> Nowdays we live in a very explosive world, and while we may not know where or when the next outburst will be, we might hope to find ways of stopping it or at any rate damping down its force.
>
> Charles S. Elton (1958)

Exotic pests frequently make the headlines because of potentially high economic losses caused by their ravages and the exorbitant costs for their control. The Mediterranean fruit fly [*Ceratitis capitata* (Wiedemann)] in California is a case in point. Despite national efforts to prevent the entry and establishment of exotic pests during the past 100 years, hundreds of exotic plant pests and diseases have become established in the United States (Sailer, 1978). During the 10-year period from 1960 to 1969, 83 new exotic species of insects and mites became established in the United States, with an average of 8.3 species per year. Furthermore, there are more than 1300 plant pests that have not yet become established but are believed to be a significant threat to American agriculture (McGregor, 1973).

Regulatory plant protection in the United States usually takes four lines of defense. Inspection of cargo, passengers, and baggage at the ports of entry has been the first line of defense to exclude exotic pests. The second involves the detection and survey of plant pests that slipped through the port inspections or that arrived by natural spread from neighboring countries. This is backed by a third line of defense, the eradication, suppression, or containment of those pests. The fourth involves the preshipment clearance and certification by inspection and treatment at the point of origin [Morschel, 1971; U.S. Department of Agriculture (USDA), 1975; Mathys and Baker, 1980].

The basic concepts and strategies formulated under the 1912 Plant Quarantine Act remain in force to guide the current regulatory plant protection programs.

However, many of these concepts and operational strategies are outmoded because of the tremendous changes that have taken place in agricultural production, demography, and international traffic. The few changes made in recent years basically have been piecemeal efforts at the operational level. Thus, fundamental problems of conventional detection and regulatory control programs have not been resolved. Historically, national efforts and resources in plant quarantine and control have been directed almost exclusively to exclusion, eradication, or suppression of exotic pests after they became established in the United States, e.g., the cereal leaf bettle [*Oulema melanopus* (Linnaeus)] and the imported fire ant [*Solenopsis geminata* (Fabricius)].

Because of its low priority in the national plant protection program, the detection of exotic pests has not received major emphasis. Furthermore, the federal leadership has always considered pest detection and survey in terms of dichotomy, i.e., exotic vs. domestic pests, perhaps because of the different jurisdictional mandates and charges of each federal agency. This dichotomous view of plant pest surveys has hindered the development of a holistic approach toward national plant protection.

The variables affecting the movements of plant pests and production costs have changed greatly during the past 50 years. For example, international trade and travel have increased and expanded in volume. New technological advances have been made in transportation and shipping. The costs of labor, energy, and materials have skyrocketed. Accordingly, the fundamental philosophy of plant protection and quarantine needs to be reviewed critically, and regulatory strategies and operational procedures modified as needed. A rational alternative to the defunct national pest survey and detection system must be developed quickly within the framework of national integrated plant protection and pest management strategies. New approaches and technology must be developed for inspection, containment, and control of exotic pests.

Our efforts to increase the production of food and fiber are continually hampered by the economic losses from plant pests and ever-increasing production costs. Economic losses to the world's total crop production due to plant pests, many of which are of foreign origin, were estimated at 35%; the losses are even higher in terms of actual production—54%, or $140 billion. In the United States, the total economic losses and added production costs were estimated to be $10–$15 billion, based on the value of the 1960 dollar (Cramer, 1967; Way, 1976). These figures are undoubtedly conservative, because crop losses and production costs are rapidly escalating. Thus, one of the best ways to keep these costs from going even higher and to minimize the additive burdens on our agricultural economy is early detection and prompt containment of exotic pests before they become permanently established and spread over large areas (Oman, 1968; Kim, 1979).

This chapter deals with biological principles of exotic pests as the basis for regulatory plant protection, and the concepts of plant protection and quarantine, with historical perspectives. It also explores organizational and operational strategies for pest detection and regulatory control and recommends rational approaches to long-term protection against exotic plant pests.

Definitions

An ''exotic pest'' is a pest species (or subspecies or other infraspecific category) which occurs commonly in a foreign country but is not known to occur naturally in the United States.

A ''new pest'' is a pest species not known to be established in the United States. Alternatively, if established, its infestation is distinctly different either in the geographic location or in the type, extent, and intensity of crop damage.

''Regulatory plant protection'' is an approach that employs a combination of biological, legal, chemical, and other appropriate techniques to prevent entry and establishment of exotic and new plant pests in a country or area and to eradicate, contain, or suppress plant pests established in a limited area [National Academy of Sciences (NAS), 1969; Council on Environmental Quality (CEQ), 1972].

II. HISTORY OF REGULATORY PLANT PROTECTION

A. Plant Quarantine and Legislation

Although extensive damage to crops by plant pests was recognized in the early days of American history, little was done about it until the late 1800s. Some concern was expressed about two insect pests introduced before and during the Revolutionary War: the Angoumois grain moth [*Sitotroga cerealella* (Olivier)] and the Hessian fly [*Mayetiola destructor* (Say)]. Although two indigenous insects—the Colorado potato beetle [*Leptinotarsa decemlineata* (Say)] in the 1860s and the Rocky Mountain grasshopper [*Melanoplus spretus* (Walsh)] in the 1870s—attracted national attention, the problems posed by exotic pests introduced with imported plant materials were not addressed (Osborn, 1937; Wiser, 1973).

In 1881 the California state legislature, prompted by the extensive damage inflicted by the San Jose scale [*Quadraspidiotus perniciosus* (Comstock)], enacted the first plant protection law to prevent the introduction and spread of insect pests. This included a system of plant inspection at ports of entry. Many states were faced with similar problems with other insects, such as the gypsy moth [*Lymantria dispar* (Linnaeus)] in New England and the boll weevil (*Anthonomus grandis* Boheman) on cotton in the southern states. Using the needed momentum provided by the California legislation, four western states had enacted general horticultural laws by 1895 (Felt, 1909). In the eastern United States, the danger from the San Jose scale caused by the free movement of nursery stock caused many states to enact legislation and inaugurate plant quarantine and inspection services (Marlatt, 1899). National programs of the U.S. Department of Agriculture (USDA) were developed for detection, quarantine, and exclusion of exotic pests during the 1890s. In 1905, the U.S. Congress passed the first comprehensive federal regulatory legislation, the *Insect Pest Act,* which enabled the federal government to regulate the importation and interstate movement of articles that might spread insect pests (Osborn, 1937; Wiser, 1973). By 1908, plant pest quarantine laws existed in 39 states (Darling, 1977; Wheeler and Nixon, 1979).

The first federal service to deal with insect problems was the U.S. Entomological Commission, headed by C. V. Riley in 1878. He later became the Chief of the Division of Entomology, USDA, and stepped down in 1879 for political reasons. The commission was initially established by an act of Congress in 1877 to study the Rocky Mountain grasshopper and subsequently became the national center of entomological activities in regard to many other economically important pests. Later, the Entomological Commission, with the reinstatement of Riley as Chief Entomologist (1881–94), was consolidated into the Division of Entomology. This was then elevated to the Bureau of Entomology in 1904 (Osborn, 1937; W. D. Rasmussen, personal communication, 1982).

In 1906, the USDA inaugurated an inspection system for plant materials imported for its own use. By 1912, the most significant legislation concerning exotic pests, the *Plant Quarantine Act,* was enacted to provide a federal inspection and quarantine system to protect American agriculture from exotic pests. In the same year, the Federal Horticultural Board, composed of representatives from the Bureau of Entomology and Plant Industry and the Forest Service, was established to administer the foreign and domestic inspections and the control and eradication of plant pests. In 1928, the Federal Horticultural Board became a part of the Plant Quarantine and Control Administration. On July 1, 1932, this was succeeded by the Bureau of Plant Quarantine and two years later was combined with the Bureau of Entomology to form the Bureau of Entomology and Plant Quarantine (Osborn, 1937; W. D. Rasmussen, personal communication, 1982).

In the 1930s, the Department of Agriculture began to consider inspection and quarantine in terms of reciprocal trade agreements with other governments. This led to an amendment of the Plant Quarantine Act in 1936 to require disinfection of plant materials and vehicles and to an informal agreement between the United States and Mexico for cooperation in eradication of the pink bollworm [*Pectinophora gossypiella* (Saunders)]. In 1942, the *Mexican Border Act* was enacted to clarify the earlier legislation for inspecting, cleaning, and disinfecting vehicles, shipments, and baggage (Wiser, 1973). Three other major pieces of legislation were passed to strengthen further federal regulatory efforts against exotic pests:

Control of Insects, Pests, and Grass Diseases Act of 1940, the *Organic Act* of 1944, and the *Golden Nematode Act* of 1948 (USDA, 1975). Noxious weeds were added to the national regulatory program under the *Federal Noxious Weed Act* of 1974.

During the mid-1950s, international travel and trade increased, and inspection at the ports of entry by U.S. Customs officials no longer provided adequate coverage. The USDA had to provide more funds and manpower to continue such activities. In 1957 the U.S. Congress enacted new legislation, the *Federal Plant Pest Act,* under which plant quarantine inspection was facilitated by providing more effective control over the movement of plants and pests. Preinspection was also instituted at some ports (e.g., Hawaii and Puerto Rico) and at Canadian rail and air terminals (Wiser, 1973).

In recent years, the responsibility for detecting and excluding or suppressing exotic and new pests has been shifted from one agency to another with reorganization of the federal government. The Bureau of Entomology and Plant Quarantine was absorbed into the Agricultural Research Service; the bureau's Plant Quarantine Division was charged with keeping invading pests out of the country, whereas the Plant Pest Control Division was charged with preventing spread and controlling and eradicating invading pests within the states and with coordinating pest detection and survey programs (USDA, 1963). These regulatory divisions were later to become the Plant Protection and Quarantine (PPQ) programs, Animal and Plant Health Inspection Service (APHIS). Thus, APHIS has the legislative mandates to "(1) prevent the entry of foreign pests into the United States, (2) control or eradicate outbreaks of those that slip in before they become established, (3) prevent the spread of foreign pests that establish colonies, and (4) suppress periodic outbreaks of native pests too widespread for farmers and ranchers to handle by themselves" (USDA, 1975). PPQ programs include plant quarantine inspections at international ports of entry, pest detection, and surveys, regulatory services, control operations, methods development, and pesticide monitoring.

In 1973, Russel C. McGregor and his Import Inspection Task Force reviewed for the USDA the problems of immigrant pests and identified the problematic areas in plant quarantine and regulatory control of exotic pests. With a request from the International Board for Plant Genetic Resources, a Task Force of the Food and Agriculture Organization of the United Nations (FAO) considered plant health and quarantine problems arising in international transfer of plant genetic resources in 1975 (Hewitt and Chiarappa, 1977).

B. Pest Detection and Surveys

The first national detection and survey program was not organized until 1921, when the federal government was asked to issue pest reports consolidating field observations from different states. Thus, the Insect Pest Survey was born as a national medium to report on insect pest conditions in its bulletin. The field reports were voluntary and of varying quality, and were received mainly from the federal and state research stations. This program neither reached a level of sophistication nor gained professional respect. During World War II, the program was used to detect possible biological warfare (R. Daum, personal communication, 1982).

In 1951, national survey and detection programs were evaluated, and a national insect survey and detection system, the Cooperative Economic Insect Survey, was developed by the Bureau of Entomology and Plant Quarantine, USDA, with empahsis placed on the state level. It had two primary goals: (1) defense against biological warfare agents, and (2) protection against the introduction of biological agents. For this program, a state survey coordinator was designated by cooperating groups in each state, and field surveys were conducted by survey entomologists. After the first survey entomologist was placed in Missouri in 1953, the program rapidly spread to other states, and the involvement of survey specialists brought renewed vigor to the program for the next few years (Gentry, 1977). The program was designed to (1) assist in protecting crops, forest, livestock, and public health, (2) aid industry in determining areas of need for supplies and equipment, (3) ensure more prompt detection of newly introduced insect pests, (4) develop a workable insect pest forecasting service, (5) develop nationwide uniformity in reporting insect conditions, (6) determine the losses caused by insects, (7) maintain records on domestic and foreign economic insects, and (8) provide a basic nationwide structure for biological warfare defense in case of a national emergency. Field observations on plant pests collected by cooperators were published through the Cooperative Economic Insect Report (CEIR), which provided communication among workers in plant protection. CEIR was discontinued with Volume 25, Number 52, in 1975, and in January 1976 the Cooperative Plant Pest Report (CPPR) succeeded CEIR to give a much broader coverage of plant pests and diseases.

CEIR had never reached its goal because of its low priority within regulatory plant protection efforts, particularly after the Bureau of Entomology and Plant Quarantine was reorganized in 1954. The emphasis of the Plant Pest Control Division and later APHIS-PPQ was placed primarily on ports-of-entry inspection and specific action programs against pests such as the gypsy moth, citrus blackfly, and pink bollworm. Thus, pest survey and detection programs became a burden to the agency. Despite the importance of detecting newly established insect pests, the role of pest detection and the quality of the cooperative program continuously declined. Under a system of cooperative agreement with the states, a budget of $500 was provided for each activity month to state cooperators; the annual federal budget for this program amounted to $200,000. This did not change, and no additional appropriations were sought for the program since its initiation in 1951 (Gentry, 1977)! With such funding levels, cooperators could not do too much and had gradually lost vigor and interest, as the federal agency tried to take more control of the program and to demand more productivity from the cooperators (Wheeler and Nixon, 1979). Consequently, state cooperators came to view pest survey and detection as a federal program and simply tried to satisfy the minimum requirements of the cooperative agreement. Finally, the entire program was terminated in 1980, with the last issue of CPPR dated October 1980.

In retrospect, the cooperative survey program was doomed to failure because the conceptual framework and organizational approach were outdated, operational strategies were outmoded, funding was inadequate, and the diverse goals of pest survey by integrated pest management (IPM) programs could not be met. Furthermore, the cooperative program that was initiated in 1951 for insect pests was not prepared to accept surveys for pathogens, nematodes, and weeds in terms of approaches, organization, manpower, and funding. This situation prompted the development of numerous specific programs, often poorly conceived and developed within the agency, without data from cooperating state agencies or the scientific community.

The federal regulatory agency has interacted primarily with the National Plant Board for the cooperative program in plant protection, although most of the actual cooperators were associated with the land grant universities. There was inadequate or often little communication among all concerned, such as the Cooperative Extension Service, Agricultural Experiment Stations, and professional societies. The spirit of cooperation and collaboration that had built an effective program had disappeared. This lack of communication also added to misunderstanding and confusion in the plant protection community.

Furthermore, very little national leadership was provided to develop and refine sampling or survey techniques and data-collecting procedures related to plant pests. For example, the APHIS-PPQ Survey Manual (rev. 1967) was never revised or updated. As a result, the quality of field data on pests submitted to and published in CPPR declined, and much of the data were useless. The quality and value of the data collected through the cooperative program varied widely by state. Some data were quantitative and reliable; others were useless. Certain real-time (urgent) data, such as detection of injurious exotic pests, must reach the action agencies of the user quickly. However, neither state pest reports nor CPPR met this need very well. Often, such data were published in CPPR a month or more after pests were detected and reported by State cooperators.

The data collected through a cooperative program had limited use without an efficient storage and retrieval system. The information published in CPPR was stored in a microfilm system that had not been updated since the mid-1970s (Daum,1982). The data collected through the state programs were stored in card files or simply in folders, and there was no effective way to retrieve and utilize them. Much of the useful data are still locked in cabinets!

The development of a pest forecasting capability was one of the ultimate goals of the survey and detection efforts, but it was never close to reality. Within the context of the cooperative program, it was not a reachable goal. Before a generalized model could be adopted for a cooperative survey program, extensive research on baseline data and systems models should have been done for several major pest species.

With the emergence of integrated pest management strategies after the mid-1970s, pest survey and detection efforts were expanded to include plant pathogens, weeds, and nematodes. In 1976, the National Plant Disease Detection and Information Program was initiated by the USDA/APHIS/PPQ. In 1977 a pilot program was started with a staff of nine plant pathologists to assess plant diseases of three major crops in 10 states in the North Central area.

As the pest survey was directed primarily to detecting exotic pests, USDA/APHIS/PPQ considered domestic pests ancillary to their program. Nevertheless, the need for quality data on domestic pests and an efficient system by which pest information could be effectively stored, retrieved, and delivered on time was appreciated among the disciplines of plant protection, particularly as IPM strategies rapidly developed. In other words, there was no national leadership in survey and monitoring systems for domestic pests, and little effort was made to integrate pest surveys and detection of exotic pests with IPM and extension programs.

The original objectives of the pest survey program, which are certainly as valid today as they were 25 years ago, are attainable with proper funding, organization, leadership, and the correct charges. However, the mission of the program was too broad and ambitious in view of the federal priority and funding level assigned to these programs.

In 1976, a task force was appointed by the Deputy Administrator, USDA/APHIS/PPQ, to review the National Cooperative Pest Survey Program, with six members representing USDA/ARS, the Agricultural Experiment Stations, Cooperative Extension Service, state regulatory agencies, and industry. The report of the task force concluded that the only successful National Cooperative Pest Survey Program is one involving maximum cooperation and input from all possible sources. Their recommendations included revitalizing the Cooperative Pest Survey with additional funding, developing specific guidelines for data collection and submission, improving the quality of CPPR, and computerizing the data storage and retrieval system.

Although the recommendations of the task force were not accepted, another pilot project was initiated in 1978 in the Northeastern Region by USDA/APHIS/PPQ, with vigorous objections from the states. A plant pest survey specialist (Plant Protection and Quarantine Programs' survey officer) was assigned to each of seven states (Delaware, Massachusetts, New Hampshire, New Jersey, New York, Pennsylvania, and Virginia) to survey and develop information on new plant pests and domestic plant pest conditions. They were not provided with adequate guidelines and directives, and the program was terminated in 1981 with little accomplished. Several other pilot projects were initiated and tried for a few years, such as the PPQ High Hazard Survey Program, which identified eight new pests exotic to the United States.

On November 30, 1977, a symposium entitled "Insect Survey and Detection: Past, Present, and Future" was held at the National Meeting of the Entomological Society of America, in Washington, D.C., to review a wide range of problems concerning pest survey and detection. In 1978 the representatives of the Entomological Society of America and the Intersociety Consortium for Plant Protection (ICPP) discussed the matter with the Deputy Administrator of the USDA/APHIS/PPQ and the National Plant Board. At the 1979 national meeting, the Entomological Society of America passed a resolution (ESA Resolution No. 7), calling for the Secretary of Agriculture to furnish the requisite leadership and staff in a cooperative National Pest Survey Program.

In 1979, USDA/APHIS/PPQ asked the ICPP to determine the information needed for plant protection. The ICPP Ad Hoc Committee identified in its final report "a great need for a centralized, national effort to collect, summarize and present plant pest information in a standardized format that will allow interchangeable use through a pest information network" (Epstein, 1980, p. 32).

Following a change of administrative leadership in USDA/APHIS/PPQ in 1980, the cooperative insect pest survey program and all other pilot projects were terminated. In the fall of 1980 a new approach to pest survey and detection was sought, and by February 1981 a cooperative National Plant Pest Survey and Detection System (NPPSDP) was conceived with emphasis on the need for pest surveys and information at the state level. Implementation began with adequate budgeting and capabilities for electronic data processing (Johnson, 1982a,b). After CPPR was discontinued, a pilot publication, "Plant Pest News," was initiated in 1981 to foster communication among workers in the plant protection community. This publication was terminated in February 1982. At present, there is no plan to publish information on plant pests (D. Barnett, personal communication, 1982), except a newsletter *Plant Pest Readout*.

C. Successes and Failures

Some quarantine activities, either alone or in combination with eradication programs, have excluded certain exotic pests. In the United States there are more than 37 plant pests documented to have been eradicated or excluded from the areas of invasion by regulatory activities, e.g., the citrus blackfly [*Aleurocanthus woglumi* (Ashby)] on citrus and mango from Florida between 1934 and 1938, again in 1976, and in Texas in 1969 and 1970; the giant African snail (*Achatina fulica* Bowdich) on many plants from Florida between 1969 and 1975; the medfly on fruits from Florida (four infestations: 1929–1930, 1956–1958, 1963, 1981), Texas in 1966, and California between 1975–1976 and 1980–1982; and the oriental fruit fly [*Dacus dorsalis* (Hendel)] from California (eight infestations from 1960 to 1974).

At the same time, despite the enforcement of quarantine, many exotic plant pests have passed through inspection and early detection networks and have become established in the United States and many other countries (McCubbin, 1954; McGregor, 1973; Gonzales, 1978). Examples in the United States are the cereal leaf beetle, golden nematode [*Globodera rostochiensis* (Woll.)], imported fire ants [*Solenopsis invicta* (Buren) and *S. richteri* (Forel)], whitefringed beetle (*Graphognathus* spp.), and witchweed [(*Striga asiatica* (Linnaeus) Kuntz]. Federal-state cooperative quarantine and control programs have been conducted against a number of exotic plant pests, including the cereal leaf beetle, the medfly, the Mexican fruit fly [*Anastrepha ludens* (Loew)], the boll weevil (*Anthonomus grandis grandis* Boheman), the golden nematode, and others (USDA, 1975). Obviously, agricultural quarantine inspections have not been an entirely effective barrier against exotic pests.

Recently, some have doubted the efficacy of regulatory and quarantine programs, and new approaches have been proposed in view of the increasing and changing volume and patterns of international trade and travel and the rapidly advancing transportation technology (Morschel, 1971; McGregor, 1973; Mathys and Baker, 1980).

III. EXOTIC COMPONENT OF THE WORLD BIOTAS

Every biota of the world has exotic components. Insect pests and diseases of cultivated plants and domestic animals make up a major part of these exotic components, and many are cosmopolitan in their distribution. Man and his activities have modified the distribution of exotic species at an alarming rate during the past few decades. Thus, the biogeography of exotic pests shows a bewildering mixture of biotas; for example, the biota of central Canada may show some similarities to those of Kenya, Peru, or Hokkaido (Japan). As species mixing continues among biotas, the different world biogeographic regions may perhaps eventually be reduced to the three major regions—*polar, temperate,* and *tropical,* or Holarctic, Holotropical, and Holantarctic. Each biota will probably have less specific diversity, but in some cases a higher density and biomass (Rapoport *et al.,* 1976; Ezcurra *et al.,* 1978).

The North American biota began to change dramatically with the arrival of

European immigrants. During the following 480 years, 1115 exotic species of insects and mites have become established in the United States, increasing the fauna by 1.3%. Since about 1920, the rate of establishment has been relatively stable, with an average of nine new insects and related anthropods added annually. More than one-half of these exotic pests are of significant economic importance (McGregor, 1973; Sailer, 1978). Of all the plant and animal species associated with certain crops, introduced species account for less than 10%. However, of the 350 agricultural pests, introduced species account for 63.3% (Simmonds and Greathead, 1977). Large numbers of different animals, including arthropods, are continuously introduced into the United States (DeBach, 1964; Laycock, 1966; Lemmon, 1968). Many animals are imported as pets (Courtenay and Robins, 1975). In 1968, 129,000 mammals, representing 302 species, and 569,000 birds were brought into the United States; in 1972 more than 100 million fishes were imported (Jones, 1970; Courtenay and Robins, 1975). Many species of arthropods are deliberately introduced to help control established exotic pests (Clausen, 1978). From about 1890 to the late 1960s, approximately 2300 species of predators and parasites were introduced worldwide to control plant pests in about 600 situations. Of these introductions, 16% were established and provided complete control of target plant pests, and about 58–60% provided some degree of control (Clausen, 1978; Hall and Ehler, 1979; Hall *et al.*, 1980).

The Hawaiian biota provides an example of species mixing (Zimmerman, 1948). The estimated number of ancestral immigrant species presumed to account for today's Hawaiian flora and fauna was 272 flowering plants, 37 ferns, 233 insects, 22 land mollusks, and seven land birds (Zimmerman, 1948; Carlquist, 1981). These immigrants originated from various continents and Pacific islands by long-distance dispersal since the islands were formed; Kauai is 5.6–3.8 million years old, and the age of Hawaii is less than 1 million years (Carson *et al.*, 1970). Additionally, since the arrival of man, many plants and animals have been introduced to the islands by human immigrants (Carlquist, 1981).

Species mixing among the world biotas is caused naturally by dispersal and migration and by man-associated movements. However, the number of immigrant insect species in a given biota is determined largely by the intensity of commerce, the degree of agricultural and industrial development, internal transportation, isolation, climate, quarantine regulations, and control programs (Ezcurra *et al.*, 1978). For example, the opening of the Suez Canal permitted the passage of many marine organisms from the Red Sea to the Mediterranean Sea (Por, 1971). Many exotic pests have been transported from one continent to another by man's carriers such as ships (Myers, 1934; Bishop, 1951) and aircraft (David, 1949), and subsequently became established in other lands (Elton, 1958). All the cosmopolitan species of rove beetles (Staphylinidae) are believed to have been spread by commerce (Moore and Legner, 1974). Since the seventeenth century, introductions and colonizations of European arthropods on the eastern seaboards of Canada and the United States have brought about the "Europeanization" of that part of the North American fauna (Lindroth, 1957, 1968). In Poland, 11–13% of the flora (250–300 species) are considered to be introduced, and about 10% of the native plant species have become extinct in some parts of Poland during the last 100 years (Kornaś, 1971).

IV. BIOLOGICAL BASIS OF REGULATORY PLANT PROTECTION

Exotic pests, like all other organisms, have an innate capacity to disperse as a means of perpetuating their own species. They are constantly moving in and out of their population area. Some of these immigrants often succeed in colonizing new areas where habitats and environmental conditions are sufficiently favorable for survival and reproduction (Baker and Stebbins, 1965; Johnson, 1969; Rabb and Kennedy, 1979). Similarly, some new pests shift to novel hosts (Templeton, 1979). Under the aegis of man, many exotic pests have gained entry and become established in the United States (McGregor, 1973; Wolfenbarger, 1975; Sailer, 1978). Subsequently, they have caused enormous economic losses in agriculture and forestry.

A. Habitat

A habitat is a place with a particular kind of environment where organisms live, and within it specific niches are occupied by different species. The physical setting of a habitat is constantly changing because of climatic variations and physiological and behavioral responses of the organisms to ever-changing inter- and intracommunity variables (Hutchinson, 1978).

Habitat quality varies over space and time, and the patchiness of a habitat may also differ from one season to the next (Taylor and Taylor, 1977). A habitat which is presently favorable to an organism may become unfavorable through time. Near the edge of a habitat, unfavorable selection pressures are intense and cause a high rate of extinction (Huffaker and Messenger, 1964; Richards and Southwood, 1968). Dispersal provides a means of escape from such unfavorable environments. In a given habitat, as the number of resident species increases, there will be more extinction and less establishment of exotic species. Extinction is likely to be lower in larger habitats where intrahabitat replacement of extinct individuals takes place (MacArthur and Wilson, 1967).

The crop community in an agroecosystem annually goes through a seasonal faunal succession. The crop habitat can be colonized only for a short period of time, the growing season. The crop develops and its quality changes with time, for example, from open to shady, hot and dry to cool and moist. With time, it becomes more complex in structure but stable in microclimate. These changes will then affect the colonization and extinction rate of arthropods in the habitat (Price, 1976).

B. Dispersal to Colonization as a Process

Immigrant populations go through three successive stages before establishing themselves in a new habitat (Fig. 1): (1) "initiation of dispersal in an old habitat (H_1)," (2) "en route," and (3) "entry and establishment in a new habitat (H_2)" (Wolfenbarger, 1975; Hughes, 1979). This process involves numerous biological and environmental parameters that influence the initiation, course, and ter-

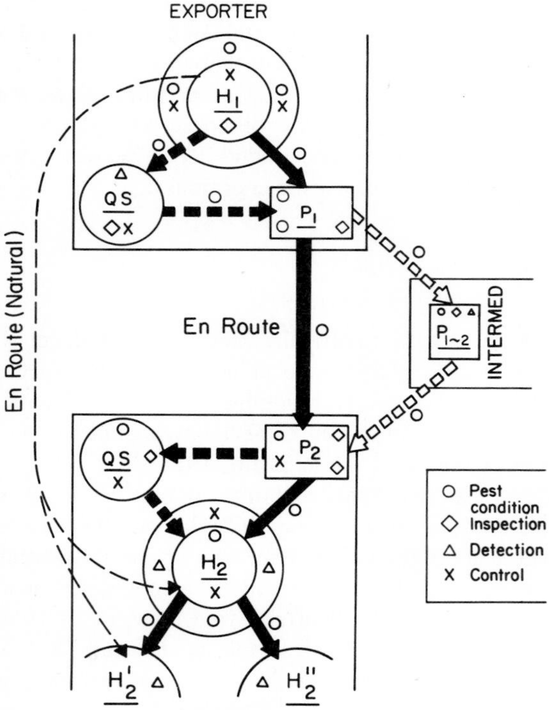

Fig. 1. Diagram showing the movement of exotic pests and plant quarantine process. H_1—habitat 1, original host crop on a farm or nursery; H_2—habitat 2, vulnerable host crop on a farm or nursery; P_1—port of export; P_2—port of entry; QS—quarantine station; INTERMED—Intermediary country or pathway.

mination of dispersal and establishment of immigrant populations (Hughes, 1979):

1. *Habitat 1:* initiation of dispersal
   a. Timing—duration of development and voltinism; suppression of migrant morph formation; reduction or density-induced deterioration in food quantity
   b. Determinants—overcrowding, food shortage, wind, moisture, temperature, light, seasonality; hunger, sex impulse; periodicity of habit, development; level of juvenile hormone; oogenesis and flight syndrome.
   c. Size of movement—population density; proportion of migratory propensity
2. *En route:* course of dispersal
   a. Physiology and reproduction—temperature, moisture, availability of food en route; age, sex, stage (phase, generation); periodicity of habit or state of hunger
   b. Direction—aidants, barriers, temperature, wind, gravity, light, medium, moisture; density level; plant host density, attraction, natural enemies
3. *Habitat 2:* entry and establishment
   Upon arrival, immigrant populations first must terminate the dispersal syndrome and adjust themselves to the environmental conditions of the new habitat. The following conditions are necessary for immigrants to become established in H_2:
   a. Cessation of dispersal—fatigue, end of reserve energy supply; reproductive development; termination of dispersal or flight syndrome; termination of hormonal activity.
   b. Establishment—favorable temperature, moisture, light, host species, host density; lack of barriers and natural enemies; flexible community structure (niche space); high immigrant density

C. Dispersal and Migration

Organisms are displaced by dispersal and migration from the original habitat (habitat 1, or H_1) to the new habitat (habitat 2, or H_2). Conditions in H_1 usually determine the initiation of dispersal movements, and those in H_2 the termination of dispersal and the success of colonization. Within the habitat, the adults move from sites of emergence to sites of feeding, oviposition, and refuge, and the immature insects also move from one site to another for feeding and refuge.

"Dispersal" and "migration" are two terms often used to describe the basic movements of organisms, and many different definitions have been proposed for these phenomena (Johnson, 1969; Kennedy, 1975; Baker, 1978; Dingle, 1981). The dictionary definitions are often quite similar. For example, the term ''dispersal'' (''dispersion'': Wolfenbarger, 1975) is defined by ''Webster's Third New International Dictionary'' (1968): ''disperse'' as ''to cause to break up and go in different ways; to cause to become spread widely; and to spread or distribute from a fixed or constant source.'' Dispersal is used here in a broad, modern sense to refer to the movement of an organism in relation to others from one place to another. This definition includes the man-associated movements of exotic or immigrant species. Thus, dispersal can be accidental (chance) movement or directional migration.

On the other hand, migration is considered an advanced or specialized form of dispersal, and thus refers to the directional movement of organisms with temporal or seasonal regularity coupled with their life history strategy (Johnson, 1969; Dingle, 1972, 1979; Kennedy, 1975; Hughes, 1979; Rankin and Rankin, 1979). Migration in an evolutionary context is a species-specific adaptive behavior especially evolved for the displacement of an individual in space relative to other individuals of the same population (Taylor, 1971; Dingle, 1981). Thus, migration is characterized by the displacement of organisms between places of birth and reproduction (Taylor and Taylor, 1977).

Taylor and Taylor (1977) considered dispersal (called ''migration'') to be a density-dependent spatial behavior resulting from both individual and population interactions. When the population unit is considered to be concentrated at the center of population, the movement of an individual must be either toward (''congregatory'') or away from (''migratory'') the center. Thus, dispersal is a redistributive movement determined by the balance between two antithetical pressures. When the environment is destabilized, selection pressure favors those individuals with a behavioral tendency to move out of the habitat (migration). Conversely, when resources are plentiful, selection favors those individuals that maximize the use of available resources and hence congregate (congregation) (Taylor, 1981a,b).

There is a spectrum of variation in dispersal and migration from chance dispersal (Carlquist, 1981) to spectacular seasonal migrations (Williams, 1958; Johnson, 1969; Orr, 1970; Schmidt-Koenig, 1975; Baker, 1978). These movements can be either short- or long-distance movements, intra- or interhabitat, and active or passive. However, it is often very difficult to determine whether dispersal is accidental, facultative, or adaptive. It is also hard to measure the factors involved in initiation, orientation, maintenance, and termination of the movement (Wolfenbarger, 1975).

Migration by flying insects is a controversial but important subject in pest control and management. Different concepts and theories about migration have been proposed by a number of workers (Williams, 1958; Johnson, 1969; Kennedy, 1961, 1975; Taylor and Taylor, 1977; Baker, 1978; Taylor, 1981a,b).

Migrations of many insects have been exhaustively documented by Elton (1958), Schneider (1962), Johnson (1966, 1969), Wolfenbarger (1946, 1975), Baker (1978), and Rabb and Kennedy (1979).

Migration usually occurs in the prereproductive period of adult insects and involves a distinct set of ecological and physiological characteristics (Johnson, 1966, 1969; Dingle, 1972, 1981; Rankin, 1978). In many insects, migration is often caused by adverse factors in the environment, such as food shortage or overcrowding. Thus, the habitat of early stages in ecological succession, such as old fields, roadsides, waste areas, and ephemeral ponds, would be the place where migration most likely occurs (Southwood, 1962). Migration is controlled by the corpus allatum, which regulates the level of juvenile hormone titers in response to the environmental stimuli of photoperiod, temperature, and food deprivation (Johnson, 1969; Dingle, 1972, 1978; Wolfenbarger, 1975; Rainey, 1978; Rankin, 1978; Rankin and Rankin, 1979).

During migration some insects move self-steered and self-propelled within the boundary layer, but others are carried by the wind when migrants are taken out of the boundary layer by temporarily dominating positive phototaxis (Johnson, 1969; Taylor, 1974). Locomotory and flight functions of adult insects are increased during migration, whereas feeding and reproductive activities are often suppressed. However, migrants are capable of modifying their physiological functions to suit the prevailing ecological and environmental conditions (Dingle, 1972).

D. Modes and Agents of Dispersal

Dispersal and colonization are determined by climate, geography, and physical barriers (Johnson, 1969; Wolfenbarger, 1975; Kennedy, 1975; Rainey, 1976; Rabb and Kennedy, 1979). In natural modes, dispersal and migration are triggered by temperature, sunshine, wind, photoperiod, and population density-dependent factors such as crowding and scarcity of food (Johnson, 1969; Dingle, 1972). However, since the days of early civilization, human activities have become dominant in the movement of exotic pests and changes in the world biota (Lindroth, 1957; Laycock, 1966; Lemmon, 1968; Jones, 1970; McGregor, 1973; Courtenay and Robins, 1975; Sailer, 1978; Clausen, 1978; Carlquist, 1981).

Many different modes of dispersal have been documented (Johnson, 1969; Wolfenbarger, 1975; Simmonds and Greathead, 1977). One obvious means is by active movement of organisms, including walking, crawling, flying, jumping, and swimming, as shown by well-documented examples in locusts, grasshoppers, and aphids (Kennedy, 1961; Hughes, 1963; Johnson, 1969). More important means of dispersal for exotic pests are passive transportation, often referred to as "hitchhiking." Plant pests can be transported by wind and air currents (Hodson and Cook, 1960), atmospheric moisture, water currents, water systems, landslides, volcanic and other geological disturbances, and animals and plants (Johnson, 1969; van der Pijl, 1969; Wolfenbarger, 1946, 1975). Some insects are phoretically transported, e.g., chewing lice (Keiran, 1975). Plant seeds, fungal spores, and even small arthropods may be passively transported by wind, air currents, rafting, or animal movements (Gressitt and Nakata, 1958; Gressitt *et al.*, 1961; Revill *et al.*, 1967; Agnew and Flux, 1970; Carlquist, 1981).

Man-associated movements of exotic species such as the cereal leaf beetle and the medfly are made either by man himself as a direct carrier; by his products of commerce such as fruits, vegetables, fibers, grains, agricultural and forest products, and other artifacts; or by his carriers such as aircraft, automobiles, ships, railways, and others (Wolfenbarger, 1975).

E. Colonization and Adaptive Strategies

The mere arrival of immigrants at new habitats does not necessarily result in successful colonization. In fact, most immigrants become extinct upon arrival because of adverse selective pressures awaiting them. Successful colonizers are usually those immigrant populations possessing adaptive strategies for survival in the different environmental conditions of new habitats. Thus, success of colonization is determined mostly by the genetic makeup of colonizers and the environmental conditions of a new habitat (Templeton, 1979; Parsons, 1982).

During colonization, immigrant populations must fit into available niches in their new habitats by coping successfully with new selection pressures (Simmonds and Greathead, 1977; Templeton, 1979), such as competition from resident organisms for food and space (Turnbull, 1967), adverse climate, attack from predators and parasites (Embree, 1979), and seasonality. This process leads to a series of genetic changes in colonizing populations. The founder effect may increase homozygosity by shifting the allele frequencies, and rapid population growth occurs among colonizers. In other words, colonizing populations can develop rapidly a new set of balanced genes suitable to the new habitat (Carson, 1968, 1975; Templeton, 1979). These genetic changes are often expressed in terms of structure and function, development, life cycle, and behavior. Such changes can be so drastic that even a new taxon can evolve from the event, perhaps complete with reproductive isolation (Templeton, 1979). Similarly, within a habitat, resident populations may shift rapidly to novel hosts because the environmental conditions have changed. Such new adaptation may also cause the evolution of new pests or taxa (Bush, 1975; Templeton and Rankin, 1978). Thus, the habitat of a climax community, where the resident species are finely tuned to a dynamic net of interrelationships, is not the most likely place for exotic species to colonize successfully because of high extinction (MacArthur and Wilson, 1967).

Colonists usually have a set of phenotypes pertaining to the r-K continuum of adaptive strategies (= ecological phenotypes) (Pianka, 1971; King and Anderson, 1971). Ecological phenotypes include tolerance of varying environmental stresses, capacity for population increase, flexible development time, and capacity for resource utilization (Dingle, 1972; Hutchinson, 1978; Parsons, 1982). Colonists are usually generalists with regard to resource utilization (Lewontin, 1965; Kennedy, 1975). Many exotic insects are usually active, vigorous competitors (Wellington, 1964), for example, the bean aphid (*Aphis fabae* Scopoli), the gypsy moth [*L. dispar* (Linnaeus)], and the European red mite [*Panonychus ulmi* (Koch)]. Thus, an understanding of the colonization potential of exotic pests can be made only by ecological and genetic assessment of the source populations.

The adaptive strategies differ depending on whether the colonists are from populations inhabiting central or marginal habitats. There is a substantial literature on the ecology and genetics of central and marginal populations, for example, Mayr (1963), Baker and Stebbins (1965), Carson (1968, 1975), Remington (1968), Southwood (1977), Wiens (1977), Clarke (1979), Jain (1979), Price (1980), and Parsons (1982). In central habitats, the populations are usually large and tend to have high chromosome and enzyme polymorphisms because they are situated in the center of distribution where the environment is most favorable for the species (Parsons, 1982). Here, the population size is primarily influenced by predation, parasitism, and competition involving density-dependent selections (Clarke, 1979).

On the other hand, populations in ecologically marginal habitats are greatly affected by climatic-environmental selection or resource limitation. As a result, the importance of density-dependent selection may be reduced, leading to considerable changes in distribution and abundance and reduction of chromosome polymorphisms and lethal and semilethal genes (Carson, 1975; Wiens, 1977). In this connection, an ecologically marginal habitat, defined by Parsons (1982, p. 117) as "one in which physical stresses tend to be variable and extreme, so that resources are unpredictable and short-lived," must be distinguished from a peripheral habitat in a geographic sense.

F. Entry and Establishment of Exotic Pests

In man-associated dispersal, the behavior of immigrant populations is considerably more difficult to predict because their origin is uncertain. Immigrants may arise from the center or the margins of population distribution. Thus, the success of colonization depends largely on the adaptive strategies of immigrant populations, which are predetermined by the ecological phenotypes of the source populations of species in H_1.

In natural dispersal and migration, the route, seasonality, and life stage of immigrant species are generally predictable. For example, adults of the desert locust [*Schistocerca gregaria* (Forsk.)] migrate seasonally in Africa, and the monarch butterfly [*Danaus plexippus* (Linnaeus)] migrates to the south in North America in the early fall. The bean aphid (*A. fabae* Scopoli) has a diel flight periodicity coupled with the emergence of the adults and thresholds of light and temperature that permit takeoff (Johnson, 1969).

In man-associated dispersal, however, immigrants may arise from source populations in any life stage (eggs, larvae, pupae, or adults) at any time. Specific seasonality, life stage, or pathways of invading immigrants cannot really be predicted unless their mode of dispersal is identified. Although a high population density at H_1 may increase the probability of infesting agents and carriers such as fruits and nursery stock, the movement of immigrant populations is primarily determined by the mode of man-associated dispersal. For example, the medfly is likely to enter the United States as larvae in contraband fruits.

Immigrant populations entering a new habitat face interspecific interactions. The arrival of colonizers may lead to an initial increase in species diversity, and the colonizers face a challenge of either competition or species packing depending largely upon their adaptive strategy involving space and food. In the face of such a challenge, specialist colonizers may become extinct or may displace some indigenous species (Turnbull, 1967), whereas generalist colonizers tend to pack themselves into a limited part of the habitat (Hutchinson, 1978). Furthermore, the outcome of the invasion is influenced by biotic and physical conditions of H_2, such as climate, predation, and parasitism (Parsons, 1982).

Successful exotic pests go through complicated processes before colonization. The balsam woolly aphid [*Adelges picea* (Ratz.)] and the European spruce sawfly [*Diprion hercyniae* (Hartig)], both parthenogenetic insects, are specialists in their feeding strategy. Both originate in Europe, where their impact on host trees is not serious; they moved into Nova Scotia in the early 1930s. *Adelges picea,* a phloem feeder, harms fir trees in two ways: by attacking twigs, which causes stunting but not mortality, and stems, which usually kills the tree (Balch, 1952). Only the first-instar nymphs can withstand freezing temperatures in the range of $-16°$ to $-25°F$. This determines the present distribution. Their habitat is relatively open. The aphid has changed the quality of the fir forest community but not the composition of its insect species (Embree, 1979). At the same time, *D. hercyniae* feeds only on the old foliage of spruce in habitats already fully occupied by indigenous species (approximately 80 defoliators). This colonizer produced a devastating effect on the forest, but once its natural enemies were established, its population decreased. The European spruce sawfly successfully fitted into the spruce habitat, where its niche has been maintained (Embree, 1979).

The medfly and the cereal leaf beetle are generalists. The medfly, distributed from central East Africa to most of the world's tropical and subtropical areas, attacks more than 260 different fruits, flowers, vegetables, and nuts. Its distribution is temperature limited, but its ability to tolerate colder climates is better than that of many other fruit flies (Weem, 1981). The medfly can be transported from one part of the world to another in a few hours as an adult or a larva in fruits or vegetables. With a eurytopic feeding strategy and relatively high reproductive and dispersal potentials, it usually succeeds in colonizing the new habitat, although most of its host trees are already infested with many indigenous pest species. Once it is established, eradication is extremely difficult and costly; its economic damage could run into billions of dollars. The 1980–1982 California Medfly Eradication Program is an example.

The cereal leaf beetle, distributed throughout southern Europe and Russia, was discovered in Berrien County, Michigan, in 1962. In Russia, Rumania, and Spain, it often causes serious injury to wheat and small grain crops. After its discovery in Michigan, the cereal leaf beetle was found in Illinois, Indiana, Ohio, Kentucky, Pennsylvania, Maryland, New York, Virginia, and West Virginia, as well as Ontario (Pfadt, 1971). Although oats and spring wheat are preferred, host plants include spring- and fall-sown small grains, cultivated and wild grasses, and a few other plants. This beetle has become successfully established in fields of small grains that are already infested with well over 100 destructive pests. With successful introductions of parasites and predators, the cereal leaf beetle is permanently fitted into the small grain community in the north central and eastern United States.

Certain organisms are naturally concentrated in a given region as the result of biotic evolution, a phenomenon referred to as the "center of diversity" (Zohary, 1970) or the "center of distribution." The center of diversity is not necessarily the center of origin in a phylogenetic sense. Organisms move from one region to another along with man, primarily through his trade and travel (Kahn, 1977). This movement could follow a direct pathway from the center of diversity to another region, or could take rather complicated pathways involving one or more countries or regions before reaching the United States (Fig. 1). The Japanese beetle and cereal leaf beetle took a simple direct pathway, but the medfly followed a complicated secondary pathway. Thus, many countries or regions serve as temporary stations for an exotic pest that has its greatest development potential and economic impact in neighboring or trading countries.

The medfly, a native of Africa, was initially introduced into Brazil in 1903. Since then it has spread and caused varying degrees of economic losses to fruit and some vegetable production in about 15 countries of Central and South America. In 1929, it penetrated the United States; infestations have been found in Florida (1929, 1956, 1964, 1981), Texas (1966), and California (1976, 1980, 1981). The populations in Central and South American countries might have evolved rapidly under different selection pressures and now may represent distinct biotypes or sibling species differing from the incipient populations in Africa (Parsons, 1982). If so, they will certainly show different biological and ecological requirements. Similar situations may be found with the oriental fruit moth [*Grapholitha molesta* (Busk)], a native of northwest China, which reached the United States in 1915 from Japan (Gonzales, 1978).

In other words, successful colonizers possess several forms of genetic adaptability inherent in their feeding and life history strategies in addition to physical and biological prerequisites. These observations emphasize the need for accurate biological data on the pest species to effectively detect and successfully control exotic pests. Accordingly, for regulatory plant protection, detection and control strategies must be developed on the basis of sound data on the mode of dispersal, the center of distribution, pathways, and adaptive strategies of exotic species, and their current population status at H_1, as well as biotic and physical descriptions and current conditions of H_2.

V. PLANT PEST INFORMATION

The large body of pest information generated and collected annually through various pest management programs is scattered and not easily accessible to the agricultural and plant protection community (Epstein, 1980). Much of this information has been fragmentary and qualitative, and is often completely useless. Thus, a national need has been repeatedly recognized: Sound biological data bases related to plant pests and their biotic and environmental factors must be developed, and pest information must be brought together for ready access by plant protection workers [Council on Environmental Quality (CEQ), 1980; Epstein, 1980].

A. Pest Information Needs

Comprehensive data on pests and related ecological and environmental factors would provide the basis for making sound management decisions. For regulatory plant protection, we must have a knowledge of the dispersal, survival, reproduction, habitat, and host plants of the pest species (Oman, 1968). When exotic pests are detected, a prompt action plan for containment can be developed only with adequate biological and ecological information on the species and their economic impact on commodity.

Needs for pest information are diverse and have become increasingly complex as the agricultural production is being intensified and international trade and traffic are rapidly increasing. Various government agencies, having specifically mandated missions and goals, require different sets of pest information. In addition, because of differing program objectives, different pest management programs and information users demand information that varies in quality and comprehensiveness. For example, each of the plant protection and quarantine programs of USDA/APHIS requires and generates specific information on exotic and new pests.

Pest management strategies require data on the distribution and abundance of endemic pest populations, new pests and their associated natural enemies, and effects on agricultural crops and forests. Current information on the activities and abundance of exotic pests in native lands is also important in predicting the level of interceptions at ports of entry in the United States.

Every plant pest has specific requirements of distribution, habitat, host plants, and climate. The distribution of plant pests and diseases specific to certain cultivated plants is related to the geography of the crops and is determined by a combination of environmental factors and the degree of euryoecity of the invading species (Rapoport *et al.*, 1976). Thus, basic information on the biology, ecology, and environment of exotic pests in their native land should provide a foundation for making better predictions about the degree of invasibility and prospective areas for the invasion and colonization.

The strategy of biological control is based on known information about the degree of specificity that exists between the pest and the biocontrol agent, as well as the distribution of both. Thus, cumulative information on biotic agents for specific pests needs to be readily available.

B. Sources of Pest Information

Pest information generated by pest management and plant protection programs is stored in many ways. Much of the background or cumulative information is documented in scientific and technical publications that can be retrieved from bibliographic resources, such as "Biological Abstracts" and the "Bibliography of Agriculture."

Records on pests and associated information accumulated in federal and state agencies and universities are often kept on cards and in files. More recently collected information on plant pests is stored in computer memories by IBM cards, tapes, and discs.

Pest information is being generated by a large number of federal agencies, such as the USDA (APHIS, ARS) and the Environmental Protection Agency (EPA), by state agencies such as the State Department of Agriculture (Plant Industry), by academic and research institutions such as Experiment Stations, the Cooperative Extension Service, and IPM programs, and by private organizations such as the pest control industry and cooperatives.

Every phase of plant protection strategy generates pest information, from pest identification and quarantines to regulatory measures and pest management. Each plant quarantine and regulatory control program requires and also generates specific pest information: international and port-of-entry operations, new pest detection and survey, emergency programs, and domestic regulatory programs.

These data are stored on microfilm or microfiche. APHIS-PPQ also maintains interception and quarantine records of plant pests; importation, introduction, and mass rearing data on biological control agents are also stored. State pest survey coordinators and survey and extension specialists also have pest information files which have been accumulated for many years. Information on the distribution, abundance, environmental conditions, life cycles, natural enemies, and effects on the host crop of endemic pests has been generated through pest surveys related to pest detection, regulatory, and IPM programs. However, the data collected through these programs vary widely in their quality and usefulness (Kim, 1979). Above all, these pest data are often inconsistent and not easily comparable because they were collected by different sampling techniques.

IPM programs throughout the United States annually generate large amounts of biological information on pests of major crops. A number of small information systems are being developed to manage properly this vast body of pest information, for example, the Kentucky IPM Data Base Management System. There are a number of national efforts to capture and coordinate this IPM-related information (Bottrell *et al.*, 1976; CEQ, 1980). Biological control programs involving exploration for natural enemies or screening of biocontrol agents, importation, introduction, mass rearing, successful establishment, and efficacy for the target pest all generate large volumes of pest information. A complete record of the introduction and establishment of biocontrol agents in this country is not readily available. The Beneficial Insects Research Laboratory and the Beneficial Insect Introduction Laboratory, among other laboratories of USDA/ARS, along with several state bureaus of plant industry, are directly involved with importation and establishment of biocontrol agents (Wheeler and Nixon, 1979). They maintain records on the importation and release of biocontrol agents in the United States.

C. Type and Content of Pest Information

In regulatory plant protection work, biological knowledge of exotic pests must be readily available so that control strategies can be quickly implemented upon the introduction of exotic pests. A modest effort was initiated to meet this need by instituting the publication of taxonomic descriptions, distribution, biology, and life cycle under the project entitled "Insects Not Known to Occur in the United States (INKTO)" in 1977, and 161 species were documented in CEIR. Subsequently, additional series were published in CEIR and CPPR. APHIS-PPQ plans to issue a separate publication on INKTO (D. Barnett, personal communication, 1982).

Indigenous plant pests may become resurgent due to (1) the establishment of a known pest in a new environment, especially without the predators and parasites that acted to keep it in balance in its native habitat; (2) the alteration of the environment by man that creates conditions favorable for sustaining high population densities of an organism; and (3) genetic mutations of an organism by natural and artificial means that make it capable of greater exploitation of the crops and animals cultivated by man (R. Daum, personal communication, 1980). Various types of biological information on the target pest are necessary to monitor the resurgence of new pests.

Although the needs and sources of pest information are different, the basic data required by all users should be included in any data base, so as to be compatible nationwide. Such a data base should be interactive and dynamic, and should be modified as the conditions change.

Changes in scientific names and in classification can affect or even change the host or commodity records, distribution data, and importance of individual pests. Likewise, the economic impact of pest species or groups can vary seasonally. Pests may become resistant to specific pesticides, or new pest management technologies may greatly affect population structure or economic threshold. Geographical distributions of pests may expand, recede, or stabilize, or they may become established in areas remote from current boundaries. Cropping, processing, grading, packaging, shipping, or storage practices can all alter pest population development and importance (Knutson and Sutherland, 1979).

The types of data generated will reflect the primary mission of the generator. All pest data have certain common items, whereas the information content of each data base is different. From the operational viewpoint, pest information may be classified into two types: cumulative (historical or background) and real-time (urgent or timely) data. In developing a pest information system, the difference and basic similarity between the two types of data base should be clearly identified. The term "cumulative" refers to the information that is recorded on paper or published in scientific journals and is not time sensitive. Real-time information, the result of immediate inquiry–response types of systems, is current information which requires immediate access by decision makers for making time-sensitive decisions. This requires a minimal time lag from the occurrence of an event to the discovery and action by decision makers. Real-time data become cumulative as they are documented and stored.

Two types of data banks may be established, one for cumulative data and another for real-time information. The cumulative data base is primarily made of alphanumeric information extracted from the literature and unpublished records. These data on pests and their hosts become the basis for risk analysis. The cumulative data base would include the following information (Epstein 1980): (1) pest—scientific name and authority, reference; (2) synonymy, reference citation; (3) host—scientific name and authority, common name, synonymy; (4) biology—life cycle, reproduction; (5) ecology—environmental requirements and constraints for dispersal and survival, habitats; (6) geographical distribution; (7)

natural enemies; (8) economic importance: loss/gain; (9) control practices; (10) modes of dispersal; (11) modes of detection; and (12) genetics—ecological phenotypes.

On the other hand, real-time data include (Epstein, 1980) (1) pest: scientific name, common name; (2) host: scientific name (family name); (3) date; (4) location/port-of-entry; (5) destination, if found in imported material; (6) intensity/abundance; (7) damage/severity; (8) area affected/size of planting; (9) type of planting; (10) sample number; (11) submitted by—person or agency; (12) identifier; (13) method of analysis; and (14) weather data.

As a starting point, the data base may include cumulative information for approximately 100 pathogens, 200 insects and related arthropods, 50 nematodes, and 75 weeds that are considered highly important economically, based on criteria developed by past studies (Arnett, 1970; McGregor, 1973). In addition to the basic data for real-time pest information, each program would add further categories to the input to meet its program needs. Genetic and ecological analyses should be made for those exotic species most likely (or most often intercepted) to become serious pests in the United States (Parsons, 1982).

Data on economic losses of agricultural resources to insect pests are numerators in benefit-cost analyses for agricultural research and regulatory activities (Norgaard, 1976). The importance of these data has long been well recognized, but no serious effort has been made to obtain truly reliable data on the economic losses in the United States. Despite the effort by FAO, research in this area has been neglected (Chiarapa, 1970).

In the past, annual estimates of economic losses and production costs were made by survey coordinators, survey entomologists, and extension specialists, based on their subjective judgments. The results of these efforts have been published in CEIR and CPPR. Data available on economic losses may have been helpful for some general purposes, but they certainly are not reliable enough to be useful. There is no easy way to obtain such data without proper research and funding. Recently, the problems of crop loss assessment were directly challenged by a scientific community (Teng and Krupa, 1980). Currently, a special effort is being made by the USDA to collect such data with proper funding and necessary research under the National Pesticide Impact Assessment Program (NAPIAP).

VI. INTEGRATED APPROACH TO PLANT PROTECTION

The introduction and establishment of exotic pests in the United States closely follows man's traffic and trading patterns, although dispersal behavior varies among species. Exotic species recognize no political or economic boundaries and prefer no particular country or state; no ports or habitats escape the invasion of exotic species. Invasion of exotic species occurs throughout the world, but colonization is limited to the habitat where environmental constraints are minimal.

All living species have different biological characteristics. Thus, every exotic species entering the United States will react differently to environmental conditions in a new habitat, including climate and changing agricultural practices. Once exotic pests become established in a crop community or habitat, they often spread rapidly to adjacent areas of similar plant communities. Economic losses from such infestation, which include crop damages and costs for eradication and regulatory control, affect a broad spectrum of the agricultural economy and society at large.

To accomplish the goals of regulatory plant protection, man-associated agents and modes that are potential carriers of exotic pests must be inspected at all ports of entry. Also, the habitats where exotic pests are most likely to colonize should be surveyed regularly to detect those species which have passed through the inspection process or been carried into the country by natural means. When important exotic or new pests are discovered in a defined area, there should be a concerted effort to eradicate or to contain and suppress them in order to prevent their spread. These efforts involve federal, state and local governments, the public, and the affected agricultural industry. Accordingly, regulatory plant protection must be a cooperative effort involving all parties concerned with plant protection, and its success depends largely upon how well the scientific, human, and financial resources are harnessed. Furthermore, plant quarantine and regulatory control programs must be considered within the total plant protection effort rather than as isolated, independent regulatory activities of the federal government.

Regulatory actions against exotic pests are usually interstate or international and therefore come under the jurisdiction of the federal government. Federal regulatory programs against exotic pests have been constrained by the sporadic and unpredictable nature of the invading pests. Under the existing program approaches and organization, it has been very difficult for APHIS-PPQ to develop long-term programs. Furthermore, the federal regulatory agency is prone to consider regulatory problems of exotic and new pests as primarily legalistic and technical problems, when in fact biological and economic considerations are most critical in regulatory plant protection. Thus, regulatory plant protection must be considered as part of the total national plant protection effort, and its programs need to be based on biological, ecological, economic, and technical as well as societal considerations.

A. Integrated System against Plant Pests

Population management of plant pests, regardless of whether it involves exotic or domestic pests, constitutes a management system based on biological information within the context of a natural ecosystem, agroecosystem, or forest ecosystem. A pest management system has the three basic components of a closed-loop system: (1) a monitoring component that collects and stores data, (2) a decision-making process that analyzes the data and determines management and control strategy, and (3) an action component that carries out the management decisions (Fig. 2). In a functioning system, the detection and monitoring component leads to the action mode via a decision-making process, in which one action loop often determines the future pest condition (Welch and Croft, 1979; Kim, 1979).

A rational pest management system requires that biological processes within the ecosystem be clearly understood, that stochastic features of monitoring and action components be accurately measured, and that the economics of monitor-

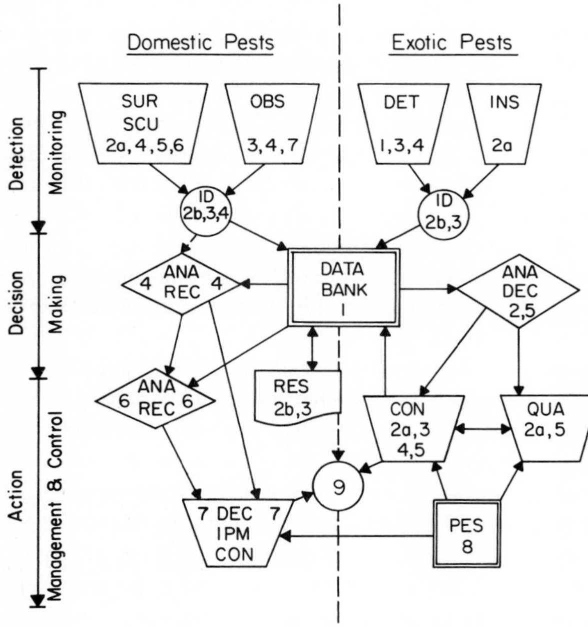

Fig. 2. Flow chart of an integrated plant pest survey and management system. 1, Pest Information Center (data bank) with Pest Management Steering Committee. 2, USDA—a. APHIS/PPQ; b. ARS/Insect Identification and Beneficial Insect Introduction Institute. 3, Agricultural Experiment Station/USDA-CSRS. 4, Cooperative Extension Service/USDA-ES. 5, State regulatory agency (e.g., Bureau of Plant Industry). 6, IPM consultant. 7, Growers. 8, Agrichemical and pesticide industries. 9, News media. Abbreviations: ANA—analysis; CON—control action; DEC—decision; DET—detection; ID—identification; INS—inspection; IPM—integrated pest management; OBS—field observation; PES—pesticides; QUA—quarantines; REC—recommendation; RES—research; SUR—survey; SCU—scouting.

ing, decision-making and action components be critically determined. The success of such a system depends on the quality of information on biological processes and economics related to plant pests (Welch and Croft, 1979). Accordingly, it is necessary for the plant protection community to develop integrated national systems against plant pests with these basic components, through which biological and economic data on plant pests are made readily available.

This pest management approach, which may be called the "Plant Pest Survey and Management System (PPSMS)," is likely to be most effective at the state or regional level, encompassing regulatory, extension, and research programs (Kim, 1979). The system should be implemented to serve and support all participating member organizations and users so that they need not change their task-driven or mandated work (Szyperski and Grochla, 1979). The primary goal of the system is to protect crops and forest resources from plant pests of immediate concern, whether they are domestic or exotic. Central to the system are (1) an efficient information storage and retrieval system, (2) strong leadership of the Pest Management Steering Committee, and (3) adequate taxonomic services (Fig. 2).

The proposed system would be managed by the state or regional coordinator under the direction of a Pest Management Steering Committee. Once it is developed, it must be funded cooperatively by federal agencies and participating state organizations. Moderate funding is needed to develop the core system. The prototype of such an information acquisition/delivery system for extension IPM has been successfully used in Michigan (Croft *et al.*, 1976). Currently, the pest detection and monitoring system similar to the one described here, and with adequate funding, is being instituted nationwide by the USDA (APHIS/PPQ) (Johnson, 1982a,b).

The participants in this system may include the federal and state regulatory agencies, Cooperative Extension Service, Agricultural Experiment Stations, and agricultural consultants and cooperatives. Once the system is established, administration and expansion of the system should be self-sufficient with the support of participating institutions. The Pest Management Steering Committee, consisting of representatives from agricultural and plant protection organizations, would chart the needs and programs of cooperative pest management for a state or region, including quarantine and regulatory control.

Within the system, an important component often overlooked or undervalued is a taxonomic service which provides accurate and timely identification of the pest species and their natural enemies. Taxonomic and diagnostic services must be developed as an integral part of the total system, without which the progress of pest management will most certainly be slow and ineffective. There are only a few adequate taxonomic service centers at the state or regional level, although sufficient systematics resources are available in some regions (Conference of Directors of Systematics Collections, 1971; Hurd, 1974, 1975; Jamnback,

1975). In the past, taxonomic component of pest management strategy have often been considered adequate without evidence. Currently, major national efforts are being made to address the questions of taxonomic services and related research in systematic biology and entomology by the Association of Systematics Collections and the Entomological Society of America. The USDA Cooperative State Research Service and the Agricultural Research Service are also conducting a study to assess and organize taxonomic diagnostic services at the state or regional level.

The first step in any pest management strategy is to know the identity of the pest species and related organisms in an ecosystem. Species identification unlocks the door to the published and unpublished information about the species. Information on the distribution, biology, natural enemies, and ecology of the pest species helps pest management workers find suitable control alternatives (Kosztarab, 1975). Accurate identification of pest species is even more critical in pest detection and quarantine programs. No regulatory action can be taken against unknown or exotic pests without species identification (Kim, 1980). In recent years, considerable concern has been expressed about the status and importance of systematics and systematics collections in relation to agriculture, biology, and environmental sciences (Belmont Writing Committee, 1973; Kim 1975a,b, 1978; Hurd, 1974, 1975; Stuessy and Thomson, 1981).

The second component of a pest management system is a decision-making process in which the data are analyzed and management and control strategies determined (Fig. 2). This process turns the biological information on pests into action. Here, the need for quality and comprehensiveness of pest information becomes apparent.

Basic information on plant pests should be centrally stored and available to all plant protection workers and organizations at the state or regional level. A centralized pest information system must be computer based and easily accessible to users (Szyperski and Grochla, 1979). Such a system would serve as the information clearing house for the state or region. This approach minimizes duplication and warrants a standardization and permanency of pest information. The data stored in the system would be of high scientific quality and would be compatible with those in other similar systems because they are collected by following nationally standardized sampling techniques and input procedures. The data in the system are stored permanently and retrieved easily on-line by workers in research, extension, regulatory, and pest management programs, as described by Brown *et al.* (1980).

At the federal level, there should be a national body, perhaps called the "National Pest Management Council," responsible for guiding national policy and programs on plant pests and related pest management. This council would consist of experts and representatives from various federal agencies concerned with plant pests and pest management, such as USDA-APHIS; Agricultural

Research Service; Extension Service; Economics, Statistics, and Cooperative Service; Forest Service; State Cooperative Research Service; Environmental Protection Agency; and Department of Defense.

The federal leadership (currently USDA/ APHIS/PPQ) should provide national standardization of sampling techniques, input formats, software development, interstate or interregional coordination, data storage for exotic pests, national and international programs and regulatory activities, and periodic publications on plant pests, their natural enemies, and pest management programs.

The Plant Pest Survey and Management System described would primarily support the first two components of the integrated system, namely, detection and monitoring and decision making. As shown in Fig. 2, the analysis and decision-making process is in the hands of an individual agency, organization, or grower. Again, the action component is in the domain of federal and state agencies or individual growers and is based on authoritative recommendations. The recommendations may be made by the Pest Management Steering Committee, National Pest Management Council, Extension Service, or pest management consultants.

B. Regulatory Plant Protection System

Regulatory plant protection is a legislative and administrative means to exclude or restrict the entry of potential exotic pests, to eradicate those detected, and to prevent further spread of plant pests already established in a limited area (NAS, 1969; McGregor, 1973). It is an ongoing process in which current action will determine a future pest situation in an agroecosystem.

Detection and regulatory control of exotic and new pests should be considered a part of the proposed Plant Pest Survey and Management System. Naturally, several aspects of the regulatory plant protection system are different from those of other pest management programs and beyond the purviews of an integrated system proposed in this paper. They include (1) port-of-entry inspections of plants and plant materials and quarantine measures, (2) preshipment inspection and clearance, (3) detection of exotic pests that pass through the inspection process or are transported by natural means, and (4) regulatory control.

Port inspection is a necessary part of regulatory plant protection. However, emphasis should be given to the worldwide movement of pests and a number of high-risk pests, since no quarantine and regulatory control program can provide complete protection against the entry and establishment of exotic pests. As discussed earlier, ecological factors of the habitat or agroecosystem relating to climate, physical condition, and biotic resistance are more important barriers to colonization than are port inspection and quarantine activities.

Detection and monitoring of exotic pests are inseparable at the operational level from pest survey and scouting in pest management programs, and a pest detection system should be an integral part of the Plant Pest Survey and Management System at the state or regional level. Thus, a pest detection and monitoring system needs to be designed to meet the two purposes already described: (1) to detect exotic species and (2) to assess the activities of both indigenous and exotic pests and store these data in a common data bank (Kim, 1979). The data in the system should be versatile and of high quality to be useful to all participating organizations. Thus, central to the success of the regulatory plant protection program is the early detection of exotic pests, including those that have slipped through the inspection process or were transported to the United States by natural means of dispersal. Information on population density and the behavior of exotic pests in their native land will facilitate their early detection in the United States.

The analysis and decision-making component of the system lies in the federal bureaucracy, usually in the hands of the Deputy Administrator of USDA/APHIS-PPQ. Here, the decision on whether to eradicate the limited infestation, to suppress further spread, to contain the population, or to ignore the infestation is made on the basis of biological, economic, legal, and political considerations.

Population management and control, which is the last component of the system, requires the direct involvement and cooperation of the state and local governments. Federal and state regulatory agencies should be prepared to launch regulatory control programs immediately after the decision is made for eradication, suppression, or containment of exotic plant pests. The success of these control programs depends on the readiness of equipment, manpower, logistics, and support funds; for that, strong emphasis must be given to research on the biology of major exotic pests, detection techniques, and regulatory control technology.

In carrying out the mission of regulatory plant protection, the objectives and tasks of every program should be clearly defined and the performance of each program should be systematically reviewed for improvement or termination based on precise patterns of crop production, damage to crops, costs of control and management, and imports, exports, and trade (Mathys and Baker, 1980).

C. International Cooperation

No country in the world escapes the invasion of exotic pests, and almost all countries have some kind of regulatory and quarantine program. An agreement to foster international cooperation in combating plant pests and diseases and preventing their spread across national boundaries was approved in 1951 at the International Plant Protection Convention by the FAO Conference (Ling, 1953; Mulders, 1977; Chock, 1979). It was modified, and 79 countries were signatories to the convention in 1979. During more than 30 years, great strides have been made in implementing the provisions of the convention, particularly in the area of legislative mandates and administrative procedures for the member countries and regional organizations. However, little has been done in the area of research on important plant pests, detection and eradication techniques, and other aspects of regulatory pest management (Mulders, 1977; Johnston, 1979). Global efforts must be focused on research and communication in regulatory plant protection.

D. Plant Pest Risk Analysis

The decision-making process in pest management programs requires the analysis of pest information to determine the magnitude of the risk to our agricultural production and natural resources. Such analysis is as useful for managing domestic pests as it is necessary for regulatory plant protection. The results of critical pest risk analysis can be used to make pest management decisions that minimize reliance on subjective and often biased criteria. Risk analyses should be made in advance for those exotic pests considered potentially serious threats to our agriculture and natural resources as relevant data become available. Such analyses will provide the basis for making sound regulatory and control decisions.

Risk analysis for plant pests and diseases follows two principles: (1) the benefits must exceed the risk, and (2) the benefits must exceed the cost (Kahn, 1977, 1979). Benefits may consist of an introduction of new crops or natural enemies, new varieties of old crops, or new genetic stocks to increase yields, quality, and protein content. Costs involve economic losses inflicted by the pest species and regulatory or control expenses, as well as the costs of manpower and materials for inspection, detection, and treatment.

Pest risk analyses provide objective reference points concerning the economic significance for the target pest species, although they cannot furnish absolute predictions. Predictive precision in pest risk analysis is the function of a conceptual model, the quality and comprehensiveness of a pest data base, and rational judgments of experts and administrators. A new effort is needed to develop a solid data base for such plant pests.

Accurate and comprehensive data on the ecology and genetics of exotic species will provide the basis to determine the kinds of pests that could find niches among major crops or habitats. Thus, pest risk analysis requires detailed biological and environmental information that goes beyond the scientific name and distribution records. The information includes various aspects of biology, endemic range, ecologic limitations, and socioeconomic parameters, as discussed earlier. Information on changes in host variety, utilization patterns, and production technology patterns must be available and subjected to assessment.

Large numbers of plant pests and diseases native to other countries are expected to be introduced into the United States (Hunt, 1946; McGregor, 1973; Reed, 1977; Simmonds and Greathead, 1977). Those exotic species not yet established in North America were enumerated and some identified by Hunt (1946) for destructive plant diseases, by Holm *et al.* (1977) and Reed (1977) for weeds, and by McGregor (1973) for insect pests. Obtaining necessary biological and environmental data is a very expensive and time-consuming task, and it is not feasible to gather such data for all of these pests and diseases. Thus, pest information should be gathered initially for a reasonable number of those species so far recognized as the most serious threats to the United States.

McGregor (1973) proposed a three-step procedure for ranking exotic pests: (1) estimate the probability of specific exotic pests becoming established in the United States, (2) evaluate the economic impact if those pests become established, and (3) multiply the first value by the second. In other words, the expected economic impact is the probability that an exotic pest will become established, multiplied by the economic impact of this occurrence. If we consider other factors of pest risk analysis, the risk value would certainly change. Much research is still needed to develop a realistic model for pest risk analysis.

VII. PLANT QUARANTINE AND INSPECTION

Every country is faced with continuous inflow of exotic pests, and continuous efforts are being made to protect crops and forest from their invasion. Many exotic species are expected continually to enter the United States. Plant quarantine and inspection programs are designed to protect the nation's agriculture and other plant resources from the attack of exotic pests without undue restriction on international commerce and travel.

Plant quarantines have not provided adequate protection against exotic pests. In fact, a comparative study of quarantine programs between the continental United States and Hawaii demonstrated that plant quarantine inspections were much less important than ecological and biological factors in preventing the colonization of exotic species (McGregor, 1973). Furthermore, there are no hard data to support the efficacy of and the argument for continuance of plant quarantine and inspection programs (Mathys and Baker, 1980). Interception records of plant pests are often used to assess the effectiveness of plant quarantine (Oman, 1968; USDA, 1974; MacLachlan, 1977). However, the number of interceptions at ports of entry does not provide a measure of exclusion capability, because the total number of plant pests and diseases entered and the proportion of interceptions are unknown. These observations can lead to the conclusion that plant quarantine and inspection programs are futile and cost-ineffective efforts in restricting the entry of exotic pests, and thus that the merit of such efforts should be questioned (McCubbin, 1954). However, there is no alternative to plant quarantine and inspection at this time. These regulatory activities must be conducted because of the high risk posed by exotic pests in the large volume of international trade and travel. High economic losses to total crop production by these exotic plant pests cannot simply be accepted as inevitable by our society (Cramer, 1967; McGregor, 1973; Way, 1976).

Plant quarantine and inspection should be established on the basis of the following principles (Morschel, 1971): (1) quarantine measures must be based on sound biological facts and principles; (2) they should be used for protection of agriculture and other plant resources, but not for the furtherance of trade; (3) they must derive from adequate legislation and mandates; (4) they should be constantly reviewed and modified or terminated as conditions change or circumstances dictate; (5) the success of programs must be considered attainable; (6) cooperation among all the parties concerned or affected is required; (7) communication must be well maintained among those responsible for the measures; and (8) these agencies must be an integral part of domestic pest management.

Basic strategies for plant quarantine have not changed much since the inception of plant quarantine inspection (McCubbin, 1954; USDA, 1975; Rohwer, 1979). The present strategy includes (1) inspection at the port of entry; (2) inspection at the point of origin, ports of dispatch, and fields in the growing season (export certification and preclearance); (3) embargoes and quarantines; (4) controlled entry (disinfection and disinfestation); and (5) postentry eradication (Mathys and Baker, 1980) (Fig. 1). The first two measures are routinely performed by plant quarantine inspectors under the legislative mandates and international agreements, and the last three activities are usually done in response to the detection of new or exotic pests and diseases.

Ships arriving at ports of entry under U.S. jurisdiction are inspected by PPQ inspectors. Inspection includes the ship's manifests, food lockers, cargoes, and passenger baggage, which are examined by U.S. Customs officers. During fiscal year (FY) 1978, more than 68,000 ships were inspected. Foreign aircraft at international airports are boarded by PPQ inspectors upon arrival to inspect crews, passengers, galleys, and cargo compartments for contraband material, garbage, and plant pests. In FY 1978 alone, approximately 301,000 aircraft and their cargoes and 20 million passengers were processed by PPQ inspectors. On the borders with Mexico and Canada, PPQ inspectors, in cooperation with U.S. Customs officials, also inspect railway cars, vehicles, and pedestrians entering the United States. At the Mexican border ports, about 46 million vehicles and railway cars were processed during FY 1978.

In 1967 the number of travelers crossing our borders was 207 million; this figure had increased to 285 million by 1979. During FY 1972, the total number of interceptions in the United States made by PPQ inspectors was more than 120,000, of which 15,844 were considered economically important and officially recorded (USDA, 1974). Included were, for example, 510 interceptions of the fruit fly, *Anastrepha mombinpraeoptans* Sein, 287 *A. ludens* (Loew) (Mexican fruit fly), 232 *Ceratitis capitata* (Wiedemann) (medfly), and 495 *Dacus dorsalis* Hendel (oriental fruit fly); citrus scales, 375 interceptions of *Parlatoria ziziphi* (Lucas) (Parlatoria scale) and 468 *Unaspis yanonensis* (Kuwana); 736 thrips [*Taeniothrips hawaiiensis* (Morgan)]; 139 khapra beetles (*Trogoderma granarium* Everts); 599 interceptions of citrus diseases, *Xanthomonas citri* (Hasse) Daws, and 694 *Guignardia citricarpa* Kiely. Most of these interceptions were found in small lots of infested plant materials carried in travelers' baggages (USDA, 1974; MacLachlin, 1977). By 1979, the number of interceptions of insects, mites, diseases, and nematodes had amounted to 18,644 (USDA, 1981).

Many countries require phytosanitary certification for importation of agricultural and horticultural products. For American exporters to meet import requirements of foreign countries, PPQ provides a service for inspection of plants and plant products, and certification of shipments being exported. During FY 1978, PPQ issued 84,000 export certificates for shipments.

Shipments of propagative plant materials enter only through designated inspection stations at major ports of entry, where they are inspected and treated if necessary. There are 14 plant inspection stations operating in the United States. Once plants are imported into the United States, certain restricted materials must be grown under controlled conditions, usually for two growing seasons, to meet postentry quarantine requirements. This ensures that the plant or its progeny is free of and will not spread plant pests and diseases.

Other activities of PPQ include foreign mail package inspection in cooperation with U.S. Customs and Postal officials. In FY 1978 about 76,000 foreign mail packages were inspected, and mail containing material dangerous to plant life was impounded or destroyed. Under the Endangered Species Act of 1973 and the Convention on International Trade in Endangered Species, PPQ regulates the movement of terrestrial plant through the plant inspection stations. Inspection personnel of the U.S. Customs Service, Department of Defense, and state regulatory agencies, under cooperative arrangements, conduct a range of plant quarantine inspections for PPQ at minor ports of entry and borders and installations far removed from PPQ-staffed locations.

Agricultural quarantine inspection programs account for approximately one-half of the APHIS/PPQ budget and almost two-thirds of the man-years. The budgeting share of this program has increased in recent years. In FY 1979, the PPQ budget for the program was \$36.2 million, with 1024 man-years, staffing 85 ports and subports. The budgetary allocation for this activity was \$38.9 million for FY 1980, \$41.4 million for FY 1981, and \$45.1 million for FY 1982, which amounts to almost 60% of the total PPQ budget.

New Approaches and Alternatives

A fundamental prerequisite for the future success of plant quarantine is to have biological and ecological data readily available for the major exotic pests of high-risk potential to North American crops and forests (Oman, 1968; Morschel, 1971). The pathways along which plant pests and pathogens enter the United States or move from one region to another must be defined and documented. These data should provide the basis for risk prediction and sound quarantine decisions.

The potential for invasion and establishment of an exotic species can be measured by matching the biological and ecological data of the species in its native land (H_1) with the biotic and environmental conditions of prospective areas or zones (H_2) in the United States that can be defined by the frequency of the interceptions in port-of-entry inspection. No one can make absolute predictions about the invasion and colonization potential of exotic pests, but such data provide the basis for a sound approach to predictive regulatory planning. As these data become available, more accurate predictions can be made on the risk potential of exotic pests and diseases for colonization and economic impact (Kahn, 1970, 1977).

Relatively few exotic pests are likely to cause large losses to U.S. agriculture and forests. Agricultural quarantine efforts should be concentrated on high-risk pests, and inspection should be directed to those commodities and conveyances most likely to harbor them. At the same time, research on risk potential and detection techniques is needed to refine the list of high-risk pests and diseases and continuously to improve inspection procedures (McGregor, 1973).

Without sufficient data on the efficacy and the problems involved in current programs, it is difficult to consider new approaches to plant quarantine and inspection. A conscious effort must be made by the federal regulatory agency to monitor, record, and conduct research on the efficacy and cost-benefits of every measure taken in regulatory plant protection. To provide research and technical support and to improve the predictive capability of regulatory pest management, there is a need to consolidate the research and technical activities of the federal regulatory agency. Toward this end, the APHIS-PPQ's Plant Importation and Technical Support Staff and its related groups within the National Program Planning Staff can be reorganized into a plant protection and quarantine research institute.

1. *Administration of Port Inspections*

Inspection of passengers' baggage and cargo at international ports of entry involves a number of federal agencies, for example, the Bureau of Customs (Department of the Treasury), APHIS (Agriculture), Food and Drug Administration (Health and Human Welfare) and Fish and Wildlife Service (Interior) after passport and document control are completed by the Immigration and Naturalization Service (Justice) and the Public Health Service (Health and Human Welfare). All of these mandated passenger controls cause delays at international ports of entry. These federal agencies are under constant pressure from transportation companies and the public to facilitate passenger control and baggage clearance processes. To improve the inspection procedures and maximize existing resources, all inspection activities should be coordinated under unified policies and guidelines developed by an interagency committee which adequately protects the pursuance of the legislative mandates of each agency. Such a committee could be called the "Interagency Inspection Services Commission," to be composed of senior-level representatives of the inspection agencies. This commission should set policies and guidelines related to port inspections to ensure that no agency unilaterally changes procedures (MacLachlan, 1977). Furthermore, a working committee should be established to provide a harmonious or unified inspection program at each port of entry. This approach would minimize the inconvenience of passenger delays and costs of the baggage and cargo clearance.

2. *Operating Procedures*

Inspection procedures should be uniform throughout the ports of entry, and the public is entitled to expect consistent quarantine and inspection procedures in an effective manner at all locations. The APHIS-PPQ Plant Quarantine Manual is too general and lacks specific instructions and guidelines as to how to proceed with carrier or cargo inspection. It should be frequently reviewed and updated. The manual should stand as an operating prescription for carrying out the goals of plant quarantine, with a step-by-step outline of procedures from the time of carrier arrival until inspection (McGregor, 1973). Instructions for inspection procedures, stored in the federal regulatory data bank, should also be accessible on-line to inspectors at the port.

3. *Ports-of-Entry Inspection*

At the ports, agricultural quarantine inspection is directed to commercial cargo and travelers' baggage, as well as carriers. Most contraband, which amounts to 80–90% of the total confiscated annually in the North American Plant Protection Organization (NAPPO) area (Canada, Mexico, and the United States), is from travelers' baggage, particularily that of airline passengers. Plant products carried by airline passengers are more likely to be carriers of pests and pathogens than are commercial sources (Anonymous, 1981). In other words, travelers' baggage represents one of the principal ways in which exotic pests enter the United States. Thus, agricultural quarantine inspection at international ports of entry is a necessary deterrent against the entry of exotic pests.

Budgetary and staff constraints prohibit the regulatory agency from inspecting all cargo and baggage. Agricultural quarantine inspection programs often resort to sampling in one form or another, and thus inspectors make daily decisions which can be important and perhaps costly without proper and standardized statistical procedures. The costs and benefits of sampling are complex and difficult to assess. Sampling techniques currently used by port inspectors are not based on a statistical design, and results from such samplings are not statistically reliable. Statistical sampling must be developed on the basis of the tolerance level, degree of assurance, and risk potential of pests and commodities in order to meet plant quarantine goals. Research is badly needed to develop statistical sampling procedures (McGregor, 1973; MacLachlin, 1977).

New approaches must be found to alleviate the problem of passenger delays and facilitate the clearance of international cargo at ports without compromising the fundamental goals of regulatory plant protection. Recently, a new system, the so-called Red Door/Green Door Procedures, has been adopted by some countries. It provides one exit for passengers with goods requiring inspection and one for those without such items (Anonymous, 1981). The APHIS-PPQ is also trying out the system in several ports of entry. The efficacy and validity of such a new system must be critically evaluated with respect to the fundamental goals of plant quarantine and inspection before its final incorporation.

The agricultural quarantine program can improve its effectiveness immediately by (1) upgrading public information and education and (2) improving detecting devices. The first can be implemented immediately, but the second will require time for research and development.

4. *Public Information and Education*

Efforts of the U.S. plant quarantine program to inform and educate the public are meager at best. The current effort began in 1960 and has not achieved the goals of creating general awareness and alerting the public to the threat of exotic plant pests and diseases. A continuous effort should be made to inform the news media adequately in order to develop public awareness in the United States. This must be done following the successful strategy used by the Smokey the Bear campaign of the U.S. Forest Service, with the endorsement of the Advertising Council. At the same time, direct messages should be given at the point of contact in order to reach the more than 200 million travelers annually entering the United States. Through the Interagency Inspection Services Commission, one unified multiagency information handout should be developed for use by carriers, terminal operators, and others on entry requirements. The quarantine regulations with respect to contraband articles carried by passengers and imported agricultural products should be strengthened, with stiff legal penalties for the violation. Together with legal deterrence, a greatly expanded effort should be made to inform the traveling public of the possible economic and environmental impact of imported contraband and agricultural products infested with plant pests and diseases (McGregor, 1973).

5. *Detection Devices*

At present, detection is made principally by visual inspection for all areas of quarantine activities. This includes examination for eggs and immature forms of insects and mites, fruiting bodies of fungi, unauthorized packing materials such as rice straw or sugarcane bagasse, hitchhiking insects, wood borers, termites, and soil that may harbor soil-borne organisms including nematodes (Kahn, 1977). There is great potential for the use of sensing devices, commonly known as "sniffers," for detecting fruits and other contraband which may carry plant pests and diseases, in addition to devices such as x-ray photography. Much contraband possesses common chemical compounds; for example, all citrus contains the hydrocarbon *d*-limonene, and all fruits contain ethyl acetate. Such sniffers would certainly reduce the load of individual inspection. Active research is needed to develop effective detecting devices, both biological and electromechanical (McGregor, 1973; MacLachlan, 1977).

6. *Source Inspection*

Containerization has revolutionized the shipping industry. International markets, ports, land and air transportation, shipbuilding, and other aspects of trade have been rapidly transformed by this new technology. The benefits of containerization are obvious for shipping, but its impact on the introduction of exotic pests is not clear. Two points, however, appear to be immediately relevant to plant quarantines: Containers (1) make cargo more difficult to inspect and (2) carry cargo unopened into the interior of the country. However, containerization provides an opportunity for increasing the selectivity of inspection with confidence that introduction will not occur during overland transport, and it makes fumigation and treatment much easier. If a source inspection system is carefully implemented worldwide, containerization may further minimize the threat of exotic pest introduction (McGregor, 1973).

Under close international cooperation and agreements, pre-export inspection and export certification at the point of origin are logically a more effective deterrent against exotic pests than the port-of-entry inspection and would minimize the quarantine risk of a given consignment (Mathys and Baker, 1980). Source inspection systems are designed to control plant pests at their origin and result in shipment of pest-free commodities between countries. They include inspection of plant materials at shipping points, export certification, and inspection at the source of spread (preclearance).

The Agricultural Source Inspection and Surveillance Technique (ASIST), proposed by McGregor (1973), is worthy of serious consideration for implementation as the use of containerization becomes prevalent. ASIST transfers much of the inspection and associated activities to the shipper and the government of the country where the material originates. The responsibility to ship pest-free commodities is shifted to the commercial interests concerned rather than to the inspectors. The incentive for the shippers is the provision of more rapid access to the markets, but for violators strong penalties and fines can be levied. As the International Plant Protection Convention recognizes (Chock, 1979), adequate control of plant pests and diseases in the country of origin is the best safeguard against their spread by plants and plant products.

7. *Controlled Entry and Quarantine*

The importation of plant germplasm is regulated by almost all countries in the world because of high pest and disease risks. Regulatory control incudes (1) absolute prohibition, (2) quarantine (Kahn, 1977; Neergaard, 1977), (3) postentry quarantine (Mathys, 1977; Berg, 1977), and (4) restricted entry. Control may involve fumigation/thermal treatments or confiscation/elimination (Kahn, 1970, 1977; MacLachlan, 1977; Mathys and Baker, 1980). The principles and methodology of plant quarantine in the international transfer of plant germplasm have been extensively reviewed (Hewitt and Chiarappa, 1977). It was concluded that safe and accelerated international transfer of germplasm or propagative units and plant material is an attainable goal with a combination of the proper application of scientific methods and the dedicated efforts of skilled professionals.

VIII. PEST DETECTION AND MONITORING

A. Plant Pest Surveys

Prerequisites to rational pest management programs are (1) early detection, (2) proper identification of pest species and associated natural enemies, and (3) collection of information on their distribution, abundance, and effects on the crop. A plant pest survey is a means to meet these requirements. The scope of surveys is limited by the specific objectives and the availability of sound techniques, funds, and manpower.

Pest surveys have been conducted in research, regulatory, and extension programs to (1) detect exotic pests or new pests and their natural enemies, (2) assess the abundance of pest species and the status of pest activities, (3) estimate damages and loss of crops by pests, (4) study population dynamics with factors affecting the life table, movement, size, structure, and quality of pest populations, (5) test the efficacy of certain control measures, (6) establish economic thresholds, and (7) develop forecasting models for pest species.

As the concepts and approaches to pest control and plant protection change, the approaches and procedures of the cooperative program naturally need to be reexamined and modified. This involves the reorganization of survey, detection, and regulatory apparatus, refined procedures, proper funding, and a resurgence of professional communication. Within the context of newly emerging pest management concepts and strategies, no single agency or organization can successfully carry out survey and detection efforts and pest control programs without cooperation from other organizations. *The ultimate key to the success of such a national program is wholehearted cooperation and collaboration among all of the individuals and organizations involved and interested in pest management and plant protection.*

B. Cooperative System for Pest Detection and Monitoring

Detection of exotic pests and diseases is done primarily for regulatory plant protection, but the assessment and monitoring of plant pest activities are conducted for pest management and pest forecasting, as well as quarantine and regulatory control. Accordingly, detection and monitoring of plant pests and diseases have operational congruence whether they are exotic or domestic. The pest data from detection and monitoring are useful to a wide range of agricultural and biological workers and organizations, including the Cooperative Extension Service, federal and state regulatory agencies, Agricultural Experiment Stations, consultants, industry, and growers. They include real-time and cumulative data. Real-time data, for example, detection of serious exotic pests such as the medfly, must reach regulatory agencies without delay, and cumulative data on pest activities must be stored and easily retrieved for use. These data become the basis of regulatory control.

The goals of pest detection and monitoring systems must be to detect exotic and new pests efficiently; to monitor accurately activities of plant pests and diseases for the purpose of pest management and control; and to recognize changing pest situations in the most cost-effective manner (CEQ, 1980). Thus, a system should be a cooperative program that maximizes the skilled manpower available for collecting field data. Detection and assessment of plant pest activities cannot be done satisfactorily by any one agency and must be considered in terms of cooperative and collaborative efforts, which mutually benefit the parties involved in pest management and plant protection.

The Plant Pest Survey and Management System (PPSMS), as described already, would be such a system that is computer-based and multipurpose. In this system, detection and monitoring of plant pests is the most basic component, along with adequate taxonomic and diagnostic services (Kim, 1979; Brown *et al.*, 1980). The objectives of the cooperative pest detection and monitoring program would be (1) to collect, store, manage, and effectively manipulate information on activities of plant pests and their natural enemies gathered through the system, (2) to provide pest information promptly to federal, state, and municipal governments, research organizations, industry, farmers, and the interested public, (3) to deliver rapidly real-time information on exotic and new pests and diseases to regulatory agencies, (4) to assist the Cooperative Extension Service programs and pest management cooperatives by providing real-time data on pest activities, and (5) to document and disseminate valuable information on pests and pest management programs (Croft *et al.*, 1976; Kim, 1979).

We must have a clear assessment of the local biota of a crop community if we are to detect exotic pests effectively. With identification manuals for domestic pests in local biota, field cooperators will be able to detect and identify exotic pests promptly. Furthermore, various mechanical sampling techniques need to be used for detection purposes at permanent sampling stations, including light traps, Malaise traps, suction traps, bait traps, and others (Peterson, 1953; Taylor, 1979).

The detection and monitoring system (Fig. 2) is made up of the three basic components: (1) data collection, (2) data bank, and (3) information delivery (Kim, 1979; Brown *et al.*, 1980). Data on pests and their activities are collected by cooperators through a pest survey network of the cooperative pest detection and monitoring system. Data collection is made by field observations or scouting, field sampling with various instruments, and trapping at permanent stations with light, suction, or other types of traps (Taylor, 1979). There are two types of cooperators: primary cooperators, who are employees of the cooperating agencies and organizations, and voluntary cooperators. Survey activities of the primary cooperators must be an official function of the cooperating agencies and organizations, which would include federal and state regulatory agencies, land grant universities (both the Agricultural Experiment Stations and the Cooperative Extension Service), agricultural cooperatives, and industry. A strong national leadership is needed to develop standard procedures in data collecting and input based on critically evaluated information needs on plant pests for the pest management community. Field observations and detection should be quantitative, standardized, and coded for damage and sampling techniques. A survey manual describing sampling techniques, coding systems, and other information on standardized procedures should be made available for all cooperators (Kim, 1979).

The software in the computer data core should include processing (input, output, and search), updating, storage, and auxiliary programs. The computer system must be flexible enough to accommodate newly developed programs, as well as the use of a library of programs for data storage, manipulation, and preparation of summary reports. Auxiliary programs may include population models, economic thresholds, economic losses, and forecasting models.

The pest information should be delivered to the users promptly upon request and in a useful format. The information delivery may be made by on-line communication, printout, or regular summary report. All of these delivery means should be available to users.

To utilize and deliver promptly the vast body of pest information, the data bank for the pest detection and monitoring system must have capabilities for (1) terminal access to stored data, (2) compiling and summarizing field data, (3) interface between the data base and other sources of on-line data, such as weather, marketing, and forecast models, and (4) interprogram data exchange. Accomplishing these tasks requires techniques of data base management (Brown *et al.*, 1980).

At the federal level, the regulatory agency (USDA/APHIS) is primarily interested in detecting exotic and new plant pests. As already discussed, detection and monitoring are the second aspect of the regulatory plant protection program, which provides the data base on exotic pests upon which regulatory decisions are made. The national program should be coordinated by the National Coordinator and his staff under the guidance of the National Pest Management Council. The National Coordinator should be responsible for (1) providing leadership in pest survey and detection; (2) collecting and storing pest information related to regulatory pest management; (3) maintaining and managing the federal pest information system; (4) coordinating interstate or interregional pest detection and monitoring programs; (5) developing and providing sound survey techniques, computer software, and other resources related to pest detection and monitoring to be used nationwide; (6) delivering important pest information to the entire plant protection community on a regular basis; and (7) serving as liaison between federal and state agencies. The national system should also be computer based so that exchange and communication of pest information between agencies and state or regional systems could be made on-line (Kim, 1979). Although timely pest information is communicated on-line, a regular summary report of the pest situation and regulatory pest management on a national level must be published in a timely manner, perhaps monthly or bimonthly.

Currently, a cooperative National Plant Pest Survey and Detection Program (NPPSDP), basically similar in approaches and concepts to the system described here and by Kim (1979), is being actively implemented by the USDA/APHIS-PPQ, with initial participation by 16 pilot states. The NPPSDP was inaugurated in 1981 following the recommendations of the ICPP Ad Hoc Committee on plant pest information needs (1980) and the PPQ Deputy Administrator's Task Force on pest survey and detection (1981). The system is administered by the National Survey Coordinator and the National Plant Pest Survey Committee. The NPPSDP is a computer-based pest survey and detection system that serves the needs of the states concerning primarily domestic pests and the needs of the federal regulatory agency specifically concerned with new and exotic pests that may enter the country. The data collection and input are to rely almost totally on ongoing surveys conducted by the states under the direction of the State Survey Committee, which has multiagency and multidisciplinary representation, and the State Survey Coordinator. PPQ Area Survey Coordinators will handle interstate pest information and those data directly relevant to regulatory plant protection for the APHIS-PPQ (Johnson, 1982a,b).

IX. REGULATORY CONTROL STRATEGIES

To minimize potentially serious economic losses, a large budgetary outlay is made annually to control these pests. For 1982 the federal regulatory control programs cooperating with state agencies cost more than $32 million for 10 pest species: e.g., $1.7 million for the boll weevil (*Anthonomus grandis grandis*

Boheman), \$5.3 million for the gypsy moth, \$6 million for the imported fire ants, \$1.02 million for the golden nematode, \$1.4 million for the Mexican fruit fly, \$8.9 million for the medfly, and \$2.4 million for the pink bollworm [*Pectinophora gossypiella* (Saunders)].

The economic impact of these exotic pests is devastating to the agricultural industry and to the nation's economy. Recent infestation of the medfly in California is a good example. The losses to the California economy due to medfly infestation are much greater than the \$1.4 billion projected, with estimated annual crop losses of \$261–567 million and \$97–330 million for production and handling costs, including chemical control and fumigation, plus \$497 million for initial investment in equipment and plants (Hess, 1981). Eradication costs since June 1980 are already more than \$70 million, and reached \$100 million before the program was completed. The California example shows that it is much more cost-effective to fight invading pests before they firmly establish themselves in the area or country.

Regulatory control strategies are a necessary part of a total plant protection and pest management program. Decisions about control strategies depend on the level of invasion and colonization and the availability of biological and ecological information about the pest species. Making a rational decision also requires information on socioeconomic impacts of the pest. Furthermore, efficient organizational structure and adequate budgetary preparedness must exist to implement swiftly pest control and management programs.

Almost any control strategy against pest species may become ineffective in time because some species may evolve resistance against it, whether chemical control, sterile male technique, or juvenile hormone (Templeton, 1979). Pest control is a man-made selection against pest species, many of which have a broad genetic adaptability. Thus, all control strategies, including regulatory control, must take an integrated, systems approach by using a combination of available techniques.

A. Eradication, Containment, and Suppression

Regulatory control strategies include eradication, containment, and suppression, each of which uses one or a combination of chemical, cultural, and biological measures (NAS, 1969). Any strategy used to control exotic pests must be effective and economical, with the fewest adverse effects on the environment and the public health.

Eradication is a pest control approach to eliminate every individual of the target species from a defined geographical area (NAS, 1969; Newsom, 1978). When eradication cannot be accomplished, the second approach to regulatory pest management is containment. The purpose of pest containment is to retard the spread of a pest species from areas where it is firmly established or where eradicative techniques are not available. A suppression program is developed as a regulatory approach when an outbreak of the pest species occurs suddenly over such wide areas that control cannot be accomplished by individual effort (NAS, 1969).

Controversy surrounding eradication as a pest management strategy is often semantic rather than scientific (Cox, 1978). Nevertheless, here are two extreme conceptual differences that are much more serious than differences in definition of terms (Newsom, 1978). There are many scientists and administrators in the plant protection community who consider eradication to be a practical approach and who believe that there is adequate technology to eliminate a large number of pest species. At the other extreme, there are those who believe that eradication is conceptually fallacious in an ecological and evolutionary context. Nevertheless, only a few seriously question the need for eradication as a control measure against exotic pest infestations in a localized area (Newsom, 1978; Rabb, 1978a; Eden, 1978). The controversy arises when eradication is considered as a control measure against long-standing pest species such as the boll worm [*Heliothis zea* (Boddie)] and the screwworm [*Cochliomyia hominivorax* (Coquerel)] (Newsom, 1978; Knipling, 1978).

Eradication, a man-made extinction of a pest species from a defined area, is not a panacea. Eradication programs cannot be successful unless the following requirements are satisfied: (1) appropriate control technique; (2) basic information on biology, ecology, behavior, physiology, and population genetics of the target species; (3) defined natural barriers; (4) cost–benefit analysis; and (5) organizational, budgetary, and procedural preparedness. Accordingly, eradication programs have achieved limited success by using effective chemical pesticides and autocidal techniques. A great majority of these efforts have been costly failures because one or more of the requirements for successful eradication were not met, as shown by eradication efforts against well-established species such as the boll weevil (Newsom, 1978; Perkins, 1980). In general, these programs usually lacked the basic biological and ecological data needed, and thus were ill-conceived and poorly executed. Necessary research for eradication-related problems has not been properly supported (Newsom, 1978; Rabb, 1978a).

Eradication of the medfly is perhaps the most successful program. The medfly was eradicated from Florida in 1929–1930, 1956–1958, 1962–1963, 1963, and August 1981, from Texas in 1966, and from California in 1975–1976. Success might have been attributed to the small areas infested and to rapid action with highly effective chemical pesticides. As the recent medfly infestation in California demonstrates, eradication is costly and much more difficult to accomplish when a large acreage is involved and action is not promptly taken.

Despite advances made in autocidal technology and population suppression of the screwworm (Bushland, 1975; Knipling, 1978), the eradication strategy with "barrier zones" against this pest has not been completely successful (Newsom, 1978). The difficulty seems to involve the genetic differences among the screwworm populations (Richardson *et al.*, 1982). The natural target populations are genetically different from those being released, and many populations which are geographically overlapping are genetically isolated. In other words, the success of screwworm eradication is naturally difficult to attain because the autocidal control technique developed with the original type of fly is being used against genetically different populations or even different species.

Eradication as a regulatory control measure is still a useful tool, since no other approach provides a swift and immediate remedial action against new exotic pests. Alternative approaches, including biological control and IPM, often require a period of from 1 to as many as 10 years for development, and these approaches are most applicable when the new exotic pest is established and has built up its population in wide areas. On the other hand, eradication as a pest management strategy cannot succeed and should not be conducted unless the invasion of the new pests can be promptly detected and regulatory action taken swiftly, with vigorous cooperative efforts, before these pests spread to a wider area.

The favorite terms related to eradication used by regulatory workers are "suppression" and "containment." These terms are confusingly applied in regulatory plant protection, sometimes by changing the objective of the control effort, the definition of the term, or both. Soon after certain pest species were declared to be contained, "suppressed," or "eradicated," they reappeared or resurged in the same area. The economic benefit from such efforts is at best difficult to estimate. For example, the 10-year suppression efforts against the boll weevil in the Texas High Plains cost $9.8 million and used 964,918 gallons of chemical pesticides (Rummel *et al.*, 1975). The economic benefits from this program were not as great as anticipated, and its worthiness is questionable. Before such eradication programs are continued or expanded and more money is committed, their "success" of such eradication programs must be critically evaluated (Perkins, 1980).

Suppression and containment as pest management strategies have highly limited application. These approaches are viable and worthy of consideration only if (1) cost–benefit analysis shows that such programs will outdo other pest management strategies in economic benefits, (2) biological and ecological constraints will contain and suppress the pest population in the affected area long enough to implement a long-term pest management strategy, and (3) research will be supported to develop a suitable pest management strategy.

B. Development of Regulatory Actions

Regulatory actions against new plant pests are necessary as long as the arrival of exotic pests and the resurgence of plant pest epidemics persist. New plant pests show a wide range of economic significance or risk potential to our agriculture and natural resources, ranging from very serious to highly insignificant. Thus, regulatory actions cannot, and need not, be taken against every new or exotic pest that is detected. Some efficient mechanism must exist in the federal and state regulatory systems to decide whether remedial action is justified against a new plant pest and what action is most appropriate within organizational, technological, and logistic constraints (R. Daum, personal communication, 1980). Action responses for a new plant pest will differ from those for others as the biology of the pest species varies. Alternative action strategies against a new plant pest are best developed on the basis of available ecological and biological information through advance, deliberate planning that takes into account the contingencies associated with the infestation, and appropriate socioeconomic and environmental safeguards.

A key to the successful implementation of regulatory action against a new plant pest is the capability of finding the pest and evaluating the infestation at an early state of invasion. A new plant pest can be promptly detected and identified through a well-organized pest survey and detection system such as PPSMS, as stated earlier. Such a system would provide a large number of trained professionals who routinely look for exotic pests and have access to the necessary taxonomic services.

When a new plant pest is detected, the information must be relayed without delay to the federal regulatory agency to evaluate and develop action plans through the national conglomerate of plant pest survey and management systems. The federal and state regulatory agencies must have the ability immediately to develop, implement, and execute the regulatory options against new plant pests. The current administrative process within the federal agency must be streamlined to have the committee of experts and regulatory administrators such as the National Pest Management Council or its experts' committee evaluate the risk potential and socioeconomic impact of the plant pest and the regulatory control alternatives. The organizational requirements must also include adequate professional manpower, proper equipment, and budgetary and administrative flexibility to execute the regulatory action. Pertinent biological and ecological information, as well as the results of the risk analysis of the plant pest, should be readily available for use through a centralized national data bank for plant pests (Kim, 1979).

C. Current System for Developing Action Plans

Every new plant pest poses different problems to U.S. agricultural crops and forests because each represents different levels of risk and economic significance. The Plant Protection and Quarantine Program, USDA/APHIS, follows two major protocols and procedures concerning the occurrence of new plant pests in the United States. When plant pests already determined to be a serious threat such as the Mexican fruit fly, khapra beetle, or witchweed are detected, prompt remedial action is implemented by PPQ line personnel without going through the time-consuming evaluation process following protocol routines established by the agency. Only those pests whose economic significance is uncertain are subjected to the evaluation process by the New Pest Advisory Group (NPAG) (USDA, 1977, 1981).

Newly detected plant pests go through one or more steps for evaluation, recommendation, and implementation in the following sequence: (1) technical evaluation, (2) regulatory action evaluation, (3) management evaluation, and (4) communication. The technical evaluation is designed to determine the biological significance of a newly reported plant pest which may pose a regional threat and whether it should be evaluated by NPAG. This evaluation by the Plant Importation and Technical Support Staff (PITSS) is usually made within 5 working days, conducted on the basis of the criteria. When a new pest is determined to be sufficiently important, it is referred to NPAG for regulatory action evaluation (USDA, 1977, 1981).

NPAG, appointed by the NPPS Director, consists of five voting members, all from the staff of NPPS, and one or more nonvoting members from PPQ or other federal agencies. It makes appropriate recommendations on program options and regulatory action, following a set of administrative protocols, to the Deputy Administrator (DA), PPQ, USDA/APHIS. The final result of the NPAG evaluation is reported to the North American Plant Protection Organization (NAPPO) and the National Plant Board Advisory Council (NPBAC), with a request for comments to the DA within 3 working days after receipt of the report. The control/management evaluation involves the action by the DA on the recommendations made by NPAG, NAPP, and NPBAC. The decision by the DA is usually made within 2 weeks after the date of transmittal. The final step involves the communication of the DA decision on new plant pests to other federal agencies, NAPPO, NPBAC, and industry groups.

D. Rapid Responses and Preparedness

The early detection of new plant pests is one of the highest priorities in regulatory plant protection (Newsom, 1978; Kim, 1979; R. Daum, personal communication, 1980). A sufficient data base on important exotic pests must exist for developing detection strategies and regulatory action plans and for evaluating pest risk analysis. Little effort has been made to develop such data bases on plant pests at either the state or federal levels. Without an efficient detection system and a pest data base, all regulatory efforts will continually be a haphazard "fire fighting" of emergency situations, responding to the new plant pest only after the problem has occurred.

Regulatory control requires a rapid response and does not permit time to develop alternative strategies. Accordingly, integrated pest management and biological control strategies are rarely suitable for regulatory control of new plant pests because they require considerable time to develop, implement, and execute. For the near future, the regulatory control strategies for most exotic pests will remain heavily dependent on different pesticides, which still are an effective and the least costly means of controlling exotic plant pests. Through data banking of the information already available in the literature and from vigorous research, alternative regulatory pest management strategies should be developed and implemented at the production level when the infestation of a new exotic pest is belatedly discovered.

The early detection of new exotic pests makes remedial action more effective because it permits the use of more vigorous and more intense control actions, which otherwise may not be acceptable sociopolitically and environmentally. Through continuous investment in regulatory research on important exotic pests and some potentially important plant pests (Newsom, 1978; Mathys and Baker, 1980; Batra, 1981), effective remedial action plans should be ready for implementation at the time of early detection, and alternative control strategies should be applicable for those exotic pests discovered to be already established in the United States.

X. CONCLUSION AND SUMMARY

Protecting agriculture and natural resources from the invasion of exotic pests is a complex and perplexing aspect of national plant protection efforts. It involves a multiplicity of parameters in legal, administrative, diplomatic, and industrial as well as scientific and technical areas. Regulatory problems are the result of biological causes no matter how exotic pests are introduced, and yet the problems are solved by human decisions based on analysis of many biological and socioeconomic factors. Thus, without a good understanding of the biology and ecology of the pest species and data on their socioeconomic impacts, rational and effective regulatory strategies cannot be implemented. Future programs will remain merely fire-fighting activities, responding only to the delayed discovery or the establishment of new plant pests. Such an approach does not afford long-term solutions to the invasion of new plant pests.

Every exotic species has distinct biological, ecological, and environmental requirements. However, there seem to exist definite dispersal patterns of certain plant pests due to the frequency and pattern of international trade and traffic (McGregor, 1973; USDA, 1975; Mathys and Baker, 1980) and weather patterns (Taylor, 1979). Information on the patterns of pest dispersal, trade and traffic, and weather, along with the host (commodity) and habitat associations of developmental stages, should be considered in developing a detection and survey program for a specific pest species. Such data must be collected and made readily available for regulatory pest management.

The most cost-effective and desirable approach is early detection and prompt eradication of new plant pests, on which national efforts should be focused and technology and systems developed. A strong emphasis should be given to research related to the biology and ecology of important exotic pests, detection techniques, pest risk analysis, and regulatory control techniques including effective pesticides and their application.

Eradication of plant pests is not a valid approach in pest management, but is applicable to certain exotic pests, when and if detected early and still confined to a limited area. More emphasis should be given to stability of production and optimization of yields with regard to pest management and plant protection (Rabb, 1978b). Regulatory control actions should be based on both biological facts and political or administrative considerations. Hence, regulatory plant protection involves a wide spectrum of scientific, governmental, and industrial sectors of society, detection and control of new plant pests, and must be conducted by the cooperative efforts of all personnel in the plant protection community. Here, the Plant Pest Survey and Management System at the state or regional level as an integrated and unified pest management approach would be an ideal way to serve all plant protection disciplines. Such a system should be as much a part of agricultural research facilities as soil-testing and milk-testing laboratories (Rabb, 1978b). The federal regulatory agency should take the lead in developing such systems throughout the nation, in standardizing sampling techniques and data base management, and in managing interstate problems and programs while meeting the regulatory mandates from inspection to control actions.

The pest management goals of farmers and producers are largely economic, and thus economic analysis can help in selecting optimal crop quality and pest management alternatives. Pest problems will change rapidly as changes in social objectives of agriculture and the economy occur, whereas the basic biology and behavior of plant pests remain mostly unchanged (Norgaard, 1976). Accordingly, economic analysis will help develop new pest management practices based on a better understanding of the economic behavior of farmers, producers, and agricultural community. Thus, economics must enter into the design of regulatory plant protection (Norgaard, 1976).

Pest risk analysis requires not only biological and ecological data bases but also economic information on crop losses. Accurate assessment of crop losses is a prerequisite to the development of plant protection and regulatory pest management strategies because it defines the pest problem. Decision makers at the policy-making level usually take action with inadequate data or without reliable information on pest-related economic losses. There are very little reliable data on crop losses by plant pests and diseases. During the last 10 years, a serious need for reliable crop loss data has been recognized, and new approaches and assessment techniques have been studied (Chiarapa, 1970; Teng and Krupa, 1980). There is an urgent need to conduct continuous research on loss assessment techniques and to have accurate assessment data on economic losses of crops and forests for making rational decisions on pest management strategies. Furthermore, in cost-benefit analysis of a pest management strategy, tradeoffs may be considered between risks and higher profits or benefits only if the probability of different levels of success or failure of the strategy is evaluated and the opportunity to change the strategy in the course of program implementation is ascertained (Norgaard, 1976).

Exotic pests do not recognize national boundaries, and no single country can protect itself from and regulate the invasion of exotic pests. The best way to achieve the goals of plant protection and quarantine is to enhance international cooperation and communication about pests, develop active regional plant protection organizations, and, above all, consolidate and strengthen pest management organizations and programs within each country.

ACKNOWLEDGMENTS

I am greatly indebted to Edgar Eckess for discussing with me many aspects of pest detection and regulatory plant protection and for giving me his invaluable views, and dedicate this work to him for his tireless efforts in promoting the improvement of plant protection and quarantine programs for many years. My sincere appreciation is also due Richard Daum, E. Eckess, and Lloyd V. Knutson for critically reading the earlier draft and making invaluable suggestions; their comments and suggestions helped greatly to improve this chapter. Thanks are also due to Stanley G. Gesell, Charles W. Pitts, Robin A. J. Taylor, Peter H. Adler, Allen Norrbom, Department of Entomology, The Pennsylvania State University; A. G. Wheeler, Jr., Pennsylvania Bureau of Plant Industry, Harrisburg, Pa.; A. H. Epstein, Department of Plant Pathology, Seed and Weed Sciences, Iowa State University; and Richard J. Sauer, Agricultural Experiment Station, University of Minnesota, for reading the manuscript and making useful suggestions. However, I am solely responsible for the contents of this chapter. Mr. Wayne D. Rasmussen, Chief, Agricultural History Branch, USDA/ESS, Washington, D.C., kindly checked the accuracy of historical dates and facts appearing in this chapter, for which I wish to acknowledge his efforts. I also express my appreciation to Judi Hicks, Thelma Brodzina, and Jackie Wolfe in my department office for their efforts in typing the final manuscript.

REFERENCES

Agnew, A. D. Q., and Flux, J. E. C. (1970). Plant dispersal by hares (*Lepus capensis* L.) in Kenya. *Ecology* **51,** 735–737.

Anonymous (1981). Position on international baggage inspection for agricultural purposes. *FAO Plant Prot. Bull.* **29**(½), 1–3.

Arnett, R. H. (1970). "Entomological Information Storage and Retrieval." The Bio-Rand Found., Inc., Baltimore, Maryland.

Baker, H. G., and Stebbins, G. L., eds. (1965). "The Genetics of Colonizing Species." Academic Press, New York.
Baker, R. R. (1978). "The Evolutionary Ecology of Animal Migration." Holms & Meier Publ., New York.
Balch, R. E. (1952). Studies of the balsam woolly aphid *Adelges picea* (Ratz.) and its effect on balsam fir, *Abies balsamea* (L.) Mill. *Can. Dep. Agric., Publ.* **807,** 1–76.
Batra, S. W. T. (1981). Biological control in agroecosystems. *Science* **215,** 134–139.
Belmont Writing Committee (1973). "America's Systematics Collections: A National Plan." Assoc. Syst. Coll., Lawrence, Kansas.
Berg, G. H. (1977). Chapter 27. Post-entry and intermediate quarantine stations. *In* "Plant Health and Quarantine in International Transfer of Genetic Resources" (W. B. Hewitt and L. Chiarappa eds.), pp. 315–326. CRC Press, Cleveland, Ohio.
Bishop, A. (1951). Distribution of barnacles by ships. *Nature (London)* **167,** 531.
Bottrell, D. B., Huffaker, C. B., and Smith, R. F. (1976). "Information Systems for Alternate Methods of Pest Control," UC/AID Pest Management and Related Environmental Protection Project Report. University of California, Berkeley.
Brown, G. C., Lutgardo, A. R., and Gage, S. H. (1980). Data base management systems in IPM programs. *Environ. Entomol.* **9,** 475–482.
Bush, G. L. (1975). Modes of animal speciation. *Annu. Rev. Ecol. Syst.* **6,** 339–364.
Bushland, R. D. (1975). Screwworm research and eradication. *Bull. Entomol. Soc. Am.* **21,** 23–26.
Carlquist, S. (1981). Chance dispersal. *Am. Sci.* **69,** 509–516
Carson, H. L. (1968). The population flush and the genetic consequences. *In* "Population Biology and Evolution" (R. C. Lewontin, ed.), pp. 123–137.
Carson, H. L. (1975). The genetics of speciation at the diploid level. *Am. Nat.* **109,** 83–92.
Chiarapa, L., ed. (1970). "Crop Loss Assessment Methods." FAO/UN, Rome.
Chock, A. K. (1979). The international plant protection convention. *In* "Plant Health." (D. L. Ebbels, and J. E. King, eds.), pp. 1–11. Blackwell, Oxford.
Clarke, B. (1979). The evolution of genetic diversity. *Proc. R. Soc. London, Ser. B* **205,** 453–474.
Clausen, C. P., ed. (1978). Introduced parasites and predators of arthropod pests and weeds: World review. *U.S. Dep. Agric., Agric. Handb.* **480,** 1–545.
Conference of Directors of Systematics Collections (CODOSC) (1971). "The Systematic Biology Collections of the United States: An Essential Resource," Part 1. N.Y. Bot. Gard., New York.
Council on Environmental Quality (CEQ) (1972). "Integrated Pest Management."U.S. Govt. Printing Office, Washington, D.C.
Council on Environmental Quality (CEQ) (1980). "Report to the President. Progress Made by Federal Agencies in the Advancement of Integrated Pest Management" (mimeo). U.S. Govt. Printing Office, Washington, D.C.
Courtenay, W. R., and Robins, C. R. (1975). Exotic organisms: An unsolved, complex problem. *BioScience* **25,** 306–313.
Cox, H. C. (1978). Eradication of plant pests—pro's and con's. *Bull. Entomol. Soc. Am.* **24,** 35.
Cramer, H. H. (1967). Plant protection and world crop production. *Pflanzenschutz-Nachr. Bayer (Ger. Ed.)* **20,** 7–524.
Croft, B. A., Howes, J. L., and Welch, S. M. (1976). A computer-based extension pest management delivery system. *Environ. Entomol.* **5,** 20–34.
Darling, T. G. (1977). International programs to prevent pest spread: U.S. activities. *In* "Selected Papers Presented at the 15th International Congress of Entomology on International Plant Pests," pp. 1–3. APHIS, USDA, Washington, D.C.
David, W. A. L. (1949). "Air Transport and Insects of Agricultural Importance," pp. 1–11. Commonw. Inst. Entomol., London.

DeBach, P., ed. (1964). "Biological Control of Insect Pests and Weeds." Van Nostrand-Reinhold, Princeton, New Jersey.

Dingle, H. (1972). Migration strategies of insects. *Science* **175,** 1327–1335.

Dingle, H., ed. (1978). "Evolution of Insect Migration and Diapause." Springer-Verlag, Berlin and New York.

Dingle, H. (1979). Adaptive variation in the evolution of insect migration. *In* "Movement of Highly Mobile Insects" (R. L. Rabb, and G. G. Kennedy, eds.), pp. 64–87. North Carolina State University, Raleigh.

Dingle, H. (1981). Ecology and evolution of migration. *In* "Animal Migration, Orientation, and Navigation" (S. A. Gauthreaux, Jr., ed.), pp. 2–101. Academic Press, New York.

Eden, W. G. (1978). Eradication of plant pests—pro. *Bull. Entomol. Soc. Am.* **24,** 52–54.

Elton, C. S. (1958). "The Ecology of Invasions by Animals and Plants." Methuen, London.

Embree, D. G. (1979). The ecology of colonizing species, with special emphasis on animal invaders. *In* "Analysis of Ecological Systems" (D. J. Horn, R. D. Mitchell, and G. R. Stairs, eds.), pp. 51–66. Ohio State Univ. Press, Columbus.

Epstein, A. H., ed. (1980). "Plant Pest Information Needs - APHIS, USDA: A Final Draft" A preliminary analysis to the Administrator, APHIS/PPQ from the Pest Survey and Detection Committee, Intersociety Consortium for Plant Protection (mimeo).

Ezcurra, E., Rapoport, E. H., and Marino, C. R. (1978). The geographical distribution of insect pests. *J. Biogeogr.* **5,** 149–157.

Felt, E. P. (1909). Insects and legislation. *J. Econ. Entomol.* **2,** 342–345.

Gentry, J. W. (1977). National programs for insect survey and detection. *Natl. Meet., Entomol. Soc. Am., 1977*. Washington, D.C. (mimeo).

Gonzales, R. H. (1978). Introduction and spread of agricultural pests in Latin America: Analysis and perspects. *FAO Plant Prot. Bull.* **26,** 41–52.

Gressitt, J. L. and Nakata, S. (1958). Trapping of air-borne insects on ships in the Pacific. *Proc. Hawaii. Entomol. Soc.* **16,** 363–365.

Gressitt, J. L., Sedlacek, J., Wise, K. A., and Yoshimoto, C. M. (1961). A high-speed airplane trap for air-borne organisms. *Pac. Insects* **5,** 549–555.

Hall, R. W., and Ehler, L. E. (1979). Rate of establishment of natural enemies in classical biological control. *Bull. Entomol. Soc. Am.* **25,** 280–282.

Hall, R. W., Ehler, L. E., and Basabri-Ershadi, B. (1980). Rate of success in classical biological control of arthropods. *Bull. Entomol. Soc. Am.* **26,** 111–113.

Hess, C. E. (1981). "Economic Impact of the Mediterranean Fruit Fly," Talk presented at the State Board of Food and Agriculture Meeting, Aug. 6, 1981, Los Gatos, California (mimeo).

Hewitt, W. B., and Chiarappa, L. (1977). "Plant Health and Quarantines in International Transfer of Genetic Resources." CRC Press, Cleveland, Ohio.

Hodson, A. C., and Cook, E. F. (1960). Long-range aerial transport of the Harlequin bug and the greenbug into Minnesota. *J. Econ. Entomol.* **53,** 604–608.

Holm, L. G., Plucknett, B. L., Pancho, G. V., and Herberger, J. P. (1977). "The World's Worst Weeds." Univ. of Hawaii Press, Honolulu.

Huffaker, C. B., and Messenger, P. S. (1964). Chapter 4. The concept and significance of natural control. *In* "Biological Control of Insect Pests and Weeds" (P. DeBach, ed.), pp. 74–117. Van Nostrand-Reinhold, Princeton, New Jersey.

Hughes, R. D. (1963). Population dynamics of the cabbage aphid, *Brevicoryne brassicae* (L). *J. Anim. Ecol.* **32,** 393–424.

Hughes, R. D. (1979). Movement in population dynamics. *In* "Movement of Highly Mobile Insects" (R. L. Rabb and G. G. Kennedy, eds.), pp. 14–32. North Carolina State University, Raleigh.

Hunt, N. R. (1946). Destructive plant diseases not yet established in North America. *Bot. Rev.* **12,** 593–627.
Hurd, P. (1974). Report of the advisory committee for systematics resources in entomology. *Bull. Entomol. Soc. Am.* **20,** 237–242.
Hurd, P. (1975). Part II. The current status of entomological collections in North America. *Bull. Entomol. Soc. Am.* **21,** 209–212.
Hutchinson, G. E. (1978). "An Introduction to Population Ecology." Yale Univ. Press, New Haven, Connecticut.
Jain, S. (1979). Adaptive strategies: Polymorphism, plasticity, and homeostasis. *In* "Topics in Plant Population Biology" (O. T. Solbrig, ed.), pp. 160–187. Columbia Univ. Press, New York.
Jamnback, H. (1975). State and private collections. *Bull. Entomol. Soc. Am.* **21,** 94–95.
Johnson, C. G. (1966). A functional system of adaptive dispersal by flight. *Annu. Rev. Entomol.* **11,** 233–260.
Johnson, C. G. (1969). "Migration and Dispersal of Insects by Flight." Methuen, London.
Johnson, R. (1982a). "History and Status of the National Plant Pest and Detection Program." USDA/APHIS-PPQ , Washington, D.C. (mimeo).
Johnson, R. (1982b). "National Plant Pest Survey and Detection Program." Eastern Plant Board Meeting, Harrisburg, Pennsylvania (mimeo).
Johnston, A. (1979). Information requirements for effective plant quarantine control. *In* "Plant Health" (D. L. Ebbels, and J. E. King, eds.), pp. 63–70. Blackwell, Oxford.
Jones, C. (1970). Mammals imported into the United States in 1968. *U.S. Dep. Int., Fish Wildl. Serv., Spec. Sci. Rep.* **137,** 1–30.
Kahn, R. P. (1970). International plant quarantine. *In* "Genetic Resources in Plants" (O. H. Frankel and E. Bennett, eds.), pp. 403–412. Blackwell, Oxford.
Kahn, R. P. (1977). Plant quarantine: Principles, methodology, and suggested approaches. *In* "Plant Health and Quarantine in International Transfer of Genetic Resources" (W. B. Hewitt and L. Chiarappa, eds.), pp. 290–307. CRC Press, Cleveland, Ohio.
Kahn, R. P. (1979). A concept of pest risk analysis. *Bull. OEPP* **9,** 119–130.
Keiran, J. E. (1975). A review of the phoretic relationship between Mallophaga (Phthiraptera: Insecta) and Hippoboscidae (Diptera: Insecta). *J. Med. Entomol.* **12,** 71–76.
Kennedy, J. S. (1961). A turning point in the study of insect migration. *Nature (London)* **189,** 785–791.
Kennedy, J. S. (1975). Insect dispersal. *In* "Insects, Science and Society" (D. Pimentel, ed.), pp. 103–119. Academic Press, New York.
Kim, K. C. (1975a). Systematics and systematics collections: Introduction. *Bull. Entomol. Soc. Am.* **21,** 89–91.
Kim, K. C. (1975b). A concluding remark. *Bull. Entomol. Soc. Am.* **21,** 98–100.
Kim, K. C., ed. (1978). The changing nature of entomological collections: Uses, functions, growth and management. *Entomol. Scand.* **9,** 146–172.
Kim, K. C. (1979). Cooperative pest detection and monitoring system: A realistic approach. *Bull. Entomol. Soc. Am.* **25,** 224–230.
Kim, K. C. (1980). Accurate insect taxonomy basic to pest management. *Sci. Agric., Pa. State Univ. Agric. Exp. Stn.* **27,** 7.
King, C. E., and Anderson, W. W. (1971). Age-specific selection. II. The interaction between r and K during population growth. *Am. Nat.* **105,** 137–156.
Knipling, E. H. (1978). Eradication of plant pests—pro. Advances in technology for insect population eradication and suppression. *Bull. Entomol. Soc. Am.* **24,** 44–52.
Knutson, L. V., and Sutherland, D. W. S. (1979). National Agricultural Pest Register (personal communication).

Kornaś, J. (1971). Changements recents de la flore polonaise. *Biol. Conserv.* **4,** 43–47.
Kosztarab, M. (1975). Role of systematics collections in pest management. *Bull. Entomol. Soc. Am.* **21,** 95–98.
Laycock, G. (1966). "The Alien Animals: The Story of Imported Wildlife." Natural History Press, Garden City, New York.
Lemmon, K. (1968). "The Golden Age of Plant Hunters." A. S. Barnes & Co., New York.
Lewontin, R. C. (1965). Selection for colonizing ability. *In* "The Genetics of Colonizing Species" (H. G. Baker and G. L. Stebbins, eds.), pp. 79–94. Academic Press, New York.
Lindroth, C. H. (1957). "The Faunal Connection between Europe and North America." Wiley, New York.
Lindroth, C. H. (1968). Distribution and distributional centers of North Atlantic insects. *Bull. Entomol. Soc. Am.* **14,** 91–95.
Ling, L. (1953). International Plant Protection Convention: Its history, objectives and present status. *FAO Plant Prot. Bull.* **1,** 65–68.
MacArthur, R. H., and Wilson, E. D. (1967). "The Theory of Island Biogeography." Princeton Univ. Press, Princeton, New Jersey.
McCubbin, W. A. (1954). "The Plant Quarantine Problem." Munksgaard, Copenhagen.
McGregor, R. C., ed. (1973). "The Emigrant Pests: A Report to Dr. Francis J. Muhern, Administrator." APHIS, USDA, Washington, D.C. (mimeo).
MacLachlan, D. S. (1977). International programs to prevent pest spread—new approaches. *In* "Selected Papers Presented at the 15th International Congress of Entomology on International Plant Pests," pp. 23–24. APHIS, USDA, Washington, D.C.
Marlatt, C. L. (1899). The laisser-faire philosophy applied to the insect problem. *USDA, Div. Entomol. Bull.* [N.S.] **20,** 5–19.
Mathys, G. (1977). Phytosanitary regulations and transfer of genetic resources. *In* "Plant Health and Quarantine in International Transfer of Genetic Resources" (W. B. Hewitt and L. Chiarappa, eds.), pp. 327–331. CRC Press, Cleveland, Ohio.
Mathys, G., and Baker, E. A. (1980). An appraisal of the effectiveness of quarantines. *Annu. Rev. Phytopathol.* **18,** 85–101.
Mayr, E. (1963). "Animal Species and Evolution." Belknap Press, Cambridge, Massachusetts.
Moore, I., and Legner, E. F. (1974). Have all the known cosmopolitan Staphylinidae been spread by commerce? *Proc. Entomol. Soc. Wash.* **76,** 39–40.
Morschel, J. R. (1971). "Introduction to Plant Quarantine." Aust. Gov. Publ. Serv., Canberra.
Mulders, J. M. (1977). The International Plant Protection Convention—25 years old. *FAO Plant Prot. Bull.* **25,** 149–151.
Myers, J. G. (1934). The Arthropod fauna of a rice-ship, trading from Burma to the West Indies. *J. Anim. Ecol.* **3,** 146–149.
National Academy of Sciences (1969). "Insect-Pest Management and Control. Principles of Plant and Animal Pest Control," Vol. 3. NAS, Washington, D.C.
Neergaard, P. (1977). Quarantine policy for seed in transfer of genetic resources. *In* "Plant Health and Quarantine in International Transfer of Genetic Resources" (W. B. Hewitt and L. Chiarappa, eds.), pp. 309–314. CRC Press, Cleveland, Ohio.
Newsom, L. D. (1978). Eradication of plant pests—con. *Bull. Entomol. Soc. Am.* **24,** 35–40.
Norgaard, R. B. (1976). II. Integrating economics and pest management. *In* "Integrated Pest Management" (J. L. Apple and R. H. Smith, eds.), pp. 17–27. Plenum, New York.
Oman, P. (1968). Prevention, surveillance and management of invading pest insects. *Bull. Entomol. Soc. Am.* **14,** 98–102.
Orr, R. T. (1970). "Animals in Migration." Macmillan, New York.
Osborn, H. (1937). "Fragments of Entomological History; Including Some Personal Recollections of Men and Events." Author, Columbus, Ohio.

Parsons, P. A. (1982). Adaptive strategies of colonizing animal species. *Biol. Rev. Cambridge Philos. Soc.* **57,** 117–148.

Perkins, J. H. (1980). Boll weevil eradication. *Science* **207,** 1044–1050.

Peterson, A. (1953). "A Manual of Entomological Techniques." Edwards, Ann Arbor, Michigan.

Pfadt, R. E., ed. (1971). "Fundamentals of Applied Entomology," 2nd ed. Macmillan, New York.

Pianka, E. R. (1971). On r- and K-selection. *Am. Nat.* **104,** 592–597.

Por, F. D. (1971). One hundred years of the Suez Canal—a century of Lessepsian migration: Retrospect and view points. *Syst. Zool.* **20,** 148–159.

Price, P. W. (1976). Colonization of crops by arthropods: Non-equilibrium communities in soybean fields. *Environ. Entomol.* **5,** 605–611.

Price, P. W. (1980). "The Evolutionary Biology of Parasites." Princeton Univ. Press, Princeton, New Jersey.

Rabb, R. L. (1978a). Eradication of plant pests—con. *Bull. Entomol. Soc. Am.* **24,** 40–44.

Rabb, R. L. (1978b). A sharp focus on insect populations and pest management from a wide area view. *Bull. Entomol. Soc. Am.* **24,** 55–61.

Rabb, R. L., and Kennedy, G. G., eds. (1979). "Movement of Highly Mobile Insects." North Carolina State University, Raleigh.

Rainey, R. C., ed. (1976). "Insect Flight," Symp. Entomol. Soc. No. 7. Blackwell, Oxford.

Rainey, R. C. (1978). The evolution and ecology of flight: "Oceanographic" approach. *In* "Evolution of Insect Migration and Diapause" (H. Dingle, ed.), pp. 31–48. Springer-Verlag, Berlin and New York.

Rankin, M. A. (1978). Hormonal control of insect migratory behavior. *In* "Evolution of Insect Migration and Diapause" (H. Dingle, ed.), pp. 1–32. Springer-Verlag, Berlin and New York.

Rankin, M. A., and Rankin, S. M. (1979). Physiological aspects of insect migratory behavior. *In* "Movement of Highly Mobile Insects" (R. L. Rabb and G. G. Kennedy, eds.), pp. 35–63. North Carolina State University, Raleigh.

Rapoport, E. H., Ezcurra, E., and Drausel, B. (1976). The distribution of plant diseases: A look into the biogeography of the future. *J. Biogeogr.* **3,** 365–372.

Reed, C. H. (1977). Economically important foreign weeds: Potential problems in the United States. *U.S., Dep. Agric., Agric. Handb.* **498.**

Remington, C. L. (1968). The population genetics of insect introduction. *Annu. Rev. Entomol.* **13,** 415–426.

Revill, D. L., Stewart, K. W., and Schlichting, H. E., Jr. (1967). Passive dispersal of viable algae and protozoa by certain craneflies and midges. *Ecology* **48,** 1023–1027.

Richards, O. W., and Southwood, T. R. E. (1968). The abundance of insects: Introduction. *Symp. R. Entomol. Soc. London* **7,** 2–7.

Richardson, R. H., Ellison, J. R., and Averhoff, W. W. (1982). Autocidal control of screwworms in North America. *Science* **215,** 361–370.

Rohwer, G. G. (1979). Plant quarantine philosophy of the United States. *In* "Plant Health" D. L. Ebbels and J. E. King, eds.), pp. 23–34. Blackwell, Oxford.

Rummel, D. R., Bottrell, D. G., Adkisson, P. L., and McIntryre, R. C. (1975). An appraisal of a 10-year effort to prevent the westward spread of the boll weevil. *Bull. Entomol. Soc. Am.* **21,** 6–11.

Sailer, R. I. (1978). Our immigrant insect fauna. *Bull. Entomol. Soc. Am.* **24,** 3–11.

Schmidt-Koenig, K. (1975). "Migration and Homing in Animals." Springer-Verlag, Berlin and New York.

Schneider, F. (1962). Dispersal and migration. *Annu. Rev. Entomol.* **7,** 223–242.

Simmonds, F. J., and Greathead, D. J. (1977). Introductions and pest and weed problems. *In* "Origins of Pest, Parasites, Diseases and Weed Problem" (J. M. Cherrett and G. R. Sagar, eds.), pp. 109–124. Blackwell, Oxford.

Southwood, T. R. E. (1962). Migration of terrestrial arthropods in relation to habitat. *Biol. Rev. Cambridge Philos. Soc.* **37,** 171–214.
Southwood, T. R. E. (1977). Habitat, the templet for ecological strategies. *J. Anim. Ecol.* **46,** 337–365.
Stuessy, T. F., and Thomson, K. S. (1981). "Trends, Priorities and Needs in Systematic Biology" A report to the Systematic Biology Program of the National Science Foundation. Assoc. Syst. Coll., Lawrence, Kansas.
Szyperski, N., and Grochla, E. (1979). "Design and Implementation of Computer-based Information Systems." Sijthoff & Noordoff, Germantown, Maryland.
Taylor, L. R. (1971). Aggregation as a species characteristics. *Stat. Ecol.* **1,** 357–377.
Taylor, L. R. (1974). Insect migration, flight periodicity and the boundary layer. *J. Anim. Ecol.* **43,** 225–238.
Taylor, L. R. (1979). Chapter 10. The Rothamsted Insect Survey—an approach to the theory and practice of synoptic pest forecasting in agriculture. *In* "Movement of Highly Mobile Insects" (R. L. Rabb and G. G. Kennedy, eds.), pp. 148–185. North Carolina State University, Raleigh.
Taylor, L. R., and Taylor, R. A. J. (1977). Aggregation, migration and population mechanics. *Nature (London)* **265,** 415–421.
Taylor, R. A. J. (1981a). The behavioural basis of redistribution. I. The △-model. *J. Anim. Ecol.* **50,** 573–586.
Taylor, R. A. J. (1981b). The behavioural basis of redistribution. II. Simulations of the △-model. *J. Anim. Ecol.* **50,** 587–604.
Templeton, A. R. (1979). Genetics of colonization and establishment of exotic pests. *In* "Genetics in Relation to Insect Management" (M. A. Hoy and J. J. McKelvey, Jr., eds.), pp. 41–49. Rockefeller Found., New York.
Templeton, A. R., and Rankin, M. A. (1978). Genetic revolutions and control of insect populations. *In* "The Screwworm Problem" (R. H. Richardson, ed.), pp. 83–111. Univ. of Texas Press, Austin.
Teng, P. S., and Krupa, S. V., eds. (1980). Crop loss assessment. *Misc. Publ.—Agric. Exp. Univ. Minn., Agric. Exp. Stn.* **7,** 1–327.
Turnbull, A. L. (1967). Population dynamics of exotic insects. *Bull. Entomol. Soc. Am.* **13,** 333–337.
U.S. Department of Agriculture (USDA) (1963). "Plant Pest Detection," ARS Spec. Rep. ARS 22-63:1–8. USDA, Washington, D.C.
U.S. Department of Agriculture (USDA) (1974). "List of intercepted plant pests, 1972," APHIS 82-4. USDA, Washington, D.C.
U. S. Department of Agriculture (USDA) (1975). "Plant Protection and Quarantine Programs." APHIS, USDA, Washington, D.C.
U.S. Department of Agriculture (USDA) (1977). "Contingency Plans for Dealing with Infestation of New Plant Pests within the United States." Plant Protection and Quarantine, National Program Planning Staff, APHIS, USDA, Washington, D.C. (mimeo).
U.S. Department of Agriculture (USDA) (1981). "Procedure for Evaluating and Reacting to Reports of New Pest Occurrence in the United States." APHIS/PPQ, USDA, Washington, D.C. (mimeo).
van der Pijl, L. (1969). "Principles of Dispersal in Higher Plants." Springer-Verlag, Berlin and New York.
Way, M. J. (1976). Entomology and the world food situation. *Bull. Entomol. Soc. Am.* **22,** 125–129.
Weem, H. V. (1981). Mediterranean fruit fly, *Ceratitis capitata* (Wiedmann) (Diptera: Tephritidae). *Entomol. Circ.* **230,** 1–8.

Welch, S. M., and Croft, B. A. (1979). "The Design of Biological Monitoring Systems for Pest Management." Wiley, New York.

Wellington, W. G. (1964). Qualitative changes in populations in unstable environments. *Can. Entomol.* **96,** 436–451.

Wheeler, A. G., and Nixon, H. F. (1979). "Insect Survey and Detection in State Departments of Agriculture," Spec. Publ. Dep. Agric., Harrisburg, Pennsylvania.

Wiens, J. A. (1977). On competition and variable environments. *Am. Sci.* **65,** 590–597.

Williams, C. B. (1958). "Insect Migration." Collins, London.

Wiser, V. (1973). 61. Inspection and quarantine work; Append. 6F. Protecting American agriculture: Inspection and quarantine. *In* "The Emigrant Pests; a report to Dr. F. J. Mulhern, Administrator (R. C. McGregor, ed.). APHIS, USDA, Washington, D.C.

Wolfenbarger, D. O. (1946). Dispersion of small organisms, distance dispersion rates of bacteria, spores, seeds, pollen and insects. Incidence rates of diseases and injuries. *Am. Midl. Nat.* **35,** 1–152.

Wolfenbarger, D. O. (1975). "Factors Affecting Dispersal Distances of Small Organisms." Exposition Press, Hicksville, New York.

Zimmerman, E. C. (1948). "Insects of Hawaii," Vol. I. Univ. of Hawaii Press, Honolulu.

Zohary, D. E. (1970). Centers of diversity and centers of origin. *In* "Genetic Resources in Plants" (O. H. Frankel and E. Bennett, eds.), IBP Handb. **11,** pp. 33–42. Blackwell, Oxford.

CHAPTER 11

Research on Exotic Insects

C. O. CALKINS

Insect Attractants, Behavior, and Basic Biology Research Laboratory
Agricultural Research Service
United States Department of Agriculture
Gainesville, Florida

I. INTRODUCTION

Insect pests have been attacking agricultural plants since man first began to cultivate crops. Only during the last 100 years have pests become critical world-wide problems. Before then, most agricultural regions of the world consisted of relatively small cultivated units interspersed among the natural vegetation. Insect outbreaks, when they occurred, tended to be confined locally except in cases of migratory pests. Since the mid-nineteenth century, because of agricultural mech-

EXOTIC PLANT PESTS AND NORTH AMERICAN AGRICULTURE

ISBN 0-12-757880-3

anization, large tracts of land have been opened up and devoted almost exclusively to the culture of a single crop (Newman, 1966). This monoculture created conditions that favored pest insects over their natural enemies, and the results were catastrophic outbreaks.

There are between 625,000 and 1.5 million insect species on earth. More than 82,000 species are known from North America north of Mexico (Hoffman and Henderson, 1966). More than 5000 different species are known to be important pests throughout the world. This is less than 1% of the species of insects present in the world. However, almost every crop grown anywhere in the world is attacked by at least one pest (Newman, 1966).

A. Examples of Exotic Insects Which Have Become Major Pests in the United States

More than 1100 species of insects and mites have invaded North America since its discovery by Columbus. Of these, 221 are not economically important. Of those classified as pests, 404 are minor and 212 are considered major, although 139 of these are not considered important pests in their countries of origin. Also, 278 of the invading species are beneficial insects, and 126 of these were deliberately introduced. The rate of entry of foreign insect species into the United States reached a high of 15 per year during the period 1910–1919. The average rate is now about 9 species per year (McGregor, 1973). Some of the major immigrant pests of U.S. agriculture are listed in Table I.

TABLE I

Foreign Pests of Agricultural Crops Now Resident in the Contiguous United States

Common name	Scientific name	Order: Family
African mole cricket	*Gryllotalpa africana palisot* de Beauvois	Orthoptera: Gryllotalpidae
Alfalfa weevil	*Hypera postica* (Gyllenhal)	Coleoptera: Curculionidae
Argentine ant	*Irklomyrmex humilis* (Mayr)	Hymenoptera: Formicidae
Asiatic garden beetle	*Maladera castanea* (Arrow)	Coleoptera: Scarabaeidae
Asparagus beetle	*Crioceris asparagi* (L.)	Coleoptera: Chrysomelidae
Black scale	*Saissetia oleae* (Oliver)	Homoptera: Coccidae
Browntail moth	*Eyproctis chrysorrhoea* (L.)	Lepidoptera: Lymantriidae
California red scale	*Aonidiella aurantii* (Maskell)	Homoptera: Diaspididae
Cereal leaf beetle	*Oulema melanopus* (L.)	Coleoptera: Chrysomelidae
Citrophilus mealybug	*Pseudococcus calceolariae* (Maskell)	Homoptera: Pseudococcidae
Citrus blackfly	*Aleurocanthus woglumi* Ashby	Homoptera: Aleyrodidae
Citrus mealybug	*Pianococcus citri* (Risso)	Homoptera: Pseudococcidae

TABLE I *Continued*

Common name	Scientific name	Order: Family
Codling moth	*Laspeyresia pomonella* (L.)	Lepidoptera: Olethreutidae
Comstock mealybug	*Pseudococcus comstocki* (Kuwana)	Homoptera: Pseudococcidae
Cottonycushion scale	*Icerya purchasi* Maskell	Homoptera: Margarodidae
European chafer	*Rhizotrogus magalis* (Razoymouwsky)	Coleoptera: Scarabaeidae
European corn borer	*Ostrinia nubilalis* (Hübner)	Lepidoptera: Pyralidae
European fruit scale	*Quadraspid iotusostraeformis* (Curtis)	Homoptera: Diaspididae
European peach scale	*Lecanium persicae* (F.)	Homoptera: Coccidae
Green bug	*Schizaphis graminum* (Rondani)	Homoptera: Aphididae
Imported cabbage worm	*Pieris rapae* (L.)	Lepidoptera: Pieridae
Japanese beetle	*Popillia japonica* Newman	Coleoptera: Scarabaeidae
Longtailed mealybug	*Pseudococcus longispinus* (Targioni-Tozzetti)	Homoptera: Pseudococcidae
Mexican bean beetle	*Epilachna varivestis* Mulsant	Coleoptera: Coccinellidae
Oriental beetle	*Anomala orientalis* Waterhouse	Coleoptera: Scarabaeidae
Oriental fruit moth	*Grapholitha molesta* (Busck)	Lepidoptera: Olethreutidae
Oystershell scale	*Lepidosaphes ulmi* (L.)	Homoptera: Diaspididae
Pea aphid	*Acyrthosiphon pisum* (Harris)	Homoptera: Aphidae
Red imported fire ant	*Solenopsis invicta* Buren	Hymenoptera: Formicidae
San Jose scale	*Quadraspidiotus perniciosus* (Comstock)	Homoptera: Diaspididae
Spotted alfalfa aphid	*Therioaphis maculata* (Buckton)	Homoptera: Aphidae
Sweetpotato weevil	*Cylas formicarius elegantulus* (Summers)	Coleoptera; Curculionidae
Whitefringed beetle	*Graphognathus* spp.	Coleoptera: Curculionidae
Yellow scale	*Aonidiella citrina* (Coquillett)	Homoptera: Diaspididae

The most important way that an insect population grows to damaging levels is by migration to a new environment, leaving its natural enemies behind. This is precisely what happens when an insect not known to occur in an area gains entry. In fact, "most losses and most insecticides used for pest control in the U.S. can be attributed to alien pests" (Knipling, 1979). Federal and state quarantines are recognized as the first line of defense for insect control by preventing encroachment of pests into areas not previously occupied.

B. Purposes of Studying Exotic Insects before Their Introduction

A great deal of basic information about exotic species can be attained by the study of model species already present. For example, the biology and behavior of the Caribbean fruit fly, *Anastrepha suspensa* (Loew), a resident of Florida, closely resembles those of the Mediterranean fruit fly, *Ceratitis capitata* (Wied.), and the Mexican fruit fly, *A. ludens* (Loew). By studying *A. suspensa* as a model species, a basic understanding of tropical tephritid behavior and biology is achieved. Other means of increasing our knowledge of exotic insects by using model species involve the understanding of insect enzyme systems, diapause, growth-regulating mechanisms, nutritional requirements, and generalized factors influencing behavior and mating strategies.

Specific information, however, can be developed only in the presence of the insect pest itself. These include the nature of host plant resistance and insecticide resistance, predatory and parasitic control agents, life history and ecological studies, host range and host relationship studies, pheromone research, dispersal characteristics, periodicity and phenology aspects, and suitable chemical control agents and methodology. Information of these types is immediately useful should a foreign pest gain entry to our agricultural systems, and could prevent the establishment of a pest species.

There is a high degree of unpredictability about the possibility of exotic insects becoming pests which may preclude compilation of lists of potential pests. However, because a large number of potential exotic pests could become immigrants, a ranking system is needed so that research priorities can be established. The U.S. Department of Agriculture's (USDA) Animal and Plant Health Inspection Service (APHIS) has compiled a "top 100" list of potential plant pests and animal diseases not occurring in the United States. In any such ranking system, strong considerations must be given to hosts, commodity pathways, world distribution, insect life cycle, seasonal and population trends, climatic requisites, frequency of infested host cargo importations, and interception frequency (McGregor, 1973). An additional list of economically less important pests should be compiled. These may occur on crops of limited importance, but pose potential problems in local or isolated environments in the United States.

A combination of lists provided by a task force commissioned by APHIS (McGregor, 1973) and a list compiled by APHIS Plant Pest Quarantine in 1975 is shown in Table II. Criteria used to develop these lists are (1) estimations of the probability that a pest will become established in the United States, (2) evaluation of the economic impact if the pest does become established, and (3) determination of the expected economic impact by multiplying criterion (1) by criterion (2).

Insects on this list probably pose the greatest potential threats to our agriculture based on the information available. However, these species should not be considered the only candidates. Other species that are relatively unimportant as pests in their native environments, as previously mentioned, may become major pests in new favorable environments (McGregor, 1973), especially if they escape most of their biotic control factors.

Additional means of identifying potential pests should be explored. Because of

TABLE II

Selected Foreign Food Crop Pests of Primary Concern to Mainland Agriculture in the United States[a]

Scientific name	Common name	Order: Family	Product attacked	Geographical distribution
Acrolepia assectella (Zeller)	Leek moth	Lepidoptera: Yponomeutidae	Seeds, flowers, vegetables	Europe, USSR, China
Agriotes lineatus (L.)	Lined click beetle	Coleoptera: Elateridae	General feeder	Europe, USSR
Agriotes obscurus (L.)	Click beetle	Coleoptera: Elateridae	General feeder	
Agriotes sputator (L.)	Click beetle	Coleoptera: Elateridae	General feeder	
Amphimallon solstitialis (L.)	Summer chafer	Coleoptera: Scarabaeidae	General feeder	Europe, USSR, Middle East, China
Anastrepha fraterculus (Wied.)	South American fruit fly	Diptera: Tephritidae	Citrus, fruits, nuts	Mexico, Central and South America
Anastrepha grandis (Macquart)		Diptera: Tephritidae	Melons, cucumbers	
Anastrepha ludens (Loew)	Mexican fruit fly	Diptera: Tephritidae	Citrus, fruits	Texas, Mexico, Central America
Argyrotaenia pulcellana (Haworth)	Leaf roller	Lepidoptera: Tortricidae	Fruit	
Aulacophora abdominalis (Fab.)		Coleoptera: Chrysomelidae	Melons, cucumbers	
Aulacophora hilaris (Boisduval)	Pumpkin beetle	Coleoptera: Chrysomelidae	Melons, cucumbers	
Austrotortrix postvittana (Wlk.)	Light-brown apple moth	Lepidoptera: Tortricidae	Apples, pears	Australia, New Zealand, Hawaii, England
Autographa gamma (L.)	Silvery moth	Lepidoptera: Noctuidae	General feeder	Europe, Asia, India, North Africa
Baris lepidii Germar		Coleoptera: Curculionidae	Fruits, carrots, beets	
Brevipalpus chilensis Baker	Chilean grape mite	Acarina: Tenuipalpidae	Fruit, nuts	South America
Busseola fusca Fuller	Maize stalk borer	Lepidoptera: Noctuidae	Corn, sorghum, sugarcane	Ethiopian region, West, East, South Africa
Calliptamus italieus (L.)	Italian locust	Orthoptera: Acrididae	General feeder	Europe

(*continued*)

TABLE II *Continued*

Scientific name	Common name	Order: Family	Product attacked	Geographical distribution
Carpomyia paradelina (Bigot)	Baluchistan melon fly	Diptera: Tephritidae	Melons, cucumbers	Pakistan
Carposina niponensis Wals.	Peach fruit moth	Lepidoptera: Carposinidae	Apples, pears, peaches	China, Korea, Japan
Ceratitis capitata (Wied.)	Mediterranean fruit fly	Diptera: Tephritidae	Fruits, certain vegetables	Mediterranean region, Africa, Central and South America, Western Australia, Hawaii
Ceratitis rosa Karsch	Natal fruit fly	Diptera: Tephritidae	Citrus, peaches	East and South Africa, Nigeria, Angola
Chloroclystis rectangulata (L.)	Green pug moth	Lepidoptera: Geometridae	Citrus, fruit	Europe, Egypt
Cnaphalocrocis medinalis (Guen.)		Lepidoptera: Pyralidae	Rice, grasses	Madagascar, India, Southeast Asia, Australia
Coenorrhinus aequatus (L.)		Coleoptera: Curculionidae	Apples, plums	
Colias lesbia F.	Lucerne caterpillar	Lepidoptera: Pieridae	Alfalfa, sugar beet	South America
Copturus aguacatae Kissinger		Coleoptera: Curculionidae	Avocado	Mexico, Central America
Cossus cossus (L.)	Goat moth	Lepidoptera: Cossidae	General feeder, fruit	Europe
Cryptophlebia leucotreta (Meyr.)	False codling moth	Lepidoptera: Tortricidae	Citrus, fruit	South, East, and West Africa
Dacus caudatus F.	Fruit fly	Diptera: Tephritidae	Citrus, melons	India, Pakistan
Dacus cucurbitae Coq.	Melon fly	Diptera: Tephritidae	Cucumber, tomato	Northern India (origin), India, Burma, Philippines, Sri Lanka, Pakistan, Okinawa, Kenya, Mauritius, Guam, Japan, Australia, Malaysia, Java, Hawaii
Dacus dorsalis Hendel	Oriental fruit fly	Diptera: Tephritidae	Citrus, melons	Pakistan, India, Southeast Asia, Taiwan, Philip-

				pines, Ryukya, North Australia, Hawaii
Dacus tsuneonis (Miy.)	Japanese orange fruit fly	Diptera: Tephritidae	Citrus	China, Japan
Dacus tyroni (Frog.)	Queensland fruit fly	Diptera: Tephritidae	Citrus, fruit	Australia
Dacus zonatus Saund.	Peach fruit fly, Punjab fruit fly	Diptera: Tephritidae	Melon, tomatoes	Pakistan, India, Bangladesh
Diabrotica speciosa Germar	Cucurbit beetle, chrysanthemum beetle, San Antonio beetle	Coleoptera: Chrysomelidae	Melons, peanuts, tomatoes	South America
Dichocrocis punctiferalis Guenee	Yellow peach moth	Lepidoptera: Pyralidae	Citrus, fruit, corn, cotton	India, Burma, Sri Lanka, China, Japan, Australia
Earias insulana (Boisd.)	Spiny bollworm	Lepidoptera: Noctuidae	Cotton	Africa, Europe, southern Asia
Epicaerus cognatus Sharp	Mexican potato weevil	Coleoptera: Curculionidae	Potatoes	Middle East, Australia
Euluia herachei (L.)	Celery fly	Diptera: Tephritidae	Celery, lettuce	
Eutetranychus orientalis (Klien)	Oriental spider mite	Acarina: Tetranychidae	General feeder	Far East
Haplorhynchites confruleus (De Geer)		Coleoptera: Curculionidae	Deciduous fruit	
Heliothis armigera (Hb.)	''American'' bollworm	Lepidoptera: Noctuidae	General feeder	Africa, Southeast, Near, and Middle East, India, Central and Southeast Asia, Japan, Australia, New Zealand
Heliothis assulta Gn.	Cape gooseberry budworm	Lepidoptera: Noctuidae	General feeder	Tropical and subtropical regions of Africa, Asia, and North Australia
Heliothis gelatopoeon (Dyar)		Lepidoptera: Noctuidae	General feeder, cotton, sorghum, flax	Argentina

(continued)

TABLE II *Continued*

Scientific name	Common name	Order: Family	Product attacked	Geographical distribution
Heliothis punctigera Wellgr.	Native budworm of Australia	Lepidoptera: Noctuidae	General feeder, cotton, tobacco, lucerne, etc.	Australia
Herpetogramma licarsisalis (Walker)	Grass webworm	Lepidoptera: Pyralidae	Rice, grasses	
Icerya aegyptiaca (Dgl.)	Egyptian fluted scale	Homoptera: Margarodidae	Citrus, avocado	Southeast Asia, India, Taiwan, Australia, Pacific Islands, Egypt, Sudan, Kenya, Tanzania, Ivory Coast
Icerya seychellarum (Westw.)	Seychelles scale, giant coccid	Homoptera: Margarodidae	General feeder, citrus	Southeast Asia, Japan, Madagascar, East Africa, Pacific Islands
Lampides boeticus (L.) (Targioni-Tozzetti)	Bean butterfly	Lepidoptera: Lycaenidae	Beans, soybean, alfalfa	Africa, Europe, USSR, China, Southeast Asia, Hawaii
Leptocorisa acuta Thunb.	Rice bug	Homoptera: Aleyrodidae	Rice, grasses	Australia, Southeast Asia, China, New Zealand, Pakistan, Tibet, India
Liriomyza bryoniae (Kaltenbach)	Leaf miner	Diptera: Agromyzidae	Potato, tomato	
Lobesia botrana (Schiff.)	Vine moth, grapeberry moth	Lepidoptera: Olethreutidae	Grapes	Southern Europe, Middle East, Japan, North Africa, Kenya
Lonchaea chalbeus Wiedemann		Diptera: Lonchaeidae	Fruits, vegetables	
Luperodes suturalis (Motschulsky)		Coleoptera: Chrysomelidae	Sugar beet, soybean, rice, sugarcane	
Lymantria monacha (L.)	Nun moth	Lepidoptera: Lymantriidae	Deciduous tree, pear, apple, plum, cherry	Europe, USSR, Turkey, Japan, Korea, Tibet
Macrosteles laevis (Ribaut)	Leafminer	Homoptera: Cicadellidae	Grape, small grain	Afghanistan
Malacosoma neustria (L.)	Lackey moth, European tent caterpillar	Lepidoptera: Lasiocampidae	Fruit trees	Europe, Turkey, Ireland

Mamestra brassicae (L.) (Tryon)	Cabbage moth, cabbage armyworm	Lepidoptera: Noctuidae	Cole crops	Europe, Asia, Middle East, Libya
Melanogromyza phaseoli	Bean fly	Diptera: Agromyzidae	Beans, legumes	Middle East, Pakistan, Egypt, Sudan
Nephotettix nigropictus (Stal.)	Green rice leafhopper	Homoptera: Cicadellidae	Rice, grasses	Subtropical and tropical Asia, southern Japan
Oligonychus gossypii (Zacher)	Mite	Acarina: Tetranychidae	General feeder	
Omophlus lepturcides (L.)	Winter moth	Lepidoptera: Geometridae	Deciduous fruit, trees	Europe, USSR, Tunisia
Ophiomyia phaseoli (Tryon)	Bean fly	Diptera: Agromyzidae	Beans, soybean	Southeast Asia, India, Pakistan, Bangladesh, Middle East, Africa, Australia, Pacific Islands, Hawaii
Orchamoplatus citri (Takahashi)	Citrus whitefly	Homoptera: Aleyrodidae	Citrus	
Orchamoplatus mammaeferus	Whitefly	Homoptera: Aleyrodidae	Citrus	
Parlatoria ziziphi (Lucus)	Black parlatoria scale	Homoptera: Diaspididae	Citrus	Southern Europe, USSR, South and North Africa, Asia, Australia, Argentina, West Indies, Hawaii
Paseolecanium bituberculatum (Targioni-Tozzetti)		Homoptera: Coccidae	Citrus, fruit	
Phyctinus callosus Boheman	Vine weevil	Coleoptera: Cucurlionidae	Grapes	
Phyllotreta nemorum (L.)	Turnip flea beetle	Coleoptera: Chrysomelidae	Vegetables	Europe, Asiatic USSR
Phytomyza horticola Gour.		Diptera: Agromyzidae	Vegetables, alfalfa, legumes, compositae	Europe, USSR, Africa, Middle East, Asia
Prays citri Mill	Citrus flower moth, rind borer	Lepidoptera: Yponomeutidae	Citrus	Mediterranean region, Australia, South and Southeast Asia
Proeulia chiliensis		Lepidoptera: Tortricidae	Fruit	South America

(continued)

TABLE II *Continued*

Scientific name	Common name	Order: Family	Product attacked	Geographical distribution
Pseudococcus saccharicola Takahashi		Homoptera: Pseudocociidae	Rice, grasses	
Pseudonapomyza spicata (Malloch)		Diptera: Agromyzidae	Corn, grasses	
Rastrococcus iceryoides Green	Mealybug	Homoptera: Pseudococcidae	Citrus, cotton	India
Rastrococcus spinosus (Robinson)	Mealybug	Homoptera: Pseudococcidae	Citrus	Indonesia
Rhagoletis cerasi (L.)	European cherry fly	Diptera: Tephritidae	Cherries, plums	Europe, Turkey, USSR, Iran
Rhagoletis lycopersella Smyth	Tomato fruit fly	Diptera: Tephritidae	Tomatoes	
Rhynchites bacchus (L.)		Coleoptera: Curculionidae	Apples, peaches	
Rhychites cupreus	Plum borer	Coleoptera: Curculionidae	Apples, peaches	
Rhychites heros Roelofs	Fruit weevil	Coleoptera: Curculionidae	Apples, peaches	Japan, Taiwan, China, Korea
Sacododes pyralis Dyar	South American cottonworm	Lepidoptera: Noctuidae	Cotton	Central America, Colombia, Venzuela, Paraguay
Silba pendula (Bezzi)		Diptera: Lonchaeidae	Beans, corn	
Spodoptera exempta (Walker)	Nutgrass armyworm	Lepidoptera: Noctuidae	General feeder	Africa south of the Sahara, Sri Lanka, Malaysia, Singapore, Indonesia, Philippines, Australia, Hawaii
Spodoptera littoralis (Boisd.)	Egyptian cottonleaf worm	Lepidoptera: Noctuidae	General feeder	South and East Asia, East Africa, Madagascar, Australia
Spodoptera litura (F.)	Tobacco caterpillar, common cutworm	Lepidoptera: Noctuidae	General feeder	Most of Asia, Australia

Spodoptera mauritia (Boisd.)	Cotton leaf roller	Lepidoptera: Pyralidae	Cotton	Tropical Africa, South and East Asia, Australia
Tetranychus viennensis Zacher	Fruit-tree spider mite	Acarina: Tetranychidae	Deciduous fruit	Europe
Timocratia haywardi Busck		Lepidoptera: Tineidae	Deciduous fruit	
Torgionia vitis Sign.	Black scale	Homoptera: Diaspididae	Alfalfa, beans	
Trogoderma granarium Everts	Khapra beetle	Coleoptera: Dermestidae	Stored products	All tropical and subtropical regions except South and Central America and Southeast Asia
Tryporyza incertulas (Walker)	Paddy borer, yellow rice borer	Lepidoptera: Pyralidae	Rice	Asia, Australia
Unaspis yanonensis (Kuwana)	Arrowhead scale	Homoptera: Diaspididae	Citrus	China, Japan
Zabrus tenebrioides Goeze	Corn ground beetle	Coleoptera: Carabidae	General feeder, small grains, corn, beets	Turkey, Iran, Ukraine, Southeast Europe

[a] From USDA, APHIS, PPQ.

our rapid transport systems, almost any insect has a chance of gaining entry. One consideration is to examine r-strategist species, those insects with life histories characterized by high biotic potential, extensive host range, good dispersal ability, and short generation time (MacArthur and Wilson, 1967). These species frequently colonize invader plant species (annuals) that are themselves r-strategists. Many of our field crops (corn, small grains, soybeans, vegetables, cotton, sugarbeets and cane, etc.) are annual plants, usually occurring only in disturbed environments. Man keeps the environment of these crop plants in a continual state of disruption, usually by plowing the soil or burning, and by controlling the invasion of unwanted plants. Because the biotic potential of r-strategist insect species is extremely high, populations can increase very rapidly under these favorable circumstances. These are the types of insect species that hold the greatest threat for field crops in the United States.

For an invading insect to become established, it must have the ability to colonize successfully any new area it enters. Certain conditions are necessary for successful establishment. The species must possess the physiological potential to survive and reproduce in the new environment, and it must be preadapted for survival. An ecological opportunity for establishment must be present, that is, there must not be too much competition for the niche. And finally, the organism must have access to the new environment (McGregor, 1973).

C. Advantages of Studying Exotic Insects in Their Endemic Range

When an insect first invades the United States, the initial strategies are to confine its spread and to eradicate it if possible. The operations resemble a war situation, and there is usually no time to do background research. Information that would support control strategies must be gleaned from the literature, usually at short notice. If control and eradication strategies fail and the pest becomes established, a flurry of research projects is initiated to answer many questions on basic biology, behavior, and control. By this time the war has been lost, and the American agricultural community resigns itself to living with the pest and the extra control expenses.

Although the benefits of quarantine are difficult to measure, we can document the consequences of not excluding a foreign pest. To use an example of the costs that occur in attempting to eradicate a very destructive pest or to live with it after its establishment, let us look at the situation involving the Mediterranean fruit fly (medfly) in California.

The medfly has been introduced into this country several times, and until recently it was fairly easily eradicated each time. In 1980, the pest was detected near San Jose, California, and immediate control efforts were initiated. The eradication efforts were finally declared successful on September 21, 1982. Although the fly was confined to a relatively small area of California, the costs of control and quarantine were phenomenal. The projected cost of the Medfly Eradication Project, headquartered at Los Gatos from July 1, 1981, to June 30, 1982, was $59,764,576. However, this figure only included aerial spraying expenses from July 1 to November 30, 1981 ($19,671,260). The cost of aerial spraying in 1982 and the loss of markets in 1981–1982 would make the total cost of the program still higher (Eula Farnham, Budget Analyst, Financial Services on the estimate of expenses of Medfly Eradication Project 1981, unpublished data in memo to Caroline Cabias, Assistant Director, Administrative Services).

If the medfly should become established in California, the costs of having to live with it are staggering. The susceptible crop areas encompass 1.4 million acres with a farm-gate crop value of $2.7 billion. The value of these commodities at the marketplace after adding the expenses for packaging, handling, shipping, and marketing is $5.4 billion, and the value of commodities shipped overseas is more than $583 million. The annual medfly control programs, including bait sprays and quarantine treatments, would cost $97–330 million per year. The initial cost of constructing refined fumigation and cold treatment facilities would be $497 million. Crop losses due to the pest, in spite of control efforts, would range from $261 to $567 million per year (C. E. Hess, unpublished data from Economic impact of the Mediterranean fruit fly, a report presented to the California State Board of Food and Agriculture, August 6, 1981). The increased handling costs and the reduced supply of fruits and vegetables also would have an impact on the consumer at both the state and national levels.

Although most foreign pests invading the United States do not generate the efforts or possess the potential for damage that the medfly does, the appearance of any important exotic pest has the potential to produce considerable expense and trauma. The threat of economic damage or the hazard of entry into the United States created by most species could be reduced by basic and applied research within the endemic range. Considerable expense involving quarantine and eradication efforts could thus be avoided.

The effective strategies for control of an invading pest is based on information developed *before* entry. This information can be obtained in two ways: (1) research in foreign countries from which the pest originates and (2) research in quarantine containment facilities in the country of potential introduction. Research conducted to reduce populations of these pests in foreign countries also would reduce the likelihood that the pests would move into international pathways. Thus, research on potential pest invaders should be encouraged and even financially supported in foreign countries that comprise the endemic range of the species.

II. RESEARCH APPROACHES

The exchange of information and the collaboration between entomologists in different countries have increased enormously in recent years. This has led to the strengthening of economic entomology as an applied science whose importance is recognized throughout the world. During the last 50 years, through international cooperation, entomologists have triumphed over many insect-borne diseases and have substantially lessened the damage caused by plant and animal pests. Some of the most outstanding successes would not have been possible without such international cooperation. For example, the worldwide campaign against malaria by entomologists and the medical profession has reduced this disease to a minor or nonexistent problem in many countries. The cooperative effort involved in the International Locust Control Program has significantly reduced the locust plagues of the past.

The exchange of basic and applied information between entomologists of various countries has accelerated the progress of research. The use of computerized literature retrieval systems, now widely employed, makes it possible to survey published scientific literature dating back to 1970 on most species being investigated. Using these retrieval systems, recent scientific literature on identified potential pests of U.S. agriculture can be collected, cataloged, and made quickly available, and many successful control programs can be implemented without undue loss of time and effort. Unfortunately, literature published before 1970 must still be searched by the older conventional methods.

Basic to the development of pest quarantine control strategies is adequate knowledge of factors such as pest biology and distribution, commodity and carrier risk, international pathways, pathway survival, and eradication and control mechanisms (McGregor, 1973). Currently, much of the knowledge of pests from remote localities has been extrapolated from closely related species or from fragmentary information. Although this technique can be used to a limited extent, pertinent, reliable information can be derived only from research conducted directly with the species in question. This type of research can be conducted most easily within the current distribution areas of the pest. This "on-site" research will accomplish a twofold objective: (1) information will be generated that will make it easier to quarantine the pest or to deal with it after it has gained entry; (2) research findings may help to control or suppress the pest within the present area of distribution, which, in effect, will reduce its chances of moving into international pathways.

A. Determination of Origins of Introductions by the Use of Genetic Markers

1. *Allele Frequency*

Gel electrophoresis was first adapted to genetic studies by Lewontin and Hubby (1966). It was used extensively in theoretical research dealing with population genetics and evolution. A large amount of genetic variation in gene products was discovered using this technique. The genetic variants, called "allozymes'' or ''electromorphs,'' can be used as markers to study the population structure and movement of some species. Comparisons between two or more populations by electrophoresis often reveal the existence of rare alleles of these allozymes (in $< 5\%$ of the population) at various loci. These rare alleles, because they do not occur in all populations and because they are frequently lost during genetic ''bottlenecking,'' frequently can be used to identify populations and gene flow (Pashley and Bush, 1979).

Emigrants from a population usually are few in number and will carry only a sample or subset of the array of rare alleles from the parent population. Subsequent emigrants from the parent population would probably not possess the same subset of alleles as other emigrants. As secondary emigrants leave the original emigrant population, the number of rare alleles would further diminish as long as no additional gene flow occurred. Thus, populations of a species may be characterized by the array of their rare alleles, and their emigration patterns may be established. This is true if gene flow is not maintained, if mutations over long periods of time do not cloud the patterns, or if man does not deliberately or accidentally provide additional introductions.

This technique was used by Pashley and Bush (1979) to determine the emigration patterns of the codling moth, *Laspeyresia pomonella* (L.), throughout the world. Populations from Austria-Hungary had twice as many rare alleles as any other population. This area of Europe was assumed to be the original range of this species. Populations in Chile and Australia have very low percentages (18% and 9%, respectively) of the total number of rare alleles, suggesting that they were colonized by small numbers of individuals. New Zealand populations, on the other hand, possess a high percentage (45%) of rare alleles in common with the Austrian-Hungarian population. This indicated that a large number of individuals were founders from the original range (Central Europe).

M. D. Huettel, research entomologist, ARS, USDA, from Gainesville, Florida (unpublished data), presently is surveying the allele frequencies of the Mediterranean fruit fly throughout the world. The medfly was thought to originate in East and South Africa. Samples from South Africa, representative of other African populations, have the greatest percentage (52%) of polymorphic loci of any population surveyed. Populations from Israel, Hawaii, and Costa Rica have only 17%, 13%, and 9% of the polymorphic loci, respectively, of field populations. These populations do not have progressively smaller numbers of the same alleles; rather, they have a few unique alleles (i.e., existing only among themselves). It should be possible to trace the origin of recent introductions as long as the introduction did not come from the original range of the species. In the case of the 1979–1980 medfly introduction in the San Jose area of California, the population possessed a rare allele that was not found in the Hawaiian population. It was present, however, in the population from Central America (Costa Rica). Thus, the introduction could not have come from Hawaii (M. D. Huettel, personal communication). This technique not only is useful in determining the origin of an infestation, but also indicates the history of the spread of certain species and identifies the pathways.

2. *Pheromone-Mediated Responses*

Another possible genetic marker for use in determining the origins of exotic pest introductions involves the response of males to ratios of pheromone blends or to different pheromones found in different strains of moths. ''Pheromone'' is defined as a substance produced by one organism and released. It is received by another individual of the same species and subsequently elicits a specific reaction (Karlson and Butenandt, 1959; Karlson and Lüscher, 1959).

Two strains of the European corn borer, *Ostrinia nubilalis* (Hübner), that responded to two different ratios of the (*Z*:*E*) isomer of 11-tetradecenyl acetate were discovered in the United States. Males from populations throughout most of the United States respond preferentially to a 97:3 *Z*:*E* isomer blend. However, males from populations located in New York and Pennsylvania responded to a 3:97 *Z*:*E* blend. Although these strains will hybridize if confined, where their ranges overlap in nature they are able to retain their identities. Efforts were made to determine the origin of these strains in Europe by Klun and associates (1975). The more common strain (97:3 *Z*,*E* responder) occurred throughout Central Europe, whereas the more rare strain (3:97 *Z*,*E* responder) seemed to be found exclusively in the Netherlands and in Italy. The investigators stated that this strain probably was introduced into the United States in shipments of broom corn from Italy in 1909–1914. The population in the Netherlands also could very well have been introduced from Italy at some time in the recent past (Klun and associates, 1975).

It appears that this technique has limited usefulness in identifying the origins of infestations of emigrant species. Other examples are almost nonexistent because sex pheromones are so critical in the reproductive process of the individual. The mutations that would result in the production of unique pheromone blends must be accompanied by receptors in the responding individuals that are tuned to those blends. This obviously has occurred in the past and can be demonstrated in closely related species. However, it is such a strong mechanism of speciation that it is doubtful whether variations in sex pheromone blends are common within a species.

3. *Cuticular Hydrocarbon Analysis*

A novel method of separating closely related subspecies or sibling species of Diptera has been developed by comparing differences in extracts of cuticular hydrocarbon complexes. Two important vectors of malaria in Africa, *Anopheles gambiae* Giles and *A. arabiensis* Patton, frequently occur sympatrically and cannot be distinguished morphologically. Extraction and analysis of cuticular hydrocarbons with gas chromatography revealed statistically significant differences in the ratios of various long-chain hydrocarbons (Carlson and Service, 1979, 1980). The technique also was applicable in separating two black fly sibling species of the *Simulium damnosum* spp. complex, *S. sirbanum* (Vajime & Dunbar) and *S. squamosum* Enderlein (Carlson and Walsh, 1981). It also was used in separating species and subspecies of the tsetse fly, *Glossina* spp. (Carlson, 1982). The utility of this technique rests with the fact that the cuticular hydrocarbons are quite stable and can even be extracted from pinned specimens several years old. Although this technique works best in separating groups at the species level, certain subspecies also can be separated. This opens the possibility that, with refinement, the method could be used for strain separation.

B. Detection Research

Early detection of an immigrant pest is critical for control and eradication strategies. Frequently, there is a period between the time an insect enters a new habitat and the time it is detected. If this interval is of sufficient length so that one or more generations are produced, the species is considered established. The longer the species is undetected, the more adapted it becomes to the new habitat and the more difficult it is eventually to eradicate.

Because scientists are constantly identifying and synthesizing new attractant materials, the use of pheromone- and food-baited traps to detect foreign pests at ports and airline terminals is increasing. However, the number of species for which pheromones or food attractants are known is still relatively small.

1. *Sex Pheromones*

Results of sex pheromone research make possible the early detection of new emigrants into pest-free areas and allow monitoring of adult eclosion, emergence from hibernation, and population buildups. The attraction response of one individual to another elicited by sex pheromones is very strong, causing responding individuals to travel relatively long distances and to enter situations such as traps that they would normally avoid. This phenomenon allows pheromone traps to be very efficient in the detection of extremely small numbers of certain pests in a quarantine situation. Such traps situated near ports of entry or at shipping terminals are apt to detect reproductively mature insects from contaminated cargoes or from the carriers themselves. Early detection might permit regulatory personnel to eradicate an invading species before it became widespread. Unfortunately, these sex pheromones are specific to species; thus, a specific pheromone would be needed for each candidate species.

Research on the use of sex pheromones is being conducted in this country and in certain foreign countries. Such research usually begins with experiments demonstrating that a species possesses a chemical that elicits a sexual response or attraction from the opposite sex. Isolation and identification of the compound(s) are necessary before the chemicals can be used. For a sex pheromone to be of value as a detection or survey tool, it must be synthesized so that it is available in usable amounts at an affordable price. Such research takes years, and because of its expense, only a few countries can afford to finance, staff, and equip laboratories necessary for programs to be successful. Unfortunately, the insects of concern to U.S. agriculture are not always found in the countries that are conducting the sophisticated research. Although field bioassays of pheromones are required where the insect is naturally present, chemical analyses and syntheses can be done at separate locations. Thus, if the field and the laboratory are not too remote from each other and communication is adequate, portions of this work could be done in separate countries.

A compendium of insect pheromones (currently being revised) was compiled by Mayer and McLaughlin (1975). Although somewhat out of date, it gives some information on the status of pheromone research throughout the world. A colloquium on management of insect pests with the use of semiochemicals, held in Gainesville, Florida, provided the most up-to-date information on the use of sex pheromones for pest detection and control (Mitchell, 1981).

2. *Food Lures*

Food lures have been the primary attractants used to detect tephritid fruit flies for many years (see the review by Chambers, 1977). With certain species, such as *A. suspensa, A. ludens,* and several *Dacus* species, food lures continue to be the only reliable means of detection. Several materials have been used, but the most efficient and inexpensive is acid-hydrolyzed proteins derived as a by-product from corn syrup. This material is used in traps for detection and in a mixture with malathion as a lethal bait.

Efforts have been and are being made to identify the active components of this material in order to enhance the attractiveness or increase the efficiency of the mixture. The combinations of sugar, yeast, and protein-containing substances yield, through fermentation, a variety of products such as alcohols, acetates, aldehydes, esters, and fatty acids (Chambers, 1977). Morton and Bateman (1981) found that the volatiles were mostly primary and secondary products of amino acid degradation. In attractancy tests with *Dacus tryoni* (Froggatt), ammonia and a feeding stimulus from the amino acids were the main attractive components. However, these components have not been tested or proven effective for other species of fruit flies.

The Japanese beetle, *Popillia japonica* Newman, is attracted to some essential oils and fruity odors. Scientists attempting to find the most active compounds have developed three highly attractive materials: genaniol, a 3:7 mixture of phenethyl propionate (PEP) and eugenol, and a 3:7:3 mixture of PEP:eugenol:genaniol (Klein, 1981). These materials are used as baits in traps for population reduction and for detection.

C. Control Program Research

1. *Genetic Control*

The use of genetic mechanisms for control of pest insect species is a relatively new concept of managing pest populations. Although this autocidal method appears to be very promising, only a few of the thousands of insect species are likely targets for its use. Fortunately, those that are good candidates are also some of the most economically important pests [e.g., several species of fruit flies, stable flies, screwworm flies, *Cochliomyia hominivorax* (Coquerel), and tsetse flies, *Glossina* spp.].

Common control strategies (insecticides, etc.) normally are used only when the pest reaches economic threshold levels. Quite often, much damage has been done by the time the population is reduced and controlled. Genetic control, on the other hand, operates most efficiently against low populations and acts to keep populations low or to eradicate them. The control mechanisms are directed at a single species. This reduces effects on the ecosystem by chemical or physical contamination and does not affect other nontarget species. This type of control is probably best suited for eradication of new introductions before they have had time to establish themselves and become widespread.

a. The Sterile Insect Technique. The Sterile Insect Technique (SIT) involves the release of sterilized male insects. Theoretically, the sterile males mate with the fertile females of the same species, and all eggs are subsequently rendered sterile. This technique works best against low populations in which the ratio of sterile to fertile males is much greater than 1:1. Sterility usually is achieved by irradiation and is technically referred to as "mutagen-induced dominant lethality."

This technique was first proposed by Bushland and Hopkins (1951) as a nonchemical means of control and eradication for the screwworm fly. It was first successfully implemented on Curaçao, on the Netherlands Antilles, and on a large scale in the southeastern United States, where an extensive isolated population of screwworms was eradicated (Baumhover *et al.,* 1955; Knipling, 1955). Since then, it has been used against the screwworm in the southwestern United States and in northern Mexico. The population has now been pushed southward to central and southern Mexico.

The technique also has been used successfully against mosquitoes in El Salvador, tsetse flies in Africa (Dame *et al.*, 1981), and stable flies (Patterson *et al.*, 1981), and against several species of fruit flies (summarized by Calkins *et al.*, 1982). It was successfully used against the medfly in Los Angeles in 1975–1976 (Harris, 1977) and again in 1980. Most of the large programs were conducted outside the United States, although in many cases much of the expertise and support came from U.S. scientists. Many of these projects also have involved collaboration between the United States, the United Nations Development Progam (UNDP), the United Nations Environment Program (UNEP), the International Atomic Energy Agency (IAEA), the Food and Agriculture Organization (FAO), and the ministries of agriculture of host countries.

The largest medfly SIT program now going on began in Mexico and Guatemala in the late 1970s under the joint sponsorship of the governments of the United States, Mexico, and Guatemala, with input from IAEA and FAO. It has achieved its goal of eradicating the pest from Mexico. A barrier of sterile flies is being maintained to prevent the spread of the fly northward from Guatemala. As the program continues, plans are being made to eradicate the population completely from Central America, provided that political situations do not interfere.

Two new medfly eradication programs have begun in Egypt and Peru using the mass-rearing techniques developed by IAEA in Vienna, Austria, and perfected in the Mexico program.

Research is continuing during the conduct of these programs in quality control of mass-reared insects, release methods, and evaluation techniques. Although SIT programs have had varying degrees of success, the technique is generally accepted as a viable means of control under proper circumstances. As it continues to be used, the knowledge gained has made the technique increasingly efficient.

Several problems are being or need to be addressed in order to improve the SIT. Efforts to improve the quality of mass-reared insects—specifically, sexual compatibility and competitiveness, motility, and the ability of released insects to locate hosts and mating areas—are being employed. Many inadequacies are presently associated with the mass-rearing, sterilization, and release of insects that will require the efforts of foreign and U.S. scientists.

b. Hybrid Sterility. Another method of genetic control of insect populations that has not been tried on a practical scale but that seems to offer good possibilities is hybrid sterility. When two closely related species mate, the resulting offspring may be sterile or have a low fecundity in one or both sexes. These genetic aberrations could be used for genetic control through release of affected progeny. However, a major problem with this approach is that these hybrids are unlikely to mate with the target species under field conditions because they often express an inappropriate mixture of the premating behavior and sex pheromones of both parental species. Hybridization between allopatric species would introduce the additional possibility that the progeny would be poorly adapted to the ecology of the target area and would be less sexually competitive (Whitten and Pal, 1974).

It is sometimes possible to use a backcrossing system to retain the characteristics of the target species while maintaining the sterility of the hybrid. This might facilitate mating between the released hybrids and the target species. This procedure presently is being investigated for the genetic control of the tobacco budworm, *Heliothis virescens* (F.) (Lester *et al.*, 1976; Makela and Huettel, 1979).

A modification of the hybrid sterility procedure that has not been attempted but that might be effective as a source of sterile hybrids for genetic manipulation would be to cross an introduced population which has been in this country for many generations with the parent stock of the same species from the country of origin or from some remote area of its natural range. The genetic makeup of the two populations could have changed during the time they were apart, enough so that genetic incompatibility may exist. This phenomenon occurred accidentally when a strain of alfalfa weevil, *Hypera postica* (Gyll.), was introduced into the eastern United States. When crossed with individuals from a population from the western United States that had been introduced 40+ years earlier (Titus, 1910), a one-way genetic incompatibility was found to be present (Blickenstaff, 1969). This phenomenon may exist among other species from early introductions and should be investigated.

c. Translocation Heterozygosity. A third genetic device that may be effective for insect control is a reciprocal chromosomal translocation. It occurs when nonhomologous chromosomes exchange sections after breakage. This breakage can be induced by irradiation or chemical disruption and takes place during meiosis. If the pieces interchanged are large, the interchanged chromosomes and their normal homologues form an interchange complex. Segregation from this complex can be adjacent, resulting in duplication-deficiency gametes or inviable gametes. Because two types of segregation occur equally, translocation heterozygotes may have low fecundity. If such males are released into a field population, they would produce only a few offspring with low fecundity. This low fecundity is a result of the chromosomal translocations (Rai *et al.*, 1974).

The synthesis of strains heterozygous for a sufficient number of translocations to render them highly sterile is quite inefficient and may not be feasible alone in a SIT program (Whitten, 1971). However, if translocations creating a low level of sterility are combined with low levels of irradiation, sufficiently high sterility would occur without reducing the competitiveness of the insect. This technique is presently being explored with the medfly by R. Steffens, University of Giessen, Giessen, Germany (personal communication).

2. *Host Plant Resistance*

The use of crop varieties resistant to insect attack is the ideal means of insect control. In some cases, no other means of protection are necessary and the threat of chemical residues is diminished. Unfortunately, it often takes from 10 to 25 years, depending on the crop, to develop suitable varieties. Much of this effort in the United States is involved in searches for resistance against established exotic pests. If the sources of such resistance already have been established, the time needed to develop resistant varieties would be greatly reduced.

During the early development of agricultural crops, the strains or varieties most susceptible to insect and disease attack were eliminated by grower selection or by failure to reproduce, much as they would have been under natural selection. Although this selection pressure usually involved only phenotypic characteristics and the selection was not intensive during many generations, a certain amount of host resistance developed under this primitive method.

With the advent of agricultural mechanization and chemical pest control, the host–pest relationship changed. Instead of small plantings of several species of food plants, extensive plantings prevailed and pest populations were artificially reduced. Plant breeders, in their quest to increase yields, inadvertently reduced insect resistance. As a result of plant protection, many plants that would have been severely damaged or eliminated were thus protected. As the percentage of these susceptible plants increased in the crop gene pool, the potential for pest damage increased. When chemical controls were no longer effective, many of the accepted plant varieties sustained heavy losses.

The defense mechanisms evolved by plants to thwart insects feeding on them include arrays of physical barriers and complexes of chemicals that render the plant repellent, toxic, unpalatable, or otherwise unsuitable for use (Beck and Schoonhoven, 1980). The evolution of chemical and physical factors in plants to deter insect feeding resulted in an evolution of enzyme systems and behavioral mechanisms in insects that tended to neutralize these defenses. This coevolutionary relationship is analogous to an arms race in which new defenses eventually are countered by appropriate offenses. In some cases, the chemical defenses allow some plant species to avoid the role of host completely for certain species of insect. In other cases, the insects not only are able to evolve neutralizing mechanisms against these defenses but may even use the chemical defense substances as "kairomones" or recognition cues (Maugh, 1982).

Chemicals for which no metabolic or nutritional function is known often have been suggested as sources of host plant resistance. These so-called secondary chemicals were termed "allelochemics" by Whittaker (1970). They are defined as nonnutritional chemicals that are produced by an individual species that affects the growth, health, behavior, or biology of another species. The effects may be behavioral as well as metabolic. These allelochemics are heritable.

To develop resistant cultivars, gene recombination by selective breeding is the most rapid method. For this technique to function properly, geneticists must have access to gene pools with genetic diversity great enough to encompass the traits desired. During the evolution of host–insect relationships, a great deal of genetic diversity developed. However, this diversity provides the raw materials for host plant resistance breeding programs. Hidden among this vast, diverse gene pool are probably the genes that impart resistance to most of the crop pests in the world. This diversity is rapidly being lost as high-yielding cultivars replace strains of more ancient cultivars.

Attempts to preserve this genetic diversity are being made. The International Board of Plant Genetic Resources (IBPGR), begun in 1973, has established priorities by regions and crops, appointed crop advisory committees for several major crops, supported development of information retrieval and documentation systems, developed guidelines for long-term cold storage of base collections, and activated portions of the global network plan. IBPGR thus is the world coordinating body for germplasm collection and maintenance activities (Harlan and Starks, 1980). A list of the major germplasm collection and maintenance laboratories is presented in Appendix 1. Material is collected from various regions of the world and is brought to these national or international genetic resource centers for cataloging and storage.

Although collections for some crops are quite extensive, most are not complete. Collections are weakest in wild races and related species; yet, these are probably the best sources of resistance to diseases, nematodes, and insects. Other sources of resistance are found in adapted cultivars, plant introductions, exotic germplasm, and even in near relatives of the cultivars. Through inbreeding, crossbreeding, and hybridization, the genes for resistance can be transferred to agronomically adapted cultivars. Unfortunately, there is some indication that larger, more complete collections are necessary to obtain insect-resistant material than to obtain disease-resistant material (Harlan and Starks, 1980).

Locating sources of insect resistance can best be accomplished by the use of networks of commodity-oriented international centers. They are strategically located and could participate in the improvement and monitoring of genetic resistance as an integral part of the overall crop improvement process. The infrastructure of facilities and services is established, and liaison with national programs and universities already exists (Ortega *et al.*, 1980). Many of these centers of crop improvement are associated with centers of germplasm laboratories (Appendix 1).

The fundamental bases of resistance to most insects are not known. Rapid plant screening for host resistance does not reveal the specific nature of resistance. If the physical and chemical bases are understood, modern chemical screening techniques might permit prompt, accurate, and inexpensive detection of these resistance factors. Rapid incorporation of these factors into existing cultivars in the United States would make them available more quickly in the event that certain exotic pest species were established.

The development of host plant resistance to insects relies on the efforts of a multidisciplinary team made up of entomologists, plant breeders, biochemists, and physiologists.

The incorporation into United States crops of resistance factors to pests not yet introduced may not be well advised or feasible. Most chemical defense factors are not directly associated with the production of biomass, as previously indicated, yet they involve rather complex metabolic processes. They create an energy drain on the plant that is of benefit only in the presence of the insect pest. If these factors are incorporated into varieties or breeding lines in the United

States in the absence of the pest, they may inadvertently be lost when plants subsequently are selected and bred for high-yield characteristics.

The development of varieties with single-gene resistance in foreign countries also may be disadvantageous to U.S. agriculture if the target pests eventually are introduced. Early development and incorporation of defense factors into varieties where intense selection pressure will be exerted against the target species could lead to adaptation of the pest or the development of biotypes.

Because many of our crops and most of our insect pests are of foreign origin, it is logical to search and do research on host plant resistance outside the country in areas indigenous to the host species. It is here that the "arms race" has had time to develop, and continued pressure by the insect has resulted in maintenance of chemical and physical resistance factors in the plant. An example of this phenomenon is the relationship of alfalfa to the spotted alfalfa aphid, *Therioaphis maculata* (Buckton). This insect is a native of the semiarid area of the Middle East. Resistance in alfalfa to the aphid varied directly with the parental source (Howe *et al.*, 1965). Varieties of Turkestan or partial Turkestan origin and African varieties all exhibited moderate to high degrees of resistance. Alfalfa types that owed their parentage to Chilean origin, where the spotted alfalfa aphid never occurred, always showed a high degree of susceptibility.

Another example is maize, which originated in Central America (Mangelsdorf, 1974). The greatest genetic diversity seems to exist in Mexico, Guatemala, and the Andean region of South America. Melhus *et al.* (1954) located some sources of resistance to several *Diabrotica* spp. in Guatemalan corn lines and speculated that Mesoamerica would be the logical area to search for resistance. Evidence has been found that *Diabrotica* also originated in this area (Webster, 1895; Smith, 1966). Thus, the insect–host plant relationship has existed for some time, and one would expect to find the greatest amount of resistance there. Branson and Krysan (1981) recently initiated a search for *Diabrotica* resistance in maize and its near relatives in Latin America.

3. *Biological Control*

When emigrant species invade a new territory, they frequently leave their controlling biotic influences behind. These influences, primarily predators, parasites, and pathogens, normally keep these insect populations at certain levels of equilibrium. In the absence of these factors, the populations are allowed to rise substantially until they reach outbreak proportions. To bring such populations down to manageable levels, it is necessary to bring their controlling biotic factors back into association with the invader. Virtually all successful classic biological control programs have resulted from this reassociation of invading species with their previous natural enemies (DeBach, 1964).

The earliest biological control project against a foreign invader by imported biotic control agents involved the cottonycushion scale, *Icerya purchasi* (Maskell), in 1888. This scale was accidentally introduced from Australia into California, where it became a serious pest on citrus. The search for its natural enemies began in the native area of the scale where the coccinellid predator, the vedalia beetle, *Rodolia cardinalis* (Mulsant), and a dipteran parasite, *Crypotochetum iceryae* (Williston), were found. After they were imported and observed to readily attack the scale, they were released into citrus groves. Within months, the scale epidemic had been reduced to noneconomic levels (van den Bosch and Messenger, 1973).

Several other predators and parasites were introduced into the United States during the late 1800s and early 1900s to combat several species of scales and mealybugs, the greenbug, *Schizaphis graminun* (Rondani), and the codling moth, *Laspeyresia pomonella* (L.). Other such introductions have continued to the present, with varying degrees of success. DeBach (1964) tabulated a total of 162 completely or partially successful biological control projects in the United States from 1890 to 1963.

Effective biological control programs against imported pest species depend on a number of factors. Foreign exploration by trained, observant entomologists is needed to locate efficient biological control organisms. These candidate agents must be reared so that large numbers of unparasitized, disease-free insects can be imported. The insects are then colonized so that enough individuals for several releases in several locations can be available. And finally, adequate evaluation of the releases must be conducted to determine whether the insects were successfully established and, if not, to understand why.

The USDA has established the European Parasite Laboratory and the Asian Parasite Laboratory (see Section III) overseas to serve as bases of operations for collecting, rearing, and identifying natural enemies of introduced insects. Entomologists employed by several states also undertake foreign explorations for natural enemies of specific pests. Considering the very small investment in money and manpower and the tremendous benefits achieved through successful introduction of natural control agents, this type of foreign research should be strongly supported.

III. RESEARCH INSTITUTIONS

Almost every country in the world is conducting agricultural research at some level of complexity. This includes plant and animal protection through insect pest and disease research. This chapter is not intended to list every entomological research facility and program outside the United States. Much of this information can be obtained through other means. For instance, the Entomological Society of America has published and is in the process of publishing a "Worldwide Directory of Institutions with Entomologists." "Part I: Latin America," prepared by

R. N. Williams, was published in 1978 in the *Bulletin of the Entomological Society of America* **24**(2), 179–193. "Part II: Western Europe, Australia, and New Zealand," prepared by C. O. Calkins, is currently awaiting publication in the *Bulletin*. Two other segments, on Asia and Africa, are now being prepared. When completed, the directory should suffice for locating almost all institutes involved in entomological research outside North America. However, because of the uniqueness of international programs and institutions, they and their roles are described briefly here.

A. Foreign Research

1. *Supportive*

a. European Parasite Laboratory. The European Parasite Laboratory, ARS, USDA, at Sevres, France, is the oldest federal biological control laboratory outside the United States. It was established in 1919 to obtain parasites and predators of the European corn borer (Drea, 1980).

During the past 60+ years, the small staff (four entomologists plus support personnel) have searched for, studied, and shipped to the United States a wide range of parasites and predators of pest species as part of the USDA biological control program. Some of the pests which were at least partially controlled by parasite introductions included the pea aphid, spotted alfalfa aphid, alfalfa weevil, cereal leaf beetle, and imported cabbageworm. During this period, natural enemies of more than 65 pest insects and 11 weeds were studied at the laboratory.

The European Parasite Laboratory continues to obtain biological control agents and to conduct basic research in Europe on agents associated with specific biological control projects. The laboratory also provides specific agents to scientists from other research organizations in North America and provides technical assistance and laboratory space to scientists conducting short-term biocontrol research in Europe (Drea, 1980).

b. Asian Parasite Laboratory. The Asian Parasite Laboratory was established in 1975 by the USDA-ARS in Sapporo, Hokkaido, Japan, in order to facilitate the search for endemic parasites of the gypsy moth, *Lymantria dispar* (L). In February 1982, the laboratory was moved to Seoul, Korea, to take advantage of a greater range of host and parasite insects. At that time, the laboratory also became engaged in the search for biological control agents of the larch case bearer, the Japanese beetle, *Popillia japonica* Newman, the Mexican bean beetle, *Epilachna varivestis* Mulsant, the European corn borer, the chestnut gall wasp, *Dryocosmus kuriphilus* Yasumatsu, and the euonymus scale, *Unaspis euonymi* (Comstock).

Material is collected and identified from the field and is frequently reared for biological and life history information. After the numbers have been increased, it is sent to the Beneficial Insects Research Laboratory, ARS, USDA, Newark, Delaware. To date, a predator and a parasite of the gypsy moth have been introduced and successfully established in Pennsylvania, and a new parasite of the chestnut gall wasp has been established in Georgia as a result of the work of this laboratory (Paul Schaeffer, Research Entomologist, Beneficial Insects Research Laboratory, ARS, USDA, Newark, Delaware, personal communication).

2. *Cooperative*

a. Binational Agricultural Research and Development Fund. In an effort to promote and support cooperative research in agriculture between the United States and Israel, a Binational Agricultural Research and Development Fund (BARD) was established in October 1977. BARD is an independent legal entity directed by a Board of Directors with equal representation from both countries.

BARD's income, from which it allocates grants for new projects, is derived from the interest earned on an endowment fund contributed by both governments. Proposals for grants are accepted from all public and private nonprofit research institutions that demonstrate the required research and development capabilities, provided the proposals meet BARD objectives and criteria. One of the supported areas of research is crop protection. This covers the development of improved protection technology such as breeding for host plant resistance and biological and integrated control. Research in the area of chemical control emphasizes environmental safety and economic feasibility. Cooperative research under this agreement involves active collaboration between Israeli and American scientists (United States–Israel Binational Agricultural Research and Development Fund, 1981).

b. Public Law 480. The USDA supports foreign research on several agricultural topics. These studies are financed through a grant program from foreign currencies that have accrued to the credit of the United States, primarily from the sales of farm products abroad under Public Law (P.L.) 480. Currently, in the case of Poland and Yugoslavia, financing of research grants is being supplemented with the currencies made available by those countries under provisions of the U.S.–Poland and U.S.–Yugoslavia Joint Boards. The monies from all such grants cannot be converted into dollars for use in the United States (Anonymous, 1980).

Several grants for entomological subjects currently are being supported in Egypt, India, Pakistan, Poland, and Yugoslavia (Appendix 3). Many of these projects are targeted for species of insects that are potential threats to U.S. agriculture, e.g., the Egyptian cottonleaf worm, *Spodoptera littoralis* (Boisd.), beet armyworm, *S. exigua* (Hübner), *Heliothis* spp., Mediterranean fruit fly, melon fruit fly (*Dacus cucurbitae* Coq.), leafhoppers, scale insects, and mites. A large percentage of the grants also involve the search for and the study of biological control organisms of exotic pests already in the United States (Anonymous, 1980).

3. *Passive*

a. International Organization of Biological Control. An organization that has done much to further cooperative entomological research among countries is the International Organization of Biological Control (IOBC), It is divided into four geographical regions for dealing with regional priorities: the East Palearctic Regional Section (EPRS), the West Palaearctic Regional Section (WPRS), the Western Hemisphere Regional Section (WHRS), and the South and East Asian Regional Sections (SEARS). It also supports projects of worldwide importance through its international and global working groups.

The International Working Group on *Ostrinia* (IWGO) brings together entomologists from 15 countries to exchange inbred lines of maize (corn) in order to test responses to *O. nubilalis* and other maize insects such as fruit flies in Europe and corn rootworms (*Diabrotica*) in North America. Selected resistant inbreds are subsequently used in producing hybrids resistant to target pests while retaining desired local adaptabilities.

The International Working Group on Fruit Flies of Economic Importance originated as a working group on tephritid fruit flies in the International Biological Programme (IBP) in 1968 (Bateman, 1976). This group became welded into a single entity with agreed-upon aims and common goals. Collaborative projects included pupal mortality factors, biological control agents, fruit-marking pheromones, color attraction, quality control, trapping techniques, and trap development. Extensive studies were also conducted jointly on microorganisms, attractants, and population genetics. Frequent communication was facilitated by development of a world list of fruit fly workers and distribution of a fruit fly bulletin to members. Workshops and meetings were organized under these auspices (Bateman, 1976). This group now forms the basis of a manpower pool for information, expertise, and action support for large control programs anywhere in the world. Members serve as consultants, experts, and collaborators for organizations such as IAEA, FAO, and APHIS. Some of the most recent programs in which members were involved included the Mexico–U.S.–Guatemala Medfly Program, the California Medfly Eradication Program, the Okinawa Prefecto Program for control of the oriental fruit fly (*Dacus dorsalis* Hendel), and the Egyptian Medfly Project.

The most recently formed group is the Global Working Group on Quality Control, which supports large programs in which insects are mass-reared for eventual release. Fruit fly programs involving release of sterile males for control and eradication were emphasized at the outset. Other programs involving Lepidoptera and biting flies and *Trichogramma* release programs soon followed.

A WPRS Working Group on Integrated Control in Orchards and a WPRS Working Group on Bruchidae are examples of regional groups that could be of benefit in gaining information on potential insect threats to the United States.

b. International Atomic Energy Agency. The International Atomic Energy Agency (IAEA) was established as a specialized agency of the United Nations in Vienna, Austria, in 1957. Its main objectives are to encourage and assist research and/or development and practical applications for peaceful uses of atomic energy. A Joint Division of Atomic Energy in Food and Agriculture was formed shortly thereafter by combining expertise from FAO and IAEA. Within this division were sections for insect and pest control and for chemical residues and pollution.

For a number of years, the Insect and Pest Control Section has been cooperating with member states in demonstrating the effectiveness of the sterile-male technique (SIT) for controlling the Mediterranean fruit fly, olive fruit fly, *Dacus oleae* (Gmelin), codling moth, and several tsetse fly species.

The section also advises governments of member states on entomological programs using atomic energy, awards fellowships for advanced study, finances research, provides consultants and experts for programs, and conducts training courses, workshops, and symposiums on entomological topics involving radiation and isotopes.

The IAEA supports a laboratory at Seibersdorf, Austria, part of which is devoted to the use of atomic energy in entomology. The main objective of the Entomology Section is to support and service the Joint Division's programs on insect control. General areas of research involve the development and improvement of mass rearing of insects, improvement of radiation techniques for insect sterility, development of laboratory methods for determining and improving insect quality control (sexual competitiveness), and development of insect shipping methodology and release techniques. A recently added area of research involves the development of isotopic techniques in insect physiology and ecology in direct support of FAO programs.

The Entomology Laboratory and the Joint Division presently (1982) are supporting the large SIT programs on medfly in Mexico–Guatemala, Peru, and Egypt and the Biological Investigations for the Control of Tsetse Fly (BICOT) program on the tsetse fly, *G. palpalis palpalis* (Robineau-Desvoidy), in Nigeria. They have also supported the USAID–Tanzania tsetse fly project by maintaining a backup colony of *Glossina morsitans morsitans* Westw. at Seibersdorf.

c. Food and Agriculture Organization. The Food and Agriculture Organization (FAO) of the United Nations was established in 1945 in Quebec, Canada. The headquarters was moved to Rome, Italy, in 1951. Its main objective is to promote research to increase food production in developing countries and the publication of information on all aspects of agriculture in support of this objective.

To identify the major factors involving crop losses, FAO maintains statistics on crop yields and losses for all countries. Statistics for crop losses caused by insect pests are essential for governments to justify expenditures for research. These statistics are also useful to APHIS to determine the most important potential pests of U.S. agriculture. FAO supports and conducts research on pest insects through the use of financial grants, research contracts, consultantships, and hiring of experts and associate experts.

One area of research on pest control heavily supported by FAO is integrated pest management (IPM), referred to by them as "integrated pest control (IPC)." This began in 1974, when a proposal was made by the FAO Panel of Experts on Integrated Pest Control to develop the FAO/UNEP (United Nations Environment Program) Cooperative Global Program for the Development and Application of Integrated Pest Control in Agriculture. This program, by providing guidance and training in research, will contribute to the development of sufficient expertise in IPM so that developing countries will be able to conduct such control programs on pests of major importance (Brader, 1979).

In its original phase, intercountry programs for three major crops (cotton, rice, and sorghum/millet) were initiated in 1978. These programs were implemented in certain regions to include countries in which these crops were of major economic importance. The research activities in each of the participating countries in the same region were designed to complement each other. The programs for cotton involved the Near East (Pakistan, Iraq, Syria, and Turkey), Africa (Egypt and the Sudan), and Latin America (Mexico, El Salvador, Guatemala, Colombia, Peru, Bolivia, and Brazil). For rice, the intercountry program comprised Indonesia, Malaysia, the Philippines, Thailand, Bangladesh, Sri Lanka, and India. The program for sorghum/millet comprised countries from the Sahel such as Gambia, Senegal, Mali, Upper Volta, Niger, Chad, Cape Verde, and Mauritania (Brader, 1979).

Other research and control programs supported by FAO include the coconut beetle (*Oryctes rhinoceros*) in Samoa and other areas of the South Pacific, the desert locust in Africa and the Near East, citrus insects, and stored-product pests.

d. International Centre of Insect Physiology and Ecology. The International Centre of Insect Physiology and Ecology (ICIPE) was established in Nairobi, Kenya, in 1970 in order to foster close cooperation with many specialized institutes in East Africa concerned with the control of insects harmful to crops, animals, and man (Pringle, 1980). Its mission was originally to provide a laboratory where a multinational group of scientists could carry out research of high quality in certain fields of insect physiology and ecology, and to train and educate future research scientists. Fields of endeavor were selected by teams of entomologists and chemists from the United States and Europe. These fields included the effects of hormones on growth and reproduction, pheromones, behavior and ecology, mosquito genetics and lethal genes, and the chemistry of plant products acting as insect hormones. Although the research was originally very basic, increasing emphasis is now being placed on integrated pest management. Target insects include the nutgrass armyworm, *Spodoptera exempta* (Walker), tsetse flies, locusts, mosquitoes, and termites. In addition, pest complexes of cotton, cocoa, coffee, sugarcane, sorghum, and others are being studied.

Funding is provided by several European countries, such as Sweden, Switzerland, the Netherlands, the United Kingdom, West Germany, and others. Several international organizations, including FAO, the World Health Organization (WHO), IAEA, UNDP, the United Nations Industrial Development Organization (UNIDO), and several university institutions also participate in funding.

B. Quarantine Research

1. *Plant Disease Research Laboratory*

The USDA currently operates the Plant Disease Research Laboratory in Frederick, Maryland, as a containment facility for the study of exotic plant diseases and their vectors. The purpose of the laboratory is to assess the potential of diseases and the vectors for economic loss if they are accidently introduced. Currently among the insect vectors being studied are the South African leafhopper (*Cicadudina mbila*), a migrating vector of maize streak virus, and a foxglove aphid (*Alacoucthum solani*) from Japan that is a vector of soybean dwarf virus, although biotypes of this aphid found in the United States do not colonize well on soybeans.

The potentially most important diseases and vectors of corn, soybeans, and wheat not yet present in the United States were chosen for initial studies. The containment facilities are under strict quarantine. Specially designed greenhouses permit simultaneous studies of several pathogens and vectors. Each greenhouse has foolproof safeguards against the escape of the test organisms.

2. *Facilities for Exotic Insect Research in the United States*

A secure facility is needed in the continental United States to conduct research on exotic nonvector insects that are serious potential pests or insects that presently are under quarantine regulation in this country, and on insects currently being studied in foreign countries with the support of USAID funds. Such a facility should be located in a geographical or climatic zone (on an island or in a very cold climate) that would preclude establishment if escape were to occur.

The mission of such a facility would be to develop basic information for the detection, control, or eradication of potential pest species. Research should concentrate on topics that cannot readily be addressed in foreign countries, such as chemical communication systems, basic behavior, basic genetics, mating behavior, ecological genetics, genetic sexing, inhibition of reproduction, host specificity, effective colonization procedures, quality control of mass-reared target insects, and development of detection devices. Such research would lead to the development of techniques and strategies to improve eradication efforts such as the sterile-insect release method.

Both the medfly and the screwworm are serious pests that currently are being mass-reared outside the contiguous 48 states in cooperation with APHIS and ARS for eradication by the sterile-male release method. These programs require such basic information to support the rearing and sterilization of high-quality insects for release that are capable of mating with their native counterparts.

Other species that might be considered for research in a quarantine facility include the frit fly, nun moth, old world bollworm, and other species of fruit flies.

IV. CONCLUSIONS

The contiguous United States is quite susceptible to pest introductions because of its size and extensive borders and because of the increase in international trade and travel. The initial appearance of any foreign pest within our borders causes consternation in the agricultural community. The strategy in dealing with such pests depends largely upon our knowledge of their biology and behavior and upon the availability of proper control techniques.

Entomological research, in general, can be conducted anyplace in the world where competent scientists and adequate equipment and facilities occur. However, the type of information needed to prevent specific pests from entering our borders or to combat them after their arrival can be procured only by investigating each species before it actually threatens to become a resident. As previously mentioned, this information cannot be acquired rapidly or easily during the pests' initial appearance but must be obtained where these species already exist prior to their encroachment.

Although there are disadvantages in compiling lists of potential pest species, it behooves us to be aware of those that pose the greatest threat and to obtain as much information about them as possible to assist us in quarantine and control campaigns.

To obtain the specific information needed about the potential pest, cooperative efforts should be established between the United States and the countries in which the pest is endemic. Although such investigations can be entirely supported by the United States with U.S. personnel, it would be more prudent to establish cooperative research projects with the candidate countries. Such research efforts will be of mutual benefit to all countries involved. Information about potentially threatening species also would be useful in solving other entomological problems. Experience gained by both parties would constitute a useable pool of knowledge. In addition, support for the training of young scientists who use these pests as their research topics would increase both the amount of critical information and the number of experts familiar with the pest. Such expertise in close proximity to these pest species could be utilized in the event that the pest invades the United States. In addition, the contacts and goodwill created by such research projects would lead to better mutual understanding of the problems associated with these pests by the respective countries.

APPENDIX 1

Major Germplasm Collection and Maintenance Laboratories and Crop Types Stored

1. Bangladesh Rice Research Institute, Dacca, Bangladesh—rice
2. Central Agricultural Station, Non Repos, Guyana—rice
3. Central Experimental Farm, Ottawa, Canada—wheat, oats, rye
4. Central Research Institute for Agriculture, Bogor, Indonesia—rice
5. Central Rice Research Institute, Cuttack, India—rice
6. Centro Internacional de Agricultura Tropical (CIAT), Cali, Colombia—cassava, beans, rice, cattle pastures (1967)
7. Centro Internacional de Mejoramiento de Maiz y Trigo (CIMMYT), El Batâh, Mexico—maize and wheat (1943)
8. Centro Internacional de Papas (CIP), Peruvian Andes—potatoes (1971)
9. Crop Research and Introduction Center, Izmir, Turkey—wheat, rye
10. Foundation of Agriculture and Plant Breeding, Waginingen, the Netherlands—wheat, rye
11. Instituto Colombiano Agropecuario (ICA), Medellin, Colombia—maize
12. Instituto Nacional de Investigaciones Agricolas (IMIA), Champingo, Mexico—maize
13. International Crops Research Institute for the Semi-Arid Tropics (ICRICAT), Hyderabad, India—sorghum, millet, peanuts (1972)
14. International Institute of Tropical Agriculture (IITA), Ibadan, Nigeria—cassava, yams
15. International Rice Research Institute (IRRI), Manila, the Philippines—rice (1959)
16. Japan National Seed Storage Facility, Hiratsuka, Japan—wheat, rice, rye
17. Laboratorio del Germoplasmo, Bari, Italy—wheat, rye
18. Ohara Institute for Agricultural Biology, Okahama University, Kurashiki, Japan—barley, rye
19. Plant Breeding Institute, Cambridge, England—wheat, rye
20. Research Institute of Cereals, Kamery, Havlivkova, Czechoslovakia—barley
21. Swedish Seed Association, Svalov, Sweden—wheat, rye
22. USDA, Beltsville, Maryland—rice, wheat, barley, rye
23. U.S. Federal Experiment Station, Mayagüez, Puerto Rico—sorghum, cassava
24. USDA Plant Introduction Station, Ames, Iowa—maize
25. U.S. Regional Soybean Laboratory, University of Illinois, Champaign, Illinois—soybeans
26. U.S. Southern Regional Plant Introduction Station, Experiment, Georgia—peanuts
27. USDA Agricultural Experiment Station, College Station, Texas—sorghum, cotton
28. National Seed Storage Laboratory, Fort Collins, Colorado,—wheat, rice, barley
29. Waite Agricultural Research Institute, Adelaid, Australia—barley, rye
30. World Collection of Wheat, Tamworth, Australia—wheat

APPENDIX 2

International Agricultural Research Centers

1. Centro Internacional de Mejoramiento de Maiz y Trigo (CIMMYT), El Batâh, Mexico—maize and wheat (1943)
2. International Rice Research Institute (IRRI), Los Banos, the Philippines—rice (1959)
3. Centro Internacional de Agriculturo Tropical (CIAT), Cali, Colombia—rice, kidney beans, cattle pasture (1967)
4. International Institute of Tropical Agriculture (IITA), Ibadan, Nigeria—sweet potato, yam, cocoyam, cassava, rice, maize, soybeans, pigeon peas, lima beans (1967)
5. Centro Internacional de Papa (CIP), Peruvian Andes—potatoes (1971)
6. West African Rice Development Association (WARDA), Liberia—rice (1971)
7. International Crops Research Institute for the Semi-Arid Tropics (ICRISAT), Hyderabad, India—ground nuts, chick peas, pigeon peas, pearl millet, sorghum (1972)
8. International Livestock Center for Africa (ILCA), Addis Ababa, Ethiopia—pastoral and mixed farming economics of Africa (1974)
9. International Laboratory for Research on Animal Diseases (ILRAD), Nairobi, Kenya—livestock diseases (1974)
10. International Board for Plant Genetics Resources (IBPGR), with FAO in Rome, Italy—conservation of crop diversity by sponsorship of international germplasm collections (1974)
11. International Center for Agricultural Research in Dry Areas (ICARDA), Aleppo, Syria—increased food production in dry subtropical and subtemperate areas
12. International Food Policy Research Institute (IFPRI), Washington, D.C.—strategies for meeting world's food needs (1975)
13. International Service for National Agricultural Research (ISNAR), The Hague, the Netherlands—assists developing countries to initiate and upgrade their agricultural research and extension services (1979)

APPENDIX 3

Examples of Entomological Studies Funded by Public Law 480 in Foreign Countries during 1980

Egypt

1. Persistence of some insecticides in semiarid conditions in Egypt
2. Surveying and ecological studies on phytophagous and predaceous pests and soil mites in Egypt, with special consideration for newly reclaimed land
3. A study of the factors influencing population density, distribution, migration, and type and degree of damage caused to cotton in Egypt by *Spodoptera littoralis* and *S. exigua*
4. Biotic factors affecting different species in the genera *Heliothis* and *Spodoptera* in Egypt
5. Biological pest control in Egypt
6. Insect-resistant packages for processed cereals

APPENDIX 3 *Continued*

7. Plant pest trapping project
8. Development of data for utilization of traps as a safeguard measure against newly introduced pests
9. Ecological studies to develop the SIT for control of the Mediterranean fruit fly, *Ceratitis capitata* (Wied.), in the government farms in the reclaimed area of the western desert of Egypt
10. The development and use of varietal resistance and other nonchemical control methods to reduce yield losses due to stalk borers in maize
11. Weed and wild plants as hosts of some economic insects and mites
12. Investigations on the role of predators and parasites in controlling dung-breeding flies
13. Integrated control of scale insects and red spider mites attacking citrus in Egypt
14. Biology and ecology of the principal natural enemies of stored-product insects in Egypt
15. Utilization of non-*Apis* bees as crop pollinators
16. Natural enemies of *Lygus* and associated plant bugs in upper Egypt

India

1. Cytotaxonomical studies on certain Indian Curculionidae (Coleoptera)
2. Control of the behavior of leafhoppers
3. Biosystematics of *Lygaeoidea* of India
4. Study of pesticide residues and monitoring of pesticidal pollution
5. Studies on the taxonomy of *Proctotrypoidea* (= Serphoidea: parasitic Hymenoptera) from India
6. Comparative study on the effects of alkylating and nonalkylating chemosterilants on the grasshoppers (*Chrotogenus* sp. and *Poecilocerus* sp.) and locust (*Schistocera gregaria*)
7. Survey for natural enemies of diaspine scale insects in South India

Pakistan

1. Melon fruit fly and its control in (NWFP) Pakistan
2. Studies on the long-term effects of gamma rays on some important pests of stored cereals
3. Investigations on the insect enemies of *Chenopodium* spp. (*album, murale,* and *botrys*) and *Kochia scoparia* in Pakistan and the supply of specific insect enemies of *Salsola* and *Halogeton* to the United States
4. Studies on potential biological control agents of white flies in Pakistan
5. Investigations on the natural enemies of selected lepidopterous pests of crucifers and feasibility studies of mass rearing and release of promising species for the control of these pests
6. Ecology of leafhopper pests of vegetables and fruit plants of Pakistan
7. Studies on honeybees and their management in Pakistan
8. Investigations on new strains of *Bacillus thuringiensis* (BT) affecting lepidopterous crop pests
9. Studies on biosystematics and control of mites of field crops, vegetables, and fruit plants in Pakistan

(*continued*)

APPENDIX 3 *Continued*

10. Studies on the storage stability of pesticides and their residues on crops in Pakistan
11. Studies on the rice insects of Pakistan, with reference to their systematics and pheromone glands
12. Development of procedures for using *Apanteles flavipes* in periodic releases for control of stalk borers feeding on sugarcane and other graminaceous plants
13. Investigations on the natural enemies of *Epilachna* spp. in Pakistan
14. Integrated control of aphids on cruciferous crops in Pakistan
15. Insect pests on stored cereal grains and their control in Pakistan
16. Studies on the biology, phenology, and field behavior of natural enemies of the pink bollworm, *Pectinophora gossypiella* (Saunders), in Pakistan
17. Survey and evaluation of the effectiveness of the parasites, predators, and pathogens of cutworms affecting fruits, vegetables, field crops, and grasses
18. Biology and ecology of the principal natural enemies of some flies that breed in dung and vegetable refuse in Pakistan
19. Investigations on nematodes parasitic on insect pests
20. Bionomics and control of coccids in Baluchistan

Poland

1. Nonchemical control methods of stored-product mites
2. Survey and study of pathogens of stored-product insects and their possible use in biological control
3. Studies on the population dynamics of the cereal leaf beetles (*Oulema* spp.) in Poland
4. Genetic basis of reproduction in the honeybee
5. The role and exploitation of predatory and parasitic insects in limiting the population of the more important pests of apple orchards
6. Development of entomophathogenic fungi for aphid control
7. Effect of calcium imbalance on the development of common stored-product arthropods

Yugoslavia

1. The revision and monographic elaboration of Cetoniinae (*Coleoptera lamellicornia*) of the Palearctic and Oriental regions.
2. Biological control of terrestrial and aquatic weeds and pests of crop plants and forests
3. The development of wheat species that are resistant to the Senn pest, *Eurygaster austriaca* Schrk., Pentatomidae, insect pests of wheat in Yugoslavia

REFERENCES

Anonymous (1980). "Active Foreign Agricultural Research Grants" USDA-IRD-OD-1011 Int. Res. Div., U.S. Dep. Agric., Washington, D.C.

Baltensweiler, W., Priesner, E., Arn, H., and Delucchi, V. (1978). Unterschiedliche Sexuallockstoffe bei Lärchen- und Arvenform des Grauen Larchenwicklers (*Zeiraphera diniana* Gn.), Lep. Tortricidae. *Mitt. Schweiz. Entomol. Ges.* **51,** 133–142.

Bateman, M. A. (1976). Fruit flies. *In* "Studies in Biological Control" (V. L. Delucchi, ed.) Int. Biol. Programme 9, pp. 11–49. Cambridge Univ. Press, London and New York.

Baumhover, A. H., Graham, A. J., Bitter, B.A., Hopkins, D. E., New, W. D., Dudley, F. H., and Bushland, R. C. (1955). Screw-worm control through release of sterilized flies. *J. Econ. Entomol.* **48,** 462–466.

Beck, S. D., and Schoonhoven, L. M. (1980). Insect behavior and plant resistance. *In* "Breeding Plants Resistant to Insects" (F. G. Maxwell and P. J. Jennings, eds.), pp. 114–135. Wiley, New York.

Blickenstaff, C. G. (1969). Mating competition between eastern and western strains of the alfalfa weevil, *Hypera postica. Ann. Entomol. Soc. Am.* **62,** 956–958.

Brader, L. (1979). Integrated pest control in the developing world. *Annu. Rev. Entomol.* **24,** 225–254.

Branson, T. F., and Krysan, J. L. (1981). Feeding and oviposition behavior and life cycle strategies of *Diabrotica:* An evolutionary view with implications for pest management. *Environ. Entomol.* **10,** 826–831.

Bushland, R. C., and Hopkins, D. E. (1951). Experiments with screwworm flies sterilized by X-rays. *J. Eco. Entomol.* **44**(5), 725–731.

Calkins, C. O., Chambers, D. L., and Boller, E. F. (1982). Quality control of fruit flies in a sterile insect release programme. *In* "Sterile Insect Technique and Radiation in Insect Control," IAEA-SM-255/37, pp. 341–355. IAEA, Vienna.

Carlson, D. A. (1982). Chemical taxonomy in tsetse flies (*Glossina* spp.) by analysis of cuticular components. *J. Insect Physiol.* (in preparation).

Carlson, D. A., and Service, M. W. (1979). Differentiation between species of the *Anopheles gambiae* Giles complex (Diptera: Culicidae) by analysis of cuticular hyudrocarbons. *Ann. Trop. Med. Parasitol.* **73,** 589–592.

Carlson, D. A., and Service, M. W. (1980). Identification of mosquitoes of *Anopheles gambiae* species complex A and B by analysis of cuticular components. *Science* **207,** 1089–1091.

Carlson, D. A., and Walsh, J. F. (1981). Identification of two West Africa black flies (Diptera: Simuliidae) of the *Simulium damnosum* species complex by analysis of cuticular paraffins. *Acta Trop.* **38,** 235–239.

Chambers, D. L. (1977). Attractants for fruit fly survey and control. *In* "Chemical Control of Insect Behavior: Theory and Application" (H. H. Shorey and J. J. McKelvey, Jr., eds.), pp. 327–344. Wiley, New York.

Dame, D. A., Lowe, R. E., and Williamson, D. L. (1981). Assessment of released sterile *Anopheles albimanus* and *Glossina morsitans morsitans. In* "Cytogenetics and Genetics of Vectors" (R. Pal, J. B. Kitzmiller, and T. Kanda, eds.), pp. 231–248. Elsevier/North-Holland Biomedical Press, Amsterdam.

DeBach, P., ed. (1964). "Biological Control of Insect Pests and Weeds." Van Nostrand-Reinhold, Princeton, New Jersey.

Drea, J. J. (1980). The European Parasite Laboratory: Sixty years of foreign exploration. *In* "Biological Control in Crop Production," pp. 107–120. Allanheld, Osmum Publishers, Granada.

Harlan, J. R., and Starks, K. J. (1980). Germplasm resources and needs. *In* "Breeding Plants Resistant to Insects" (F. G. Maxwell and P. J. Jennings, eds.), pp. 253–276. Wiley, New York.

Harris, E. J. (1977). The threat of the Mediterranean fruit fly to American agriculture and efforts being made to counter this threat. *Proc. Hawaii. Entomol. Soc.* **22,** 475–480.

Hoffman, C. H., and Henderson, L. S. (1966). The fight against insects. *In* "Protecting Our Food: The Yearbook of Agriculture," pp. 26–38. U.S. Dep. Agric., Washington, D.C.

Howe, W. L., Kehr, W. R., and Calkins, C.O. (1965). Appraisal for combined pea aphid and spotted alfalfa aphid resistance in alfalfa. *Res. Bull.—Nebr., Agric. Exp. Stn.* **221,** 1–31.

Karlson, P., and Butenandt, A. (1959). Pheromones (ectohormones) in insects. *Ann . Rev. Entomol.* **4,** 39–58.

Karlson, P., and Lüscher, M. (1959). "Pheromones," a new term for a class of biologically active substances. *Nature (London)* **183,** 55–56.

Klein, M. G. (1981). Mass trapping for suppression of Japanese beetles. *In* "Management of Insect Pests with Semiochemicals" (E. R. Mitchell, ed.), pp. 183–190. Plenum, New York.

Klun, J. A., and Associates (1975). Insect sex pheromones; intraspecific pheromonal variability of *Ostrinia nubilalis* in North America and Europe. *Environ. Entomol.* **4,** 891–894.

Knipling, E. F. (1955). Possibilities of insect control or eradication through the use of sexually sterile males. *J. Econ. Entomol.* **48,** 459–462.

Knipling, E. F. (1979). The basic principles of insect population suppression and management. *U.S., Dep. Agric., Agric. Handb.* **512.**

Lester, M. L., Martin, D. F., and Parvin, D. W. (1976). Potential for suppressing tobacco budworm (Lepidoptera: Noctuidae) by genetic sterilization. *Tech. Bull.—Miss., Agric. For. Exp. Stn.* **82,** 1–9.

Lewontin, R. C., and Hubby, J. L. (1966). A molecular approach to the study of genetic heterozygosity in natural populations. II. Amount of variation and degree of heterozygosity in natural populations of *Drosophila pseudoobscura. Genetics* **54,** 559–604.

MacArthur, R. H., and Wilson, E. O. (1967). "The Theory of Island Biogeography." Princeton Univ. Press, Princeton, New Jersey.

McGregor, R. C., ed. (1973). "The Emigrant Pests: A Report to Dr. Francis J. Mulhern, Administrator." APHIS, USDA, Washington, D.C.

Makela, M. E., and Huettel, M. D. (1979). Model for genetic control of *Heliothis virescens. Theor. Appl. Genet.* **54,** 225–233.

Mangelsdorf, P. C. (1974). "Corn: Its Origin, Evolution and Improvement." Belknap Press, Cambridge, Massachusetts.

Maugh, T. H. (1982). Exploring plant resistance to insects. *Science* **216,** 722–723.

Mayer, M. S., and McLaughlin, J. R. (1975). An annotated compendium of insect sex pheromones. *Fla., Agric. Exp. Stn., Monogr. Ser.* **6,** 1–91.

Melhus, I. E., Painter, R. H., and Smith, F. O. (1954). A search for resistance to the injury caused by species of *Diabrotica* in the corns of Guatemala. *Iowa State Coll. J. Sci.* **29,** 75–94.

Mitchell, E. R., ed. (1981). "Management of Insect Pests with Semiochemicals." Plenum, New York.

Morton, T. C., and Bateman, M. A. (1981). Chemical studies on proteinaceous attractants for fruit flies, including the identification of volatile constituents. *Aust. J. Agric. Res.* **32,** 905–916.

Newman, L. H. (1966). "Man and Insects." Natural History Press, Garden City, New York.

Ortega, A., Vasal, S. K., Mihm, J., and Hershey, C. (1980). Breeding for insect resistance in maize. *In* "Breeding Plants Resistant to Insects" (F. G. Maxwell and P. R. Jennings, eds.), pp. 371–419. Wiley, New York.

Pashley, D. P., and Bush, G. L. (1979). The use of allozymes in studying insect movement with special reference to the codling moth, *Laspeyresia pomonella* (L.) (Olethreutidae). *In* "Movement of Highly Mobile Insects" (R. L. Rabb and G. G. Kennedy, eds.), pp. 333–341. North Carolina State University, Raleigh.

Patterson, R. S., LeBrecque, G. C., Williams, D. F., and Weidhaas, D. E. (1981). Control of the stable fly, *Stomoxys calcitrans* (Diptera: Muscidae), on St. Croix, U.S. Virgin Islands, using integrated pest management measures. *J. Med. Entomol.* **18,** 203–210.

Pringle, J. W. S. (1980). Science of a pest: Research on the African armyworm at the International Centre of Insect Physiology and Ecology, Nairobi. *In* "Insect Biology in the Future" (M. Locke and D. S. Smith, eds.), pp. 925–945. Academic Press, New York.

Rai, K. S., Lorimer, N., and Hallinan, E. (1974). The current status of genetic methods for controlling *Aedes aegypti. In* "The Use of Genetics in Insect Control" (R. Pal and M. J. Whitten, eds.), pp. 119–132. Elsevier/North-Holland, Amsterdam.

Smith, R. F. (1966). Distribution patterns of selected western North American insects; the distribution of diabroticites in western North America. *Bull. Entomol. Soc. Am.* **12,** 108–110.
Titus, E. G. (1910). The alfalfa leaf weevil. *Utah, Agric. Exp. Stn., Bull.* **110,** 1–72.
United States-Israel Binational Agricultural Research and Development Fund (1981). "Guidelines and Regulations for Grant Applicants and Recipients." U.S. Dep. Agric., OICD, Agric. Res. Cent., Beltsville, Maryland.
van den Bosch, R., and Messenger, P. S. (1973). "Biological Control." Intext Educational Publishers, New York.
Webster, F. M. (1895). On the probable origin, development, and diffusion of North American species of the genus *Diabrotica. J.N.Y. Entomol. Soc.* **3,** 158–166.
Whittaker, R. H. (1970). The biochemical ecology of higher plants. *In* "Chemical Ecology" (E. Sondheimer and J. B. Simeone, eds.), pp. 43–70. Academic Press, New York.
Whitten, M. J. (1971). Use of chromosome rearrangements for mosquito control. *In* "Sterility Principle for Insect Control or Eradication," pp. 339–411. IAEA, Vienna.
Whitten, M. J., and Pal, R., eds. (1974). "The Use of Genetics in Insect Control." Elsevier/North-Holland, Amsterdam.

CHAPTER 12

Research on Exotic Plant Pathogens

P. LAWRENCE PUSEY

Southeastern Fruit and Tree Nut Research Laboratory
Agricultural Research Service
United States Department of Agriculture
Byron, Georgia

and

CHARLES L. WILSON

Appalachian Fruit Research Station
Agricultural Research Service
United States Department of Agriculture
Kearneysville, West Virginia

I. INTRODUCTION

In defense of our nation militarily, it is important to know where the enemy is and what his strengths and weaknesses are. This same type of knowledge is needed to stop or arm ourselves against possible invasions by exotic pathogens.

EXOTIC PLANT PESTS AND
NORTH AMERICAN AGRICULTURE

ISBN 0-12-757880-3

Sound information on which to base judgments concerning the potential of exotic pathogens has been generally lacking (Thurston, 1973; McGregor, 1978). As a result, quarantine laws have been based largely on authority, without scientific support or verification. There is little question about the need for research in this area if we are to continue to have quarantine programs.

Can we really anticipate plant pathogen introductions? Is it possible to prevent the introduction of the next Dutch elm disease fungus or citrus canker bacterium? Perhaps plant pathologists have not fully addressed these questions. Since the onus to deal with this problem has been primarily on regulatory people, researchers have not been fully challenged to apply their skills. Because of poor communications between researchers and regulatory agencies, the full force of our present technology has not been realized.

Not only is research a prerequisite to adequate programs for detecting and stopping dangerous pathogens, but it may also be applied to the development of control strategies as insurance against possible introductions. In some cases, efforts should be directed predominantly toward control strategies if the risk of a pathogen's introduction is very high and the probability of preventing it appears to be low.

With increased odds of pathogen introductions and a more sophisticated technology, it would behoove us to continue the battle against these enemies of our agricultural plants. This chapter is an attempt to define the problem, discuss pest research, and develop a rational framework within which meaningful research questions can be asked in the future.

II. HOW WELL CAN WE PREDICT?

Schoulties *et al.* (Chapter 6, this volume) have addressed the question, where are the threats?, and Shrum and Schein (Chapter 15, this volume) discuss predictive models. Since these topics will ultimately require an input of data accumulated through research, it seems necessary that we consider them as well. We may reiterate what these other authors have expressed or differ with their views.

Entomologists appear to have a better understanding of the nature of their pest introductions at this point than plant pathologists. This is partially because of the greater genetic stability of their populations and the greater ease in identifying biotypes. It may be instructive to learn that "the behavior of two-thirds of the important pest immigrants came as a surprise to entomologists" (McGregor, 1973, p. 23). This statement applies equally to the minor pest species. Only 18% of the immigrant species that proved to be either important or minor pests in the United States would have been expected to behave as they did.

McGregor (1973) lists 551 foreign plant pathogens which pose a risk to our agriculture. These were selected primarily on the basis of the greater economic value of their hosts. Of the 49 which are included in the 100 most dangerous exotic pests and diseases, 7 are given a 25–99% probability of becoming established in the United States within 3 years, and 37 are given a 15–24% probability of becoming established in 4–6 years. Although it has been more than 6 years since this report, its accuracy is difficult to assess because percentages are used, and some pathogens may have become established but await discovery. It appears, however, from the available information that the above estimates for these pathogens are high. This is not to say that few dangerous pathogens have been introduced during the elapsed time, but only that few of the new introductions are among those predicted.

Although increased knowledge of exotic pathogens should improve predictability and lead to a better defense against possible threats, we are aware of limitations and realize that not all occurrences can be predicted. When unexpected diseases appear and the economic risk is high, our only recourse is to act quickly in assessing the problem and developing control measures.

Often, destructive pathogens do not "flag" themselves in their natural habitats, where they may exist in a balanced relationship with their host. They may go unnoticed but become destructive when introduced into another area of the world. Classic examples are the fungi causing Dutch elm disease and chestnut blight, both believed to have originated in East Asia. Chinese and Japanese trees have greater resistance to these diseases than do the American species.

Exotic pathogens may not only threaten species related to plants in their area of origin, but could unexpectedly affect plant species in the United States which are quite unrelated. This could easily happen with viruses, many of which are known to have wide host ranges. CMV is known to infect plant species in more than 67 families representing both dicotyledons and monocotyledons, and both herbaceous and woody plants (Waterworth and Kaper, 1980). Viruses can be latent in one host but very damaging to another.

Also posing a threat is the reintroduction of pathogens that are more aggressive or virulent than strains of the same species already established. A major new outbreak of Dutch elm disease in Western Europe was recognized in the late 1960s, when elms were dying in large numbers in parts of southern England (Gibbs, 1978). According to research by Brasier and Gibbs (1973), this epidemic was caused by a very aggressive strain of the Dutch elm fungus, which was probably introduced into Britain on diseased logs of rock elm imported from Canada.

Even less foreseeable is the possibility that newly introduced strains could sexually mate with established or endemic strains of a pathogen to form recombinants with greater virulence or aggressiveness. Other ways in which genes from newly introduced entities might combine with those of endogenous or established entities include parasexual mechanisms in fungi, conjugation or the transferral of genes in bacteria via bacteriophages or plasmids, and the combining of genes from different strains of viruses or viroids.

Viruses could also be introduced along with associated satellite viruses or nucleic acids capable of regulating disease severity. An interesting case is the cucumber mosaic virus (CMV) with its associated ribonucleic acid-5, designated CARNA 5 (Kaper and Tousingnant, 1977). It has been shown that CARNA 5 is a low-molecular-weight RNA, encapsidated with CMV protein, and dependent for its replication on, but not part of, the viral genome (Diaz-Ruiz and Kaper, 1977; Kaper *et al.*, 1976). Infection of tomato plants with CARNA 5 along with CMV results in a lethal necrotic disease (Kaper and Waterworth, 1977). However, in the absence of CARNA 5, CMV in tomato usually causes the familiar fernleaf syndrome, often accompanied by mosaic (Gibbs and Harrison, 1970). Although experiments with tomato demonstrated with CARNA 5 can dramatically increase disease severity caused by CMV, more recent studies have shown that CARNA 5 has an opposite effect on disease in certain other crops (Waterworth *et al.*, 1979). It has been suggested that satellite entities may have originated from host plants (Waterworth *et al.*, 1979). If this is the case, such an entity could originate in nonnative plants in the United States, such as exotic weeds, or it could originate in native plants and have an exotic virus for a helper.

In addition, introduced pathogens of minor importance could conceivably interact with established pathogens to form a synergistic combination that may threaten U.S. crops. A number of disease complexes involving the combination of nematodes with other pathogens are now known. A destructive disease complex of wheat in India is caused by the bacterium *Corynebacterium tritici* in association with the seedgall nematode *Anguina tritici* (Midha and Swarup, 1974). Neither organism has so far been reported in the United States.

The possibilities are, of course, unlimited, and many types of occurrences cannot be anticipated. However, there may be things we can do to determine where some of the major threats are. Obviously, we can detect and examine only a small fraction of the exotic pathogens threatening U.S. plants; therefore, criteria are needed to narrow the field of potential threats. Later, we will make suggestions on what these criteria might be.

III. POTENTIAL OF EXOTIC PATHOGENS

Research on exotic plant pathogens has been conducted, for the most part, after the fact, or following their discovery in the United States. Many classic examples could be given. A recent introduction that has been cause for concern is the corn cyst nematode, *Heterodera zeae,* found for the first time in the United

States (Sardanelli *et al.*, 1981). The parasite was previously known only from India, Egypt, and Pakistan. Research plans include (1) improving the description of *H. zeae* to make identification easier, (2) detection surveys in corn-growing areas, and (3) studies on host range and pathogenicity.

To the present, only a few studies have been conducted by American researchers on specific foreign pathogens not yet reported in the continental United States. Included are investigations on the causal agents of soybean rust and diseases of corn such as downy mildew, rust, and maize streak. Most of the work has been conducted in containment facilities at the Plant Disease Research Laboratory in Frederick, Maryland.

The fungus *Phakopsora pachyrhizi* is the causal organism of soybean rust, a serious disease in the Eastern Hemisphere. In eastern Australia, Thailand, and Taiwan, soybean rust is generally considered to be the disease of soybean having the greatest economic impact (Bromfield, 1976). Studies on colonization and uredospore production of *P. pachyrhizi* have led to a greater understanding of the epidemiology of this disease (Marchetti *et al.*, 1975; Melching *et al.*, 1979). The taxonomy of the pathogen has been reviewed (Bromfield, 1976) and the histology of the pathogen–suscept relationship investigated (Bonde *et al.*, 1976). The recent discovery of *P. pachyrhizi* in Puerto Rico was disturbing, and host range studies indicated that Puerto Rican cultures of the fungus have a wide range of pathogenicity among legume species (Vakili and Bromfield, 1976; Bromfield *et al.*, 1980). However, these cultures were found to be less virulent than cultures from the Orient.

The fungi that cause downy mildew of corn are among the most destructive pathogens in the tropics and subtropics. These fungi, which belong to the family Peronosporaceae, originated in the Eastern Hemisphere. None originated on corn; however, they possess the ability to attack corn or change rapidly to become virulent to corn (Frederiksen and Renfro, 1977). The most important corn downy mildew species are in the genus *Peronosclerospora.* Morphological studies have been made by U.S. researchers to clear up some of the confusion existing within this genus (Schmitt *et al.*, 1979). *Peronosclerospora sacchari* causes sugarcane downy mildew, a serious disease on both sugarcane and corn in the Orient. In Taiwan from 1960 to 1964, 70% of the popular corn hybrid Tainan No. 5, grown in the Chiayi-Tainan area, was affected (Sun *et al.*, 1976). *Peronosclerospora sorghi* is the causal fungus of sorghum downy mildew, a serious problem to corn and sorghums. Although this disease has been known in Asia and Africa for several decades, it was not reported in the United States until 1961, when it was observed at two locations in Texas (Frederiksen *et al.*, 1973). It has since spread as far as southern Indiana (Warren *et al.*, 1974) and Illinois (Frederiksen and Renfro, 1977). Most disease loss to date has occurred in the South on grain sorghum and sorghum–sudan grass hybrids (Bonde, 1980).

Temperature and moisture conditions that have been shown to favor infection by *P. sacchari* (Bonde and Melching, 1979) are common during the growing season in much of the U.S. corn belt. In a host range study, 20 new suscepts of *P. sacchari* were identified within the genera *Andropogon, Bothriochloa, Eulalia, Schizachyrium,* and *Sorghum.* Since several of these newly identified hosts are common perennial weeds and/or grasses in the United States, they might serve to allow *P. sacchari* to survive the winter and act as an inoculum source the following spring to infect corn.

In studies with *P. sorghi,* optimum temperature for germination and germ-tube growth was lower for an isolate from Texas than one from Thailand (Bonde *et al.*, 1978; Bonde, 1980). The lower optimum for the Texas isolate may represent an adaptation to the more temperate environment of the continental United States. In a host study with the Texas and Thai strains (Bonde and Freytag, 1979), several new hosts for the Texas strain within the genus *Sorghum* were determined. However, since none of these hosts are common in the United States, they are not epidemiologically important here. No additional hosts for the Thai strain were found. According to Bonde (1980), it appears doubtful that perennial hosts of this strain exist in the United States which would allow the pathogen to survive the absence of corn. This, in addition to the lack of formation of oospores by the Thai strain (Pupipat, 1976; Bonde, 1980), suggests that the Thai strain of *P. sorghi* is unlikely to be a threat to the United States.

Isolates of *Peronosclerospora* from various parts of the world continue to be studied at the Plant Disease Laboratory. This is warranted since some corn downy mildew species and biotypes exotic to the United States are considerably more virulent to corn than the American strain of *P. sorghi* (Bonde, 1980).

Among the pathogens responsible for destructive epidemics in the past are the rust-causing fungi. Except for a few instances, corn has not yet suffered extensive and serious attacks of rust. Three important species of fungi which cause rusts of corn have been studied at the Plant Research Laboratory (Melching, 1975). These are *Puccinia sorghi, P. polysora,* and *Physopella zeae. Puccinia sorghi* causes common rust and occurs essentially everywhere corn is grown. *Puccinia polysora,* the causal agent of southern corn rust (or American rust), occurred only sporadically in the United States until 1972 and was restricted to Central America and the Caribbean area until 1949. *Physopella zeae,* which causes tropical corn rust, has never been observed in the United States and has remained in Mexico and Central America, the Caribbean area, and the northern portions of South America. Thus far, the two species occurring in the United States have caused only minor damage to corn. Melching (1975) showed that moisture and temperature requirements for infection with *P. polysora* and *P. zeae* were more narrow than those for *P. sorghi.* This explains in part why *P. sorghi* is the most widespread. It appeared from experiments, however, that *P. polysora* was the most potentially damaging of the three fungi on the commercial hybrids tested. Most commercial hybrids grown in the United States express nonspecific resistance to these pathogens which appears adequate at the present time.

Maize streak virus (MSV) is a pathogen of corn present in Africa, Mauritius, Madagascar, India, and probably elsewhere in Southeast Asia. In recent years, MSV has become a serious threat to corn production in West Africa (Fajemisin, 1977). The disease is characterized by chlorotic streaking on newly emerging leaves, stunting, and eventually poor yield. The most important vector of this virus is *Cicadulina mblia,* a leafhopper found in the tropics and subtropics, but not in the United States. If *C. mbila* and MSV become established in North America, it probably will be restricted to the warmer regions of the southern United States and Mexico. Most of our corn belt lies north of this area, so that the *C. mbila*–MSV system is unlikely to pose a serious threat. One danger is the possibility that migratory domestic vectors could pick up the virus in the southern regions and move it across our corn belt. Graham (1979) tested seven species of leafhoppers, two species of planthoppers, and three species of aphids for their ability to transmit MSV, but none were shown to be vectors.

It is apparent that most of the nonintroduced pathogens being studied by U.S. researchers cause diseases in tropical and subtropical areas of the world. Since the major portion of the United States lies in the temperate zone, the authors question whether these organisms are actually among those exotics posing the greatest danger.

IV. STOPPING THE WOULD-BE INVADERS

A. Pathways

In numerous instances, knowledge is lacking on the survival pathway of pathogens. With the increased volume and speed of transportation in world trade and international travel, the odds are increased that propagules of plant pathogens will survive transit and escape into new geographical areas. Air travel is especially important to this problem. The fast rate at which aircraft move from one country to another increases the likelihood that short-lived spores of pathogens or insects-carrying plant viruses will make the trip.

Another important factor which is becoming more obvious (McGregor, 1973, 1978) is the scientist himself. Although having good intentions, he may still be an important avenue of entry for exotic pathogens. This is somewhat like the situation in science fiction in which the aim of the scientist is to benefit man, but instead he creates a monster or new bacterium that plagues the world. Often, biologists wish to use exotic plants or pathogens as objects of experimentation. Frequently, information involving the particular species has been accumulated by foreign scientists and can serve as a starting point for the American scientists' research. At other times, the traveling scientist is tempted to collect plants which are in some way peculiar or interesting or which may have use in a breeding program back home. The importance of utilizing resistance genes from foreign germplasm collections or plants native to other world areas in order to combat exotic pathogens cannot be minimized. On the other hand, it is essential that researchers take precautions and proceed through proper quarantine channels in acquiring plant material. Otherwise, they may introduce far greater problems than they can ever solve.

B. Ports of Entry

Pathogens reaching U.S. ports via travelers are likely to be associated with plants or plant commodities. Researchers are now working on various devices that can detect contraband fruits, vegetables, and other plant matter (U.S. Department of Agriculture, 1981). Possible developments, which are discussed more fully in Chapter 10, include odor-detecting instruments known as "sniffers," electromagnetic devices, and imaging by x-ray.

An alternative to the detection and confiscation of plant material may be the control of pathogens by irradiation. In 1970, Agricultural Quarantine Inspection considered the possibility of utilizing irradiation to eliminate the threat of contraband in passenger baggage at Honolulu Airport (McGregor, 1973). A procedure using radiation was found to be effective, but was too costly and time-consuming. Nevertheless, this or other similar methods could have potential in the future.

The import of agricultural cargoes must be closely regulated. Because there is neither sufficient funding nor trained manpower to inspect all pest hosts or even a large proportion of those allowed to enter, representative sampling must be performed. In addition, methods of treating commodities to eliminate pathogens may be developed.

C. Quarantine Stations

Plant genera that are normally prohibited from entering the United States from specified countries are sometimes accepted for breeding or other scientific purposes. On arrival, such material is subject to inspection and possible chemical treatment. For example, seeds of soybean and garden bean from Africa and Australia are chemically treated because of soybean rust. Inspection and treatment are not enough, however, since the plants may contain latent viruses or incubating rust fungi (Kahn *et al.*, 1979). The plants must be indexed or tested at approved quarantine facilities. The principal location for testing is the U.S. Plant Quarantine Facility at the U.S. Plant Introduction Station, Glenn Dale, Maryland. Cooperative facilities have been established with the Science and Education Administration at Beltsville, Maryland, and Davis, California, and with the University of California at Davis and Riverside.

At the quarantine facilities, nonviral pests such as insects, fungi, or bacteria on introductions are found infrequently because (1) the foreign collector usually selects healthy-looking plant parts; (2) the shipment is small; and (3) items are often dipped or fumigated to kill surface-infesting organisms (Kahn, 1977; Waterworth and White, 1982). Detection and identification of internal latent pathogens, such as viruses, viroids, mycoplasmas, and spiroplasmas, require much greater effort. Procedures used include mechanical transmission to herbaceous indicators, graft transmission to varieties that are sensitive to particular viruses, graft transmission to plants of the same species to determine whether a viruslike symptom is transmissible, serological methods, and electron microscopy. One or more of these methods may be used depending on the plant genus. The percentage of introductions infected with viruses or mycoplasmas has varied according to the crop, ranging from 2% for cacao to as high as 68% for apples (Waterworth and White, 1982). Certain genera, such as apples, potatoes, and grapes, are often infected with two or three pathogens (Kahn *et al.*, 1979; Waterworth, 1980).

Plants are still the only "eyes" we have for the detection of many viruses or viruslike agents. Bioassay procedures, involving either mechanical or graft transmission, are the methods most frequently used at the quarantine locations. These procedures may be slow and require considerable space, thus imposing limitations on the importation of germplasm.

Serological detection methods, which have the advantage of speed and specificity, may be used to a greater extent in the future for detection of exotic plant pathogens. The enzyme-linked immunosorbent assay (ELISA) and radioimmunosorbent assay (RISA) techniques have proved valuable in plant virus detection (Clark and Adams, 1977; Ghabrial and Shepherd, 1980). Recently the use of immunosorbent assay has been expanded to detect spiroplasmas (Clark *et al.*, 1978), bacteria (Kishinevsky and Bar-Joseph, 1978; Vruggink, 1978; Stevens and Tsiantos, 1979; Nome *et al.*, 1980), and fungi (Casper and Mendgen, 1979; Savage and Sall, 1981). Immunofluorescence, another method, has been shown useful for detecting fastidious bacteria (French, 1974; French *et al.*, 1978; Lee *et al.*, 1978; Brlansky *et al.*, 1982).

Electron microscopy is commonly used to identify plant viruses. Rod-shaped virions can often be observed in crude sap preparations by negative staining. The immune electron microscopy method of virus detection and identification is receiving increased attention (Milne and Luisoni, 1977). The technique, now applied in a number of ways, involves observation of specific binding of antigen and antibody in the electron microscope. Immune electron microscopy was nearly twice as sensitive as ELISA in detecting arabis mosaic, prunus necrotic ringspot, and strawberry latent ringspot viruses in extracts from rose (Thomas, 1980).

Recently, a new technique was developed for the detection of the potato spindle tuber viroid (PSTV) (Owens and Diener, 1981). Recombinant DNA technology made it possible to produce radioactively labeled strands of deoxyribonucleic acid (DNA) that are complementary to the ribonucleic acid (RNA) of the viroid. In the test procedure, PSTV, if present in samples from potato, is bound to a nitrocellulose membrane. When complementary DNA is added later, it will hybridize with viroid RNA present on the membrane. The viroid RNA, hybridized with the labeled DNA, can then be detected indirectly using autoradiography. Possibly the technique can be applied to the detection of other disease agents.

If all available materials of certain plant cultivars or clones contain disease agents, methods may be employed to exclude the pathogens. A variety of procedures has been used, including thermotherapy, meristem tip culture, nucellus or ovule culture, and shoot-tip grafting *in vitro*. The various methods of thermotherapy that have proven successful include preconditioning to a warm temperature prior to heat treatment, use of hot and moist air, use of hot water, specialized heat-treatment chambers, and small heat-treating plant enclosures. Thermotherapy has been shown to eliminate many pathogens from citrus budwood (Roistacher, 1977). It has also been used alone or in combination with meristem-tip culture to eliminate viruses in ornamentals (Lawson, 1981). Pathogen-free citrus selections have been obtained by nucellar embryony or nucellar tissue culture; however, these plants tend to be excessively vigorous, thorny, and late bearing, and have fruit which may vary in quality and likeness from the original (Weathers and Calavan, 1959; Bitters *et al.*, 1972). Shoot-tip culture *in vitro* has not been proven to be applicable to trees, although it could be in the future. Meanwhile, a technique has been developed for grafting the small apical shoot of citrus to the top of a decapitated seedling grown *in vitro* (Murashige *et al.*, 1972). An improved version of the technique (Navarro *et al.*, 1975) was shown to be effective in eliminating a number of citrus viruses (Roistacher *et al.*, 1976).

V. PREPARING FOR INVASIONS

Through research it may be possible to develop strategies of combating exotic pathogens that are anticipated or are likely to be introduced. Simons and Browning, who discuss this subject in Chapter 16 of this volume, stress diversity as insurance against loss.

We agree with Simons and Browning that a diversity of resistant genotypes in agriculture can be an effective safeguard against exotics. Primary questions might be: Does resistance exist? If so, where? This questioning relates to Schmidt's (1978) comment that diversity in itself is no safeguard, but must be of a kind that is functional against pathogens. If resistance is not present in existing cultivars, possibly it can be found in wild relatives of the cultivated species and perhaps where the host or pathogen originated.

Little work has been done by U.S. scientists to screen or breed for resistance to diseases not yet found in the United States. Diseases which have been given some attention include soybean rust (Bromfield, 1976; Bromfield and Hartwig, 1980), sugarcane downy mildew of corn (Schmitt *et al.*, 1977), maize streak virus (Damsteegt, 1980), and yellow wilt of sugarbeets (Gaskill and Ehrenfeld, 1976). In some cases, strains or races within a pathogen species have complicated breeding of resistant genotypes (Bromfield, 1976; Schmitt *et al.*, 1977).

Great gains may be made through cooperative arrangements with other countries. In 1961, through cooperation between the U.S. Department of Agriculture (USDA) and the Taiwan Agricultural Research Institute, the entire USDA soybean germplasm collection was planted in Taiwan, where it was subjected to rust from natural infection. Two U.S. accessions from Japan were reported to have some resistance to rust, and these were used in Taiwan in subsequent crosses. The Beet Sugar Development Foundation, an association of U.S. and Canadian beet sugar companies and other organizations, cooperates with the USDA, Dutch and Chilean seed companies, and an Argentine experiment station in conducting breeding research to develop sugarbeets resistant to a disease called ''yellow wilt'' (Gaskill and Ehrenfeld, 1976).

When sources of resistance have been identified, it is important that the material not be lost, but maintained in germplasm collections. It will then be available if introductions occur.

To prepare for possible invasions, it would be well for us to examine present control schemes in the United States for their effect on dangerous exotic pathogens. Would fungicides now in use, for instance, reduce diseases caused by certain exotic fungi? We may also learn from research conducted by foreign scientists in areas where the pathogens are currently pests. Bear in mind, however, that studies in another part of the world may not necessarily be applicable to conditions in the United States.

Biological control has received increased attention. Hyperparasites or microbial antagonists may keep a pathogen in check on the continent where it is endogenous or has existed long enough to become part of a balanced system. The observant scientist may discover phenomena or interrelationships that could have practical application to the control of exotic diseases. An interesting case which illustrates this point concerns chestnut blight. Europeans were alarmed when blight was reported in northern Italy in 1938. Later, Biraghi, an Italian plant pathologist, found trees that seemed unusually healthy after repeated attacks by the blight fungus (Biraghi, 1953). Cankers on these trees had apparently healed, and the fungus was restricted to the outer layer of bark.

Biraghi's work attracted the attention and imagination of a French mycologist,

Grente, who visited Italy in the late 1950s and collected bark samples from healing trees. From these he isolated forms of the blight fungus with reduced virulence, which he referred to as "hypovirulent." These hypovirulent forms cured existing blight when they were inoculated into cankers (Grente, 1965). Grente and Sauret (1969) speculated that cytoplasmic determinants are transferred through hyphal anastomosis, and convert virulent strains to hypovirulent ones. This hypothesis was supported by Puhalla and Anagnostakis (1971) in tests with strains identifiable by nuclear genes. Viruses have been suggested as the determinants of hypovirulence since double-stranded ribonucleic acid (dsRNA), the genetic material of most fungal viruses, is present in hypovirulent strains (Day *et al.*, 1977). Also, the dsRNA has been associated with club-shaped viruslike particles in at least one strain (Dodds, 1980).

In recent years, hypovirulent strains of the chestnut blight fungus have been found in the United States. The effectiveness of these strains in controlling blight of individual trees depends on their vegetative compatibility with the disease-causing virulent strains (Anagnostakis, 1977). Natural spread of hypovirulence, which occurs in Italy, has not been shown to occur in the United States. Despite the difficulties so far encountered with this new approach in disease control, there is now new hope for the American chestnut.

VI. WHAT SHOULD OUR FOCUS BE?

It would be impossible to assess adequately all exotic pathogens that could potentially threaten the United States. The idealistic approach would be to direct our attention to the most dangerous exotics. However, this approach brings us back to the question addressed at the beginning of this chapter and in Chapter 6 of this volume: Where are the exotic disease threats? Obviously, we will never be able to respond with great accuracy. Nevertheless, our best efforts to identify dangerous pathogens should help us to minimize economic losses in the United States. In many cases, we may be buying time.

Criteria need to be established which will aid in focusing our energies on the greatest potential problems. Wilson (1982) has proposed that certain ecological principles be used in setting priorities. These are as follows:

1. Continuous and perennial ecosystems are more threatened by exotics than are discontinuous, annual agroecosystems (annual crops).
2. Organisms from large land mass areas (e.g., Eurasia) are more apt to replace native ones in smaller land mass areas if introduced.
3. Organisms are limited in their distribution and spread by the climatic conditions in regions where they evolved.

If these are valid statements, in general, then we would expect the greatest threats to come from exotic pathogens of forests, rangelands, and orchards in the Eurasia land mass area with climates comparable to ours.

The risk to perennial plants is high primarily because of their longer life span and their slower replacement rate. The time required for trees to reach maturity or an age at which they are of value to man, whether it is for shade, timber, food, or recreation, is several to many years. In the event of an epidemic, even if control measures are developed and total recovery is achievable, this could take years, decades, or even centuries in the case of forest trees. The losses incurred might be incalculable. Certainly, our most destructive and costly epidemics have been caused by exotic pathogens affecting forest, shade, and orchard trees. Included are Dutch elm disease, chestnut blight, white pine blister rust, and citrus canker. These diseases are often given as classic examples when pointing out the potential destructiveness of exotic pathogens. It is not surprising then that the first nine exotic plant pathogens listed by McGregor (1973) as being the most dangerous to the United States are pathogens of trees.

A severe epidemic of an annual food crop, on the other hand, may be limited to a single season if research scientists respond quickly and work together in finding a solution. It may be that resistant germplasm is available, and this could be used in the ensuing years. The southern corn leaf blight epidemic of 1970, which was responsible for a 15% loss of the corn crop and a monetary loss of $1 billion, essentially ran its course in 1 year and ended. All corn lines carrying Texas cytoplasm male sterility (Tcms) were shown to be susceptible to a new strain of the fungus *Helminthosporium maydis*. When this was understood, preventive measures were taken, including the planting of corn not possessing Tcms.

Perennials are also vulnerable since pathogens can survive in their tissues over winter and persist indefinitely. In our vast forests, exotic pathogens might survive and spread for years before being detected. By this time, they may be widely distributed.

Research directed toward disease control generally progresses more slowly for trees than for agronomic crops. This is another reason why a higher priority should be given to the quarantine of exotic tree pathogens. Many diseases of perennial tissues advance slowly, and host susceptibility is frequently dependent on seasonal cycles. To investigate canker or vascular diseases of trees, it may be necessary to make observations and record data over periods of 1 year or longer. The age of trees may also be a factor. Studies on Dutch elm disease have been impeded by resistance in juvenile trees (Nestor and Feldman, 1951; Arisumi and Higgins, 1961), which has minimized their value in greenhouse and growth chamber experiments and delayed the time for determining the results of breeding and selection programs.

Another hurdle in tree research is the frequent difficulty in obtaining genetic uniformity of test plants. Systems for propagating plant clones are also important for maintenance and distribution of resistant germplasms. A considerable amount of effort has been made to develop vegetative propagation methods for trees, including grafting, rooting cuttings, layering, and tissue and organ culture. What works for one species or variety may not work for another.

Furthermore, breeding for resistance in trees is a much slower process than for other plants since the time required for trees to begin flowering is longer. In breeding programs, the rotation age for the American chestnut is 5–6 years, and for apples and pears it is 4–8 years. This is drastically different from grain and vegetable crops, for which two or three generations per year is not uncommon. Tree-breeding programs generally require larger commitments of personnel and funds for a long period of time. The investment of time and money in such programs should be a consideration in evaluating the potential economic impact of exotic pathogens.

Second, it is a well-established principle of zoogeography that an organism from a larger land mass area will tend to replace an organism in a smaller land mass area when the two species are competing for the same ecological niche. A case in point is the high rate of colonization of new species of organisms in Hawaii. In the period 1942–1972, the rate of colonization of insects and mites in Hawaii per 1000 square miles was 40 species, 500 times the rate of colonization for the continental United States (McGregor, 1973).

It may be argued whether the invading pathogen is replacing another organism occupying a particular niche or creating a new niche. At any rate, the introduced pathogen intercepts the energy flow to other organisms, whether they are parasites, saprophytes, or resident organisms which depend on the tree for their food source.

The most devastating epidemics of trees in the United States have been caused by pathogens believed to have originated in the East Asia land mass area. A number of these pathogens, including the causal agents of Dutch elm disease and white pine blister rust, caused epidemics in Europe before they reached North America. In view of past experience, disease outbreaks in Europe involving pathogens unknown to North America should be a warning to us, and should prompt immediate investigatory and quarantine action. Unfortunately, though, the recent gain in commerce and relations between the United States and the People's Republic of China has increased the probability of pathogens being introduced directly into the United States from East Asia. Since Chinese officials are also concerned about the potential danger if disease agents are introduced into their country from North America, perhaps scientists and regulatory officials of both countries can cooperate in efforts to reduce the threat.

Third, organisms generally have a limited distribution that is determined by climatic conditions. With exceptions, pathogens endemic to tropical regions are not likely to threaten plants in temperate regions, and pathogens of temperate regions are not likely to threaten plants in the tropics. *Peronosclerospora sorghi,* one of the downy mildew fungi of corn originating in tropical South Asia, has been responsible for some disease loss in the United States, but this has been restricted mainly to the south on grain sorghum and sorghum-sudangrass (Bonde, 1980). Perennial hosts, which would allow strains of *P. sorghi* to survive over winter, are either uncommon or have not been found (Bonde and Freytag, 1979). The causal agent of southern corn rust, *P. polysora,* was introduced into the United States from Central America and the Caribbean area decades ago, but has so far caused only minor damage to corn. Even viruses tend to be limited by climatic conditions, although in most cases this appears to be related to the distribution of their vectors and not to their own temperature requirements. The primary vector of MSV, a pathogen threatening corn production in parts of Africa and Southeast Asia, is not likely to become established in temperate regions of North America. Also, none of our domestic insects have so far been shown to transmit this virus (Graham, 1979). Maize rayado fino virus, a corn virus formerly known to occur only in Central and South America, was recently reported in Texas and Florida (Bradfute *et al.,* 1980). As long as the virus maintains its limited host range and is efficiently transmitted only by *Dalbulus* species of leafhoppers, its distribution in the United States is not likely to expand (Nault *et al.,* 1980).

It is not suggested that the above principles be rigidly adhered to, but that they be used as guidelines in establishing priorities and in making the best investment of our time and resources. There may be some serious threats that do not come under the general principles outlined here.

REFERENCES

Anagnostakis, S. L. (1977). Vegetative incompatability in *Endothia parasitica. Exp. Mycol.* **1,** 306–316.

Arisumi, T., and Higgins, D. J. (1961). Effect of Dutch elm disease on seedling elms. *Phytopathology* **51,** 847–850.

Biraghi, A. (1953). Possible active resistance to *Endothia parasitica* in *Castanea sativa. Proc. Congr. Int. Union For. Res. Organ., 11th, 1953* pp. 643–645.

Bitters, W. P., Murashige, T., Rangan, T. S., and Nauer, E. (1972). Investigations on establishing virus-free citrus plants through tissue culture. *Proc. Conf. Int. Organ. Citrus Virol., 5th, 1969* pp. 267–271.

Bonde, M. R. (1980). Research program to determine threat of maize downy mildews to American agriculture. *Prot. Ecol.* **2,** 223–230.

Bonde, M. R., and Freytag, R. E. (1979). Host range for an American isolate of *Peronosclerospora sorghi. Plant Dis. Rep.* **63,** 650–654.

Bonde, M. R., and Melching, J. S. (1979). Effects of dew-period temperature on sporulation, germination of conidia, and systemic infection of maize by *Peronosclerospora sacchari. Phytopathology* **69,** 1084–1086.

Bonde, M. R., Melching, J. S., and Bromfield, K. R. (1976). Histology of the suscept-pathogen relationship between *Glycine max* and *Phakopsora pachyrhizi*. *Phytopathology* **66,** 1290–1294.

Bonde, M. R., Schmitt, C. G., and Dapper, R. W. (1978). Effects of dew-period temperature on germination of conidia and systemic infection of maize by *Sclerospora sorghi*. *Phytopathology* **68,** 219–222.

Bradfute, O. E., Nault, L. R., Gordon, D. T., Robertson, D. C., Toler, R. W. (1980). Identification of maize rayado fino virus in the United States. *Plant Dis.* **64,** 50–53.

Braiser, C. M., and Gibbs, J. N. (1973). Origin of the Dutch elm disease epidemic in Britain. *Nature (London)* **252,** 607–609.

Brlansky, R. H., Lee, R. F., Timmer, L. W., Purcifull, D. E., and Raju, B. C. (1982). Immunofluorescent detection of xylem-limited bacteria *in situ*. *Phytopathology* **72,** 1444–1448.

Bromfield, K. R. (1976). World soybean rust situation. *World Soybean Res., Proc. World Soybean Res. 1975* pp. 491–500.

Bromfield, K. R., and Hartwig, E. E. (1980). Resistance to soybean rust and mode of inheritance. *Crop Sci.* **20,** 254–244.

Bromfield, K. R., Melching, J. S., and Kingsolver, C. H. (1980). Virulence and aggressiveness of *Phakopsora pachyrhizi* isolates causing soybean rust. *Phytopathology* **70,** 17–21.

Casper, R., and Mendgen, K. (1979). Quantitative serological estimation of a hyperparasite: Detection of *Verticillium lecanii* in yellow rust infected wheat leaves by ELISA. *Phytopathol. Z.* **94,** 89–91.

Clark, M. F., and Adams, A. N. (1977). Characteristics of the microplate method enzyme-linked immunosorbent assay for the detection of plant viruses. *J. Gen. Virol.* **34,** 475–483.

Clark, M. F., Flegg, C. L., Bar-Josef, M., and Rottem, S. (1978). The detection of *Spiroplasma citri* by enzyme-linked immunosorbent assay (ELISA). *Phytopathol. Z.* **92,** 322–337.

Damsteegt, V. D. (1980). Investigations of the vulnerability of U.S. maize streak virus. *Prot. Ecol.* **2,** 231–238.

Day, P, R., Dodds, J. A., Elliston, J. E., Jaynes, R. A., and Anagnostakis, S. L. (1977). Double-stranded RNA in *Endothia parasitica*. *Phytopathology* **67,** 1393–1396.

Diaz-Ruiz, J. R., and Kaper, J. M. (1977). Cucumber mosaic virus associated RNA 5. III. Little or no sequence homology between CARNA 5 and helper RNA. *Virology* **80,** 204–213.

Dodds, J. A. (1980). Association of type 1 viral-like dsRNA with club-shaped particles in hypovirulent strains of *Endothia parasitica*. *Phytopathology* **70,** 1217–1220.

Fajemisin, J. M. (1977). Maize streak and other maize virus and virus-like diseases in West Africa. *Proc. Maize Virus Dis. Colloq. Workshop, 1976* pp. 16–19.

Frederiksen, R. A., and Renfro, B. L. (1977). Global status of maize downy mildew. *Annu. Rev. Phytopathol.* **15,** 249–275.

Frederiksen, R. A., Bockholt, A. J., Clark, L. E., Casper, J. W., Craig, J., Johnson, J. W., Jones, B. L., Matocha, P., Miller, F. R., Reyes, L., Resenow, D. T., Tuleen, D., and Walker, H. J. (1973). Sorghum downy mildew: A disease of maize and sorghum. *Tex., Agric. Exp. Stn.,* [*Monogr.*] **2,** 1–32.

French, W. J. (1974). A method for observing rickettsialike bacteria associated with phony peach disease. *Phytopathology* **64,** 260–261.

French, W. J., Stassi, D. L., and Schaad, N. W. (1978). The use of immunofluorescence for the identification of phony peach bacterium. *Phytopathology* **68,** 1106–1108.

Gaskill, J. O., and Ehrenfeld, K. R. (1976). Breeding sugarbeet for resistance to yellow wilt. *J. Am. Soc. Sugar Beet Technol.* **19,** 25–44.

Ghabrial, S. A., and Shepherd, R. J. (1980). A sensitive radioimmunosorbent assay for the detection of plant viruses. *J. Gen. Virol.* **48,** 311–317.

Gibbs, A. J., and Harrison, B. D. (1970). Cucumber mosaic virus. No. 1. *In* ''Descriptions of Plant Viruses,'' pp. 1–4. Commonw. Mycol. Inst., *Assoc. Appl. Biol. Kew, Surrey, England.*

Gibbs, J. N. (1978). Intercontinental epidemiology of Dutch elm disease. *Annu. Rev. Phytopathol.* **16,** 287–307.

Graham, C. L. (1979). Inability of certain vectors in North America to transmit maize streak virus. *Environ. Entomol.* **8,** 228–230.

Grente, J. (1965). Les formes hypovirulentes d'*Endothia parasitica* et les espoirs de lutte contre le chancre du chataignier. *C. R. Hebd. Seances Acad. Agr. France* **51,** 1033–1037.

Grente, J., and Sauret, S. (1969). L' ''hypovirulence exclusive'' est-elle controlee par des determinants cytoplasmiques? *C. R. Hebd. Seances Acad. Sci., Ser. D* **268,** 3173–3176.

Kahn, R. P. (1977). Plant quarantine: Principles, methodology, and suggested approaches. *In* ''Plant Health and Quarantine in International Transfer of Genetic Resources'' (W. B. Hewitt and L. Chiarappa, eds.), pp. 289–307. CRC Press, Boca Raton, Florida.

Kahn, R. P., Waterworth, H. E., Gillespie, A. G., Jr., Foster, J. A., Goheen, A. C., Monroe, R. L., Povich, W. L., Mock, R. G., Luhn, C. F., Calavan, E. G., and Roistacher, C. N. (1979). Detection of viruses or virus-like agents in vegetatively propagated plant importations under quarantine in the United States, 1968–1978. *Plant Dis. Rep.* **63,** 775–779.

Kaper, J. M., and Tousignant, M. E. (1977). Cucumber mosaic virus associated RNA 5. I. Role of host plant and helper strain in determining amount of associated RNA 5 with virions. *Virology* **80,** 186–195.

Kaper, J. M., and Waterworth, H. E. (1977). Cucumber mosaic virus associated RNA 5: Causal agent for tomato necrosis. *Science* **196,** 429–431.

Kaper, J. M., Tousignant, M. E., and Lot, H. (1976). A low molecular weight replicating RNA associated with a divided genone plant virus: Defective or satellite RNA? *Biochem. Biophys. Res. Commun.* **72,** 1237–1243.

Kishinevsky, B., and Bar-Joseph, M. (1978). *Rhizobium* strain identification in *Arachis hypogaea* nodules by enzyme-linked immunosorbent assay (ELISA). *Can. J. Microbiol.* **24,** 1537–1543.

Lawson, R. H. (1981). Controlling virus diseases in major international flower and bulb crops. *Plant. Dis. Rep.* **65,** 780–786.

Lee, R. F., Feldman, A. W., Raju, B. C., Nyland, G., and Goheen, A. C. (1978). Immunofluorescence for detection of rickettsialike bacteria in grape affected by Pierce's disease, almonds affected by almond leaf scorch, and citrus affected by young tree decline. *Phytopathol. News* **12,** 265 (Abstr.).

McGregor, R. C. (1978). People-placed pathogens: The emigrant pests. *In* ''Plant Disease: An Advanced Treatise'' (J. G. Horsfall and E. B. Cowling, eds.), Vol. 2, Chapter 18, pp. 383–397. Academic Press, New York.

McGregor, R. C. (1973). ''The Emigrant Pests,'' A report to F. J. Mulhern, Administrator of the Animal and Plant Health Inspection Service (unpublished).

Marchetti, M. A., Vecker, F. A., and Bromfield, K. R. (1975). Uredial development of *Phakopsora pachyrizi* in soybeans. *Phytopathology* **65,** 822–823.

Melching, J. S. (1975). Corn rusts: Types, races, and destructive potential. *Proc. 30th Annu. Corn Sorghum Res. Conf.* pp. 90–115.

Melching, J. S., Bromfield, K. R., and Kingsolver, C. H. (1979). Infection, colonization, and uredospore production on Wayne soybean by four cultures of *Phakopsora pachyrizi,* the cause of soybean rust. *Phytopathology* **69,** 1262–1265.

Midha, S. K., and Swarup, G. (1974). Factors affecting development of earcockle and tunda diseases of wheat. *Indian J. Nematol.* **2,** 97–104.

Milne, R. G., and Luisoni, E. (1977). Rapid immune electron microscopy. *Methods Virol.* **6,** 265–281.

Murashige, T., Bitters, W. P., Rangan, T. S., Nauer, E. M., Roistacher, C. N., and Holliday, P. B. (1972). A technique of shoot apex grafting and its utilization towards recovering virus-free citrus clones. *HortScience* **7,** 118–119.

Nault, L. R., Gingery, R. E., and Gordon, D. T. (1980). Leafhopper transmission and host range of maize rayado fino virus. *Phytopathology* **70,** 709–712.

Navarro, L., Rostacher, C. N., and Murashige, T. (1975). Improvement of shoot-tip grafting *in vitro* for virus-free citrus. *J. Am. Soc. Hortic. Sci.* **100,** 471–479.

Nestor, E. C., and Feldman, A. W. (1951). Dutch elm disease in young elm seedlings. *Phytopathology* **41,** 46–51.

Nome, S. F., Ratur, B. C., Goheen, A. C., Nyland, G., and Docampo, D. (1980). Enzyme-linked immunosorbent assay for Pierce's disease bacteria in plant tissues. *Phytopathology* **70,** 746–749.

Owens, R., and Diener, T. O. (1981). Sensitive and rapid diagnosis of potato spindle tuber viroid disease by nucleic acid hybridization. *Science* **213,** 670–672.

Puhalla, J. E., and Anagnostakis, S. L. (1971). Genetics and nutritional requirements of *Endothia parasitica. Phytopathology* **61,** 169–173.

Pupipat, V. (1976). Corn downy mildew research at Kasetsart University. *Kasetsart J.* **10,** 106–110.

Roistacher, C. N. (1977). Elimination of citrus pathogens in propagative budwood. I. Budwood selection, indexing, and thermotherapy. *Proc. Int. Soc. Citric., 1977* Vol. 3, pp. 965–972.

Roistacher, C. N., Navarro, L., and Murashige, T. (1976). Recovery of citrus selections free of several viruses, exocortis viroid, and *Spiroplasma citri* by shoot-tip grafting *in vitro. Proc. Conf. Int. Organ. Citrus Virol., 7th, 1976* pp. 186–192.

Sardanelli, S., Krusberg, L. R., and Golden, A. M. (1981). Corn cyst nematode, *Heterodera zeae,* in the United States. *Plant Dis.* **65,** 622.

Savage, S. D., and Sall, M. A. (1981). Radioimmunosorbent assay for *Botrytis cinera. Phytopathology* **71,** 411–415.

Schmidt, R. A. (1978). Diseases in forest ecosystems: The importance of functional diversity. *In* "Plant Disease: An Advanced Treatise" (J. G. Horsfall and E. B. Cowling, eds.), Vol. 2, Chapter 14, pp. 287–315. Academic Press, New York.

Schmitt, C. G., Scott, G. E., and Freytag, R. E. (1977). Response of maize diallel cross to *Sclerospora sorghi,* cause of sorghum downy mildew. *Plant Dis. Rep.* **61,** 607–608.

Schmitt, C. G., Woods, J. M., Shaw, C. G., and Stansbury, E. (1979). Comparison of some morphological characters of several corn downy mildew incitants. *Plant Dis. Rep.* **63,** 621–625.

Stevens, W. A., and Tsiantos, J. (1979). The use of enzyme-linked immunosorbent assay (ELISA) for the detection of *Corynebacterium michiganense* in tomatoes. *Microbios* **10,** 29–32.

Sun, M. H., Chang, S. C., and Tseng, C. M. (1976). Research advances in sugarcane downy mildew of corn in Taiwan. *Kasetsart J.* **10,** 89–93.

Thomas, B. J. (1980). The detection by serological methods of virus infecting the rose. *Ann. Appl. Biol.* **94,** 91–101.

Thurston, H. D. (1973). Threatening plant diseases. *Annu. Rev. Phytopathol.* **11,** 27–53.

U.S. Department of Agriculture (1981). Contraband foods. *Agric. Res.* **30**(4), 4–7.

Vakili, N. G., and Bromfield, K. R. (1976). Phakopsora rust on soybean and other legumes in Puerto Rico. *Plant Dis. Rep.* **60,** 995–999.

Vruggink, H. (1978). Enzyme-linked immunosorbent assay (ELISA) in the serodiagnosis of plant pathogenic bacteria. *Proc. Int. Conf. Plant Pathog. Bact., 4th* pp. 307–310.

Warren, H. L., Scott, D. H., and Nicholson, R. L. (1974). Occurrence of sorghum, downy mildew of maize in Indiana. *Plant Dis. Rep.* **58,** 430–432.

Waterworth, H. E., and White, G. A. (1982). Plant introductions and quarantine: The need for both. *Plant Dis.* **66,** 87–90.

Waterworth, H. E., Kaper, J. M., and Tousignant, M. E. (1979). CARNA 5, the small cucumber mosaic virus-dependent replicating RNA, regulates disease expression. *Science* **204,** 845–847.

Weathers, L. G., and Calavan, E. C. (1959). Nucellar embryony—A means of freeing citrus clones of viruses. *In* "Citrus Virus Diseases" (J. M. Wallace, ed.), pp. 197–201. University of California, Div. Agric. Sci., Davis.

Wilson, C. L. (1982). Criteria for evaluating exotic pathogens. *Phytopathology* **72,** 712.

CHAPTER 13

Research on Exotic Weeds

DAVID T. PATTERSON

Agricultural Research Service
United States Department of Agriculture
and Department of Botany
Duke University
Durham, North Carolina

I. INTRODUCTION

Weeds are among the most important of all crop pests. They compete directly with the crop for the same limited environmental resources upon which plant growth depends. In addition to competing directly with the crop for light, water, and soil nutrients, weeds may damage crops mechanically through choking, twining, or smothering growth, or chemically through allelopathy, the production of toxic or inhibitory compounds that affect neighboring plants.

Weeds interfere with harvesting and reduce crop quality. Some are toxic to livestock, whereas others furnish reservoirs for other crop pests. In contrast to the epidemic nature of many other crop pests, weeds are almost always present at levels sufficient to reduce crop yield in the absence of some control measures.

In the United States, annual crop losses and control costs of weeds exceed those of insects, plant diseases, or nematodes (Klingman and Ashton, 1975).

Many of the worst agronomic weeds in the United States are exotics, with

EXOTIC PLANT PESTS AND
NORTH AMERICAN AGRICULTURE

ISBN 0-12-757880-3

origins in Europe, Africa, Asia, or Central and South America. For example, of the 37 worst soybean weeds listed by Wax (1976), 23 are exotic (McWhorter and Patterson, 1980). Of the 25 most frequently reported weeds or weed complexes in all crops in the United States in 1968 (USDA, 1972), 19 are of European or Eurasian origin. Fogg (1956) and Reed (1971) list exotic origins for many common weeds in the United States.

The impact of exotic weeds may be caused largely by the absence of the natural controls present in their lands of origin. In addition, the long history of agriculture in Europe and Asia has provided more opportunity for the evolution of agronomic weeds that readily infest crops when introduced to new areas (Baker, 1974; Fogg, 1956). The repeated extensive glaciation of Europe also resulted in naturally disturbed habitats conducive to the evolution of weeds (Fogg, 1956).

The most devastating agronomic weeds share many characteristics that distinguish them from the rest of the world's flora. For example, of the 37 worst soybean weeds (Wax, 1976), 14 are C_4 plants, 14 are monocots, 10 have vegetative as well as sexual reproduction, and 20 are allelopathic (McWhorter and Patterson, 1980).

Of the 25 most frequently reported agronomic weeds in the United States (U.S. Department of Agriculture, 1972), 9 are monocots, 8 are C_4 plants, and 10 have vegetative reproduction.

Of the 18 worst weeds in the world (Holm *et al.*, 1977), 13 are monocots, 11 have vegetative reproduction, 6 produce rhizomes, and 14 are C_4 plants (McWhorter and Patterson, 1980). Clearly, there is a disproportionate representation of C_4 plants and monocots among the most important weeds, in comparison to the composition of the world's flora.

There are many other characteristics that contribute to the impact of agronomic weeds (Baker, 1974; Hill, 1977; King, 1966a; Patterson, 1982b). These are as follows:

1. Rapid initial growth rate and expansion of the root system and leaf canopy, resulting in rapid exploitation of available environmental resources
2. Special ability to interfere with the growth of neighboring plants, with twining, climbing growth, and mechanical damage to other plants
3. High seed output under optimum conditions, with some seed produced even under unfavorable conditions
4. Morphological and/or physiological similarity to crops
5. Breeding system that allows both self-pollination and outcrossing.

The potential threat to U.S. agriculture from weeds not yet introduced is considerable. For example, of the 6741 species reported by Holm *et al.* (1979) as significant agronomic weeds, 4678 are not known to be present in the contiguous United States. Reed (1977) lists some 1200 species of noxious weeds either absent from the United States or present in only limited areas.

A major goal of exotic weed research must be to establish methods for screening and evaluation to predict the potential agronomic, ecological, and economic impact of exotic weeds in the United States. Only through such research can priorities for prevention, control, or eradication be established and contingency plans developed, in advance of the appearance of such weeds at economically damaging levels (Parker, 1977).

This review examines the areas of research needed to reach this goal of predicting the impact of exotic weeds. Such research involves physiological and ecological studies of individual weeds, as well as studies of the actual and potential interactions of exotic weeds with crop plants, weeds, and other crop pests.

II. RESEARCH ON INDIVIDUAL SPECIES

Ecophysiological and autecological research on individual species of exotic weeds can be of considerable value in predicting their potential geographic distribution and impact in the United States.

For example, itchgrass (*Rottboellia exaltata* L. f.), an exotic weed native to tropical Asia, is currently present in the United States only in limited areas in Florida and Louisiana. Patterson *et al.* (1979) grew itchgrass in controlled-environment greenhouses of the Duke University Phytotron and determined its responses to 36 combinations of day and night temperature, ranging from 17/11°C to 32/26°C. They found that itchgrass grew vigorously in temperature regimes simulating the warmest 2 months of the growing season throughout most of the major agronomic regions of the United States.

In subsequent studies, itchgrass grew to maturity and produced viable seed in controlled-environment chambers programmed to simulate summer temperatures as far north as southern Wisconsin (Patterson and Flint, 1979). However, itchgrass was found to be extremely sensitive to simulated short-term chilling events typical of the early growing season for corn in both the South and the Midwest and for soybean in the Midwest. In comparison with the growth of adapted varieties of soybean and corn, the growth of itchgrass was markedly reduced by chilling. These findings led to the conclusion that itchgrass was unlikely to be a serious early season competitor with corn or with soybeans outside the South (Patterson and Flint, 1979). Even so, the definite potential for the extensive further spread of itchgrass in the United States certainly warrants increased efforts to contain and eradicate this exotic weed.

Similar studies of the exotic weed cogongrass [*Imperata cylindrica* (L.) Beauv.] showed that its sensitivity to cool temperatures renders it unlikely to become a serious problem outside the Gulf Coast states (Patterson *et al.*, 1980).

The temperature responses of the exotic parasitic weed witchweed [*Striga asiatica* (L.) O. Ktze.] have been studied in detail since its discovery in the coastal plain of North and South Carolina in the 1950s (Nelson, 1958; Robinson, 1960; Patterson *et al.*, 1982). The results of these studies indicate that neither minimum soil temperatures during the winter nor soil and air temperatures during the growing season will prevent the spread of witchweed into the midwestern corn belt in the United States.

Studies of the effects of environmental variables other than temperature also can contribute to our prediction of the potential impact of exotic weeds. For example, Patterson (1979) reported that the growth and photosynthetic rates of itchgrass were greatly reduced by artificial shading simulating light levels occurring under the canopy of many row crops. However, some adaptation to shading occurred in this species, with shaded plants developing more leaf area per unit of plant weight. The responses of itchgrass to shading indicate that it could survive long exposure to dense shade and still retain the capacity for high rates of photosynthesis and growth if subsequently exposed to full sunlight.

Similar studies with the exotic perennial weeds cogongrass (Patterson, 1980) and purple nutsedge (*Cyperus rotundus* L.) (Patterson, 1982a) indicated that these plants also adapted to shading, at the whole-plant level, by increasing the partitioning of plant biomass into leaves and the distribution of leaf biomass as leaf area. The ability of purple nutsedge to produce both rhizomes and tubers even when grown in dense shade helps explain its persistence in agronomic crops. Similar findings with johnsongrass [*Sorghum halepense* (L.) Pers.] help explain its competitiveness with row crops (McWhorter and Jordan, 1976).

Significant adaptation of the photosynthetic apparatus to low light occurs in the widespread exotic weed oriental bittersweet (*Celastrus orbiculatus* (Thunb.) (Patterson, 1975). In contrast, Sakhalin knotweed (*Polygonum sachalinense* F. Schmidt), an exotic currently limited to abandoned homesites and highway and railroad rights-of-way, lacks the ability to acclimate photosynthetically to low light (Patterson *et al.*, 1977).

Future variations in another environmental variable, the atmospheric CO_2 concentration, may influence the impact of exotic weeds as well as other crop pests. A controlled-environment study by Patterson and Flint (1980) showed that the increases in the CO_2 concentration projected for the next 50–100 years should decrease the competitiveness of C_4 weeds with C_3 crops and increase the competitiveness of C_3 weeds with C_4 crops. Since many of our worst exotic weeds are C_4 plants, the direct effect of increased CO_2 levels in the atmosphere should be to lessen the weed problem. However, possible attendant climatic changes (increased global temperature and aridity) may increase the impact of C_4 weeds. This could occur because C_4 plants characteristically are better adapted to hot, dry conditions than are C_3 plants (Teeri and Stowe, 1976). Further research certainly is needed to evaluate the potential effects of global atmospheric CO_2 enrichment on weed–crop interactions.

The use of limited-access, controlled-environment facilities, or phytotrons, in studying the environmental responses of exotic weeds under simulated natural conditions has several advantages. The most obvious advantage is that the use of such facilities greatly reduces the risk of spreading the exotic weeds into areas not already infested. This risk would be much greater, for example, if even carefully controlled field plantings were made at various locations to test the ability of an exotic weed to grow and reproduce. Further, such controlled-environment studies can be accomplished more rapidly and less expensively than the extensive field studies that would be required to obtain similar information.

Of course, the purposeful introduction of an exotic noxious weed into the United States to evaluate its environmental responses would entail some risk, even if the plants were grown under carefully controlled conditions. Thus, for the many noxious weeds not yet known to be present in the United States, other means of predicting potential geographical distribution and impact are desirable. An obvious alternative would be to study the temperature responses of the candidate weed in controlled-environment facilities located in an area where it is already present. This should be possible for weeds present in Europe, Japan, New Zealand, and Australia, where adequate phytotron facilities exist. For the many weedy species presently restricted elsewhere, other alternatives should be sought.

One promising strategy is the use of the "indicator species-ecological amplitude" concept as applied by Duke (1976). Environmental information is compiled for sites throughout the world where economically important plants (both crops and weeds) occur naturally. These data are used to estimate the ecological amplitude of each species with respect to soil pH, annual precipitation, and mean annual temperature. Information on life zone, habit, and chromosome number is also included in a computerized description of each species. Knowledge of the ecological amplitude of an exotic weed should aid in predicting its performance in a new area to which it is introduced.

Another approach which should prove useful in predicting the potential range and impact of exotic noxious weeds in the United States is the "agroclimatic analogue" concept of Nuttonson (1965). Nuttonson has used crop phenological data and records of temperature, humidity, and rainfall from diverse sites throughout the world to develop climatic analogues for locations in the major agronomic regions of the United States. According to Nuttonson (1965, p. 3), "the identification of agroclimatically similar areas and the indication of the prevalence of a given plant or disease in these areas, may, among other things, be useful as a warning of the likelihood of the appearance of a given pest or disease . . . in areas where they had not previously been found.''

Both Nuttonson and Duke have emphasized the limitations of these methods for estimating the potential range and impact of exotic organisms in a new environment. Their approaches certainly hold promise as initial screening steps to help single out the most serious potential problems among exotic weeds not present in the United States.

However, the optimum conditions for a particular species cannot necessarily be predicted from its natural distribution. Competitive pressure, pests and diseases, limitations to dispersal, and other biotic factors may restrict or limit the distribution of an organism to a narrow range within its potential ecological amplitude. Perhaps the best-known example of this phenomenon is Monterey pine (*Pinus radiata* D. Don.). In its natural distribution, this species is a narrow endemic in California. However, in the rather different climates of Australia, New Zealand, South America, and Africa, where it has been introduced, Monterey pine is an extremely productive and valuable timber tree. For example, Jackson (1965) reported that annual volume increments in Monterey pine in New Zealand are up to twice as great as those in its native range in California. The performance of this species in these new environments could not have been predicted from its apparent ecological amplitude in its native range or through use of the agroclimatic analogue technique. In a phytotron study of this species, Hellmers and Rook (1973) grew Monterey pine in 30 combinations of day and night temperature, ranging from 17/5°C to 32/29°C. They found that temperature regimes simulating growing season conditions in New Zealand were very favorable for the growth of Monterey pine.

In addition to ecophysiological research to evaluate the environmental responses of exotic weeds, there is a definite need for taxonomic and genecological studies of important exotic weed species (McNeil, 1976). For example, intraspecific variability in herbicide susceptibility has been demonstrated in many species of native and exotic weeds (Jensen *et al.*, 1977; McWhorter, 1971b; Thomas and Murray, 1978). Such variability may help account for the occasional failures of certain herbicides to control specific populations of weeds (King, 1966b; Holliday and Putwain, 1980). Herbicide-resistant species or biotypes may pose particularly serious threats for U.S. agriculture where little hand weeding is practiced. Such species might not be recognized as serious problems in their native lands if hand weeding and mechanical cultivation keep them in check.

Increased knowledge of the genetics and breeding systems of exotic weeds may help document species which could exchange genes with important crop plants. Such genetic exchange can result in new weeds, which, because of their morphological and physiological similarity to the crop, are extremely difficult to control. Weed–crop hybridization has generated "new" weeds in the genera *Sorghum, Beta, Cucumis,* and *Raphanus* (McNeil, 1976).

The introduction of exotic species also provides the opportunity for genetic exchange with native, noncrop plants. Particularly aggressive or weedy new species that may have arisen in this manner include *Spartina* X *townsendii* and *Salvinia molesta* (Parker, 1977).

The development and importance of weed ecotypes have been reviewed by King (1966b), McNeil (1976), Parker (1977), Baker (1974), and others. In addition to variability in herbicide susceptibility, weed ecotypes may differ in numerous morphological characteristics and physiological responses. Few exotic weeds in the United States have been surveyed adequately for the presence of ecotypes. However, McWhorter (1971a) was able to distinguish 37 ecotypes of johnsongrass in collections from local populations throughout the United States. In subsequent work, McWhorter (1971b) found significant variation in the susceptibility to dalapon herbicide among 55 ecotypes of johnsongrass from the United States and abroad.

The many ecotypes of johnsongrass now present in the United States may have resulted from recent evolution in this country, from the many separate introductions over the past 150–200 years, or, more likely, from a combination of the two. However, the relative importance of these factors is difficult to assess, even though the history of johnsongrass in the United States has been thoroughly studied (McWhorter, 1971c).

For less well documented exotic weeds, the sources of present variability are even more obscure. However, significant morphological and/or physiological variation among distinct populations of exotic weeds in the United States seems to occur in most species studied (e.g., Patterson, 1973, 1980; Patterson *et al.*, 1980). Such variation, however, does not necessarily constitute ecotypic differentiation.

In addition to taxonomic, biosystematic, and genecological studies, research to discover improved methods of detecting exotic weeds at ports of entry and of surveying for newly established populations of exotic weeds in the United States is essential.

The survey, control, containment, quarantine, and eradication program initiated by the USDA in response to the discovery of witchweed in the United States in the mid-1950s serves as the best example of an integrated research and control effort aimed at a specific exotic weed (Shaw *et al.*, 1962; Eplee, 1979). This program has been responsible for preventing the further expansion of witchweed over a large part of the southeastern, southern, and midwestern areas of the United States.

With the passage and subsequent funding of the Federal Noxious Weed Act of 1974 (PL 93-629), similar approaches to other exotic weeds are now possible. The USDA Technical Committee to Evaluate Noxious Weeds is responsible for evaluating and recommending specific exotic weeds for regulation under PL 93-629 (Gunn *et al.*, 1981). Surveying and cataloging of such weeds is an essential part of exotic weed research. B. A. Auld and his associates in Australia have developed simulation models to estimate the rates of spread of introduced noxious weeds (Auld *et al.*, 1979; Auld and Coote, 1980). This approach should be useful for establishing priorities in exotic weed control programs.

III. INTERACTION OF EXOTIC WEEDS WITH OTHER ORGANISMS

Research on the interactions of exotic weeds with other organisms can contribute to two major goals:

1. Predict the total potential impact of the exotic weed in selected agroecosystems
2. Provide a data base for the development of integrated weed and pest management systems

Examples of ecological research aimed at predicting the direct impact of exotic weeds on U.S. crops include studies of the competitive and/or allelopathic interactions of weeds and crops.

Zimdahl (1980) reviewed the competitive effects of a number of native and exotic weeds on crop growth and yield. Crop yield losses from heavy infestations of weeds may be as great as 90%, but losses in farmers' fields are generally less. Yield losses vary greatly with weed species, weed density, and the duration of weed competition. Weeds emerging late in the growing season seldom compete effectively with a crop, but even these weeds may interfere with harvesting and reduce yield quality.

The competitive effects of many major exotic weeds are reasonably well documented (Zimdahl, 1980; Holm *et al.*, 1977). However, knowledge of the effects of specific weed population densities on yields of major crops is by no means complete. Therefore, evaluation of exotic weed–crop competitive interactions under cultural conditions typical of the United States should be part of a comprehensive research program to evaluate exotic weeds.

The role of allelopathy (Rice, 1974) in accounting for the impact of weeds on crop growth and yield is not as well established as that of competition. However, many of the major exotic weeds in U.S. agriculture are known to have allelopathic properties (McWhorter and Patterson, 1980). Weeds in this category include field bindweed (*Convolvulus arvensis* L.), crabgrass [*Digitaria san-*

guinalis (L.) Scop.], foxtail (*Setaria* spp.), jimsonweed (*Datura stramonium* L.), johnsongrass, lambsquarters (*Chenopodium album* L.), wild mustard (*Brassica* spp.), purple nutsedge, and velvetleaf (*Abutilon theophrasti* Medic.). Of course, many other noxious weeds are not known to be allelopathic. Nevertheless, screening programs for predicting the potential impact of exotic weeds should include bioassays for allelopathic properties.

In addition to chemically inhibiting the growth of neighboring plants, many other exotic weeds are poisonous to humans, livestock, poultry, or wildlife (King, 1966a; Hardin, 1966; Schmutz *et al.*, 1968). Notable exotic weeds in this category include halogeton [*Halogeton glomeratus* (M. Bieb.) C. A. Mey.], Russian thistle (*Salsola kali* L. var. *tenuifolia* Tausch), jimsonweed, and showy crotalaria (*Crotalaria spectabilis* Roth). Of these species, at least one, crotalaria, was introduced purposely for use as a green manure and orchard cover crop. Nonpalatable weeds, even though not toxic to livestock, may significantly reduce the value of pastures and rangelands.

In addition to their harmful direct effects through competition, allelopathy, animal toxicity, and reduction of crop harvestability and quality, exotic weeds may interact with other crop pests. Weeds may serve as alternate hosts for crop pathogens, insect pests, and nematodes and as additional reservoirs for crop pathogens and their insect vectors (King, 1966a).

For example, johnsongrass, in addition to being a highly competitive weed in corn, serves as an overwintering host for two viruses that attack corn. Johnsongrass also is a host for the insect vectors of these viruses and for a nematode that infests corn (Bendixen *et al.*, 1979).

Bendixen *et al.* (1981) have provided a comprehensive annotated bibliography of weeds as hosts of insects which attack various crop plants. An earlier bibliography dealt with weeds as hosts of crop-damaging nematodes (Bendixen *et al.*, 1979).

The potential effects of exotic weeds on the feeding patterns, predator–prey relationships, and population dynamics of other crop pests are largely unknown. Ecological research in these areas certainly should contribute to our predictions of the potential impacts of exotic weeds.

Exotic weeds, upon introduction into a new area, may have significant effects on the weed flora already present. Such effects could include competitive and allelopathic interactions which suppress weeds already present. As already pointed out, genetic exchange with closely related native or previously introduced exotic plants may result in better-adapted weeds with more noxious characteristics. So-called ecological shifts in the weed flora may result from direct competitive or allelopathic effects of the newly introduced exotic weeds or from cultural or chemical control programs undertaken to combat the new weeds (McWhorter and Patterson, 1980).

IV. RESEARCH ON CONTROL OF EXOTIC WEEDS

There are some 180 basic herbicides available today for chemical weed control. Thus, the chances are very remote that any exotic weed will be totally resistant to the chemical controls now available. However, exotic weeds often are tolerant of the herbicides commonly used in the crops they infest, and this can increase their potential economic impact. For example, the exotic weed itchgrass has the potential to create serious weed control problems in corn production, because it is resistant to atrazine (Parker, 1977). Thus, should widespread infestations of itchgrass develop in the 70 million acres of corn grown in the United States, the increased cost to farmers for new chemical control measures would be considerable. Other exotic weeds that create serious problems because of their resistance to commonly used herbicides include prickly sida (*Sida spinosa* L.) and sicklepod (*Cassia obtusifolia* L.).

On the other hand, the intensive mechanical cultivation practiced in many row crops in the United States may reduce the impact of exotic weeds that cause serious problems in other parts of the world. For example, cogongrass, which ranks as the seventh worst weed in the world (Holm *et al.*, 1977), has failed to become a major weed in agronomic crops along the U.S. Gulf Coast because it is easily controlled by cultivation (Dickens, 1974). Certainly, any proposed screening of exotic weeds should include some evaluation of the ease or difficulty of their control with the chemical and cultural practices currently used in the major crops likely to become infested in the United States.

Exotic weeds are often cited as particularly susceptible to biological control (Wilson, 1964). The greatest successes in biological control of weeds have followed the introduction of agents that are present in the land of origin of the weed (Holloway, 1964). As was pointed out earlier, one of the reasons for the aggressiveness of exotic weeds is the absence of the natural control agents that hold them in check in their native habitats.

Biological control of weeds has met with limited success in agronomic crops (Holloway, 1964). However, the use of native or introduced plant pathogens for control of agronomic weeds is promising (Wilson, 1969). Pathogen propagules are more adaptable to mass production, storage, encapsulation, and mechanical application in row crops than are those of insects. Pathogens also have the advantage of being less affected by commonly applied crop protection chemicals. Recent work with *Alternaria macrospora* Zimm. as a biocontrol agent for spurred anoda [*Anoda cristata* (L.) Schlect.] in cotton is exemplary (Walker and Sciumbato, 1979; Walker, 1981).

Where complete eradication of an exotic weed of limited distribution is required, biocontrol obviously is less feasible than conventional chemical control. However, contingency plans for control of exotic weeds probably should incorporate as many methods as possible.

V. CONCLUSIONS

Exotic weed research encompasses many disciplines. To reach the major goal of predicting with some certainty the threat and potential impact of a specific foreign weed in the United States will require a comprehensive research effort in both field and laboratory. Therefore, the selection of exotic weeds for detailed study should be done with care. As we have pointed out in this review, there are certain plant characteristics and other factors that may help us to recognize the most serious threats among the many foreign weeds not yet present or widely distributed in the United States. Thus, a plant having the characteristics or properties of physiological and/or morphological similarity to a major crop, resistance or tolerance to herbicides commonly used in that crop, difficulty of control by mechanical cultivation, adaptive seed dormancy and dispersal mechanisms, C_4 photosynthesis, rapid growth rates, exploitive use of environmental resources, allelopathic inhibition of neighboring plants, phenotypic plasticity, capacity for rapid evolution, and special competitive abilities should attract our attention as a serious potential problem. If a weed with a number of these properties has populations near ports or shipping points from which agricultural products are exported to the United States, if it has propagules that are likely to contaminate ballast or packing material, and if it exists in an area climatically analogous to a major U.S. agronomic area, the potential threat is probably very great.

Perhaps consideration of these factors may aid weed scientists and research administrators in singling out specific exotic weeds for detailed study.

REFERENCES

Auld, B. A., and Coote, B. G. (1980). A model of a spreading plant population. *Oikos* **34,** 287–292.

Auld, B. A., Coote, B. G., and Menze, K. M. (1979). Dynamics of plant spread in relation to weed control. *Proc. Asian-Pac. Weed Sci. Soc. Conf. 7th, 1979* pp. 399–402.

Baker, H. G. (1974). The evolution of weeds. *Annu. Rev. Ecol. Syst.* **5,** 1–24.

Bendixen, L. E., Reynolds, D. A., and Riedel, R. M. (1979). "An Annotated Bibliography of Weeds as Reservoirs for Organisms Affecting Crops. I. Nematodes," Res. Bull. 1109. Ohio Agric. Res. Dev. Centr., Wooster, Ohio.

Bendixen, L. E., Kim, K. U., Kozak, C. M., and Horn, D. J. (1981). "An Annotated Bibliography of Weeds as Reservoirs for Organisms Affecting Crops. IIa. Arthropods, Res. Bull. 1125. Ohio Agric Res. Cent. Wooster, Ohio.

Dickens, R. (1974). Cogongrass in Alabama after sixty years. *Weed Sci.* **22,** 177–179.

Duke, J. A. (1976). Perennial weeds as indicators of annual climatic parameters. *Agric. Meterol.* **16,** 291–294.

Eplee, R. E. (1979). The *Striga* eradication program in the United States of America. *Proc.—Int. Symp. Parasit. Weeds, 2nd, 1979* pp. 269–272.

Fogg, J. M., Jr. (1956). "Weeds of Lawn and Garden." Univ. of Pennsylvania Press, Philadelphia.

Gunn, C. R., Ritchie, C. A., and Poole, L. (1981). "Exotic Weed Alert. Noxious Weed Printout." U.S. Dept Agric., Beltsville, Maryland.

Hardin, J. W. (1966). Stock-poisoning plants of North Carolina. *N.C. State Univ. Agric. Exp. Stn., Bull.* **414** (rev.).
Hellmers, H., and Rook, D. A. (1973). Air temperature and growth of radiata pine seedlings. *N. Z. J. For. Sci.* **3,** 271–285.
Hill, T. A. (1977). "The Biology of Weeds." Arnold, London.
Holliday, R. J., and Putwain, P. D. (1980). Evolution of herbicide resistance in *Senecio vulgaris:* Variation in susceptibility to simazine between and within populations. *J. Appl. Ecol.* **17,** 779–791.
Holloway, J. K. (1964). Projects in biological control of weeds. *In* "Biological Control of Insect Pests and Weeds" (P. DeBach, ed.), pp. 650–670. Van Nostrand-Reinhold, Princeton, New Jersey.
Holm, L. G., Plucknett, D. L., Pancho, J. V., and Herberger, J. P. (1977). "The World's Worst Weeds. Distribution and Biology." Univ. Press of Hawaii, Honolulu.
Holm, L. G., Pancho, J. V., Herberger, J. P., and Plucknett, D. L. (1979). "A Geographical Atlas of World Weeds." Wiley, New York.
Jackson, D. S. (1965). Species siting: Climate, soil and productivity. *N. Z. J. For.* **10,** 90–102.
Jensen, K. I. N., Bandeen, J. D., and Souza Machado, V. (1977). Studies on the differential tolerance of two lambsquarters selections to triazine herbicides. *Can. J. Plant Sci.* **57,** 1169–1177.
King, L. J. (1966a). "Weeds of the World. Biology and Control." Leonard Hill, London.
King, L. J. (1966b). Weed ecotypes—a review. *Proc. Northeast. Weed Control. Conf.* **20,** 604–611.
Klingman, G. C., and Ashton, F. M. (1975). "Weed Science: Principles and Practices." Wiley, New York.
McNeil, J. (1976). The taxonomy and evolution of weeds. *Weed Res.* **16,** 399–413.
McWhorter, C. G. (1971a). Growth and development of johnsongrass ecotypes. *Weed Sci.* **19,** 141–147.
McWhorter, C. G. (1971b). Control of johnsongrass ecotypes. *Weed Sci.* **19,** 229–233.
McWhorter, C. G. (1971c). Introduction and spread of johnsongrass in the United States. *Weed Sci.* **19,** 496–500.
McWhorter, C. G., and Jordan, T. N. (1976). The effect of light and temperature on the growth and development of johnsongrass. *Weed Sci.* **24,** 88–91.
McWhorter, C. G., and Patterson, D. T. (1980). Ecological factors affecting weed competition in soybeans. Proc.—*World Soybean Res. Conf., 2nd, 1979* pp. 371–392.
Nelson, R. R. (1958). The effect of soil type and soil temperature on the growth and development of witchweed (*Striga asiatica*) under controlled soil temperatures. *Plant Dis. Rep.* **42,** 152–155.
Nuttonson, M. Y. (1965). "Global Agroclimatic Analogues for the Southeastern Atlantic Region of the United States and an Outline of its Physiography, Climate and Farm Crops." Am. Inst. Crop Ecol., Washington, D.C.
Parker, C. (1977). Prediction of new weed problems, especially in the developing world. *In* "Origins of Pest, Parasites, Disease and Weed Problems" (J. M. Cherrett and G. R. Sagar, eds.), pp. 249–264. Blackwell, Oxford.
Patterson, D. T. (1973). The ecology of oriental bittersweet, *Celastrus orbiculatus,* a weedy introduced ornamental vine. Ph.D. Thesis, Duke University, Durham, North Carolina.
Patterson, D. T. (1975). Photosynthetic acclimation to irradiance in *Celastrus orbiculatus* Thunb. *Photosynthetica* **9,** 140–144.
Patterson, D. T. (1979). The effects of shading on the growth and photosynthetic capacity of itchgrass (*Rottboellia exaltata*). *Weed Sci.* **27,** 549–553.
Patterson, D. T. (1980). Shading effects on growth and partitioning of plant biomass in cogongrass (*Imperata cylindrica*) from shaded and exposed habitats. *Weed Sci.* **28,** 735–740.

Patterson, D. T. (1982a). Shading responses of purple and yellow nutsedges (*Cyperus rotundus* and *C. esculentus*). *Weed Sci.* **30,** 25–30.

Patterson, D. T. (1982b). Effects of light and temperature on weed/crop growth and competition. *In* "Role of Biometeorology in Integrated Pest Management" (J. L. Hatfield, ed.), pp. 407–420. Academic Press, New York.

Patterson, D. T., and Flint, E. P. (1979). Effects of simulated field temperatures and chilling on itchgrass (*Rottboellia* exaltata), corn (*Zea mays*), and soybean (*Glycine max*). *Weed Sci.* **27,** 645–650.

Patterson, D. T., and Flint, E. P. (1980). Potential effects of global atmospheric CO_2 enrichment on the growth and competitiveness of C_3 and C_4 weed and crop plants. *Weed Sci.* **28,** 71–75.

Patterson, D. T., Longstreth, D. J., and Peet, M. M. (1977). Photosynthetic adaptation to light intensity in Sakhalin knotweed (*Polygonum sachalinense*). *Weed Sci.* **25,** 319–323.

Patterson, D. T., Meyer, C. R., Flint, E. P., and Quimby, P. C. Jr. (1979). Temperature responses and potential distribution of itchgrass (*Rottboellia exaltata*) in the United States. *Weed Sci.* **27,** 77–82.

Patterson, D. T., Flint, E. P., and Dickens, R. (1980). Effects of temperature, photoperiod and population source on the growth of cogongrass (*Imperata cylindrica*). *Weed Sci.* **28,** 505–509.

Patterson, D. T., Musser, R. L., Flint, E. P., and Eplee, R. E. (1982). Temperature responses and potential for spread of witchweed (*Striga lutea* = *S. asiatica*) in the United States. *Weed Sci.* **30** 87–93.

Reed, C. F. (1971). "Common Weeds of the United States." Dover, New York.

Reed, C. F. (1977). Economically important foreign weeds. Potential problems in the United States. *U.S., Dep. Agric., Agric. Handb.* **498.**

Rice, E. L. (1974). "Allelopathy." Academic Press, New York.

Robinson, E. L. (1960). Growth of witchweed as affected by soil types and soil and air temperatures. *Weeds* **8,** 576–581.

Schmutz, E. M., Freeman, B. N., and Reed, R. E. (1968), "Livestock-Poisoning Plants of Arizona." Univ. of Arizona Press, Tucson.

Shaw, W. C., Shepherd, D. R., Robinson, E. L., and Sand, P. F. (1962). Recent advances in witchweed control. *Weeds* **10,** 182–192.

Teeri, J. A., and Stowe, L. C. (1976). Climatic patterns and the distribution of C_4 grasses in North America. *Oecologia* **23,** 1–12.

Thomas, S. M., and Murray, B. G. (1978). Herbicide tolerance and polypoidy in *Cynodon dactylon* (L.) Pers. (Gramineae). *Ann. Bot.* (*London*) [N.S.] **42,** 137–143.

U.S. Department of Agriculture (1972). "Extent and Cost of Weed Control with Herbicides and an Evaluation of Important Weeds, 1968," ARS-H-1. U.S. Dept. Agric., Beltsville, Maryland.

Walker, H. L. (1981). Factors affecting biological control of spurred anoda (*Anoda cristata*) with *Alternaria macrospora. Weed Sci.* **29,** 505–507.

Walker, H. L., and Sciumbato, G. L. (1979). Evaluation of *Alternaria macrospora* as a potential biocontrol agent for spurred anoda (*Anoda cristata*): Host range studies. *Weed Sci.* **27,** 612–614.

Wax, L. M. (1976). Difficult-to-control annual weeds. *World Soybean Res., Proc. World Soybean Res. Conf., 1975* pp. 420–425.

Wilson, C. L. (1969). Use of plant pathogens in weed control. *Annu. Rev. Phytopathol.* **7,** 411–434.

Wilson, F. (1964). The biological control of weeds. *Annu. Rev. Entomol.* **9,** 225–244.

Zimdahl, R. L. (1980). "Weed-Crop Competition—A Review." Int. Plant Prot. Cent., Oregon State University, Corvallis.

CHAPTER **14**

Biological Control: Exotic Natural Enemies to Control Exotic Pests

L. E. EHLER

Department of Entomology
University of California
Davis, California

and

L. A. ANDRES

Biological Control of Weeds Laboratory
Agricultural Research Service
United States Department of Agriculture
Albany, California

EXOTIC PLANT PESTS AND
NORTH AMERICAN AGRICULTURE

ISBN 0-12-757880-3

I. INTRODUCTION

Fortunately, not all of the immigrant organisms arriving at North American shores are destined to become pests. Of those immigrants that become established, a certain percentage can be classified as economically unimportant and others as minor pests, whereas still others may enhance the beauty and productivity of the environment (Sailer, 1978). In fact, organisms in the last category are sought throughout the world with the intention of introducing them to control other immigrants that have become pests. This activity, known as "classical biological control," is engaged in by a cadre of scientists in a number of countries who continue to search for and introduce a variety of natural enemies from one area to the next.

Unfortunately, not all of the intentionally introduced species become established, nor, once established, do they always effect the desired level of control. The inability to predict the outcome of these planned introductions has been the bane of biological control specialists down through the years. In many respects, this unpredictability parallels the problems faced by federal and state quarantine officials whose job is to identify and screen out those species of the immigrant horde that are destined to become pests. Conceivably, the observations, preintroduction studies, and experiences that biological control workers have had with their intentionally introduced natural enemies may be of some value to those who study immigrant pests. In biological control of weeds, for example, an introduced herbivore must establish itself, build up a population, and go on to inflict severe damage on its weed host. These are precisely the steps that an accidentally imported herbivorous insect species must complete before it will become a major pest.

In this chapter, we will briefly describe the theory and practice of classical biological control and then consider some of the factors which can influence the degree of success of a biological control project. The latter analysis is intended to be illustrative rather than exhaustive; however, it should be sufficient to identify some of the similarities between immigrant pests and introduced natural enemies.

II. THEORY AND PRACTICE OF CLASSICAL BIOLOGICAL CONTROL

Current theory suggests that in the native home of a given plant or plant-feeding insect, we should expect to find a complex of natural enemies and/or other environmental factors capable of maintaining the plants or insects at relatively low levels. This is usually termed "natural control," and it serves as the basic premise of classical biological control—the intentional introduction of exotic natural enemies to control introduced or naturalized pest species. In other words, crop-feeding insects which are accidentally introduced into new areas without their adapted natural enemies, and which subsequently become pests, can then be controlled through intentional introduction of the appropriate natural enemies. Conversely, weed plant species can be effectively controlled by the intentional introduction of host-specific, weed-feeding insects.

Most natural enemies employed in biological control are either parasites, predators and pathogens of arthropod pests, or phytophagous arthropods and pathogens associated with weeds. In this chapter, we will deal largely with predators and parasites of insect pests and phytophagous insects used in weed control. These are the natural enemies which are most often employed and the ones which have yielded the greatest number of successes in classical biological control.

The basic steps in importing a biological control agent were outlined by Zwölfer *et al.* (1976):

1. *Select the area of exploration.* Identify the pest and determine its native range and point of origin. Match the climates between the target area and the native areas of the pest.
2. *Provide an inventory of natural enemies.* Collect material in a variety of habitats and at varying host population densities.
3. *Analyze pest-regulating factors.* While collecting, assay the effects of and interactions with natural enemies; assess other factors that may be limiting the host.
4. *Study promising candidates.* Assemble information on host-finding and selection processes, adaptation to climate, reproductive capacity, potential impact, synchronization with the host, etc.
5. *Import.* Select organisms and ship in the proper stages, all of which are disease and parasite free.

Upon receipt of a foreign shipment, the material must be transferred to a certified quarantine facility and held until any unwanted organisms (e.g., hyperparasites, plant pathogens) have been excluded and appropriate studies carried out. Ideally, the candidate biological control agent is cultured for at least one generation in the quarantine laboratory. At the appropriate time, a pure culture of the natural enemy is removed from quarantine and either mass-cultured in the insectary and/or released in the field. Follow-up studies concerning establishment and ecological impact of established species are routinely carried out.

A. Establishment of Natural Enemies

Just how successful have we been in our efforts to establish exotic natural enemies, and what can we learn from these efforts that may help in understanding and stemming the flow of exotic pests? Summations of biological control attempts (e.g., Clausen, 1978; Luck, 1982; Julien, 1982; and others) are helpful in providing a broad overview of biological control activities, but often leave much to be desired in terms of useful information for an analysis of success or failure. However, despite this caution and lack of adequate information on the circumstances surrounding so many of the biological control attempts, we have chosen to derive what we can from a summary of these efforts.

1. *Arthropod Pests*

Working with Clausen's (1978) summary of approximately 600 biological control projects (1890–1968), Hall and Ehler (1979) determined the rates (= percentages) of establishment of parasites and predators introduced for the control of insects and arachnids. Relevant published and unpublished data are summarized in Table I. In this case, only introductions against pests in the orders

TABLE I

Rates of Establishment of Exotic Parasites and Predators Introduced to Control Exotic Pest Insects in the Orders Lepidoptera, Coleoptera, and Homoptera, ca. 1890–1968

Region	Order of target pest	Number of introductions[a]	Establishment	
			Number	Percent
World[b]	Lepidoptera	628	172	27.4
	Coleoptera	364	85	23.4
	Homoptera	819	351	42.9
	Totals	1811	608	$\bar{x}$ = 33.6
United States[c] (contiguous 48 states)	Lepidoptera	238	55	23.1
	Coleoptera	125	23	18.4
	Homoptera	298	79	26.5
	Totals	661	157	$\bar{x}$ = 23.8
California[c]	Lepidoptera	34	7	20.6
	Coleoptera	14	4	28.6
	Homoptera	272	57	21.0
	Totals	320	68	$\bar{x}$ = 21.3
Hawaii[c]	Lepidoptera	52	21	40.4
	Coleoptera	33	12	36.4
	Homoptera	58	43	74.1
	Totals	143	76	$\bar{x}$ = 53.1

[a] One introduction equals release of one species in a given geographic area (e.g., state or country); hence, the number of introductions will be greater than the number of species released.

[b] Data from Hall and Ehler (1979).

[c] Unpublished data of R. W. Hall and L. E. Ehler.

Lepidoptera, Coleoptera, and Homoptera are presented; this is because the pests in question are virtually all pests of plants and because most of the introductions listed in Clausen (1978) dealt with pests in these taxa.

On a worldwide basis, 33.6% of the introductions resulted in establishment. In other words, about two-thirds of all introduction attempts were failures. In the 48 contiguous U.S. states, the percentage of establishment was even lower (23.8%). For the two states which have been most active in classical biological control, the percentage of establishment for California (21.3%) was much lower than that for Hawaii (53.1%). The rate of establishment for all parasites and predators introduced to control all insect pests (phytophagous plus others) in Hawaii up to 1978 was 38.9% (Ota *et al.*, 1981). The fact that the number of natural enemies which fail to become established is greater than the number which do establish is hardly a startling revelation (cf. Clausen, 1956; Wilson, 1965; Turnbull, 1967; van den Bosch, 1968; Carter, 1970; Stehr, 1974; Beirne, 1975; Ehler, 1976; Hall and Ehler, 1979; Ota *et al.*, 1981). However, whether or not this record of establishment is a poor one is largely a philosophical issue.

2. *Weeds*

Updating Julien's (1982) world catalogue of biological control attempts against weeds, but focusing attention only upon the 51 species of arthropods intentionally introduced to North America against 25 species of weeds, we find that 38 of these (74%) became established in one or more areas of their host's range. The disparity in percentage of establishment for entomophagous arthropods (24%) and weed-feeding arthropods (74%) in North America may be explained, in part, by the detailed background studies that are required to clear a plant-feeding arthropod (in the view of some, a crop-threatening arthropod) for importation. The biology, host range, host plant relationships, and habitat requirements are all studied before each natural enemy is released. Also, the limited number of importations of weed-feeding arthropods has often allowed weed researchers to make many of their own releases and to work more closely with cooperators in achieving establishment. The releases are, for the most part, in areas free from insecticides. On the other hand, the greater flow of entomophagous parasites and predators does not always allow similar care in making releases, and in many instances, insecticides and other cultural practices can hamper establishment.

B. Impact of Established Natural Enemies

The standard classification of success in biological control is based largely on the necessity of chemical or other control of the target pest following establishment of natural enemies. According to DeBach (1964b), *partial* successes occur when control measures remain necessary, but either the intervals between treatments are lengthened or outbreaks occur less frequently. Partial successes may also occur when complete control is obtained in a small portion of the pest's range. In the present case, partial successes will not be emphasized; the level of biological control of these pests is often so low that to refer to them as "successes" is misleading. *Substantial* successes include cases in which occasional insecticidal treatments or other controls are necessary or in which economic savings are less pronounced because the crop is less important or the crop area is restricted (DeBach, 1964b). *Complete* successes occur when biological control is obtained against a major pest of a major crop over an extensive area, so that other controls are rarely (if ever) required (DeBach, 1964b). Although the degree of success in classical biological control is measured in economic terms, a project may be successful, ecologically speaking, when the natural enemy establishes itself, impacts on the host, and enters into balance with its host and the environment. Unfortunately, this balance may not always result in the desired level of economic control.

How well a pest organism thrives is inversely proportional to the degree of stress under which it lives. When existing stresses are great or approach the limit for pest survival, the additional stress imposed by one or more immigrant natural enemies can cause a spectacular collapse of the pest population. This is especially the case in biological control of weeds. For example, the rosette-defoliating action of *Chrysolina quadrigemina* Suffrian larvae on *Hypericum perforatum* L. limited root development and ultimately the plant's ability to obtain needed moisture in the dry California summers, resulting in 99% control of the weed in that state (Huffaker, 1957). Where summer rainfall is less limiting or fall temperatures limit the beetle's cycle, control of the weed has not been as spectacular. Burdon *et al.* (1981) noted that the presence of competitive pasture plants hastened the control of *Chondrilla juncea* L. with biotic agents in Australia, the grasses adding another element of stress on the weed. Thus, the heterogeneity of the environment can set the checkerboard pattern of partial, substantial, and complete control often observed in the case of weeds.

1. *Insect Pests*

Classical biological control of insect pests has been actively practiced for about the last 100 years. During this period, there have been more than 2000 introductions of parasites and predators (worldwide) directed for control of more than 100 species of pests in more than 600 situations. The results of these efforts are summarized in the following reviews: Bay *et al.* (1976), Bennett *et al.* (1976), Beirne (1975), Caltagirone (1981), Clausen (1956, 1978), Commonwealth Institute of Biological Control (1971), DeBach (1964a,b, 1971, 1974), Greathead (1971, 1976), Hagen and Franz (1973), Hagen *et al.* (1976b), Hall and Ehler (1979), Hall *et al.* (1980), Laing and Hamai (1976), Legner *et al.* (1974), Luck (1982), MacPhee *et al.* (1976), McGugan and Coppel (1962),

McLeod (1962), Ota *et al.* (1981), Rao *et al.* (1971), Sweetman (1958), Turnbull and Chant (1961), Turnock *et al.* (1976), van den Bosch *et al.* (1976), Waters *et al.* (1976), and Wilson (1960). These efforts have resulted in successes of varying degrees in a multitude of situations.

The approximately 2300 introductions directed against insect pests provided complete biological control in about 100 instances (DeBach, 1971; Laing and Hamai, 1976; Hall *et al.*, 1980). These include instances in which the same pest was controlled, usually by the same enemy or complex of enemies, in more than one place. Substantial control was obtained in more than 140 cases (Laing and Hamai, 1976). However, whereas there are about 100 cases of complete biological control on record, the rate of complete success is not as high as we might expect. In this regard, about 16% (n = 602) of all attempts (i.e., projects) resulted in complete biological control (Hall *et al.*, 1980).

Successful biological control (i.e., substantial to complete) has been achieved against insect pests in the orders Orthoptera, Hemiptera, Homoptera, Coleoptera, Lepidoptera, Diptera, and Hymenoptera. Such control was obtained against pests on the following: alfalfa, apple, avocado, banana, bread fruit, carrot, citrus, coconut, coffee, crucifers, eucalyptus, fig, grape, guava, larch, mango, mulberry, oak, oleander, olive, papaya, passion fruit, pear, persimmon, pigeon pea, pine, plum, sugarcane, spruce, taro, tea, walnut, and wheat, plus various forest, shade, and fruit trees (cf. Laing and Hamai, 1976). Success was also obtained in the following categories: on islands and continents; against native and exotic pests; using parasites, predators, and both parasites and predators; with single-species and multiple-species introductions; in the tropics, subtropics, and temperate zone (cf. Laing and Hamai, 1976). Thus, the empirical record of success in classical biological control of insects is a diverse one and clearly attests to the robustness of this method. We thus concur with DeBach (1971, p. 231), who provided the following admonition: "No geographical area or crop or pest insect should be prejudged as being unsatisfactory for biological control attempts. The wide variety of successful results now obtained from importation of natural enemies suggests that nearly anything is possible."

2. *Weeds*

In the preface to Julien's (1982) worldwide summary of attempts at biological control of weeds, Harris noted that 192 exotic organisms were introduced into various countries against 86 weed species, totaling 525 attempts at biological control. In this instance, each attempt signifies the introduction into a country of a weed control candidate species against the target weed host. In only 117 of these attempts did the organisms inflict major damage to the weed. That is, 60–75% of the attempts were failures. Focusing just on the 51 arthropods introduced to North America for weed control, of the 38 that became established, only 15 (39% of those established; 29% of those introduced) provided control somewhere within the range of the host weed. Control in this case refers to the actual reduction in abundance of the weed (whether of economic significance or not) rather than just damage. Thus, under what are supposedly ideal conditions, the release of selected and tested arthropods directly onto their hosts resulted in a surprisingly high degree of failures. It is a wonder that unplanned immigrant phytophagous species ever establish themselves as pests.

III. FACTORS AFFECTING SUCCESS IN CLASSICAL BIOLOGICAL CONTROL

Biological control requires that the region in which the natural enemy operates shall provide the conditions and requisites necessary for the expression of the intrinsic capacity to control the host. If these requirements are met, biological control can operate either in a natural or an artificial habitat (Wilson and Huffaker, 1976). Below, we list some of the conditions and requisites that have been observed as affecting the establishment, buildup, and impact of natural enemies. This list is illustrative only.

A. Host Compatibility

A compatible host is one that allows the natural enemy to develop from egg to adult. Natural enemies that have coevolved with a particular host seem more likely to become established on that host in areas where it has become naturalized as a pest. Hall and Ehler (1979) noted that parasites and predators introduced to control exotic pests became established at a significantly higher rate (34.4%, n = 2163) than did parasites and predators introduced against native pests (25%, n = 132). In these cases, it is reasonable to assume that (1) most of the natural enemies introduced against exotic pests actually coevolved with the particular pest, and (2) essentially all of the enemies introduced for control of native pests had no such coevolutionary history. In other words, attempting to reestablish a natural (coevolved) host–enemy relationship in a new place should be easier than establishing a new host–enemy relationship between two organisms which have not coevolved. Thus, introducing exotic enemies to control native pests has probably contributed to establishment failure. However, the factors which confer immunity on the noncoevolved host may be so subtle that the only way to determine if there is a host incompatibility problem is to proceed with the introduction.

Canadian leafy spurge (*Euphorbia* sp.) proved toxic to the larvae of the introduced natural enemy *Chamaesphecia tenthrediniformis* (Schiffermueller) (P. Harris, personal communication, cited in Julien, 1982), bringing to light the fact that the Canadian spurge, on which the enemy was released, actually differed from the spurge on which it had coevolved and was collected in Europe. Based on their examination of a limited number of herbarium sheets of leafy spurge, Dunn and Radcliffe-Smith (1980) concluded there are five morphologically separable taxa of this weed present in North America. Host incompatibility was also cited in the failure of *C. empiformis* Esper to establish on *E. cyparissias* L. in Canada (Peschken, 1979). The failure of the seed head fly [*Urophora sirunaseva* (Hering)] to become established on *Centaurea solstitialis* L. in California was also attributed to host incompatibility. The fly would oviposit on the young buds, and the eggs would hatch, but the larvae would not develop (D. M. Maddox, personal communication). Similarly, *Aceria chondrillae* (G. Can.), an eriophyid mite imported from Greece (via established colonies in Australia) to control *Chondrilla juncea* L., failed to form galls when placed on California plants. Mites subsequently obtained from *C. juncea* in southern Italy readily attacked the naturalized North American *Chondrilla* (Sobhian and Andres, 1978) and have since become established in California, Idaho, and Washington.

The high degree of specificity exhibited by a number of weed-feeding arthropods makes imperative the positive identification and matching of the pest weed with its foreign counterpart if we expect the arthropods to become established and control the pest. Detailed taxonomic research is needed on many of our weeds, including chemotaxonomic, electrophoretic, and micromorphological studies.

Sands and Harley (1981) discuss the incompatibility between candidate biological control agents and target weeds, noting that incompatibility can be due either to the phenotypic plasticity or the genetic diversity of the weed, in either case resulting in plants different than that on which the natural enemies evolved. Burdon *et al.* (1981) suggest that the greater the genetic variability inherent in the target weed, the less the likelihood of successful biological control. They compare the reproductive mode of previous target weeds, noting that sexually reproducing vs. apomictic plants are likely to have a wider range of genetic variability that will tend to make them incompatible or will allow them to occupy a wider range of habitats beyond the scope of the natural enemies. However, their view should best be regarded as a hypothesis.

Host plant mineral deficiencies in *Opuntia* species precluded the buildup of the introduced *Cactoblastis cactorum* (Berg) on these cacti in Australia. Applying fertilizer resolved this problem, allowing the moth to develop (Wilson, 1960). Mineral imbalances in alligatorweed, *Alternanthera philoxeroides* (Martius) Grisebach, in the laboratory reduced both the attractiveness of the plant to its natural enemy (*Agasicles hygrophila* Selman and Vogt) and the reproductive rate of the insect (Maddox and Rhyne, 1975).

Host defenses can also adversely affect an introduced natural enemy. For example, *Bathyplectes curculionis* (Thomson) is often encapsulated in larvae of the Egyptian alfalfa weevil in California (van den Bosch, 1964; van den Bosch and Dietrick, 1959; Salt and van den Bosch, 1967). In other words, hemocytes of the host larva cluster on the parasite's egg and finally encapsulate it. This eventually leads to the death of the embryo. Encapsulation is also a mechanism whereby a host can develop resistance to its parasite; however, there is no convincing evidence in the literature that this has actually occurred and thus led to pest outbreaks. Presumably, as the host develops some form of resistance (e.g., encapsulation), the parasite evolves new avenues for overcoming this resistance. These and other questions related to host resistance to parasites were discussed by Huffaker *et al.* (1971), van den Bosch (1971), Messenger and van den Bosch (1971), and Huffaker *et al.* (1976).

B. Climate

Many established species are poorly adapted to climatic conditions and simply are not able to maintain the target pest at low levels. Given such circumstances, introduction of a different race of the parasite or predator in question and/or additional species of enemies—all better adapted to the regional climate—can lead to improved levels of biological control. Both of these strategies have been used in the past and eventually resulted in cases of complete biological control. For example, the French race of *Trioxys pallidus* (Hal.), a parasite of walnut aphid [*Chromaphis juglandicola* (Kalt.)], was poorly adapted to the Central Valley of California; however, an Iranian race of *T. pallidus* was well adapted to the valley and has provided generally outstanding biological control of walnut aphid (Frazer and van den Bosch, 1973; van den Bosch *et al.*, 1979). Similarly, *Aphytis paramaculicornis* DeBach and Rosen, a parasite of olive scale [*Parlatoria oleae* (Colvee)], provided good but not complete biological control of the scale in California (Huffaker *et al.*, 1962). In this case, the hot, dry summers in the San Joaquin Valley adversely affected the parasite. Later, a second species (*Coccophagoides utilis* Doutt) was imported which could compensate for the ineffectiveness of *A. paramaculicornis* at certain times of the year (Kennett *et al.*, 1966; Huffaker and Kennett, 1966). Complete biological control resulted. Additional case histories plus other aspects of climatic limitation to biological control were reviewed by Messenger (1970, 1971) and Messenger and van den Bosch (1971).

Climate can also hinder the establishment and buildup of weed-feeding insects. Dry climate precluded the establishment of *Chrysolina varians* (Schaller) on the naturalized weed *Hypericum perforatum* L. in Canada. The beetle is a native in the more northern areas of Europe, where summers are generally moist (Harris and Peschken, 1971). Similarly, *Zeuxidiplosis giardi* Kieffer failed to survive winter temperatures in British Columbia following population buildup during the summer of release (Harris and Peschken, 1971). In California, *Leucoptera spartifoliella* Hubner initially became established only on Scotch broom,

Cytisus scoparius (L.) Link, growing in shaded sites in the Sierra foothills; this was due to high first instar larval mortality on sunny sites. After about 10 years, the number of moths on broom increased dramatically in sunny and shady areas alike, apparently due to an adaptation of this insect to the climate (L. A. Andres, unpublished notes).

Recognition of the importance of climate as it may affect the outcome of biological control efforts has spurred researchers to seek out natural enemies initially in those areas of the pests' native range which are climatically similar to the problem area and to release all natural enemies in varying climatic situations. However, much more research in the area of climate matching needs to be done.

C. Natural Enemies

Hyperparasites, particularly obligate secondary ones, could theoretically prevent an introduced primary parasite from effecting satisfactory levels of biological control. However, such secondary parasites are routinely screened out in the quarantine laboratory. To our knowledge, no obligate secondary parasite has ever been intentionally released in classical biological control. [*Quaylea whittieri* (Girault), an obligate secondary parasite, was introduced into California in the mistaken belief that it was a primary parasite.] Nevertheless, some relevant empirical data are available. For example, *Cyzenis albicans* (Fall.), a tachinid parasite of the winter moth, *Operophtera brumata* L., is not an important mortality factor in at least one location in the native home of the winter moth (England); this is largely due to hyperparasitism and predation of host pupae containing the parasite (Hassell, 1969a,b; East, 1974). However, when *Cyzenis* was released in Canada for control of the winter moth (i.e., free of its endemic natural enemies), it and another parasite [*Agrypon flaveolatum* (Grav.)] provided outstanding biological control (Embree, 1971; Hassell, 1980). In contrast, *Trioxys pallidus* (Haliday), the exotic aphidiid parasite of walnut aphid, has provided outstanding levels of biological control of the aphid in California despite comparatively high levels of hyperparasitism by several species of presumably native secondary parasites (Frazer and van den Bosch, 1973; van den Bosch *et al.*, 1979). Although the data relative to importation of secondary parasites are equivocal, the conventional practice of not introducing such species remains sound (see also Bartlett and van den Bosch, 1964; Doutt and DeBach, 1964; van den Bosch, 1971; Doutt *et al.*, 1976; Huffaker *et al.*, 1976; Caltagirone and Huffaker, 1980; Rosen, 1981). This policy is generally supported by theoretical models (Hassell, 1978; May and Hassell, 1981; Luck *et al.*, 1981), although certain models reveal that introduction of the appropriate secondary parasite could, in fact, stabilize an unstable host–parasite relationship (Beddington and Hammond, 1977; Luck *et al.*, 1981).

It is well known that certain species of ants are part of a mutualistic relationship with honeydew-producing Homoptera such as aphids, soft scales, mealy bugs, and whiteflies (Nixon, 1951; Way, 1963). The ants protect the honeydew producer from its natural enemies and, in return, have access to a copious amount of food in the form of honeydew. The most notorious ant with respect to disrupting biological control in the United States is the Argentine ant, *Iridomyrmex humilis* Mayr, itself an exotic species. This ant has had an adverse effect in a number of biological control projects (cf. DeBach *et al.*, 1951; Bartlett, 1961) and remains a potential problem in many areas. Chemical or mechanical control may be needed in order to prevent ant-introduced outbreaks of honeydew-producing Homoptera in some cases.

Natural enemies have also affected weed control projects. In Canada, *Altica carduorum* Guerin-Meneville suffered high egg mortality from insect and acarine predators, which precluded its establishment on Canada thistle, *Cirsium arvense* (L.) Scopoli (Peschken *et al.*, 1970). Indigenous grasshoppers and crickets preyed on the pupae of *Cystiphora schmidti* Rubsamen along the stems of *Chondrilla juncea* in California. This predation, along with attack by indigenous parasites transferring from native gall midges, reduced the overwintering number of the introduced *C. schmidti,* limiting its spring buildup and subsequent impact on the host weed (L. A. Andres, unpublished notes). Ant predation on the larvae of *Hyles euphorbiae* L. precluded establishment of this insect on *Euphorbia esula* and *E. cyparissias* at all but one site in Canada (Harris and Alex, 1971).

Goeden and Louda (1976) describe a number of other instances in which predators, parasites, and pathogens have been observed in association with weed-feeding arthropods. However, the affect of these organisms on the outcome of the project is not always obvious, and the authors note the difficulty and lack of experimentation in separating biotic interference from other factors precluding the establishment or increase of introduced weed control agents.

D. Habitat

The nature of the habitat can have some influence on the outcome of a natural enemy introduction. For example, Hall and Ehler (1979) showed that the rate of establishment of predators and parasites on oceanic islands (40%, n = 827) was significantly higher than that for continents (30.4%, n = 1468). The data in Table I also attest to this—i.e., data for Hawaii vs. data for the contiguous 48 states. Probably a major reason for the higher rate of establishment on oceanic islands involves climate. That is, these islands, by and large, are tropical (e.g., Hawaii, Bermuda, Fiji) and therefore have a more salubrious climate. Such a climate presumably enhances the establishment of introduced enemies compared to the more harsh, and highly seasonal, climates associated with most of the continental areas under consideration.

The stability of the habitat of the target insect pest can also influence the rate of establishment. According to Hall and Ehler (1979), the rate of establishment of natural enemies introduced into unstable environments (28%, n = 640), such as annual and other short-duration crops, was somewhat lower than that of enemies introduced into (1) habitats of intermediate stability (32.2, n = 916), such as orchard crops, and (2) highly stable environments (36.4%, n = 535), such as forests and rangelands.

In California, *Aphidius smithi* Sharma and Subba Rao, an imported parasite of pea aphid, *Acyrthosiphon pisum* (Harris), is severely affected by harvesting of hay alfalfa at approximately 30-day intervals; however, when hay alfalfa is strip-cut (cf. Stern *et al.*, 1964), a comparatively stable environment is created and the parasite is more effective in suppression of the aphid (van den Bosch *et al.*, 1966, 1967; van den Bosch and Stern, 1969). The importance of habitat stability in biological control must be emphasized. The available evidence indicates that certain kinds of predators and parasites are adapted to stable environments (e.g., orchards), whereas other kinds of natural enemies are adapted to unstable habitats (e.g., annual crops) (Force, 1972; Ehler, 1977; Ehler and Miller, 1978; Horn and Dowell, 1979; Bisabri-Ershadi and Ehler, 1981). Thus, it behooves practitioners of classical biological control to consider the nature of the habitat of the target pest when choosing natural enemies for importation.

With respect to weed control, any factor which drastically disturbs the pest–natural enemy habitat will alter the outcome of the project. Cultivation and harvesting contribute to the instability of the weed habitat and can disrupt developing weed-feeding biological control arthropods. This action is especially harmful to insects with prolonged life cycles (e.g., the univoltine *Rhinocyllus conicus* Froelich on musk thistle, *Carduus nutans* L.), as compared to those that complete their growth quickly (e.g., *Microlarinus* spp. weevils, which attack the annual *Tribulus terrestris* L.). Sheep-grazing a field of *Senecio jacaobaeae* L. destroyed the developing *Hylemya seneciella* Meade larvae in flower heads, precluding establishment of the fly at one site in California (R. B. Hawkes, personal communication). Similar catastrophic events (e.g., herbicide applications to plots, trampling of release sites by cattle, flooding) have taken their toll, adding to the percentage of biological control failures. These setbacks need be only temporary, but often, due to limited time and resources, researchers and cooperators alike are unable to make follow-up introductions until some years later.

The lack of some key ecological requisite in the habitat can also cause the failure of an introduced predator or parasite to become established. Such requisites could be as obscure as a symbiote (e.g., Hagen *et al.*, 1976a) or as commonplace as an alternate host (van den Bosch, 1968; Eikenbary and Rogers, 1973; Hoy, 1976). There is ample empirical evidence that the lack of these and other requisites has led to failures in establishing natural enemies, thus contributing to the low rate of establishment.

E. Competition

Interspecific competition among introduced natural enemies, leading to competitive exclusion, has apparently contributed to the low rate of establishment. DeBach (1965), and especially Turnbull (1967), suggested that introduced parasites and predators might be prevented from establishing due to competition from previously introduced incumbent species associated with the target pest. Van den Bosch (1968) vigorously refuted this hypothesis, particularly as put forth by Turnbull (1967); however, when adequate empirical data became available, Turnbull's claims were upheld. Ehler and Hall (1982) provided an empirical test of the competitive-exclusion hypothesis by comparing the rate of establishment for single-species releases (in the absence of incumbent species) with the rates of establishment for the following categories of multiple-species introductions (MSI):

1. Simultaneous MSI in the absence of exotic incumbent species
2. Sequential MSI involving either (a) single-species releases or (b) simultaneous release of two or more species, in each case in the presence of one or more exotic incumbent species

Their results consistently showed that the rate of establishment was inversely related to (1) the number of species released simultaneously and (2) the number of exotic incumbent species present. Ehler and Hall (1982) were also able to discount the possibility that lower establishment rates in the case of MSI were due to the release of fewer individuals per species. In other words, competitive exclusion of introduced natural enemies has probably occurred and contributed to the low rate of establishment.

Competition from other species of natural enemies can also lessen the impact of an exotic enemy. However, whether or not such competition can actually prevent the entire enemy complex from controlling the target pest has not been demonstrated under field conditions. Moreover, current empirical and theoretical evidence indicates that an increase in the number of species of natural enemies should lead to (1) a reduction in the amount of control effected by an individual species (due to competition) and (2) an increase in the amount of control by the aggregate of species (Ehler, 1982a). Nevertheless, competitive inhibition among established enemies which actually leads to lower levels of biological control should be considered a possibility which is worthy of further research. In addition, the possibility of competitive exclusion of an introduced natural enemy (either by species released simultaneously or incumbent species) which is actually capable of providing control of the target pest should also be investigated (see Ehler and Hall, 1982). It should not be assumed, *a priori,* that a natural enemy capable of controlling a target pest *will always become established* in the presence of exotic incumbent species and/or species released simultaneously.

It is not clear what role competitive exclusion has played in the outcome of weed control projects, but caution should be exercised when more than one arthropod species is to be introduced, particularly when these species attack the same plant structure. Concern over the potential importance of competition led Zwölfer (1973) to study interactions of the insects infesting the heads of *Carduus nutans* L. in Europe. He concluded that if the insects were to be used for biological control purposes, it would be best for the sequence of introduction to proceed first with the intrinsically inferior weevil, *Rhinocyllus conicus*. If this insect proved ineffective, the more intrinsically superior insects could follow without danger of competitive exclusion.

F. Genetics

Other possible explanations for the low rate of establishment of predators and parasites involve genetic phenomena. For example, it is known that the crosses between two strains of a given species can actually result in reduced fitness (compared to fitness of the parents) in the progeny. This hybrid dysgenesis (cf. Bregliano *et al.*, 1980) or negative heterosis (cf. Legner, 1972) could thus lower the fitness of individuals to be released in cases in which geographic races or strains of a particular species are allowed to interbreed in the laboratory (see also Force, 1967). Reduced fitness may also be brought about by prolonged culture in the laboratory (e.g., Legner, 1979). However, deterioration of stocks after prolonged culture need not necessarily be genetic in nature (cf. Jones *et al.*, 1978). Another possible explantion for failure to establish involves "inbreeding depression" (Sailer, 1978; see also Peters, 1977). According to Sailer (1978), a population of an introduced species may, because of inbreeding, become homozygous for certain deleterious recessive alleles; the attendant depression of fitness may then result in extinction of the population. At present, these genetic explanations for failures in establishment should best be regarded as hypotheses. To our knowledge, there has been no convincing validation of any of these hypotheses—i.e., in terms of actually causing the failure of an introduced biological control agent to become established. These and other aspects of genetics of colonization of natural enemies were discussed by Force (1967), Callan (1969), van den Bosch (1971), Messenger and van den Bosch (1971), Messenger *et al.* (1976), Mackauer (1976), and Hoy and McKelvey (1979).

Weed workers are also concerned with the genetic qualities of the arthropods introduced for biological control. In reviewing this question, Myers and Sabath (1981) noted that successful introductions have been achieved with organisms having both high and low levels of genetic variability. Further, if we are ever to resolve the question of genetic fitness, more consistent data-recording procedures must be followed in carrying out and reporting biological control introductions so that analyses can be made. The authors suggest that the best source of insects for biological control introductions are expanding populations either in areas where the insects are native or in other areas where they were formerly introduced. They state that "these insects will be large, healthy, have high fecundities and will be loaded with genetic and phenotypic variance. All of these characteristics should maximize the probability of successful establishment" (p. 99).

G. Host Tolerance

Establishment and buildup are but two of the keystones to successful biological control with immigrant natural enemies. The third is the capacity of the natural enemy to control, damage, or in some way stress the pest to the extent that it cannot maintain itself under existing environmental conditions. This phenomenon is especially relevant to biological control of weeds. For example, *Coleophora parthenica* Meyrick established well on *Salsola australis* R. Brown in California but apparently has had little or no impact on the pest plant. Larvae of *C. parthenica* tunnel the stems of *Salsola,* but the tunneling focuses in the pith and rarely destroys any of the vascular tissues (Pemberton, 1980). Similarly, *Leucoptera spartifoliella* larval tunneling on the new annual twigs of Scotch broom, *Cytisus scoparius,* causes blotching of the twigs, but it is not until the final larval instar (during March–April), just prior to blossoming, that some of the vascular tissues are severed. This damage has not yet proven significant in the control of Scotch broom in California (L. A. Andres, unpublished notes).

Obviously, in the biological control of weeds, the practitioner tries to select those insects which cause significant damage. In an attempt to systematize this selective process, Harris (1973a) listed and rated the desirable qualities to be sought out in a biological control arthropod. Although these guidelines have provided an interesting exercise, almost any type of feeding by adults or immatures can be significant if there is enough of it, if it occurs at the right time, and if it is complemented by other environmental stresses. To determine whether feeding is significant or not may require the actual sectioning of infested plants, examining the structures destroyed, and determining the roles these tissues play in the plant's life cycle.

H. Host Phenology

The timing and extent of feeding are as important as the type of feeding in the biological control of weeds. Plants go through a sequence of growth stages during which the role and relative importance of specific tissues and structures will change. For example, in most instances foliage is necessary for photosynthesis and the accrual of energy for growth, reproduction, and winter reserves. Defoliation by an arthropod when the plant's carbohydrate reserves are low is often more stressful to the plant than defoliation later in the cycle or at a time when the plant can recover or compensate for the foliar loss (Harris, 1973b).

Many of the arthropods used in weed control have only one generation a year. Thus, their proper synchrony of periods of activity and rest to the presence of particular host structures or stages is critical to the success of the project. The larvae of *Tyria jacobaeae* are synchronized to the mature *Senecio jacobaea* rosette and bolting flower stalk. In eastern Canada, where defoliation of the stems is supplemented by early frosts, plant death often results. In the West, however, the extended growing period, due to the absence of early frosts and an abundance of moisture, allows the plant to recover from *Tyria* defoliation with minimal impact (Harris *et al.*, 1978).

Lack of phenological synchrony between an insect pest and its enemy can also lessen the impact of the natural enemy. For example, the tachinid parasite *Hyperechteina aldrichi* Mesnil is fully synchronized with its host [*Popillia japonica* (Newman)] in the native home (Japan); however, in New Jersey, the adult flies emerge several weeks too early in the spring (vis-à-vis host emergence) and presumably die before peak host emergence. About 10% of the emerging hosts are eventually parasitized (Clausen, 1978). Alternate hosts can also be important in maintaining synchrony between hosts and enemies (e.g., Eikenbary and Rogers, 1973).

IV. SUMMARY AND CONCLUSION

The lessons we have learned in classical biological control of exotic insects and weeds should be of interest to those concerned with resolving (or preventing) immigrant pest problems. Host incompatibility has been a major factor in the failure of exotic, host-specific, weed-feeding arthropods to become established and/or flourish in North America. In this regard, the development and use of plant incompatibility in the form of plant resistance and diversity of crop varieties as a deterrent to establishment of exotic phytophagous pests should be explored. Predation by indigenous natural enemies has apparently precluded the establishment of exotic weed-feeding insects in several instances, suggesting that indigenous natural enemies—particularly general predators—should be encouraged to reduce the establishment of immigrant plant pests in those areas where exotic pests are frequently detected. Habitat instability has presumably hindered establishment of natural enemies of weeds and thus may be of value in reducing the establishment of potential exotic pests. For example, scheduled crop rotations could interrupt life cycles and prevent permanent establishment of an exotic herbivore. Of course, these suggested ecological actions should be of value not only against immigrant pests but against indigenous pests as well. In view of our inability to predict which exotic immigrants will become pests, any action that will close the environment to pest establishment and buildup will be helpful.

Despite the fact that biological control does work and has yielded complete and lasting control in a variety of instances, there are many aspects of this practice which require additional research. Foremost among these is the need for preintroduction investigations designed to determine the optimal release strategy for a given situation. In biological control of weeds, detailed and extensive preintroduction studies to ascertain the degree of host specificity of candidate natural enemies and the safety and wisdom of their introduction to a new area have long been standard operating procedures. Some attention has been directed toward assessing which of the weed natural enemies would be the most effective, but success in this area is still quite limited. In contrast, workers in biological control of arthropod pests have seldom (if ever) employed comparable preintroduction investigations (that is, apart from detecting and screening out obligate secondary parasites). In fact, the necessity and value of preintroduction investigations in classical biological control of arthropod pests is a controversial issue at present. There are many points of view on this subject; the two extremes are as follows:

1. Conduct thorough preintroduction investigations in order to determine the best species or combination of species to be released (the predictive approach). Proponents include Turnbull and Chant (1961), Watt (1965), Zwölfer (1971), Pschorn-Walcher (1977), Ehler (1982a,b), and Waage and Hassell (1982).
2. Release all suitable species of natural enemies with the hope that the best species or combination of species will be sorted out in the field (the empirical approach). Proponents and/or critics of the predictive method include Huffaker *et al.* (1971), Simmonds (1972), and van Lenteren (1980).

It should be noted that these authors do not necessarily endorse the wholesale use of either the empirical or the predictive method. It is our belief that, whereas the empirical method has produced a number of outstanding successes, it should continue to be practiced. However, following the suggestion of Myers and Sabath (1981), a systematized method of recording data is needed if we wish to develop a predictive model to improve the rates of establishment and success in biological control of weeds. With respect to insect control, the lack of an overall theoretical framework for carrying out preintroduction investigations is still one of the major impediments to the implementation of the predictive method (Ehler, 1982b). Let us hope that, by the year 2000, we will be well on our way to supplanting the empirical approach with the predictive one. With the proper theoretical tools, we can greatly increase both the rate of establishment and the rate of success in classical biological control of arthropod pests and weeds.

REFERENCES

Bartlett, B. R. (1961). The influence of ants upon parasites, predators, and scale insects. *Ann. Entomol. Soc. Am.* **54,** 543–551.

Bartlett, B. R., and van den Bosch, R. (1964). Foreign exploration for beneficial organisms. *In*

"Biological Control of Insect Pests and Weeds" (P. DeBach, ed.), pp. 283–304. Van Nostrand-Reinhold, Princeton, New Jersey.

Bay, E. C., Berg, C. O., Chapman, H. C., and Legner, E. F. (1976). Biological control of medical and veterinary pests. *In* "Theory and Practice of Biological Control" (C. B. Huffaker and P. S. Messenger, eds.), pp. 457–479. Academic Press, New York.

Beddington, J. R., and Hammond, P. S. (1977). On the dynamics of host-parasite-hyperparasite interactions. *J. Anim. Ecol.* **46,** 811–821.

Beirne, B. P. (1975). Biological control attempts by introductions against pest insects in the field in Canada. *Can. Entomol.* **107,** 225–236.

Bennett, F. D., Rosen, D., Cochereau, P., and Wood, B. J. (1976). Biological control of pests of tropical fruits and nuts. *In* "Theory and Practice of Biological Control" (C. B. Huffaker and P. S. Messenger, eds.), pp. 359–395. Academic Press, New York.

Bisabri-Ershadi, B., and Ehler, L. E. (1981). Natural biological control of western yellow-striped armyworm, *Spodoptera praefica* (Grote), in hay alfalfa in Northern California. *Hilgardia* **49**(5), 1–23.

Bregliano, J. C., Picard, G., Bucheton, A., Pelisson, A., Lavige, J. M., and L'Héritier, P. (1980). Hybrid dysgenesis in *Drosophila melanogaster. Science* **207,** 606–611.

Burdon, J. J., Marshall, D. R., and Groves, R. H. (1981). Aspects of weed biology important to biological control. *Proc. Int. Symp. Biol. Control Weeds, 5th, 1980* pp. 21–29.

Callan, E. McC. (1969). Ecology and insect colonization for biological control *Proc. Ecol. Soc. Aust.* **4,** 17–31.

Caltagirone, L. E. (1981). Landmark examples in classical biological control. *Annu. Rev. Entomol.* **26,** 213–232.

Caltagirone, L. E., and Huffaker, C. B. (1980). Benefits and risks of using predators and parasites for controlling pests. *Ecol. Bull.* **31,** pp. 103–109.

Carter, W. (1970). The dynamics of entomology. *Bull. Entomol. Soc. Am.* **16,** 181–185.

Clausen, C. P. (1956). Biological control of insect pests in the continental United States. *U.S. Dep. Agric., Tech. Bull.* **1139,** 1–151.

Clausen, C. P., ed. (1978). Introduced parasites and predators of arthropod pests and weeds: A world review. *U.S. Dep. Agric., Agric. Handb.* **480,** 1–545.

Commonwealth Institute of Biological Control (1971). Biological control programmes against insects and weeds in Canada, 1959–1968. *Commonw. Inst. Biol. Control, Tech. Commun.* **4,** 1–266.

DeBach, P., ed. (1964a). "Biological Control of Insect Pests and Weeds." Van Nostrand-Reinhold, Princeton, New Jersey.

DeBach, P. (1964b). Successes, trends, and future possibilities. *In* "Biological Control of Insect Pests and Weeds" (P. DeBach, ed.), pp. 673–713. Van Nostrand-Reinhold, Princeton, New Jersey.

DeBach, P. (1965). Some biological and ecological phenomena associated with colonizing entomophagous insects. *In* "The Genetics of Colonizing Species" (H. G. Baker and G. L. Stebbins, ed.), pp. 287–303. Academic Press, New York.

DeBach, P. (1971). The use of imported natural enemies in insect pest management ecology. *Proc. Tall Timbers Conf. Ecol. Anim. Control Habit. Manage.* **3,** 211–233.

DeBach, P. (1974). "Biological Control by Natural Enemies." Cambridge Univ. Press, London and New York.

DeBach, P., Fleschner, C. A., and Dietrick, E. J. (1951). A biological check method for evaluating the effectiveness of entomophagous insects. *J. Econ. Entomol.* **44,** 763–766.

Doutt, R. L., and DeBach, P. (1964). Some biological control concepts and questions. *In* "Biological Control of Insect Pests and Weeds" (P. DeBach, ed.), pp. 118–42. Van Nostrand-Reinhold, Princeton, New Jersey.

Doutt, R. L., Annecke, D. P., and Tremblay, E. (1976). Biology and host relationships of parasitoids. *In* "Theory and Practice of Biological Control" (C. B. Huffaker and P. S. Messenger, eds.), pp. 143–68. Academic Press, New York.

Dunn, P. H., and Radcliffe-Smith, A. (1980). The variability of leafy spurge (*Euphorbia* spp.) in the United States. *Res. Rep.—North Cent. Weed Control Conf.* **37,** 48–53.

East, R. (1974). Predation on the soil-dwelling stages of the winter moth at Wytham Woods, Berkshire. *J. Anim. Ecol.* **43,** 611–626.

Ehler, L. E. (1976). The relationship between theory and practice in biological control. *Bull. Entomol. Soc. Am.* **22,** 319–321.

Ehler, L. E. (1977). Natural enemies of cabbage looper on cotton in the San Joaquin Valley. *Hilgardia* **45,** 73–106.

Ehler, L. E. (1982a). Ecology of *Rhopalomyia californica* Felt (Diptera: Cecidomyiidae) and its parasites in an urban environment. *Hilgardia* **50**(1), 1–32.

Ehler, L. E. (1982b). Foreign exploration in California. *Environ. Entomol.* **11,** 525–530.

Ehler, L. E., and Hall, R. W. (1982). Evidence for competitive exclusion of introduced natural enemies in biological control. Environ. Entomol, **11,** 1–4.

Ehler, L. E., and Miller, J. C. (1978). Biological control in temporary agroecosystems. *Entomophaga* **23,** 207–212.

Eikenbary, R. D., and Rogers, C. E. (1973). Importance of alternate hosts in establishment of introduced parasites. *Proc. Tall Timbers Conf. Ecol. Anim. Control Habit. Manage.* **5,** 119–33.

Embree, D. G. (1971). The biological control of the winter moth in eastern Canada by introduced parasites. *In* "Biological Control" (C. B. Huffaker, ed.), pp. 217–226. Plenum, New York.

Force, D. C. (1967). Genetics in the colonization of natural enemies in biological control. *Ann. Entomol. Soc. Am.* **60,** 722–729.

Force, D. C. (1972). r- and K-strategists in endemic host-parasitoid communities. *Bull. Entomol. Soc. Am.* **18,** 135–137.

Frazer, B. D., and van den Bosch, R. (1973). Biological control of the walnut aphid in California: The interrelationship of the aphid and its parasite. *Environ. Entomol.* **2,** 561–568.

Goeden, R. D., and Louda, S. M. (1976). Biotic interference with insects imported for weed control. *Annu. Rev. Entomol.* **21,** 325–342.

Greathead, D. J. (1971). A review of biological control in the Ethiopian region. *Commonw. Inst. Biol. Control, Tech. Commun.* **5,** 1–162.

Greathead, D. J., ed. (1976). A review of biological control in western and southern Europe. *Commonw. Inst. Biol. Control, Tech. Commun.* **7,** 1–182.

Hagen, K. S., and Franz, J. M. (1973). A history of biological control. *In* "History of Entomology" (R. F. Smith, T. E. Mittler, and C. N. Smith, eds.), pp. 433–476. Annu. Rev., Inc., Palo Alto, California.

Hagen, K. S., Bombosch, S., and McMurtry, J. A. (1976a). The biology and impact of predators. *In* "Theory and Practice of Biological Control" (C. B. Huffaker and P. S. Messenger, eds.), pp. 93–142. Academic Press, New York.

Hagen, K. S., Viktorov, G. A., Yasumatsu, K., and Schuster, M. F. (1976b). Biological control of pests of range, forage and grain crops. *In* "Theory and Practice of Biological Control" (C. B. Huffaker and P. S. Messenger, eds.), pp. 397–442. Academic Press, New York.

Hall, R. W., and Ehler, L. E. (1979). Rate of establishment of natural enemies in classical biological control. *Bull. Entomol. Soc. Am.* **25,** 280–282.

Hall, R. W., Ehler, L. E., and Bisabri-Ershadi, B. (1980). Rate of success in classical biological control of arthropods. *Bull. Entomol. Soc. Am.* **26,** 111–114.

Harris, P. (1973a). The selection of effective agents for the biological control of weeds. *Can. Entomol.* **105,** 1495–1503.

Harris, P. (1973b). Weed vulnerability to damage by biological control agents. *Proc. Int. Symp. Biol. Control Weeds, 2nd, 1971* pp. 29–39.

Harris, P., and Alex, L. J. (1971). *Euphorbia esula* L., leafy spurge and *E. cyparissias* L., cypress spurge (Euphorbiaceae). *Commonw. Inst. Biol. Control, Tech. Commun.* **4,** 83–88.

Harris, P., and Peschken, D. (1971). *Hypericum perforatum* L., St. John's Wort (Hypericaceae). *Commun. Inst. Biol. Control., Tech. Commun.* **4,** 89–94.

Harris, P., Thompson, L. S., Wilkinson, A. T. S., and Neary, M. E. (1978). Reproductive biology of tansy ragwort, climate and biological control by the cinnabar moth in Canada. *Proc. Int. Symp. Biol. Control Weeds, 4th 1976* pp. 163–173.

Hassell, M. P. (1969a). A study of the mortality factors acting upon *Cyzenis albicans* (Fall.), a tachinid parasite of the winter moth *Operophtera brumata* (L.). *J. Anim. Ecol.* **38,** 329–339.

Hassell, M. P. (1969b). A population model for the interaction between *Cyzenis albicans* (Fall.) (Tachinidae) and *Operophtera brumata* (L.) (Geometridae) at Wytham, Berkshire. *J. Anim. Ecol.* **38,** 567–576.

Hassell, M. P. (1978). "The Dynamics of Arthropod Predator-Prey Systems." Princeton Univ. Press, Princeton, New Jersey.

Hassell, M. P. (1980). Foraging strategies, population models and biological control: A case study. *J. Anim. Ecol.* **49,** 603–628.

Horn, D. J., and Dowell, R. V. (1979). Parasitoid ecology and biological control in ephemeral crops. *In* "Analysis of Ecological Systems" (D. J. Horn, R. D. Mitchell, and G. R. Stairs, eds.), pp. 281–307. Ohio State Univ. Press, Columbus.

Hoy, M. A. (1976). Establishment of gypsy moth parasitoids in North America: An evaluation of possible reasons for establishment or non-establishment. *In* "Perspectives in Forest Entomology" (J. F. Anderson and H. K. Kaya, eds.), pp. 215–232. Academic Press, New York.

Hoy, M. A., and McKelvey, J. J., Jr., eds. (1979). "Genetics in Relation to Insect Management." Rockefeller Foundation, New York.

Huffaker, C. B. (1957). Fundamentals of biological control of weeds. *Hilgardia* **27,** 101–157.

Huffaker, C. B., Kennett, C. E. (1966). Studies of two parasites of olive scale, *Parlatoria oleae* (Colvée). IV. Biological control of *Parlatoria oleae* (Colvée) through the compensatory action of two introduced parasites. *Hilgardia* **37,** 283–335.

Huffaker, C. B., Kennett, C. E., and Finney, G. L. (1962). Biological control of olive scale, *Parlatoria oleae* (Colvée), in California by imported *Aphytis maculicornis* (Masi) (Hymenoptera: Aphelinidae). *Hilgardia* **32,** 541–636.

Huffaker, C. B., Messenger, P. S., and DeBach, P. (1971). The natural enemy component in natural control and the theory of biological control. *In* "Biological Control" (C. B. Huffaker, ed.), pp. 16–67. Plenum, New York.

Huffaker, C. B., Simmonds, F. J., and Laing, J. E. (1976). The theoretical and empirical basis of biological control. *In* "Theory and Practice of Biological Control" (C. B. Huffaker and P. S. Messenger, eds.), pp. 41–78. Academic Press, New York.

Jones, S. L., Kinzer, R. E., Bull, D. L., Ables, J. R., and Ridgway, R. L. (1978). Deterioration of *Chrysopa carnea* in mass culture. *Ann. Entomol. Soc. Am.* **71,** 160–162.

Julien, M., ed. (1982). "Biological Control of Weeds: A World Catalogue of Agents and Their Target Weeds." Commonw. Inst. Biol. Control, Slough, U. K.

Kennett, C. E., Huffaker, C. B., and Finney, G. L. (1966). Studies of two parasites of olive scale, *Parlatoria oleae* (Colvée). III. The role of an autoparasitic aphelinid, *Coccophagoides utilis* Doutt, in the control of *Parlatoria oleae* (Colvée). *Hilgardia* **37,** 255–282.

Laing, J. E., and Hamai, J. (1976). Biological control of insect pests and weeds by imported parasites, predators, and pathogens. *In* "Theory and Practice of Biological Control" (C. B. Huffaker and P. S. Messenger, eds.) pp. 685–743. Academic Press, New York.

Legner, E. F. (1972). Observations on hybridization and heterosis in parasitoids of synanthropic flies. *Ann. Entomol. Soc. Am.* **65,** 254–263.

Legner, E. F. (1979). Prolonged culture and inbreeding effects on reproductive rates of two pteromalid parasites of muscoid flies. *Ann. Entomol. Soc. Am.* **72,** 114–118.

Legner, E. F., Sjögren, R. D., and Hall, I. M. (1974). The biological control of medically important arthropods. *CRC Crit. Rev. Environ. Control* **4,** 85–113.

Luck, R. F. (1982). Parasitic insects introduced as biological control agents for arthropod pests. *In* "CRC Handbook of Pest Management in Agriculture" (D. Pimentel, ed.), Vol. II, pp. 125–284. CRC Press, Boca Raton, Florida.

Luck, R. F., Messenger, P. S., and Barbieri, J. (1981). The influence of hyperparasitism on the performance of biological control agents. *In* "The Role of Hyperparasitism in Biological Control: A Symposium" (D. Rosen, organizer), pp. 34–42. University of California, Div. Agric. Sci., Berkeley, California.

McGugan, B. M., and Coppel, H. C. (1962). A review of the biological control attempts against insects and weeds in Canada. Part II. Biological control of forest insects, 1910–1958. *Commonw. Inst. Biol. Control, Tech. Commun.* **2,** 35–216.

Mackauer, M. (1976). Genetic problems in the production of biological control agents. *Annu. Rev. Entomol.* **21,** 369–385.

McLeod, J. H. (1962). A review of the biological control attempts against insects and weeds in Canada. Part I. Biological control of pests of crops, fruit trees, ornamentals, and weeds in Canada up to 1959. *Commonw. Inst. Biol. Control, Tech. Commun.* **2,** 1–34.

MacPhee, A. W., Caltagirone, L. E., van de Vrie, M., and Collyer, E. (1976). Biological control of pests of temperate fruits and nuts. *In* "Theory and Practice of Biological Control" (C. B. Huffaker and P. S. Messenger, eds.), pp. 337–358. Academic Press, New York.

Maddox, D. M., and Rhyne, M. (1975). Effects of induced host-plant mineral deficiencies on attraction, feeding and fecundity of the alligatorweed flea beetle. *Environ. Entomol.* **4,** 682–686.

May, R. M., and Hassell, M. P. (1981). The dynamics of multiparasitoid-host interactions. *Am. Nat.* **117,** 234–261.

Messenger, P. S. (1970). Bioclimatic inputs to biological control and pest management programs. *In* "Concepts of Pest Management" (R. L. Rabb and F. E. Guthrie, eds.), pp. 84–99. North Carolina State Univ. Press, Raleigh.

Messenger, P. S. (1971). Climatic limitations to biological control. *Proc. Tall Timbers Conf. Ecol. Anim. Control Habit. Manage.* **3,** 97–114.

Messenger, P. S., and van den Bosch, R. (1971). The adaptability of introduced biological control agents. *In* "Biological Control" (C. B. Huffaker, ed.), pp. 68–92. Plenum, New York.

Messenger, P. S., Wilson, F., and Whitten, M. J. (1976). Variation, fitness, and adaptability of natural enemies *In* "Theory and Practice of Biological Control" (C. B. Huffaker and P. S. Messenger, eds.), pp. 209–231. Academic Press, New York.

Myers, J. H., and Sabath, M. D. (1981). Genetic and phenotypic variabiliy, genetic variance, and the success of establishment of insect introductions for the biological control of weeds. *Proc. Int. Symp. Biol. Control Weeds, 5th, 1980* pp. 91–102.

Nixon, G. E. J. (1951). "The Association of Ants with Aphids and Coccids." Commonwealth Institute of Entomology, London.

Ota, A., Mau, R. F. L., Funasaki, G., and Ikeda, J. (1981). A summary of classical biological control in Hawaii—up to 1978. (Unpublished manuscript.)

Pemberton, R. W. (1980). The impact of the feeding of *Coleophora parthenica* Meyrick (Lep: Coleophoridae) on the tissues, physiology and reproduction of its host, *Salsola australis* R. B. (Chenopodiaceae). Ph.D. Dissertation, University of California, Berkeley.

Peschken, D. (1979). Biological control of weeds in Canada with the aid of insects and nematodes. *Z. Angew. Entomol.* **88,** 1–16.

Peschken, D., Friesen, H. A., Tonks, N. V., and Banham, F. L. (1970). Releases of *Altica carduorum* (Chrysomelidae: Coleoptera) against the weed Canada thistle (*Cirsium arvense*) in Canada. *Can. Entomol.* **102,** 264–271.

Peters, T. M. (1977). The biology of invasions. *Minn. Agric. Exp. Stn., Tech. Bull.* **310,** pp. 56–72.

Pschorn-Walcher, H. (1977). Biological control of forest insects. *Annu. Rev. Entomol.* **22,** 1–22.

Rao, V. P., Ghani, M. A., Sankaran, T., and Mathur, K. C. (1971). A review of the biological control of insects and other pests in southeast Asia and the Pacific region. *Commonw. Inst. Biol. Control, Tech. Commun.* **6,** 1–149.

Rosen, D. (Organizer) (1981). "The Role of Hyperparasitism in Biological Control: A Symposium." University of California, Div. Agric. Sci., Berkeley, California.

Sailer, R. I. (1978). Our immigrant insect fauna. *Bull. Entomol. Soc. Am.* **24,** 3–11.

Salt, G., and van den Bosch, R. (1967). The defense reactions of three species of *Hypera* (Coleoptera, Curculionidae) to an ichneumon wasp. *J. Invertebr. Pathol.* **9,** 164–177.

Sands, D. P. A., and Harley, K. L. S. (1981). Importance of geographic variation in agents selected for biological control of weeds. *Proc. Int. Symp. Biol. Control Weeds, 5th, 1980* pp. 81–89.

Simmonds, F. J. (1972). Approaches to biological control problems. *Entomophaga* **17,** 251–264.

Sobhian, R., and Andres, L. A. (1978). The response of the skeletonweed gall midge, *Cystiphora schmidti* (Diptera: Cecidomyiidae), and gall mite, *Aceria chondrillae* (Eriophyidae) to North American strains of rush skeletonweed (*Chondrilla juncea*). *Environ. Entomol.* **7,** 506–508.

Stehr, F. W. (1974). Release, establishment and evaluation of parasites and predators. *Proc. Summer Inst. Biol. Control Plant Insects Dis., 1972* pp. 124–136.

Stern, V. M., van den Bosch, R., and Leigh, T. F. (1964). Strip cutting of alfalfa for lygus bug control. *Calif. Agric.* **18**(4), 5–6.

Sweetman, H. L. (1958). "The Principles of Biological Control." W. C. Brown, Dubuque, Iowa.

Turnbull, A. L. (1967). Population dynamics of exotic insects. *Bull. Entomol. Soc. Am.* **13,** 333–337.

Turnbull, A. L., and Chant, D. A. (1961). The practice and theory of biological control of insects in Canada. *Can. J. Zool.* **39,** 697–753.

Turnock, W. J., Taylor, K. L., Schröder, D., and Dahlsten, D. L. (1976). Biological control of pests of coniferous forests. *In* "Theory and Practice of Biological Control" (C. B. Huffaker and P. S. Messenger, eds.), pp. 239–311. Academic Press, New York.

van den Bosch, R. (1964). Encapsulation of the eggs of *Bathyplectes curculionis* (Thomson) (Hymenoptera: Ichneumonidae) in larvae of *Hypera brunneipennis* (Boheman) and *Hypera postica* (Gyllendal) (Coleoptera: Curculionidae). *J. Insect Pathol.* **6,** 343–367.

van den Bosch, R. (1968). Comments on population dynamics of exotic insects. *Bull. Entomol. Soc. Am.* **14,** 112–115.

van den Bosch, R. (1971). Biological control of insects. *Annu. Rev. Ecol. Syst.* **2,** 45–66.

van den Bosch, R., and Dietrick, E. J. (1959). The interrelationships of *Hypera brunneipennis* (Coleoptera: Curculionidae) and *Bathyplectes curculionis* (Hymenoptera: Ichneumonidae) in southern California. *Ann. Entomol. Soc. Am.* **52,** 609–616.

van den Bosch, R., and Stern, V. M. (1969). The effect of harvesting practices on insect populations in alfalfa. *Proc. Tall Timbers Conf. Ecol. Anim. Control Habit. Manage.* **1,** 47–54.

van den Bosch, R., Schlinger, E. I., Lagace, C. F., and Hall, J. C. (1966). Parasitization of *Acyrthosiphon pisum* by *Aphidius smithi,* a density-dependent process in nature (Homoptera: Aphidae) (Hymenoptera, Aphidiidae). *Ecology* **47,** 1049–1055.

van den Bosch, R., Lagace, C. F., and Stern, V. M. (1967). The interrelationship of the aphid, *Acythosiphon pisum,* and its parasite, *Aphidius smithi,* in a stable environment. *Ecology* **48,** 993–1000.

van den Bosch, R., Beingolea G., O., Hafez, M., and Falcon, L. A. (1976). Biological control of insect pests of row crops. *In* "Theory and Practice of Biological Control" (C. B. Huffaker and P. S. Messenger, eds.), pp. 443–456. Academic Press, New York.

van den Bosch, R., Hom, R., Matteson, P., Frazer, B. D., Messenger, P. S., and Davis, C. S.

(1979). Biological control of the walnut aphid in California: impact of the parasite, *Trioxys pallidus*. *Hilgardia* **47,** 1–13.

van Lenteren, J. C. (1980). Evaluation of control capabilities of natural enemies: Does art have to become science? *Neth. J. Zool.* **30,** 369–381.

Waage, J. K., and Hassell, M. P. (1982). Parasitoids as biological control agents—a fundamental approach. *Parasitology* **82,** 241–268.

Waters, W. E., Drooz, A. T., and Pschorn-Walcher, H. (1976). Biological control of pests of broad leaved forests and woodlands. *In* "Theory and Practice of Biological Control" (C. B. Huffaker and P. S. Messenger, eds.), pp. 313–336. Academic Press, New York.

Watt, K. E. F. (1965). Community stability and the strategy of biological control. *Can. Entomol.* **94,** 887–895.

Way, M. J. (1963). Mutualism between ants and honeydew-producing Homoptera. *Annu. Rev. Entomol.* **8,** 307–344.

Wilson, E. O. (1965). The challenge from related species. *In* "The Genetics of Colonizing Species" (H. G. Baker and G. L. Stebbins, eds.), pp. 7–24. Academic Press, New York.

Wilson, F. (1960). A review of the biological control of insects and weeds in Australia and Australian New Guinea. *Commonw. Inst. Biol. Control, Tech. Commun.* **1,** 1–102.

Wilson, F., and Huffaker, C. B. (1976). The philosophy, scope and importance of biological control. *In* "Theory and Practice of Biological Control" (C. B. Huffaker and P. S. Messenger, eds.), pp. 3–15. Academic Press, New York.

Zwölfer, H. (1971). The structure and effect of parasite complexes attacking phytophagous host insects. *In* "Dynamics of Populations" (P. J. den Boer and G. R. Gradwell, eds.), pp. 405–418. Cent. Agric. Public Doc., Wageningen.

Zwölfer, H. (1973). Competitive coexistence of phytophagous insects in the flower heads of *Carduus nutans* L. *Proc. Int. Symp. Biol. Control Weeds, 2nd, 1971* pp. 74–80.

Zwölfer, H., Ghani, M. A., and Rao, V. P. (1976). Foreign exploration and importation of natural enemies. *In* "Theory and Practice of Biological Control" (C. B. Huffaker and P. S. Messenger, eds.), pp. 189–207. Academic Press, New York.

CHAPTER 15

Prediction Capabilities for Potential Epidemics

R. D. SHRUM

Plant Disease Research Laboratory
Agricultural Research Service
United States Department of Agriculture
Frederick, Maryland

and

R. D. SCHEIN

Department of Plant Pathology
The Pennsylvania State University
University Park, Pennsylvania

ISBN 0-12-757880-3

I. INTRODUCTION

A. Approach

The theme of this volume, the danger posed by exotic pests and diseases, assumes there is indeed a threat to U.S. agriculture. History (Chapters 2–4) documents the danger. Still other threats exist (Chapters 5–7). Efforts to prevent their introduction are not totally successful (Chapters 9 and 10), so we do research (Chapters 11–13) on the justified assumption that if we knew more we could do more. As we face the future, we see opportunities (Chapter 14) and ways to protect ourselves (Chapters 16 and 17).

The questions which are left unanswered are the subjects of this chapter: Which threats are most likely to be imported, and how soon? If they are introduced, where will they be most troublesome and how much damage will they do? Such questions border on the imponderable. Prediction is still a risky business. And, one might ask, why predict? Why not just watch and wait? Those questions, too, must be addressed.

In reading this chapter, the reader must understand that we do not profess clairvoyance. In recent years, our field—plant disease epidemiology—has come a long way. A new methodology, or manner of approach to a problem, has been developed. We ask questions and make analyses that are different from those of the usual plant pathologist. The answers we get can be used in specially developed computer programs to yield predictive insights. As yet we are not very good, but there have been successes and the future looks encouraging. When we are done with this chapter, we will not have shown that we can at this moment provide an agricultural DEW line—an early warning system. What we hope to have shown is how we approach this rather nebulous problem and how we try to predict impact. We will have summarized the state of our art, shown its weaknesses and its strengths. Then the reader may ask the critical question: Is prediction worthwhile?

B. Importance of the Problem

At this point, the reader must have seen that many people think there is a serious problem. Chapters 5–7 make this abundantly clear. The threats exist, and some are imminent. Not one or two, but dozens of diseases seem poised to pack their bags for America and a perpetual feast on U.S. agriculture. Faced with such an armada, should we capitulate? Not without a fight. Machiavelli taught us to know our enemy, his strengths and his weaknesses, in order to construct an efficient offense.

C. Questions

The first question is, which pest will attack first? How do we identify major threats, those high epidemic potential diseases of major crops? Can ecologic analogs help us predict the probability of introduction? Can we predict impact, a probable range of epidemic potential, for the several kinds of likely pathogens causing both foliar and root diseases? Are there some diseases which may, though of slow progression, result after a period of years in major problems such as citrus decline?

If the answers to such questions are positive, then prediction has a value. It helps us prepare our defenses, plan our research, allocate our resources, strengthen quarantine, and carry out long-range breeding programs.

II. A TWO-PART PROBLEM

Two kinds of prediction are needed. The first task is to estimate the probability of introduction of the pathogen. Having thus established a list of likely emigrés, the more complex though probably less problematic task is to predict the impact of introduction. The two problems are quite different, and so are discussed separately.

A. Predicting Introduction

The McGregor group (1973) identified 551 specific pathogens which they believed would pose serious problems if introduced into the United States. They estimated that 13 of these would have very high impact and 36 more were of possibly high impact. They guessed that within 6 years from the time of their report, 44 of these pathogens would have made the trip. Of the 7 plant pathogens expected by 1976, none is known to have arrived by 1982. We must therefore apply McGregor's disclaimer: We probably do not know enough to make such predictions. Others have treated this topic in various ways (Mathys and Baker, 1980; Spaulding, 1961; McGregor, 1978; Thurston, 1973).

Most would agree that we must first identify major threats. That is the easiest part. Crop production statistics show that seven crops constitute 75% of our agricultural output. To these might be added a few specialty crops of very high local importance, e.g., mushrooms in Pennsylvania, mint in Oregon, strawberries in California, and tobacco in the Southeast. Next comes identification of highly destructive diseases or specific pathogen races of these same crops in other places. The third step is to evaluate the likelihood that the pathogens will be transferred. Rarely, transfer may not be a man-induced process, such as transport of spores of a particular race of rice blast fungus across the Gulf of Mexico from northern South America or Cuba. More often, transfer will be by human beings, so that one must look at interregional commerce. Imported primary agricultural produce poses a risk, but our history shows that nursery stock, bulbs, corms, seeds, and other living parts are much more likely to transfer a pathogen effectively. Travelers, too, may bring in uninvited pathogens, but as far as we know, there is no documented instance in which this has occurred.

To be transferred effectively, a pathogen must arrive in a viable state and in sufficient quantity to establish itself; hence, the efficiency of transfer on planting materials is a useful determinant of the potential for introduction. White pine blister rust arrived in America on infected planting stock; white rust of chrysanthemum invaded Europe via infected South African cuttings while still in the latent period; the disease and the pathogen were not visible, but the disease was later reported from the several places where the cuttings were bought. Chestnut blight, too, probably arrived on planting stock, but there is no documentation. We are still arguing about tobacco blue mold's introduction to Europe.

Survival is different for different pathogens. White pine blister rust would not have had such a high impact if its alternate hosts (*Ribes* spp.) had not been so common, for this fungus must pass from one host to another in order to survive. This is not so with many other pathogens, for instance, chestnut blight, autoecious rusts, or rusts that can survive in the uredial state.

Obviously, then, one can estimate in advance the likelihood of effective transfer if one knows the life cycle requirements of the pathogen and the most likely carriers.

There is still the question of the environment. As a generality, it is easy to assume that maize, cotton, or potato climates are similar, no matter where they occur. On the other hand, one may assume that different climates may preclude transfer.

Such generalities may not hold in terms of certain critical details. Potatoes of Andean origin flourish in mild, wet northern Europe and in Mediterranean and desert Israel, for example. It is irrigation in Israel that allows potato culture. Whether late blight disease is economically important in Israel depends on the availability of moisture for the critical germination and penetration phases, and this availability varies widely depending on the microclimate and the method of irrigation (Rotem and Palti, 1969). Estimates of the potential for successful transfer of a major crop pathogen to the United States can be evaluated only with knowledge of crop microclimate and pathogen requirements.

B. Predicting Impact

Predicting the impact of a potential pathogen is by far the larger, more complex task; all the same, it is a quantifiable task, whereas predicting introductions is less so. But impact assessment is, in the last analysis, a socioeconomic matter with which plant pathologists can deal only as part of an assessment team On the biological–environmental side is a network of interrelated functions and phenomena. What follows is a general description of how the plant pathologist approaches his part of the impact assessment process.

The objective is to develop the capability for a quick response to as yet unknown (unspecified) pathogens in as yet unknown (undefined) cropping regions. Until introductions can be predicted reliably, the key to developing impact predictions must be flexibility to whatever poses the most imminent threat.

Interactions

Obvious contributors to impact are the size of the crop and its value. The plant pathologist, in making an estimate of the potential loss to a crop from an absent pathogen, considers many other factors. It is helpful at first to divide these variables into three groups: (1) intrinsic (or genetic) capabilities of the host and the pathogen, (2) interactions of the crop and pathogen with the physical environment, and (3) effects of cultural practices. Ultimately, the interactions among elements of these three groups must be integrated into a dynamic model, which, with different values supplied for elements to "drive" it, will predict probable rates of disease. In final form, it should fix loss values for different terminal disease levels.

That is a big order, and accomplishing it is at present beyond our ability. The tools are still being invented. We will now describe an approach to this process of development so that the reader can assess the eventual likelihood of success in prediction.

a. Intrinsic Variables. Each host species has large genetic variability. Disease intensity in wild populations usually oscillates at rather low levels, with genetic diversity of the host population presenting an array of resistances to the also diverse pathogen population. Unlike modern cultivars, ancient landrace cereals were not pure, possessed considerable genetic diversity, and so gave disease protection which ensured species preservation over evolutionary time. Modern crop cultivars, and clonally propagated crops of whatever date of origin, have much less genetic resistance diversity. Metaphorically, they present an army which is specialized against a particular enemy which has limited weapons but is highly effective when any of these weapons are not resisted.

In predicting impact, we must consider a large number of variables and develop what military people call "scenarios"—a range of schematic descriptions of likely developments. The game is played with certain rules and boundaries. Once established, these must be acknowledged and updated continuously. We are our own umpires, and we must progress in light of new knowledge, whether it originates from conventional experiments with the organism and epidemics or with model gaming.

We must also acknowledge that pathogens are genetically quite plastic. A population of a pathogen will respond to pressures—host resistance, fungicides, cultural practices, and the physical and biological environments. We can expect that an exotic pathogen of a crop will have the genetic plasticity to attack the crop. We ourselves control the genetics of the hosts. Historically we have preferred pure cultivars and clonal propagation. As noted, both of these limit genetic diversity of resistance within a stand, a region, or a nation. Genetic uniformity is a weakness in our agricultural system; the word "vulnerability" is often used. Because natural populations, ancient landrace agriculture, and mixed cropping systems all have greater resistance diversity, some modern systems, such as multilines, attempt to mimic these older systems.

The pathogens, of course, have different kinds of life cycles; an exact understanding of these is critical. Of particular importance in considering exotic diseases is how the pathogen exists between cropping seasons. Can it withstand hot summers or cold winters? Does it have alternate obligate or nonobligate hosts, which may act as reservoirs of inoculum for the next season? Do we have plants which could act as alternate hosts, or can a pathogen survive as a saprophyte? All of these factors obviously have a potential which must be estimated in order to assess probable impact.

b. Environmental Variables. Most serious, economic plant diseases depend on weather in greater or lesser degree (Beaumont, 1947; Bourke, 1957; Waggoner, 1968). We can rather easily assess general weather and climatic similarities between areas from which a pathogen may be introduced and U.S. agricultural areas with the same crop. But whether or not the pathogen will be an economic threat depends on the microclimate, the environment of the phyllosphere and rhizosphere (Baker, 1966; Brooks, 1963; Waggoner, 1965).

c. Cultural Practices. Culture can modify these microenvironments. It is in the crop environment that epidemics occur, and to assess impact we must make estimates or measurements of that environment. Uniformity of cultivars, density of planting, and plant and crop structure all have their effects on temperature and humidity, light, soil moisture levels, and pathogen dissemination.

At the next larger scale, the seriousness of an epidemic may depend on the size of fields, their aggregation, the number of crops per year, or the length of the entire production season in the region where plantings are made at several times.

Finally, in regard to overseasoning and the inoculum that initiates an epidemic, we must consider the likely level of the initial inoculum and whether it can be easily reduced by eliminating weed hosts, by rotating crops, or by applying seed and soil treatments.

As a generality, we can say that agricultural intensification—striving for greater productivity and profit through use of pure lines, dense stands, fertilizers, short rotations, multiple crops per year, and monoculture—tends to make epidemics more severe. The impact of a potential pathogen is apt to be proportional to the intensity of the agricultural system, i.e., much greater in highly intensive than in more extensive types of agriculture. California farms, New York or Pennsylvania orchards, and Connecticut tobacco fields are more seriously threatened than mixed-crop dairy farms of Pennsylvania or New York.

C. What Does the Modeler Use for Data?

As will be seen in Section III a simulation model is driven by equations and computer algorithms which calculate the rates and ends of various subprocesses. Rates of change are, of course, regulated by the interactions of the host, pathogen, and environment. A model is supposed to have reality, i.e., it should tell us what can realistically be expected as controlling variables change. Reality comes, at least in part, from using good research results to drive and to constrain the model.

Do such data exist? Emphatically, yes! Are these data adequate? Not entirely. Models must be able to run on currently available data, but ideally, they need to be flexible so as to accommodate better data as they are reported. In this sense, models are never completed because we continue to learn things about the host and the pathogen—and some are important enough to be included as predictors.

The best available data on rates and end states come from well-designed experiments in greenhouses and growth chambers. More often than not, these data were not produced by epidemiologists. They come from good, standard, classic plant pathology; indeed, such data do not have to be new to be relevant. Modelers often "mine" the literature. The epidemiologist–modeler makes his extra contribution by stringing together data which quantify such subprocess responses to make an estimate of the success of a whole cycle or interacting chain of cycles.

A common, and sometimes just, criticism of such work has to do with reality. The experimental regimen, and therefore the conclusions, are not representative of farmers' fields. Growth chambers, with all variables other than the experimental one standardized, produce neat data largely because field variation and covariation are eliminated. The small numbers of plants in the experiments, their physiology, density, etc., are hardly like a field. One cannot assume that the process will act the same at the larger scale.

The modeler must be aware of all such problems; reliability and validity are addressed in Section III,C. In practice, he has used the most apt data available. Usually, he has been aware of the imprecision, unreality, disproportionate scale, absence of estimates of covariation, incompleteness, faulty resolution, etc. of such data. Recently, as the perspective of epidemiology has reached more researchers, data of greater acuity have been produced and modelers themselves have undertaken experiments to provide to-the-point data. Ultimately, the model is only as good as the data on which it is based (Waggoner and Horsfall, 1968).

The ultimate test of a quantitative model is its goodness of fit to the real or field situation. As time passes, modelers do two things: They use better data which give more realistic predictions, and they modify their models. This last item points to a special problem—what kind of model to use. As this is explained in the next section, a case will be made for generic over specific models, i.e., for flexible models which can be updated and adapted readily to a wide range of diseases. Can a single model or model type be used for a variety of pathogens and environments, and easily adapted to changing cultural practices or different management strategies? Given the relatively few practitioners and the large number of potential threats, such capability seems necessary if models are to serve the purposes addressed in this chapter.

III. MODELING

A. General Approach: Types of Models

Models are abstract representations of a real system that exists. The types of models are as varied as this definition is broad. Everyone is a modeler. Everyone carries a model of a sunset in his head. Every pathologist carries a mental image of a disease cycle or an epidemic. These are *mental* models. Each one is somewhat different from all others because each thinker has a different set of experiences and data from which to construct mental images. Each bit of research, each discovery, and each new finding changes one's perception of the system under study. To help communicate one's image to others, mental models are converted into word models, pictures, graphs, flow diagrams, drawings, and other types of *physical* models which help document an individual's perceptions of the concepts involved. Pathologists make physical models in the form of *qualitative* flow diagrams of disease cycles and *quantitative* graphs and tables of individual life cycle processes. Epidemiologists construct *mathematical formulae* and *computer program* models to document quantitative aspects of disease cycles and epidemics. Some of these quantitative models can be used to make predictions of system responses and epidemic development. Various types of decision aids will be discussed in this section.

Any system can be modeled from a number of perspectives, and each perspective by a number of approaches and types of models. The perspectives discussed in this chapter concern pest populations, with a view toward predicting crop damage potential. The specific objective is to predict pest spread and intensity. The scope in determining the approach and type of model to use is somewhat limited by constraints of available technology, data, and time—availability of environmental data in the crop, region, state, or nation where predictions are needed, quantitative data on the organisms involved, and time required for constructing the model.

B. Prediction Activity

Development of epidemics is determined by the environment, pathogen, suscept, and man's actions; but individual relationships are complex and, even when understood, are difficult to monitor in the field. Reliability and adequacy of prediction are limited by the least sufficient phase of the three-phase prediction activity: (1) developing fundamental knowledge of the biologic system under study (polishing one's mental model); (2) representing that knowledge by automated decision aids (constructing and integrating quantitative physical models); and (3) monitoring and otherwise measuring the individual biotic and abiotic factors which collectively propel the suscept-pathogen machinery (environmental driving functions). In developing each phase, there are built-in deficiencies. Even with diligent effort, mental models always less than fully represent real-world systems; physical models always less than fully represent mental models; and the measured environment (necessary as inputs for driving the constructed physical model) less than fully or accurately represents the total environment experienced by real organisms.

But the use of elaborate physical models to organize available information becomes more and more necessary as science progresses. As information accumulates, data must either be comprehensively summarized in a form which abstractly represents the real systems under study or left in a less well summarized form but put into minds large enough to keep both the overview of the system and the quantitative detail which accurately represents real system functioning. Quantitative system models put quantitative detail into an interactive system structure.

The process for learning about real-world phenomena which generates this modeling data is a stepwise progression of cycling between mental and various types of physical models. Once a phenomenon has been visualized, verbalized, and perhaps debated, hypothetical models are formulated and tested (hypothesis testing); these lead to new data. The newly garnered information leads to updated mental models, new hypotheses, and additional tests. As this reductionist cycle of scientific discovery continues, qualitative perceptions become more quantitative. As more hypotheses are tested and finer details of system function become known, data accumulate to the point where it becomes increasingly difficult to keep track of all the available details. The observer's mental accounting capabilities become swamped with information, and meaningful calculations on the integrated system become impossible. Attempting an unaided disease prediction involves a balance between developing a mental overview of the epidemic, which loses account of some of the quantitative detail for subsystem processes, and concentrating on known quantitative data for one or more of the parts, which soon loses the necessary overview. Thus, "number crunchers" and models become a necessity; with them, all the information can be utilized.

C. Judging Model Value

Before discussing what kinds of models to use, let us briefly develop some concepts and definitions for evaluating models. How do we judge whether a model has prediction value? When is a model useful, when is it reliable, and when is it adequate for the needs at hand? As illustrated in Fig. 1, for a model to be considered "useful" it must predict at least some system behavior correctly. Two other aspects of its value must be considered: (1) "reliability," the proportion of the model's predictions which match real-system behavior; (2) "adequacy," the proportion of real-system behavior which can be correctly predicted by the model. These are different, and the differences are important. This description requires further explanation and some examples.

Picture some real-world epidemic system and consider the properties which represent it. We may consider the color of the fungus mycelium, measure its diameter, count spore production, determine lesion numbers or size, measure physiological aberrations, evaluate the amount or condition of host tissue, and consider many other properties. Each such property describes some aspect of the real epidemic. However, when disease prediction is the important objective, only some of these properties are of direct importance. Now, let us consider Figs. 1 and 2. Let P in these diagrams represent the set of *all* system properties and let S be a *subset* of P for which we have some direct interest. In this case, let's arbitrarily choose the weekly disease severities (for the 20 weeks of May through September) as the system properties of direct interest (i.e., S, in this case represents the ratio of diseased to healthy tissue—severity—on a weekly basis). This is the subset of system properties that our model should correctly predict. In these diagrams the model's output, or predicted values, are represented by M; and Q represents the intersection of M and S, those predictions of the model (M) which are correct (i.e., in S). As illustrated in Fig. 1, there is a 20% overlap of M and S, so that the model has predicted four of the 20 weekly severities correctly. How valuable is that? Such a capability is not without usefulness, especially if the same values would not be expected otherwise. The model has predicted at least some of the desired values correctly, but it is not strictly valid because incorrect values are sometimes predicted as well. To determine its relative value, one must judge if the capability for correctly predicting four of 20 of the desired severities is sufficiently adequate for the needs at hand, and whether four of 20, as a success rate, is reliable enough to be depended on.

Adequacy is a measure of how many of the 20 desired severities are correctly

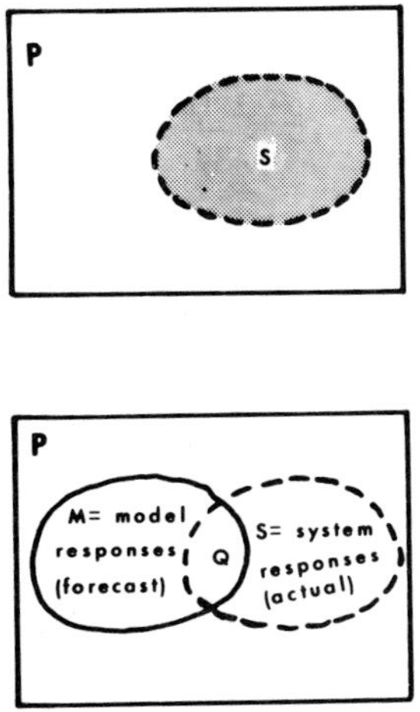

P = THE SET OF ALL PROPERTIES IN A SYSTEM (SOME REAL-WORLD SYSTEM)

S = THE CHOSEN SET OF OBSERVABLE PROPERTIES IN THAT REAL-WORLD SYSTEM

M = THE SET OF MODEL (SIMULATED) RESPONSES FOR THE CHOSEN SET (SEE ABOVE) WHEN THE MODEL IS SUBJECTED TO THE SAME CONDITIONS IMPOSED ON THE REAL SYSTEM.

Q = A SUBSET OF M AND OF S. THOSE MODEL RESPONSES THAT MATCH THE RESPONSES OF THE REAL SYSTEM (I.E., CORRECT PREDICTIONS.)

Useful= A useful model predicts some system behavior correctly

Validation= A valid model has no behavior which does not correspond to system behavior

Reliability = The proportion of model behavior which matches actual system behavior (i.e., Q/M)

Adequacy= The proportion of actual system behavior which can be forecast with the model (i.e., Q/S)

Fig. 1. Schematic to illustrate various aspects of model value. (From Mankin *et al.*, 1975.)

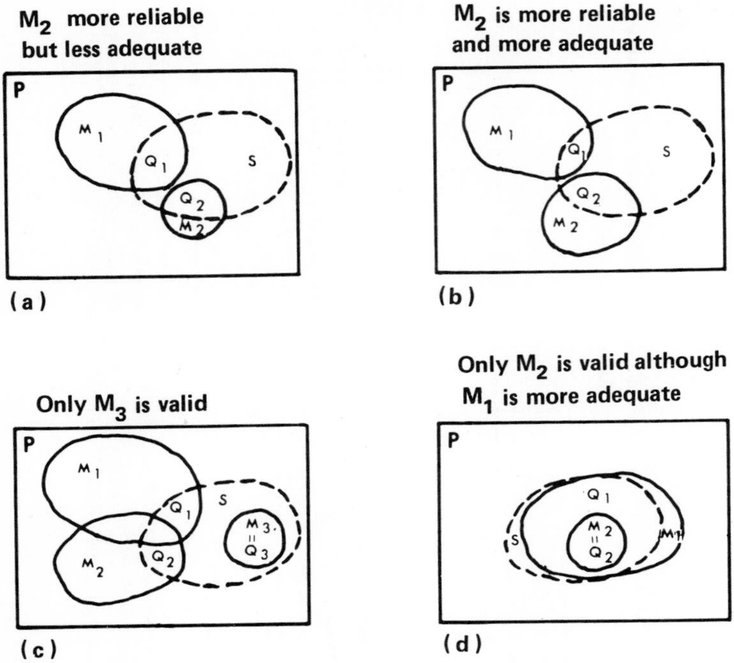

Fig. 2. Model adequacy, reliability, and validity. (From Mankin *et al.*, 1975).

forecast by the model (i.e., what proportion of S is covered by Q). Reliability, on the other hand, is a measure of how many of the model's predictions can be depended upon (i.e., what proportion of M is represented by correct predictions, Q). The diagrams of Fig. 2 show that even a useful model may be reliable but inadequate, or adequate but unreliable. The two are not necessarily maximized at the same time because they represent different prediction attributes. The nature of the application determines which is most important at any given time; thus, the value may differ for each of the model's different uses.

Figure 2a illustrates the predictions of competing models M_1 and M_2. Although model M_2 was designed to forecast only five of the desired 20 severities, the predictions of M_2 are correct a greater proportion of the time (it is more reliable); however, M_2 predicts fewer of the 20 severities than does M_1 (it is less adequate). For some needs, M_2 is superior because its predictions are more reliable; for others, M_1 is superior because it is capable of predicting more of the 20 desired values. Figure 2b shows model M_2 to be both more reliable and more adequate than M_1. In this case, M_2 is clearly superior to M_1. Figures 2c and 2d show some additional scenarios to illustrate the relationships among the qualities of adequacy, reliability, and validity. Figure 3 immediately stresses this new

ENVIRONMENTAL CONGRUITY = RELATIVE CHANGE IN THE MODEL'S RELIABILITY/ADEQUACY AS ENVIRONMENTAL DRIVING FUNCTIONS CHANGE (I.E., HOW DO Q/M AND Q/S VARY WITH ENVIRONMENTS FROM DIFFERENT YEARS AND LOCATIONS?)

TIME-COURSE CONGRUITY = RELATIVE CHANGE IN THE MODEL'S RELIABILITY/ADEQUACY AS THE TIME SCENARIO IS VARIED (I.E. HOW DO Q/M AND Q/S VARY WHEN, FOR EXAMPLE, THE HOST'S AGE WHEN ATTACKED, IS VARIED?)

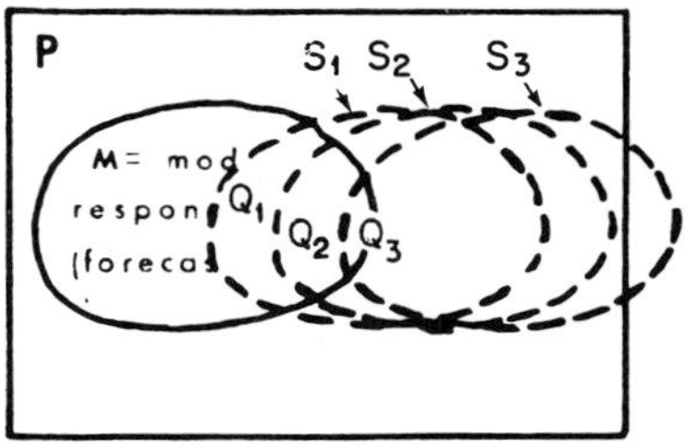

Fig. 3. The value of a model (its adequacy, reliability, and validity) will change when the modeled system itself is changed.

understanding by emphasizing that model reliability and adequacy may change from one set of environments (or one set of time scenarios) to another. There is no guarantee that an adequate or reliable prediction model will function adequately or reliably when (as is the topic of this text) the pathogen under scrutiny is introduced to a new climate with new cropping practices. This is as expected. It was pointed out in Chapter 6, that only 18% of the immigrant pests reported on by McGregor (1973) responded as they were expected to do. Some were unexpectedly destructive, and others were unexpectedly nondestructive in their new habitat. In a new habitat, it is possible that the modeled system itself has changed—different competitors, different day lengths, etc. The model needs to be adjusted to these new conditions.

Section III,E suggests some model design rationale to accommodate that eventuality, and to avoid the concern that in such a case one might have sacrificed the time and effort already devoted to constructing the prediction model. Some models cannot be adjusted.

D. What Kind of Model to Use

Model design sets model function, so the array of possible modeling objectives must be studied before we can attempt to establish a model's preferred design.

Early disease management decisions are important if explosive pathogens are to be controlled; thus, the modeling effort must be started in anticipation of the arrival of the pathogen. In the unusual situation posed by exotic pathogens, modeling in anticipation of an organism's actual arrival, there are many unknowns which cloud the modeling (or other prediction) effort. For which organisms, cropping systems, environments, cultivars, races, etc. will models be necessary? Which of hundreds of potentially destructive organisms will become established, and when and where might they attack first? The dilemma: When lead time is available for experimentation and model construction, the specific system in need of modeling remains unknown; and when the system is known (an introduction has occurred), prediction response time is very short.

Which models should be built in anticipation of an introduction; and if the wrong organisms and/or environments are guessed, is a quick response to those that are introduced precluded? If we correctly anticipate the organisms to be introduced, the problems of developing an adequate, reliable model are big enough; but with the large number of potential introductions and the difficulty of predicting which specific organism will enter first, or where, we must at least consider whether the modeling effort already expended would improve our capabilities for a timely response in the likely event that anticipatory guesses will sometimes be wrong. Can we prepare for unexpected introductions while still preparing for expected ones? We think so.

The choice of modeling approaches ranges from doing nothing to constructing individual holistic models, one for each potential introduction and environment. Both extremes are impractical, but constructing one or a few mechanistic models, each capable of being applied to any of a wide range of hosts, pathogens, environments, new information, etc. would maximize the response options while minimizing the modeling effort. The choice is between models of individual diseases, already prefit to current data for a specific disease and environment; or generalized models of epidemics, readily adjustable to the needs that become known once a specific introduction occurs or becomes imminent (Shrum, 1975a,b).

Mechanistic prediction aids (models) are normally data limited, and their prediction capabilities can be expanded as new information is gathered. Holistic ones are often model limited, so that adjustments to updated information are precluded except through model reconstruction—which takes times and sacrifices the data sets originally used to construct the model (Butt and Royle, 1974). When modeling is done in anticipation of the need, the choice between rigid and flexible models is especially important: (1) constructing a specific model involves the gamble that modeling lead time will not have been wasted by modeling the wrong disease situation and thus leaving us wholly unprepared to respond to the unanticipated introduction; (2) any model can be unreliable when used under system constraints or with environments which differ from the site where the original modeling data were derived (Fig. 3), and holistic models are inflexible to needed adaptations for accommodating factors such as new cultural practices, new varieties, different competitors, climate, etc.; and (3) the ability to make retrospective adjustments, to continue improving prediction as the epidemiologic data base expands, in order to meet the developing threat, is a compelling one.

E. How Genetic, Environmental, and Other Variables Are Estimated, Quantified, and Used in the Modeling Effort

Here we will summarize some basic matters which will allow the reader to understand what goes into a simulator of impact. We will discuss how a model can be constructed to represent the chain of events—the rates and states which together determine disease development—and illustrate how much of the necessary work can be completed before the identity of the target organism is fully known.

Details relating to this core of epidemiology are mostly beyond the scope and intent of this chapter. Readers are referred to standard works on the subject (van der Plank, 1963; Zadoks and Schein, 1979).

Systems analysts think in terms of processes and subprocesses. Any system can be divided into processes, and a process can be subdivided into its component subprocesses; usually, subprocesses can be further segmented. Plant disease is a state resulting from disease production processes. Physiologists, on the one hand, and epidemiologists, on the other, would use different words and different subprocesses to describe achievement of the same state. Quantitative epidemiologic simulation models are necessarily based on the epidemiologist's view, and epidemiology is normally focused on organism, population, and ecosystem processes.

Explosive plant disease, exemplified by rusts, mildews, and blights, develop each cropping season by repetitions of a cycle of events, the disease monocycle (Fig. 4).

A single effective spore leads to a lesion or pustule, which produces many spores. A small proportion of these new spores will initiate new cycles, each of which will culminate in a population of lesions producing many more spores. Such a chain of monocycles leads to exponential growth of the pathogen, and therefore of the disease. This process is summarized in Fig. 5.

When a disease goes through several cycles of the pathogen, this larger pro-

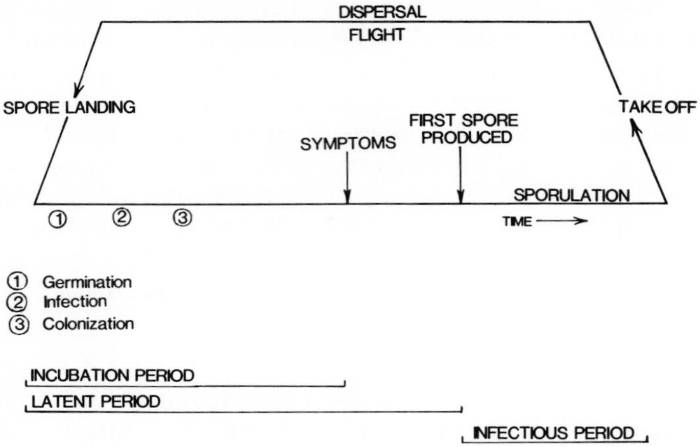

Fig. 4. Events, processes, subprocesses, and periods of the epidemic monocycle.

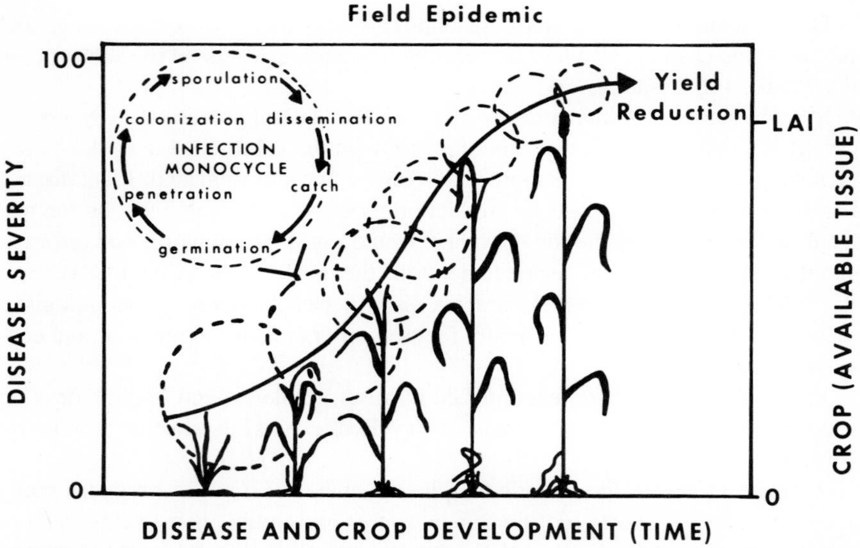

Fig. 5. Unequally overlapping, freely interacting monocycles of infection. The rate of disease increase is dependent on two things: (1) the rate of cycle turning, and (2) the amount of disease (number and size of infectious organisms) resulting from each turn. (Used with permission from Teng, 1981.)

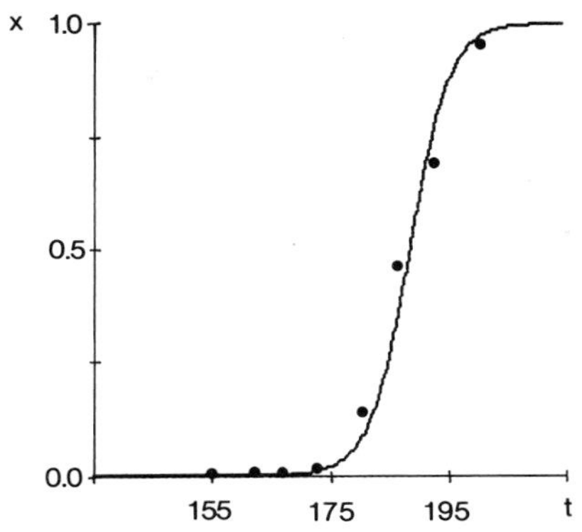

Fig. 6. Sigmoid (linear–linear) disease progress curve.

cess is called "polycyclic development." It is implicit in epidemiologic reasoning that the success of each cycle has an effect on the beginning state of the next one. Similarly, each subprocess within the monocycle affects each that follows it. In Fig. 4 we have labeled some of the important subprocesses, to which we will return in a moment in discussing components analysis.

Estimating and expressing disease progress in a population of hosts has for some years been done as follows: If disease intensity is measured several times during a season, a sigmoid curve (Fig. 6) would result. Disease is expressed as a proportion, x, of diseased tissue or of total numbers of plants affected.

Of course, as time passes, the host grows too, as is shown pictorially in Fig. 5. One convenient way to express host tissue increase for foliar pathogens is the leaf area index (LAI). Since x is the proportion of diseased tissue, the apparent growth rate of disease is slowed by increases in available host tissue. Apparent disease increase is also slowed by the fact that tissue, once infected, cannot be infected again. As the epidemic matures, its growth rate is slowed by the decreasing availability of uninfected tissue. The disease progress curve illustrated in Fig. 6 can be transformed to determine a rate of disease progress, r. The widely accepted equation of van der Plank (1963) for doing this is:

$$r = \frac{1}{t_2 - t_1}\left(\log_e \frac{x_2}{1 - x_2} - \log_e \frac{x_1}{1 - x_1} \right)$$

in which t_2 and t_1 are two times at which disease intensity is measured by some standard or experimental scale; x_2 and x_1 are the proportions of tissue diseased at those two times. The expressions $(1 - x_2)$ and $(1 - x_1)$ are corrections for the fact that, as more tissue is diseased, it becomes more difficult for a pathogen to find a new site to infect. The expression $\left(\log_e \frac{x}{1 - x} \right)$ is called a "logit";

with inexpensive calculators, it can be determined quickly; commonly, a value, logit x, is taken from a table of logits. In the latter case, r can be estimated by:

$$r = \frac{\text{logit } x_2 - \text{logit } x_1}{t_2 - t_1}$$

The dimensions of r are a relative rate, tissue diseased/total tissue per time unit. The time unit can be a day, week, month, season, or year. In explosive diseases, such as rusts, disease increase per day is the common measure. Common values are:

Phytophthora infestans	0.16–0.42 per day
Puccinia striiformis	0.10–0.27 per day
Swollen shoot virus, cacao	0.30 per year

An extension of this process is to make disease measurements at several times, convert x to logit x, and use linear regression to fit a line. The slope of the regression line is a weighted estimate of r. The result of either technique is a straight-line graph (Fig. 7), the slope of which is r, the rate of disease increase.

If we use these techniques to compare two isolates on the same cultivar, two cultivars against the same isolate, or the effects of two environments or two fungicide treatments, a graph such as Fig. 8 usually results.

The point to be made here is that r, the apparent infection rate, is an estimate of how fast a disease progresses and can be a predictor of final, or harvest, disease and of disease loss. It may be useful as a quick estimator of the potential impact of an introduced pathogen.

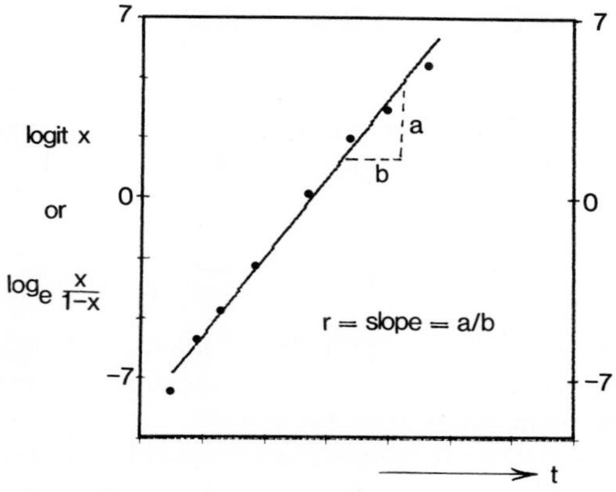

Fig. 7. The disease progress curve is straightened by plotting logit x against time.

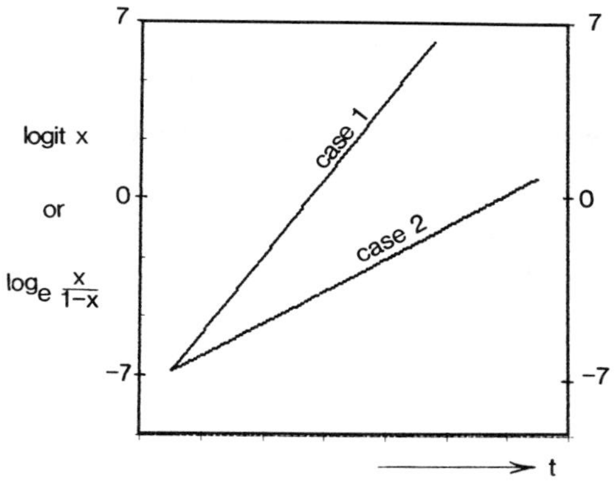

Fig. 8. Use of transformed disease progress curves to compare effects.

A critical question now arises: What controls *r?* For each subprocess of the disease cycle, different environmental variables affect "success" or "outcome," and so the rate of disease production in each monocycle is a result of how the whole sequence of processes fared and how the environmental counterforces were balanced (Fig. 9).

During a single season, an explosive disease goes through several cycles of the pathogen. This polycyclic development is a chain of cycles, the success of each monocycle affecting the production of the next. Each monocycle can be subdivided into states and subprocesses such as spore deposition, germination frac-

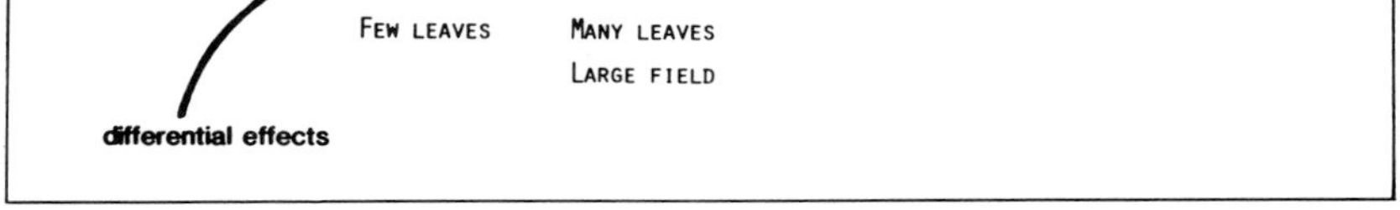

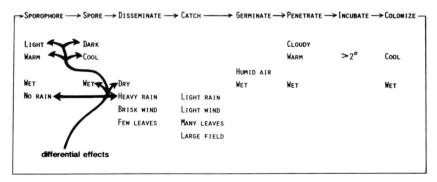

Fig. 9. With differential effects on sequential processes, the timing of environmental episodes is all-important. Thus, dry before wet weather may help a wet-weather pathogen if dissemination is the most limiting state in the sequence at that time. Wet before dry weather may be important if spore production is limiting. There is no way of making a reliable evaluation (prediction) of the influence of dryness without knowing the relative degree to which each in the sequence of states is limiting at the time of dryness, wetness, warmth, etc. The law of limiting factors applies (Blackman, 1905).

tion, infection success or efficiency, or sporulation; and the time needed to achieve a particular state varies with each turn of the cycle. We can measure or calculate the rate at which these successive states are achieved, as well as the success value of each state as affected by the host, pathogen, and environment. The cumulative product of these successes determines how much pathogen will be available for the next cycle. Analysis to determine to what degree different states influence epidemic dynamics is called "components analysis" (Zadoks, 1972; Zadoks and Schein, 1979).

Omitting much important detail (for which, see Zadoks and Schein, 1979), we can summarize these important aspects with the epidemiologic quintuplet, five quantified parameters with distinct dimensions:

- x_0: The amount of initial disease, equal to the amount of *effective* inoculum at the beginning of the season.
- p: The length of the latent period, or the time between penetration and the beginning of sporulation.
- N: The number of spores produced per unit area of host, or per lesion, per day.
- i: The length of the infectious or sporulation period. The product of number/day (N) times i is the total sporulation per lesion and can be extended to the total per leaf, per area, or per field.
- E: The effectiveness of the inoculum, i.e., what proportion of that produced actually causes new lesions in the next cycle. This can be divided into:
- P: The proportion of all inoculum that reaches susceptible sites and not nonhosts, the soil, etc., and
- Q: The proportion of P that actually brings about new lesions.

Each of these factors is affected by the host, pathogen, and environment. They are a basis of comparative epidemiology (Kranz, 1974b; Zadoks and Schein, 1979), and so can be the basic parameters used to estimate the potential impact of an introduced pathogen. Any interaction of the host, pathogen, and environment that shortens the latent period (p), lengthens the infectious period (i), or increases the rate of inoculum production (N), its effectiveness (E), or initial disease (x_0) will lead to a greater rate of disease development (r). Conversely, longer p, shorter i, or reduced N, E, or x_0 tend to reduce the rate of disease progress.

In simulation, to be discussed next, these and other parameters can be made to vary with the dynamic agroenvironment rather than remain constant. Differential equations involving three or more varying parameters are often difficult to construct, but computer technology provides for dynamic simulation which allows the dynamic interaction of rates and states of system processes to be more reflective of changing environments. Many variables and parameters can be varied simultaneously and/or sequentially as changing environments dictate.

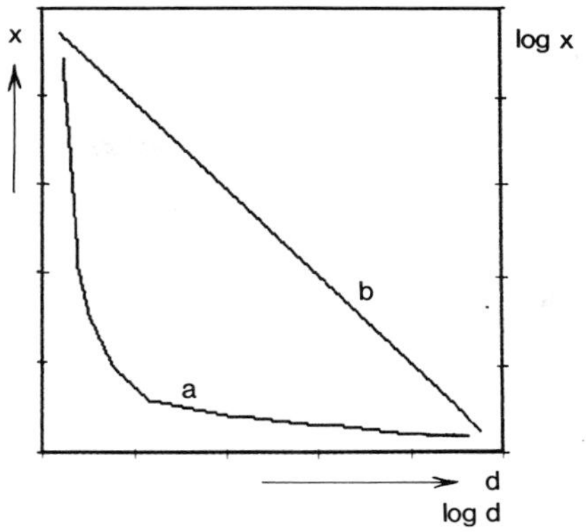

Fig. 10. Gradient of disease with distance: line a, linear; line b, log–log.

One other epidemiologic variable is the distance that disease is likely to spread, and the gradient of that spread from the source to its effective endpoint. A graph of spread usually takes the form of Fig. 10, line *a*, a negative exponential curve. It can be straightened by plotting the log of both the *x* and *y* axes (line *b*) (Gregory, 1968). The slope of that line is an estimate of the amount of disease likely to occur at specific distances, given particular wind speeds, spore sizes, and source strengths.

Presented in this way, complex though it is, epidemic analysis seems much more simple and straightforward than is really the case. Consider this single complication. A single spore can cause a lesion. When that lesion begins to sporulate, it may produce thousands of spores for days or even weeks. New infection, and therefore new cycles, can be initiated on any subsequent day when the environment is right. So, the chain of monocycles is not a chain of simple repetitions, but a very complex one of overlapping cycles. If the infectious period lasts, say, 21 days and the latent period is 7 days, then spores from the third cycle will be available for as long as 2 weeks before the second cycle is completed, by which time the third and fourth cycles may already be underway. This muddiness, which is beyond traditional computation, can be precisely accounted for by computer simulation.

The key idea is that a potential epidemic is viewed as a system which can be subdivided into comprehensible parts, often comparable to the parts and processes studied by the bench scientist. The system boundaries referred to earlier are chosen to correspond with known environmental influences on the epidemic, so that a system environment can be defined—the environment can influence the system, but the reverse is normally not true. The epidemic is defined by four types of factors (for a fuller discussion, see Zadoks and Schein, 1979):

1. A set of state variables which characterize (represent) the state of the system (how many spores, how much disease, how much host, etc.) at any time
2. A set of definitions governing the rates of transition from one state variable to the next according to the influence of changing time, distance, or environment (e.g., the number of germinated spores is a function of temperature, free moisture, etc. and available ungerminated spores)
3. A set of initial conditions to start the system working (how much host, how much inoculum, etc.)
4. A set of constraints to prevent unreality (never more than 100% of the potential or less than 0% of any factor calculated)

If the system state of interest, set 1, is X_t, the simulator must have equations or algorithms which calculate this state at any time. Set 2 from the above is computed by measuring the response of *p, i, N,* and *E* to relevant environmental factors. The initial conditions, set 3, involve x_0. The computer is programmed to cycle through a series of statements in order to calculate the state variables (set 1) in accordance with a transition matrix (set 2) and the constraints (set 4). When a computer cycle (solution interval) is completed, i.e., a discrete time interval has passed, the state of the system (set 1) has been updated to reflect the effect of environmental pressures during that solution interval. Environmental variables are changed to represent measured field conditions for the next solution interval, and transition rates (set 2) are updated accordingly. Then the cycle of calculations is repeated, the computer evaluating the change of disease with time and distance according to sequentially changing states of such factors as *p, i, N,* and *E,* the gradient of dispersal, and the growth of the host as environment and time exert their continuing influences.

F. Flexible Model

Figure 11 demonstrates a simplified form of one such model. It accepts various forms and types of epidemiologic data for foliar, polycyclic pathogens. Hopefully, the pathologist is not limited by the model—only by the available information and the difficult judgments on which of various conflicting data should be used and how. This generalized mechanistic model is designed to accommodate spatial, temporal, and environmental interactions, and thus allows broad flexibility to new data on disease cycle interactions, regardless of the form, or source and purpose, of data origination.

How is this an advantage? First, the generalized model can be used as a system to organize available data. Then, as additional data are accumulated or previous studies are refined, the model (this system of information) can be updated to accommodate this new knowledge. Intensified research is the common response

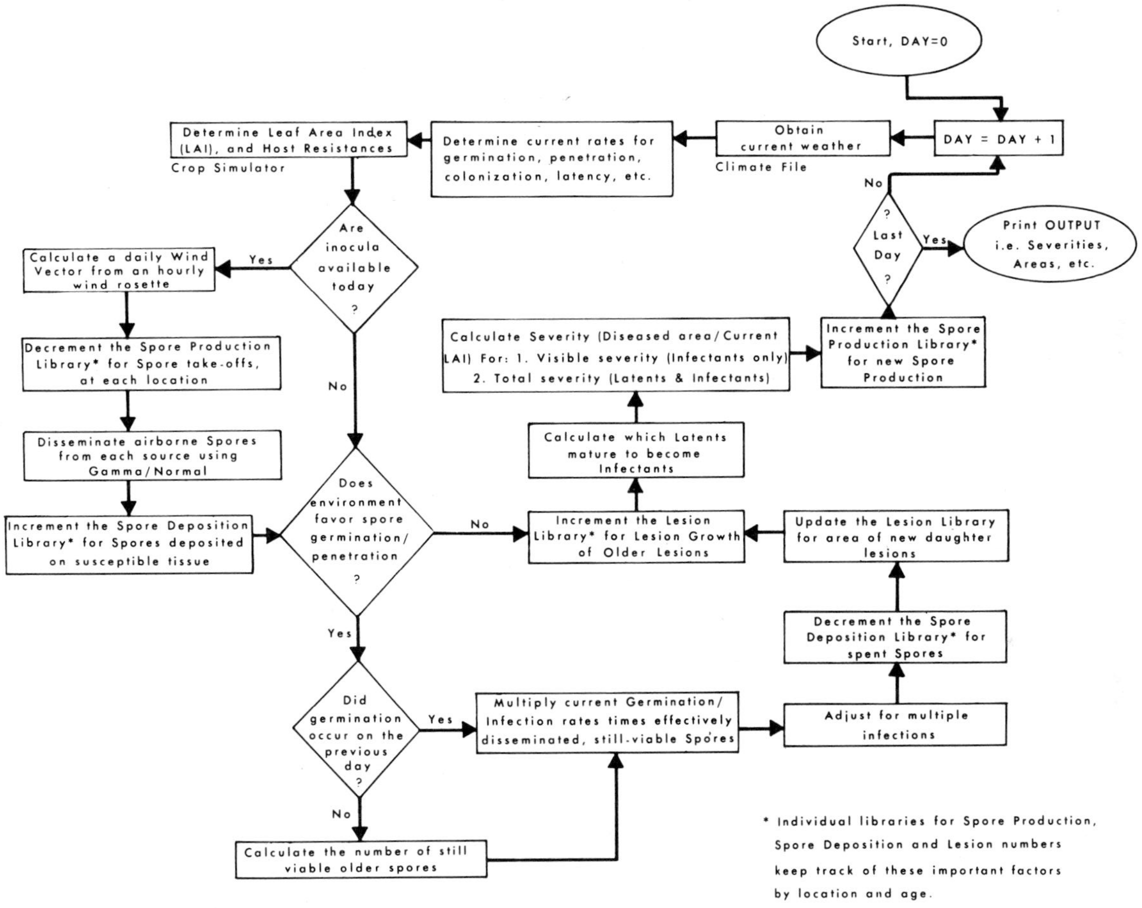

Fig. 11. Flow chart showing how a single flexible model can be used for a broad range of foliar pathogens.

once a potentially destructive introduction occurs or becomes imminent. If data do not currently exist for, say, colonization rate of the organism, a best estimate or data from a similar pathogen can be used until that deficiency is corrected by new research. When the new data is generated, it can be substituted without disrupting the remainder of the model. The same process can be used for any other epidemiologic parameters involved. This phase of modeling is discussed in Section III,B under developing fundamental knowledge. An additional advantage is that much of the necessary modeling work can be done in anticipation of the introduction. The development of physical models, discussed in Section III,B, requires much testing (debugging), even before specific data are applied to it, to ensure that the logical flow presumed to occur in the model is actually taking place. Just as epidemic systems have complex interrelationships, the simulator models which attempt to mimic them are complex and must undergo extensive testing of extreme conditions to ensure that unforeseen (unrealistic) magnifications or miniaturizations do not occur as the result of spurious calculations due to logic errors in the computer model. This phase of model testing can be carried out to any degree that judgment and time will allow. Thus, this developmental effort is valuable regardless to which disease the model is eventually applied. Dealing with unknowns has a likely potential for testing extremes. The more complete the testing for logical errors in the model structure, the shorter the modeling response time and the more likely that prediction deficiencies will result from shortfalls of the available data and not of the model itself.

Figure 12 shows some of the results from tests of the logic in the EPIDEMIC model. Here various biological and environmental conditions for stripe rust of wheat are under test. As an example, Fig. 12a shows model predictions for severity on the ordinate, time on the abscissa, and increasing levels of initial inoculum, INITIA (X0), as one proceeds into the plane of the paper. The environmental data driving the model were derived from measurements in an Oregon wheat field where the original severity data were collected. Figure 12b is the same as Fig. 12a, except that initial inoculum is constant at 10^4 and lesion growth rate (SIZINC) has been varied through the indicated range of values. Figure 12c shows the effects of large changes in spore production (KSPORE) as INITIA and SIZINC are kept constant at standard values. Figure 12d shows epidemic response when all conditions and biological parameters except temperature are at their standard or measured values. Progressing from front to rear are five epidemics for which each hourly temperature, originally measured in the wheat field, has, respectively, −7, −3, 0, +3, and +7°F added to it. Thus, the middle curve represents the Oregon wheat field measured, and the other curves represent cooler or warmer climes. Such gaming allows an evaluation of the model's realism. In this case, the high temperature responses appear to be unrealistically favorable for this cool-weather pathogen of wheat, so perhaps the high

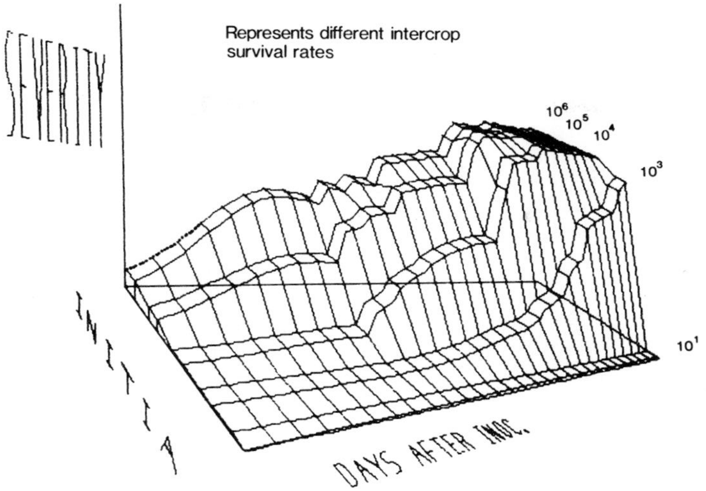

- a -

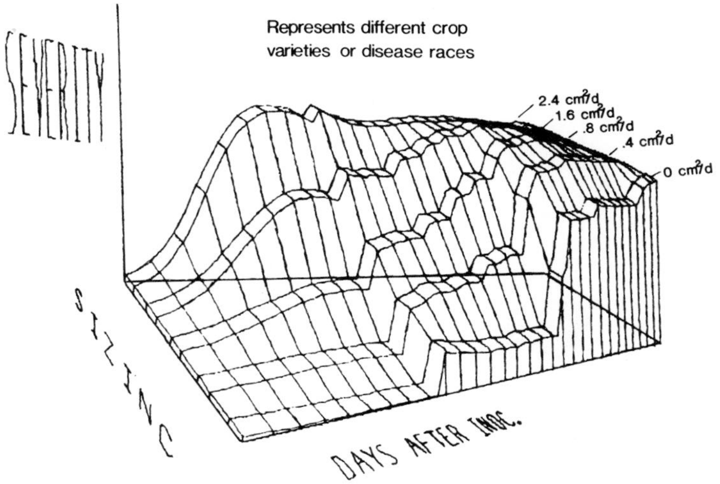

- b -

Fig. 12. Simulator gaming has two aspects: (1) to illustrate a flexible model's broad capabilities, and (2) to serve as an initial test of model realism (*continued*).

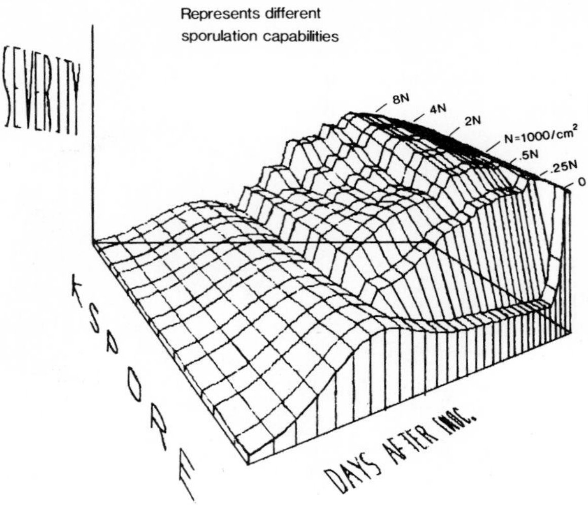

– c –

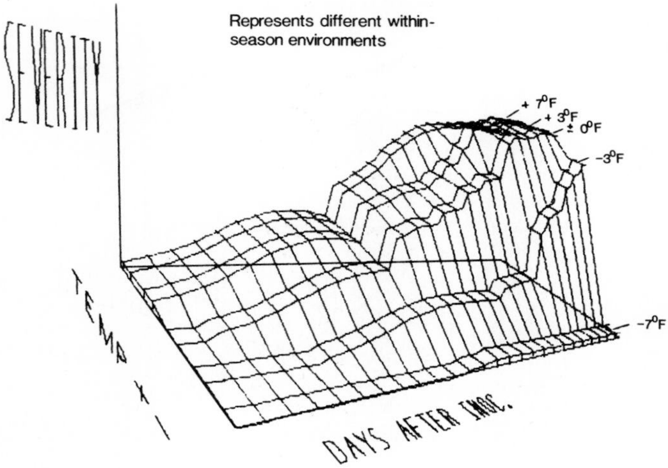

– d –

Fig. 12. *Continued*

temperature thresholds which normally stop stripe rust epidemics are not properly adjusted within the model. This can be checked (experimentally, if need be) and adjusted to order. The model can be tested through a full range of conditions for any of hundreds of potentially influential factors and its output evaluated for realism. When failures occur, they can be traced within the model and repaired even before the model needs to be used.

The response to anticipated introductions can be pretested by fitting the generalized model to the suspected target organisms and evaluating its prediction capabilities where the disease already occurs. If adjustments for new climate, new competitors, or new cultural practices are needed when the introduction actually occurs (as is illustrated in Fig. 3), these can be more simply and easily accommodated. Capability for spatial, temporal, and environmental flexibility within the model ensures that adaptability.

REFERENCES

Aldwinckle, H. S. (1975). Effect of leaf age and inoculum concentration on the symptoms produced by *Gymnosporangium juniperi-virginiane* on apple. *Ann. Appl. Biol.* **80,** 147–153.

Baker, D. N. (1966). The microclimate in the field. *Trans. Am. Soc. Agric. Eng.* **9,** 77–81.

Beaumont, A. (1947). The dependence on the weather of the dates of outbreak of potato blight epidemics. *Trans. Br. Mycol. Soc.* **31,** 45–53.

Blackman, F. F. (1965). Optimal and limiting factors. *Am. Bot. (London)* **19,** 281–295.

Bourke, P. M. A. (1957). The use of synoptic weather maps in potato blight epidemiology. *Ir. Dep. Ind. Commun. Mater. Serv., Tech. Note* **23,** 1–35.

Brooks, F. A. (1963). Biometerological data interpretation to describe the physical micro-climate. *Phytopathology* **53,** 1203–1209.

Burgess, L. W., and Griffin, D. M. (1967). The influence of diurnal temperature fluctuations on the growth of fungi. *New Phytol.***67,** 131–137.

Butt, D. J., and Royle, D. J. (1974). Multiple regression analysis in epidemiology of plant diseases. *In* "Epidemics of Plant Diseases: Mathematical Analysis and Modeling" (J. Kranz, ed.), pp. 78–114. Springer-Verlag, Berlin and New York.

Corneli, E. (1932). On mildew warnings in central Italy. *Riv. Patol. Veg.* **22,** 1–9.

de Wit, C. T., and Goudrian, J. (1978). "Simulation of Ecological Processes," 2nd ed. Wiley, New York.

Ellingboe, A. H. (1968). Inoculum production and infection of foliage pathogens. *Annu. Rev. Phytopathol.* **6,** 317–330.

Eyal, Z., Clifford, B. C., and Caldwell, R. M. (1968). A settling tower for quantitative inoculation of leaf blades of mature small grain plants with uredospores. *Phytopathology* **58,** 530.

Great Britain Ministry of Agriculture, Fish and Food (1976). "Plant Pathology Manual of Plant Growth Stages and Disease Assessment Keys." Middlesex, England.

Gregory, P. H. (1968). Interpreting plant disease dispersal gradients. *Annu. Rev. Phytopathol.* **6,** 189–212.

Hastings, A. (1981). Population dynamics in patchy environments. Modeling and differential equations in biology. *Lect. Notes Pure Appl. Math.* **58,** 217–224.

Hirst, J. M., and Stedman, O. J. (1956). The effects of height of observation in forecasting potato blight by Beaumont's method. *Plant Pathol.* **5,** 135–140.

Hyre, R. A. (1957). Forecasting downy mildew of lima beans. *Plant Dis. Rep.* **41,** 7–9.

Jowett, D., Haning, B. C., and Browning, J. A. (1973). Nonlinear disease progress curves. *Int. Cong. Plant Pathol., 2nd, 1973* Abstract 0244.

Kiyosawa, S. (1972). Mathematical studies on the curve of disease increase. *Ann. Phytopathol. Soc. Jpn.* **38,** 30–40.

Klinkowski, M. (1971). Epidemics and pandemics of plant pathogenic disease agent and its relation to men. *Math. Naturwiss. Kl.* **51**(4), 49–53.

Kranz, J., ed. (1974a). "Epidemics of Plant Diseases: Mathematical Analysis and Modeling." Springer-Verlag, Berlin and New York.

Kranz, J. (1974b). Comparison of epidemics. *Annu. Rev. Phytopathol.* **12,** 355–374.

Kranz, J., Mogh, M., and Stumpf, A. (1973). EPIVEN: Ein simulator fur apfelschorf. *Z. Pflanzenkr. Pflanzenschutz* **80,** 181–187.

Krause, R. A. (1974). Blitecast: A computerized forecast of potato late blight—Implementation and evaluation. *In* "Modeling for Pest Management: Concepts, Techniques, and Applications," pp. 187–189.

Krause, R. A., and Massie, L. B. (1975). Predictive systems modern approaches to disease control. *Annu. Rev. Phytopathol.* **13,** 31.

Leben, C., and Daft, G. C. (1968). Cucumber anthracnose: Influence of nightly wetting of leaves on number of lesions. *Phytopathology* **58,** 264–265.

McCoy, R. E. (1972). Epidemiology of chrysanthemum *Ascochyta* blight. Ph.D. Thesis, Cornell University, Ithaca, New York.

McDonald, G. I., Hoff, R. J., and Wykoff, W. R. (1981). Computer simulation of white pine blister rust epidemics. I. Model formulation. *USDA For. Serv. Res. Pap. INT* **INT-258,** 1–136.

McGregor, R. C. (1973). "The Emigrant Pests (Report to the Administrator, Animal and Plant Health Inspection Service)." U.S. Dept. Agric., Washington, D.C.

McGregor, R. C. (1978). People-placed pathogens: The emigrant pests. *In* "Plant Disease: An Advanced Treatise" (J. G. Horsfall and E. B. Cowling, eds.), Vol. 2, pp. 383–396. Academic Press, New York.

Mankin, J. B., O'Neill, R. V., Shugart, H. H., and Rust, B. W. (1975). The importance of validation in ecosystem analysis. *In* "New Directions in the Analysis of Ecological Systems" (G. S. Innis, ed.), Simul. Counc. Proc. Ser., Vol. 5, Part I, pp. 63–71. Soc. Comput. Simul., La Jolla, California.

Massie, L. B. (1973). Modeling and simulation of sourthern corn leaf blight disease caused by race T. of *Helminthosporium maydis* Nisik & Miyake. Ph. D. Thesis, Pennsylvania State University, University Park.

Mathys, G., and Baker, E. A. (1980). An appraisal of the effectiveness of quarantines.*Annu. Rev. Phytopathol.* **18,** 85–101.

May, R. M., and Oster, G. F. (1976). Bifurcations and dynamic complexity in simple ecological models. *Am. Nat.* **110,** 573–599.

Minkevich, I. I. (1976). Selection and utilization of disease forecasting models for plant protection. *In* "Modeling for Pest Management: Concepts, Techniques, and Applications" (R. L. Tummala, D. L. Haynes, and B. A. Croft, eds.), pp. 171–175. Michigan State Univ. Press, East Lansing.

Platt, R. B., and Griffiths, J. F. (1964). "Environmental Measurement and Interpretation." Van Nostrand-Reinhold, Princeton, New Jersey.

Raeuber, A. (1957). Investigation of the dependence of potato blight on weather with reference to a *Phytophthora* warning service. *Abh. Meteorol. Hydrol. Dionstes D.D.R.* **6,** 38.

Rafila, C. (1962). A new device for automatic recording of meteorological factors determining infection of cultivated plants by parasitic fungi. *Culegere Lucr. Meteorol. Inst. Meteorol. Hidrol.* pp. 405–413.

Rapilly, F., and Jolivet, E. (1976). Construction d'un modèle (EPISEPT) permettant la simulation d'une epidemic de *Septoria nodorum* sur blé. *Rev. Stat. Appl.* **24,** 31–60.

Richter, J., and Haussermann, R. (1975). An electronic scab warning apparatus. *Hortic. Abstr.* **46,** 5395.

Rotem, J., and Palti, J. (1969). Irrigation and plant diseases. *Annu. Rev. Phytopathol.* **7,** 267–288.

Rotem, J., Palti, J., and Lomas, J. (1971). Epidemiology and forecasting of downy mildews and allied fungi in an arid climate with and without irrigation. *Spec. Publ.—Agric. Res. Organ., Volcani Cent. (Bet Dagan, Isr.).*

Scarpa, M. J., and Raniere, L. C. (1964). The use of consecutive hourly dewpoints in forecasting downy mildew of lima beans. *Plant Dis. Rep.* **48,** 77–81.

Schein, R. D. (1964). Design, performance, and use of a quantitative inoculator. *Phytopathology* **54,** 509–513.

Schrodter, H., and Ullrich, J. (1966). Investigations on the biometeorology and epidemiology of *Phytophthora infestans* (Mont.) de Bry. on a mathematical-statistical basis. *Phytopathol. Z.* **54,** 87–103.

Schurer, K., and Van der Wal, A. F. (1972). An electronic leaf wetness recorder. *Neth. J. Plant Pathol.* **78,** 29–32.

Shrum, R. D. (1975a). Simulation of stripe rust of wheat using EPIDEMIC-a flexible plant disease simulator. *Prog. Rep.—Pa., Agric. Exp. Stn.,* **347,** 1–81.

Shrum, R. D. (1975b). EPIDEMIC-A flexible plant disease simulator Ph.D. Thesis, Pennsylvania State University, University Park.

Skellam, J. G. (1971). Some philosophical aspects of mathematical modeling in empirical science with special reference to ecology. *In* "Mathematical Models in Ecology" (J. N. R. Jeffers, ed.), pp. 13–28. Blackwell, Oxford.

Smith, R. S. (1964). Effect of diurnal temperature fluctuations on linear growth rate of *Macrophomina phaseoli* in culture. *Phytopathology* **54,** 849–852.

Spaulding, P. (1961). Foreign diseases of forest trees of the world. *U.S. Dept. Agric., Agric. Handb.* **197.**

Strand, M. A., and Roth, L. F. (1976). Simulation model for spread and intensification of western dwarf mistletoe in thinned stands of ponderosa pine saplings. *Phytopathology* **66,** 888–895.

Tanner, C. B. (1963). Plant temperatures. *Agron. J.* **55,** 210–211.

Teng, P. (1981). University of Minnesota, St. Paul, Minnesota (personal communication).

Thurston, H. D. (1973). Threatening plant diseases. *Annu. Rev. Phytopathol.* **11,** 27–52.

Van Arsdel, E. P., Tullis, E. C., and Panzer, J. D. (1958). Movement of air in a rice paddy as indicated by colored smoke. *Plant Dis. Rep.* **42,** 721–725.

van der Plank, J. E. (1963). "Plant Diseases: Epidemics and Control." Academic Press, New York.

Virginia Polytechnic Institute (1981). "General Description and Operation of the Agro-environmental System: Crop Management Modeling." Virginia Polytechnic Inst., Blacksburg.

Waggoner, P. E. (1965). Microclimate and plant disease. *Annu. Rev. Phytopathol.* **3,** 103–26.

Waggoner, P. E. (1968). Weather and the rise and fall of fungi. *In* "Biometeorology" (W. P. Lowry, ed.), pp. 45–66. Oregon State Univ. Press, Corvallis.

Waggoner, P. E., and Horsfall, J. G. (1968). Assessing an epidemic by computer. *Proc. Natl. Acad. Sci. U.S.A.* **61,** 1157–1158.

Waggoner, P. E., and Horsfall, J. G. (1969). EPIDEM: A simulator of plant disease written for a computer. *Bull.—Conn. Agric. Exp. Stn., New Haven* **698.**

Waggoner, P. E., and Parlange, J.-Y. (1975). Slowing of spore germination with changes between moderately warm and cool temperatures. *Phytopathology* **65,** 551–553.

Waggoner, P. E., Horsfall, J. G., and Lukens, R. J. (1972). EPIMAY. A simulator of southern corn leaf blight. *Bull.—Conn. Agric. Exp. Stn., New Haven* **729,** 1–84.

Wallin, J. R., and Shaw, R. H. (1953). Studies of temperature and humidity at various levels in crop cover with special reference to plant disease development. *Iowa State Coll. J. Sci.* **28,** 261–267.

Wang, J. Y. (1981). A computerized weather monitoring unit for farm operation. *Interciencia* **6,** 254–257.

Yarwood, C. E. (1972). Dark therapy of bean rust. *Phytopathology* **62,** 1139–1140.

Zadoks, J. C. (1972). Methodology of epidemiological research. *Annu. Rev. Phytopathol.* **10,** 253–276.

Zadoks, J. C., and Schein, R. D. (1979). ''Epidemiology and Plant Disease Management.'' Oxford Univ. Press, London and New York.

Zadoks, J. C., Klomp, A. O., and Van Hoogstraten, S. D. (1969). Smoke puffs as models for the study of spore dispersal in and above a cereal crop. *Neth. J. Plant Pathol.* **75,** 229–232.

CHAPTER 16

Buying Insurance against Exotic Plant Pathogens

M. D. SIMONS

Agricultural Research Service
United States Department of Agriculture
and Department of Plant Pathology
Iowa State University
Ames, Iowa

and

J. A. BROWNING

Department of Plant Sciences
Texas A&M University
College Station, Texas

I. INTRODUCTION

Other chapters in this volume have dealt with cataloging exotic pathogens and insects, stopping them at our shores, predicting their behavior, or eradicating them should they achieve a beachhead. This chapter considers what we might do

EXOTIC PLANT PESTS AND
NORTH AMERICAN AGRICULTURE

ISBN 0-12-757880-3

to insure ourselves against severe damage by the exotic disease-causing organism that invades our territory and then cannot be eradicated. The well-documented examples of once exotic pathogens that have become established in North America, as presented by Horsfall in Chapter 1, leave no doubt that contending with well-established exotic pathogens constitutes an area of great importance.

McGregor (1978) spelled out the magnitude of the problems involved in attempting to intercept plant pathogens under current conditions. The vast amounts of baggage entering this country every year from all areas of the world simply cannot be inspected adequately by available personnel. The situation is almost as bad for commercial cargo. Statistics strongly suggest that our inspection efforts over the years have done little or nothing to reduce the rate of introduction of exotic pathogens. The point, for our purposes, is that, as we seem unable to exclude exotic pathogens from our country, we should seriously consider the subject of insurance against severe losses from those that will inevitably enter and become established. We obviously need all the insurance we can afford. This insurance could take several forms, and a discussion of it could be organized in various ways. Here we will consider the subject under the categories of diversity, types of resistance, and utilization of information on diseases in their present geographic ranges.

II. NATURAL DIVERSITY AND DISEASE LOSS

In the centers of origin of hosts and pathogens in nature, host–pathogen systems coevolved over geologic time to their present states of equilibrium in which the parasites do not exterminate their hosts and the hosts have not developed resistance mechanisms sufficiently effective to eliminate the pathogens; both survive—in fact, belong—in the ecosystem. Furthermore, there is evidence that the hosts ordinarily suffer only minor damage in many host–pathogen systems in natural populations.

Harlan (1956) made this generalization clearly a generation ago. He commented that one would expect that where diseases are severe, the resistant plants in the population would have a very strong selective advantage and would increase in frequency. Population studies in barley and alfalfa, however, indicate that this is often not the case. Resistance seems to be maintained in the population at a low frequency in both the presence and the absence of the pathogen. Van der Plank (1968) also observed that after prolonged coevolution, neither highly virulent races nor highly resistant hosts dominate the populations. In fact, there are even some hosts with few or no resistance genes and some races of the pathogens with few or no virulence genes.

This observation was confirmed in the wild populations of small grain progenitors in Israel (Browning, 1974; Segal *et al.*, 1980). In carefully conducted studies there with natural populations of the wild oat *Avena sterilis* and its fungal parasite *Puccinia coronata,* it was found that resistant genotypes in the host did not predominate in the population, nor did susceptible genotypes drop out (Wahl, 1970). From the standpoint of the host, both virulent and avirulent races were maintained in the fungus population. Dinoor (1969) included the function of the alternate host, *Rhamnus palestina,* and observed that indigenous populations of the fungus and of its two hosts were both variable and stable. Variability in the gramineous host was reflected in that of the parasite, and a diverse population of both was thus preserved.

Segal *et al.* (1980) recently reviewed the literature on disease in several natural ecosystems. A number of such systems have been studied in sufficient detail to permit meaningful conclusions. These include, in addition to the cereals in Israel, the rusts (*Puccinia* spp.) and powdery mildew (*Erysiphe graminis*) on *Triticum* species in the Caucasus; *Helminthosporium* and rusts (*Puccinia* spp.) on *Tripsacum,* teosinte, and maize in Mexico and Central America; rusts (*Puccinia* spp.) on the wild grasses of the Great Plains of North America; late blight (*Phytophthora infestans*) on *Solanum* species in Mexico; *Venturia* on wild pears in central Asia; and mildews (*Plasmopora viticola* and *Uncinula hecator*) on wild grapes in the eastern United States. In all these natural systems, the general picture is the same. Disease organisms are present in a diversity of forms, but the hosts, which are also present in diversity, are very seldom damaged significantly.

In a discussion of the relationship of diversity and diseases in forest ecosystems, Schmidt (1978) noted that epidemics are not infrequent in forest ecosystems, but that they are limited in time and space. They are thus limited not by diversity as such, but by *functional* diversity. The host components of such diversity include specific and general resistance, tolerance, number of species, horizontal and vertical dispersion of host plants, etc. Pathogen components include latent and infectious periods, genetic potential for infection, dependence on insects, etc.

The foregoing and other work pertaining to the relevance of naturally occurring host–pathogen systems to practical plant disease control was summarized and interpreted by Browning (1974) and Segal *et al.* (1980). Susceptible host plants and avirulent pathogen races are not eliminated from natural populations because selection pressure on both host and pathogen is minimized. This low level of selection pressure is the result of the buffering action of different types of resistance in the host population and of homeostatic tendencies in the pathogen population. These populations interact in an environment that includes diverse plant species intermixed as integral parts of the same natural ecosystem. A more detailed discussion of these protective factors will be presented later in the chapter.

III. WHEN DIVERSITY IS LACKING

Although great emphasis has recently been placed on the potential dangers of a lack of diversity in our cultivated crops, such concern is by no means new. Writing a decade before the American Revolution of a severe epidemic of wheat stem rust (*Puccinia graminis*) in Italy, Tozzetti (1767, p. 24) said, "It is something worth pondering, that in this Calamitous Year, Sowings of Rye only, or of Segalato, this is to say of Wheat and Rye, were immune from rust, and I have understood that in the Valdinievole, those who had sown Segalati, had a very beautiful Crop, in which the Wheat was the finest to be seen in Tuscany. The same thing happened in the Vecciati, that is to say, Wheat sown along with Vetch. It is not so easy to render a reason, why Wheat growing seeded with Rye, or with Vetch, was not damaged by the rust, while a Field of Wheat alone, standing between one of Rye, and one of Vetch, yielded scarcely any seed, and that the most miserable."

Writing in the 1930s, Hartley (1939, p. 9) noted that forest pathologists have "long viewed with alarm the planting of pure stands of single species." He believed that genetic uniformity of the host would favor the buildup of specialized strains of parasites, and to support this view, he cited practical experience with such clonal cultures as Lombardy poplar avenues, rubber plantations, fruit trees, roses, potatoes, bananas, sugarcane, and creeping bentgrasses. He proposed that planting should consist of mixtures of desirable clones rather than of blocks of a single clone. Such mixtures would be better able to adapt themselves to disease attack than the pure-clone plantings.

Stevens (1948) also considered the potential for damage from disease in clonal and self-pollinating crops. Like the authors of the present chapter, he was interested in defense measures that could be employed as insurance against losses caused by new pathogens. He accepted as fact the proposition that any particular strain of a pathogen is likely to spread less rapidly when in contact with a heterogeneous host where some plants are resistant than in homogenous stands of plants. Any such new pathogen, he believed, would be less destructive in a mixed stand.

Browning and Frey (1969) established the philosophical basis for diversity and reviewed the literature to that time. A contemporary summarization of the historical relationships between uniformity of agricultural crops and the principal aspects of the major technological revolution in farming methods that has taken place in the last century has been provided by Marshall (1977). This revolution has led to a dramatic increase in the level and efficiency of agricultural production. It also has led to an equally dramatic decrease in the amount of diversity in the agricultural plants we depend on. This decrease has occurred at all levels of diversity. First, the number of species used by man as major sources of food has decreased greatly. At the same time, the spatial distribution of the remaining species has changed. In primitive agriculture, different crops were often grown in multispecific mixtures. Modern agriculture is perceived to require that such mixtures give way to monocultures. The loss of diversity within individual crops has been even more complete than at the specific level. Before the advent of modern methods, there was great diversity both within and among primitive crop varieties or land races. These have been replaced by a relatively small number of widely adapted, genetically homogeneous cultivars. In extreme, but by no means rare, cases many millions of acres have been planted to varieties all plants of which are genetically identical in disease resistance.

Following the great southern corn leaf blight pandemic in the United States in 1970, the potential danger inherent in genetic uniformity of important crop plants was the subject of much interest and discussion. According to a detailed report by the National Academy of Sciences (Anonymous, 1972), genetic uniformity very likely played a crucial role in many of the great classical epidemics of history. These would include the Panama wilt (*Fusarium oxysporum*) of bananas in the Caribbean in the early 1900s, the great potato blight (*Phytophthora infestans*) epidemic in Ireland in the 1840s, and certainly the wheat rust (*Puccinia graminis*) epidemics of the twentieth century in North America. So great was the danger to man inherent in the genetic uniformity of crops that Harlan (1971) published a paper appropriately entitled "The Genetics of Disaster," and Browning (1971) editorialized on "Corn, Wheat, Rice, Man: Endangered Species."

An examination of selected epidemics will now be undertaken to illustrate the relationship between genetic homogeneity and vulnerability to damage by plant disease organisms. In the United States, oats has the dubious distinction of furnishing back-to-back examples of such epidemics. The crown rust (*P. coronata*) disease of oats has been recognized as a potentially serious production problem since pioneer times. Hence, it was not surprising that the discovery that the Victoria oat, introduced from South America into the United States in the late 1920s, was an excellent source of resistance to the crown rust fungus was looked upon as the answer to the crown rust problem. All that was required was to transfer the gene for crown rust resistance from the unadapted Victoria oat to oat varieties adapted to different areas of the United States. This was accomplished, and the new resistant oat varieties were accepted enthusiastically by farmers. It was estimated that by 1945, 90% of the oat acreage of the Corn Belt and 75% of the oat acreage of the entire United States was sown to oats carrying this single gene for crown rust resistance (Murphy, 1952). At about the same time, a new disease of oats caused by a previously unnoticed, unidentified fungus, *Helminthosporium victoriae,* was discovered. This fungus was highly destructive to all oats carrying the Victoria gene for resistance to the crown rust fungus, but was innocuous on oats lacking this gene. It was estimated (Pady *et al.*, 1947) that oats with the Victoria gene were planted on an astonishing total of 35 million acres in the United States in 1946. Also in 1946, the Victoria blight disease caused severe losses in many of the major oat-producing areas of the country. Iowa and Illinois, two of the largest oat-producing states at that time, suffered losses estimated at 25% and 20%, respectively.

Acreage of oat varieties carrying the Victoria gene declined as rapidly as it had increased, with Victoria oats being replaced by varieties that carried a gene for crown rust resistance derived from the Bond variety, which had been introduced from Australia. These varieties came to dominate the scene as rapidly as the Victoria varieties had a few years before. By 1951 (Murphy, 1952), most of the oat acreage in the United States was planted to varieties that carried the Bond gene for crown rust resistance. In Iowa a single variety, Clinton, representative of this group, made up over 90% of the acreage in 1948. It had been known for some years that crown rust races capable of parasitizing the Bond-type resistance existed in the United States, but they were rare. However, these previously rare races increased in prevalence in synchrony with the increase in popularity of Bond varieties with growers. As a result, they caused estimated losses in Iowa of 18% of the crop in 1950, 20% in 1951, and 30% in 1953 (Sherf, 1954).

Another striking example of the potentially disastrous consequences of extreme genetic uniformity is furnished by the southern corn leaf blight (*Helminthosporium maydis*) epidemic of 1970, as mentioned earlier in this section. A special issue of the *Plant Disease Reporter* was devoted to describing and documenting this epidemic (Anonymous, 1970). In the years before 1970, hybrid seed corn producers had developed a technique for producing seed that involved the use of a cytoplasmic male sterility factor rather than manually detasseling the female plants. By 1970, about 85% of the hybrid seed corn produced in the United States was produced on female plants having this single, uniform type of cytoplasm (Ullstrup, 1972). Prior to this time, southern corn leaf blight had been regarded as a relatively unimportant disease of corn. In 1970, however, a new race of the pathogen that had recently appeared, and that was highly pathogenic to corn with the male-sterile cytoplasm, caused severe damage throughout much of the United States. In the South, many fields were a total loss. In some regions of the Corn Belt, yields were reduced by 50% and test weights were reduced to 45–50 pounds per bushel. Losses in two of the important corn-growing states in the Corn Belt, Illinois and Indiana, were 20–30% statewide. Losses nationwide were estimated at nearly $1 billion, representing a loss of food many times greater than occurred in the devastating potato blight epidemics in Ireland in the 1840s. Ullstrup (1972, p. 46) concluded from his analysis of the southern corn leaf blight pandemic that "Never again should a major cultivated species be molded into such uniformity that it is so universally vulnerable to attack by a pathogen, an insect, or environmental stress. *Diversity must be maintained in both the genetic and cytoplasmic constitution of all important crop species*" (author's emphasis).

A more recent example of the potential vulnerability of large areas planted to genetically uniform crops occurred in Sonora, Mexico, in the 1976–77 growing season. About 75% of the wheat acreage of this important wheat-growing area was planted to the single pure-line variety, Jupateco 73. A combination of circumstances led to relatively severe leaf rust (*Puccinia recondita*) infection on very young plants, clearly pointing to a disastrous epidemic later in the season. The disaster was averted, at great cost, by heroic efforts to obtain and apply eradicant and systematic fungicides. Controlled experiments using fungicides on commercial field-size plots showed that yield losses were excess of 50% in some cases (Dubin and Torres, 1981).

Other examples of severe epidemics could be cited, but the three just discussed illustrate the critical relationship between genetic homogeneity and potential vulnerability to severe disease damage. In these and other cases, the pattern is the same, at least for field crops. First, large acreages are planted to genetically homogeneous varieties. The plants on such acreages then serve as ideal, highly selective substrates for the increase of any pathogenic form capable of parasitizing the particular host genotype involved. Under such favorable conditions, the reproductive potential of most highly epidemic plant pathogens is so great (more than 50% per day) that, in a very short time, virulent strains can increase from being minor constituents of the pathogen population to a position of domination and destruction.

The natural phenomenon of "resistant" plants allowing only virulent strains of a pathogen to reproduce has been a part of agriculture from earliest times, but only with the growing of pure, homogeneous stands—made possible with weed control, pure-line selection, crossing to incorporate single resistance genes, and using these varieties over wide areas—did it become a serious problem. Thus, in a very real sense, the above-cited epidemics were all caused by *man;* they did not occur in natural ecosystems or in primitive agricultural systems, but only in modern, intensive, Western-type agriculture following man's manipulation of genetic material and cultural practices. Johnson (1961) coined the term "man-guided evolution" to describe the increase of pathogenic forms in response to changing host varieties. He concluded that in a typical oligogenically protected, unstable cereal monoculture, the lack of host variability is reflected in the parasite. This severe limitation in the variability of the parasite population maximizes its destructiveness to the crop.

To keep a sense of balance, it seems appropriate at this point to mention that Way (1979) has sounded a note of caution regarding indiscriminant reliance on diversity per se as a solution to all problems of disease and pest control. He noted, for example, that increasing diversity can sometimes increase rather than decrease pest problems. Also, robust monocultures do sometimes exist in nature and thus cannot be regarded as purely man-made phenomena. Although playing the role of devil's advocate, Way did not minimize the potential importance of diversity as a fundamental principle of crop protection, but believed that each crop and cropping system must be viewed as a unique situation. The quality, he believed, and not the amount of diversity was what mattered. This suggests Schmidt's (1978) *functional* resistance, mentioned earlier.

If diversity per se does not necessarily insure a crop against severe disease loss, what of the converse: Does a *lack* of diversity always mean vulnerability to loss? More than a decade ago, before the southern corn leaf blight epidemic and before the wheat leaf rust scare in Mexico, Watson (1970) reviewed the problem of changing virulence and population shifts in plant pathogens. He stated that, as general information has accumulated and as many crops have been evaluated in the field, it has been found that in all but the most exceptional cases, single genes are inadequate for protection against plant pathogens. Thus, although diversity may not guarantee crop protection over time, uniformity of resistance, at least of the single-gene kind, will almost certainly result in vulnerability.

Finally, what does the lack of diversity have to do with the subject of this chapter, namely, insurance against damage by exotic plant pathogens that might become established in the United States, or in any other new area for that matter? In the first place, we can only guess at what the new pathogen might be. It could be a new disease-causing organism, for example, *Endothia parasitica,* which was introduced around 1900 and caused the devastating chestnut blight, or it could be *Phakopsora pachyrhizi,* the causal agent of soybean rust, which fortunately has not yet been introduced. Or it could be a new race of an old pathogen, such as a new and damaging race of the wheat stem rust fungus. The foregoing discussion has indicated that genetic diversity of the host generally produces satisfactory resistance to all pathogens as they occur on wild hosts in their centers of origin and, conversely, that lack of diversity leads to vulnerability. These generalizations seem to be applicable to all pathogens, known and unknown, that might threaten a crop species. Thus, if we could achieve diversity in our cultivated crops that approximates the diversity found in nature, we should at the same time achieve insurance against significant damage from any new pathogen that managed to establish itself on this continent. In view of the past record of successful introductions of serious plant pathogens from abroad, such insurance would certainly be well worth the time and trouble it would cost.

IV. DIVERSITY IN AGROECOSYSTEMS AS INSURANCE

We have seen that the diversity of hosts and pathogens found in nature offers promise of protection from severe disease damage. Therefore, we consider it as one of the fundamental elements of insurance against loss from the introduction and establishment of the unwanted and potentially damaging exotic pathogen.

We will not consider what either has been or could reasonably be done to achieve useful diversity in agroecosystems.

It is common knowledge that the agriculture of primitive societies was characterized by great diversity of crop species, and that such agricultural systems ordinarily suffered little serious damage from diseases. Unfortunately, scientifically documented studies that might confirm this are rare. Several informative studies that might confirm this are rare. Several informative examples, however, can be cited. P. J. Keane (personal communication) observed in New Guinea that pure varieties of taro (*Colocasia esculenta*) planted in pure stands at an agricultural experiment station were severely damaged by *Phytophthora colocasiae*. Native farmers in the vicinity, who interplanted taro with other crop species in their gardens, did not experience severe disease problems and obtained satisfactory yields of taro (Fig. 1).

When monocultures of beans (*Phaseolus vulgaris*) are grown under conditions of modern, intensive agriculture in tropical America, diseases often are a serious problem (N. G. Vakili, personal communication). In primitive agroecosystems in the same area, beans are often grown intercropped with corn (''milpa''). Bean diseases under this system are common, but severe damage over an entire field

Fig. 1. Primitive farm in New Guinea showing banana, taro, coconut, sweet potato, sugarcane, cocoa, and aibika (a green vegetable) growing intermixed. (Photo by P. J. Keane.)

rarely occurs. This effective, practical degree of control is attributed, first, to the fact that the beans themselves consist of a mixture of genotypes having varying resistances and, second, to the physical effects of growing in a mixed stand with corn. The corn, for example, tends to restrict the movement of air and thereby also restricts the movement of aphids, which in turn reduces the incidence of virus diseases. The presence of the corn also results in less dew on the bean plants, which reduces the severity of foliar diseases.

Glass and Thurston (1978) discussed slash-and-burn agriculture in relation to crop protection. They noted that, overall, slash-and-burn agriculture is feasible as long as sufficient land is available to allow for long fallow periods, perhaps up to 20 years. The people who practice slash-and-burn agriculture grow a great variety of crops in their individual fields, with sometimes as many as 40 species growing together. This provides a degree of protection, because pathogens and insects are seldom able to build up to destructive levels on the isolated plants of each species. Similarly, Anderson (1952) described the gardens grown by Guatemalan Indians as consisting of apparently planless mixtures of a large number of different fruits and vegetables. He concluded that plants of the same sort were so isolated from one another by intervening vegetation that pathogens and insects could not readily spread from plant to plant.

According to Borlaug (1959), corn rust (*Pucinnia sorghi* and *P. polysora*) is an unimportant disease in tropical Mexico despite ideal environmental conditions for development of epidemics and the constant presence of low levels of infection. Apparently, the great diversity of the host and the maintenance of this diversity by frequent outcrossing of the Mexican open-pollinated corn varieties provide protection from destructive epidemics.

Observations on somewhat more advanced systems of agriculture can also be informative. Thus, Aiyer (1949, p. 470) noted that multispecific, or mixed, cropping is carried on systematically as a very common practice in India. It was based on economic considerations, being regarded as insurance "against the inclemencies of the weather, pests and diseases." He discussed at some length the relationship of mixed cropping with pathogens and insects, noting that in India they are many and destructive, and that remedial measures are few. Of special interest to us is his statement that the "insurance supported by mixtures in respect of wilts and rusts especially is worthy of special notice" (p. 473). He clearly recognized the potentially important relationship of mixtures to the severity of an epidemic, stating that an immune variety of the same crop interposed in rows or broadcast with a susceptible variety breaks or checks the progress of the disease. As an example, he cited the value of growing together two varieties of cotton, one of which was resistant to one disease and susceptible to another, whereas the second showed the reverse reactions.

In conclusion, it seems safe to say that primitive agriculture was characterized by diversity and that this diversity contributed to a generally satisfactory degree of disease control. When we attempt to extrapolate this to modern, intensive agriculture, however, it is immediately apparent that this level of diversity is incompatible with modern agriculture. The question, then, is not so much whether diversity is desirable, but rather how much is necessary and whether that level can be attained in a manner compatible with the requirements of modern agriculture. Within the last three decades, a considerable amount of both theoretical and practical research has been applied toward the solution of this problem. As is often the case, early thinking on the subject evolved slowly, over a period of time; we will pick it up in the early 1950s with the work of Jensen (1952). He noted that the risk of serious losses in oats due to the rusts (*P. graminis* and *P. coronata*) had been created by the fact that the fungi could move freely from oat field to oat field without being checked by genetic barriers in the form of resistant varieties. Diversification of varieties, he suggested, would reestablish barriers which had existed within and between regions not so many years ago, but which by that time had been virtually eliminated. Specifically, he suggested a form of intravarietal diversity in which the cultivated unit would consist of a blend of pure lines of different resistance genotypes. The lines would be chosen from an oat-breeding project for uniformity of appearance but diversity for resistance to diseases. Each line should contribute additional desirable genetic factors without detracting from phenotypic uniformity. The component lines could be changed from year to year to meet changes in pathogen populations. Jensen believed that variety blends should have longer varietal life, greater stability of production, and greater protection against disease than would pure-line varieties. Information on geographical distribution and trends in populations of the rust fungi would serve as a guide to ensure that the variety blend was at all times balanced with respect to resistance to the potentially dangerous forms of a given target pathogen.

This general concept was developed further by Borlaug (1953, 1965), who assumed that if a wheat variety is to have a chance of remaining rust resistant indefinitely, it must be so constituted that its resistance can be modified to meet the changing prevalence of different rust races. He proposed a composite variety that would be produced by backcross methods. The backcross lines that make up the variety would be developed by backcrossing a current commercial variety to a number of sources of different types of resistance. The resulting lines would be increased separately and mixed mechanically to form the commercial variety that would be released to farmers. This multiline variety would be morphologically uniform but, as the rust races changed, individual lines constituting the variety could be removed or replaced. The plan could also be adapted for developing resistant hybrid wheat varieties.

The multiline hypothesis has been subjected to experimental scrutiny by several investigators in the last decade. Cournoyer (1970), one of the first of these, used 50 × 50 foot field plots planted to oats consisting of various mixtures of crown rust-resistant near-isogenic lines. Spores were trapped daily adjacent to the plots over the period of disease development. The final cumulative spore counts showed that incorporation of resistant plants decreased the number of spores produced. A greater proportion of resistant plants resulted in the production of fewer spores. Since spore production and disease severity are directly correlated, she concluded that her results supported the multiline hypothesis that mixtures of near-isogenic lines effectively buffer the host population against the pathogen population. Politowski and Browning (1978) reported field experiments showing that the commercially available cultivars, Multiline M73 and Multiline E74, developed much less rust than susceptible pure-line varieties—in fact, only slightly more than resistant lines. These relationships were substantiated by calculating the amount of reduction in yield and seed weight attributable to the infection in the same plots.

The two oat multiline studies just summarized showed that multilines gave good control of crown rust in Iowa, where the disease season is severe but short, never more than a few weeks. Oats are grown over a much longer disease season in the southern United States, and Browning and Frey (1981) reported results of multiline experiments carried out at two locations on the southern coastal plain of Texas. Crown rust overwinters in the uredial stage on oats in this area, and severe epidemics, initiated by natural inoculum, developed in the large isolated field plots that were used. Relative urediospore yields from multilines and from susceptible isoline checks were the same in Texas as in Iowa, showing that multilines will protect oats from crown rust in both short and long disease seasons, even under conditions very favorable for the disease. At the location where the environment was relatively less favorable for disease, crown rust still killed susceptible pure-line check varieties prematurely except in small plots protected with 11 weekly applications of fungicide. By contrast, both the multiline and an artificial varietal mixture in the ratio of one resistant to two susceptible pure lines had no more rust than did the susceptible, fungicide-sprayed plots. Thus, in this environment, with only one-third of the plants resistant, protection was equivalent to that given by 11 fungicidal sprays!

Groenewegen and Zadoks (1979) also strongly recommended abandoning the concept of the pure-line variety that relies on a single resistance gene, and replacing it with either general-resistance (which we will discuss later) or "within-field" diversity. They regarded general resistance as effective, but noted that its use in breeding programs is arduous. By contrast, within-field diversity could be achieved with relative ease and speed. The multiline variety, they noted, provides only one possible way of achieving diversity. There is actually a continuous range from mixtures of species all the way to mixtures of isogenic lines. They recognized five points along this continuum: (1) mixtures of species; (2) mixtures of varieties; (3) mixtures of related lines; (4) multilines; and (5) mixtures of isogenic lines. The difference between the last two was admittedly more or less theoretical. These levels of diversity have various advantages and disadvantages. Mixtures of species and of varieties can be readily adjusted to control nontarget diseases should the need arise, but if quality is a major concern, marketing the grain harvested from such fields may be a problem. Multilines and mixtures of isolines will yield homogeneous grain, but there is little opportunity to control nontarget diseases.

An examination of some specific experimental studies involving varietal mixtures is helpful at this point. Wolfe and Barrett (1980) considered the use of simple mixtures of barley varieties for the control of powdery mildew (*E. graminis*). They found that mixtures of three varieties, each of which carried a different major gene for mildew resistance and which were suceptible in the field to different races of mildew, developed much less mildew when grown as mixtures than occurred on the same varieties when grown alone. The mean leaf infection rate of one representative mixture was 6.5%, whereas that of its three components when grown in pure stands was 20.3%. The mean infection rate of all four of the mixtures used in the study was 5.1%, as compared with 14.3% for all the varieties individually. In further experiments in which conditions were relatively well controlled in the field, the epidemic was effectively reduced in mixtures regardless of the structure of the starting pathogen population. With a three-variety scheme planted at three locations, one variety was markedly susceptible at one location, another variety was markedly susceptible at a second location, and all three showed considerable susceptibility at the third location. However, plots sown to a mixture of the three varieties showed only about one-fourth, one-third, and three-fifths as much infection, respectively, as the means of the individual plots at the three locations. The plots at the first two locations were 2 hectares in size, whereas those at the third location consisted of only 3 m^2. The lesser effect of the mixture in the small plots was attributed to stray inoculum.

Using stripe rust (*Puccinia striiformis*) of wheat and powdery mildew (*E. graminis*) of wheat and barley, Priestley and Byford (1980) and Priestley and Wolfe (1980) started with the assumption that the effectiveness of varietal diversification in reducing pathogen spread in the field depends on the extent to which noncorresponding virulence can survive in the pathogen population. They presumed that the inoculum generated by a variety possessing some particular resistance gene will consist largely of rust isolates having the corresponding virulence. However, a certain proportion may also possess noncorresponding virulence. If the latter proportion is small, the inoculum will be largely ineffective on adjacent fields or plants of varieties susceptible to the noncorresponding virulence. If this proportion is large, of course, varietal diversification will be less effective. To illustrate, they reported the results of tests in which adult plant reactions were determined during the period 1975–79. Inoculum of stripe rust was collected from varieties possessing resistance genes R6, R13, or R14. In all cases, infection levels resulting from inoculation of varieties from inoculum from the same varieties were substantially higher than levels resulting from inoculum from varieties with different specific resistances. For example, isolates collected on varieties with R14 resistance resulted in a leaf infection rate of 14% on Hobbit, the only variety in the test with this resistance. Using the same isolates, average infection rates of 2.3% and 0.6% were obtained on varieties with R13 and R6 resistances, respectively.

The same investigators noted that varietal diversification is also valuable insurance against severe damage to an entire acreage by sudden outbreaks of disease on previously resistant or partially resistant varieties. Growing a range of varieties lessens the chance that all will be susceptible to any new pathogenic race of disease. This benefit would apply regardless of any reduction in pathogen spread.

Efforts to implement varietal diversification at the practical production level in the United Kingdom were described by Priestley and Bayles (1980a,b). Farmers were encouraged to grow a number of varieties possessing different specific resistances in order to reduce pathogen spread. The United Kingdom Cereal Pathogen Virulence Survey provides farmers with the information they need by annually devising varietal diversification schemes to reduce the spread of stripe rust in wheat and powdery mildew in barley. The general principle is to place varieties with similar resistances in the same "diversification group" (DG). In practice, DGs are given numbers representing their resistance to stripe rust and letters representing resistance to powdery mildew. The risks associated with various combinations of varieties are designated by "y" (risk of yellow rust spreading) and "m" (risk of mildew spreading). To use the table in which these data are summarized, the grower is instructed to decide upon a first choice of variety and locate its DG number from a listing made available for this purpose. Then he has only to find this number under the "chosen DG" on the left-hand side of the table. Reading across the table gives the risk of pathogen spread when the chosen DG is grown adjacent to varieties in each of the diversification groups in terms of a risk of yellow rust (y), a risk of mildew (m), or "+," which indicated a low risk of spread of either pathogen. The scheme can be applied in actual practice by sowing varieties from different DGs in adjacent fields. Diversity can also be achieved either by sowing small areas within fields to different pure-line varieties or by sowing variety mixtures. Priestley and Bayles noted that, on theoretical grounds, increasing the intimacy of the component varieties would be expected to increase the effectiveness of the diversification.

This program has been well accepted by farmers in the United Kingdom. A carefully conducted survey showed that in 1979, more than half of those farmers who replied to a questionnaire on the subject were using the varietal diversification scheme designed to reduce the spread of yellow rust in winter wheat. About one-fourth were using the comparable spring barley scheme.

Currently, research on the theoretical aspects of varietal diversification is being emphasized, with modeling studies being used to predict the course of events that might be expected when a pathogen population is confronted with some sort of varietal diversification scheme. A few examples will be cited. Using a "population-genetic" approach, Kiyosawa (1977) concluded that for multiline varieties, the proportion of susceptible plants necessary to maintain equilibrium of the races comprising the fungus population is very large, unless multiplication rates are considerably lower in virulent races than in avirulent races, or unless field susceptibility (disease severity when attacked by virulent races) of resistant varieties is considerably higher than that of susceptible varieties. This suggests that the use of a multiline variety for purposes of stabilizing race frequencies in pathogen populations is not practical. Is should be emphasized, however, that the goal of using multiline varieties is usually to control disease in a given year, not to stabilize the pathogen population. This is true because multilines generally are advocated to confront the highly epidemic continental pathogens, not residual pathogens. It is noteworthy, however, that *E. graminis* is a residual pathogen in the United Kingdom, yet has responded well to control by diversity.

Along somewhat similar lines, Trenbath (1977) reported that his modeling studies suggested that the rate of "breakdown" of resistance genes would be about the same whether the genes were used in a multiline, combined in a single variety, or used as single varieties each with one gene to be employed in sequence. Wolfe's (1979) studies of varietal mixtures to control powdery mildew of barley suggested that the population dynamics of the pathogen showed that either simple or complex races may come to dominate the pathogen population. The direction of change is, he believed, unpredictable. The evolution of complex races may be slow, but to guard against such undesirable shifts, he recommended a dynamic program of utilizing diversity of resistance as well as seed-applied systemic fungicides.

The concept of "cross protection" has been in the literature for many years, and research is now underway to determine the possible application of it to the problem of explaining the disease protection that is observed in multiline varieties. To illustrate, Johnson and Allen (1975) inoculated wheat seedlings with a nonvirulent race of stripe rust. They found that this delayed sporulation of a virulent race and decreased the total quantity of spores produced. Similar findings were obtained with bean rust (*Uromyces appendiculatus*). They suggested that such effects of induced resistance could retard the development of rust in multiline varieties in the field.

In concluding this discussion of multilines and varietal mixtures, it seems safe to say that the final answer is not yet in on the value of diversity per se, at least insofar as we can control it via multilines or by similar schemes. However, there is much evidence to indicate that such diversity will provide very substantial protection against any race or disease that might be introduced unexpectedly and threaten a major crop. Thus, we believe that such diversity should constitute a major part of our insurance against the introduction and establishment of exotic pathogens. True multilines are excellent for control of specific pathogens; however, three-way mixtures of unrelated varieties also portend control of nontarget pathogens of special importance in considering insurance.

The basic schemes that we have just examined for the utilization of resistance genes to exploit the natural advantages of diversity generally involve the use of only a single resistance gene in any given plant, whether this involves multiline varieties or a mixture of pure-line varieties. An alternative scheme is to "stack," or pyramid, two or more genes for resistance in a single plant. The concept was clearly spelled out by Watson and Singh (1952) years ago and has received considerable attention since then. In considering stem rust of wheat in Australia, they noted that new rust races appeared to have increased their range of pathogenicity in steps, each step involving a single new gene in the host. There was no evidence of a change involving two host genes simultaneously. Therefore, they proposed to establish a breeding program that would incorporate pairs of genes for resistance into a basic genotype that is well adapted to local conditions. Such "doubly resistant" varieties should remain resistant in the field for longer periods than varieties having only single resistance genes because, in theory, simultaneous mutations at two loci would be required in the pathogen to enable it to overcome the resistance. Empirically, it can be stated that this approach has been successful in controlling wheat stem rust in Australia for many years. Critical experimental data on the subject are scarce. Observations made by Simons *et al.* (1957) suggest that this approach is no certain panacea. During the early 1950s, two crown rust resistance genes, which came from the varieties Landhafer and Santa Fe, were being widely used in breeding programs in the United States. Initially, no race was known that could parasitize either of these two sources of resistance. But in the mid-1950s, races started to appear that were virulent on not just one or the other of them but on both, and such races soon dominated the fungal population. No race was ever seen during this period that could parasitize only one or the other. The origin of such "doubly virulent" races is still a matter of conjecture, but there is no doubt that they appeared and negated in one step the resistance of any variety that carried the Landhafer and Santa Fe resistance genes in combination. In the context of this chapter, stacking of resistance genes provided no insurance against the appearance and spread of the new pathogenic race.

The foregoing discussion has presupposed the presence of resistance genes, but obviously, such genes must be located by the pathologist before they can be used. Increasingly, they are being located in the wild relatives or exotic varieties of our major crop species. It is necessary, therefore, to consider the problem of transferring such disease resistance genes to agronomically adapted varieties without carrying along undesirable traits from the unadapted parents. In general, the consensus is that obviously deleterious characters can be separated from the desired genes, but this may be a laborious task. On the other hand, there is also a possibility that the resistance genes may have some beneficial effects in addition to the resistance. Frey and Browning (1971), for example, found that major genes for crown rust resistance could be successfully transferred from the wild oat species *A. sterilis* to cultivated oats, and in some cases yields of derived, resistant lines were higher than those of the recurrent parent even in the absence of rust. Brinkman and Frey (1977) followed up on this study with growth analysis studies carried out on isolines showing such yield deviations. The yield differences were attributable to the presence of "yield genes" from the *A. sterilis* parent that were tightly linked to the gene for crown rust resistance. These yield genes conditioned such growth traits as leaf area duration, leaf area ratio, and growth rate. In a related study, Simons (1979) found that, averaged over a number of cultivated-type derived lines, there was a measurable loss of yield associated with the transfer of a certain major gene for crown rust resistance to cultivated oats. Presumably, this was due to the effect of deleterious *A. sterilis* genes that were linked to the crown rust resistance gene. There were individual resistant lines, however, that yielded as well as sister lines that lacked the crown rust resistance gene. Minor genes conditioning small, quantitative amounts of field resistance or tolerance were also transferred from the *A. sterilis* parent in the process of transferring the major gene, which should make the resistance broader, more durable, and therefore of more interest as insurance against the unknown pathogen of the future.

To conclude this section, a pertinent question might be: Where do these genes come from, or where should we go to search for plants that might carry them? This is not the place for an in-depth answer, but suffice it to say that much work on the subject indicates that the geographical centers of origin of the host species will be the most fruitful place to search for them (Vavilov, 1957; Browning, 1974). Baker and Cook (1974) added a new dimension to the potential importance of centers of origin when they suggested a possibly important role for antagonists of plant pathogens in the control of plant disease. Such antagonists could also probably be best located in the geographical centers of origin of the host species in question. Utilization of such antagonists would serve as additional insurance against severe damage by exotic disease organisms, should such organisms become established in new areas.

V. FUNGICIDES, DIVERSITY, AND INSURANCE

The appearance of strains of plant pathogenic fungi with resistance to fungicides is now known to be a common phenomenon of great practical importance in controlling many major plant diseases (Dekker, 1976). A recent report of this phenomenon (Davidse *et al.*, 1981) suggests that the acquisition by pathogenic fungi of resistance to fungicides has ramifications that extend to host diversity; thus, the phenomenon has a very real link to the concept of insurance against exotic pathogens. The fungicide metalaxyl was registered in the Netherlands for use on potato late blight (*P. infestans*) in 1979, and was used on about 5% of the potato acreage in that year. In 1980 it was used on about 50% of the acreage. Development of resistant strains of the late blight fungus was extremely rapid, and these strains caused extensive damage in 1980. Presumably, their appearance and increase were associated with the almost exclusive use of metalaxyl on one-half of the potato acreage. Once the resistant mutants arose, they multiplied with great rapidity on potatoes on which there was no competing strain. In theory, the buildup of such a fungicide-resistant strain would be slowed in a diversified host population, because any pathogenic strain that developed fungicide resistance would not be able to parasitize host genotypes having resistance to it. The necessity for a pathogen simultaneously to overcome a specific resistance gene and a systemic fungicide applied to one of three components of a three-way variety mixture keeps the barley powdery mildew fungus off guard in the United Kingdom (Wolfe and Barrett, 1980).

VI. INSURANCE VALUE OF DIFFERENT TYPES OF RESISTANCE

The classification and characterization of different kinds and degrees of reaction of host plants to disease organisms has been a part of plant pathology probably since the beginning of this science. Van der Plank's work (1963, 1968), however, gave new meaning and significance to the subject, and at the same time initiated a great deal of theoretical and applied research in the area. He divided host plant resistance into two rather clear-cut categories, vertical and horizontal, based primarily on the specificity of the host–parasite interaction. Where a high degree of specificity was involved, he called the resistance "vertical." In vertical resistance, there is an interaction between cultivars of the host and races of the pathogen. One cultivar will be resistant to certain races of the pathogen but susceptible to others; a second cultivar may show opposite reactions to the same races. In contrast, lack of host–parasite interaction characterizes "horizontal" resistance. Horizontal resistance is spread evenly against all races, van der Plank (1963, 1968) said. A horizontal line results when the reaction of a cultivar is plotted against races of the pathogen, hence the term "horizontal." The insurance value of such resistance is obvious.

Horizontal resistance does not mean that all variants of the pathogen are equally pathogenic, or that all cultivars are equally resistant. Van der Plank's analysis of data published by Paxman (1963) illustrates this point. Paxman had studied the rate of spread of mycelium of the potato late blight fungus (*P.*

infestans) in potato tubers. Pathogenic races of the fungus, distinguished by patterns of pathogenicity toward potato varieties with vertical resistance, were used to inoculate potato varieties that had no vertical resistance. Analysis of variance of the rates of spread of the mycelium showed that both the varieties and the races differed. However, there was no significant race–variety interaction. The races did not interact with the varieties; the resistance, therefore, was horizontal.

According to van der Plank, vertical resistance is ordinarily characterized also by simple inheritance, furnishing a high degree of protection from those pathogenic variants to which it is effective, and, most important, being subject to "breakdown" because of the relative ease with which new pathogenic variants arise and parasitize the previously resistant genotype. Vertical resistance is perhaps best exemplified by the reactions of small grains to their rust pathogens. Single genes each conditioning a high degree of resistance have been used repeatedly and extensively, and they have also broken down repeatedly upon the advent of new races. Parenthetically, however, we should note that examples can be cited in which simply inherited resistance has held up very well over long periods of time (Simons, 1972b).

Horizontal resistance is usually polygenically inherited and often gives a lesser degree of protection than does vertical resistance. Central to van der Plank's thinking is the assumption that horizontal resistance, unlike vertical resistance, is not subject to breakdown; it is stable and enduring over time.

We must now consider briefly some nomenclatural and classification problems that have arisen as a result of van der Plank's work, and the work of others that has stemmed from it. Caldwell (1968) recognized van der Plank's concept of horizontal resistance, at least in broad outline, but preferred to label the resistance "general," believing that this term was easier to understand and also had precedence over horizontal resistance. He defined general resistance as resistance that experience and adequate testing had shown to confer enduring and stable protection against a pathogen or disease. For general resistance to be effective over time, natural variants of a pathogen must be unable to compensate for the restrictions to their penetration, development, or dispersion that such resistance imposes. This can usually be determined only by prolonged testing.

Van der Plank's very stimulating but simplistic concepts of resistance combined genetic control and epidemiologic result, and exceptions to the general trend caused problems. In an effort to clarify these nomenclatural problems, Browning *et al.* (1977) pointed out that much of the confusion was due to failure to distinguish between underlying genetic mechanisms of resistance and the effect of resistance on epidemics. They proposed retaining "specific" and "general" to describe genetic concepts and suggested the use of new terms in the epidemiologic sense that would imply nothing genetically. These terms would describe only the epidemiologic action of the resistance, however it was controlled genetically. "Discriminatory" resistance is characterized by its reduction of incoming inoculum by recognizing incompatible pathogen genotypes and discriminating selectively against them. "Dilatory" resistance is that which delays the rate of progress of the epidemic through its action on all components of the pathogen population. Since dilatory resistance delays all variants of a pathogen, it is obviously of interest as insurance against exotic or unknown pathogenic forms.

Parlevliet (1981) and Zadoks and Parlevliet (1979) also believed that van der Plank's classification of disease resistance as vertical and horizontal was simplistic. Although recognizing that evolution of a host and pathogen must be analyzed and described in a generalized way, they emphasized that there are vast numbers of host–pathogen interactive systems that are in various stages of coevolution. Vertical and horizontal resistance were seen as the two extremes of a wide range of possibilities. There is, they believed, a fluid transition between the extremes as horizontal and vertical resistance were envisioned by van der Plank. Their modeling studies illustrate this point. They set up a model pathosystem that consisted of host genotypes and pathogen genotypes that could be varied for quantitative response such as spore production. The hypothetical data were then subjected to analysis of variance in which the main effects were presumed to measure horizontal compatibility and interaction effects were presumed to indicate differential or vertical compatibility. The results suggested that both main effects and interaction effects can often be significant.

Quite a different view was recently postulated by Browning (1980). Using certain basic evolutionary assumptions, he emphasized two systems that govern host–pathogen interactions. The first, incompatibility, corresponding roughly to specific or vertical resistance, functions to protect populations of plants rather than individual plants. The mechanism hypothesized to achieve this in nature was the maintenance of homeostatic host–pathogen balance by controlling aggressiveness in the pathogen. This type of resistance has the same effect when used in man-made diverse populations such as multilines. The second system, which corresponds roughly to horizontal or general resistance, protects the individual plant. In theory, this type of resistance could be used alone in agroecosystems, but greater insurance would be achieved if it were used in conjunction with the first system, which is the situation found in nature.

Unfortunately, the simple fact that a given example of disease resistance is inherited polygenically does not guarantee that it will be stable over time. In Indiana (Caldwell, 1968), a gradual loss of resistance to leaf rust (*P. recondita*) was observed in widely grown wheat varieties derived from the Chinese Spring variety. The resistance of Chinese Spring is typical of resistances that are regarded as horizontal or general in that it is a mature plant with a moderate type of resistance under polygenic control. Segregated populations from appropriate crosses were shown to vary continuously in infection severity from 0% to 100%.

From 1961 to 1965, rust populations of gradually increasing virulence finally overcame this polygenic, mature plant resistance.

More recently, Johnson (1978) also recognized the objections to van der Plank's terms of horizontal and vertical resistance and coined the term "durable resistance," which has become very popular in Europe. "Durable" completely avoids the genetic and epidemiologic problems associated with "vertical" and "horizontal"; it carries no message but a historical one, i.e., that a given type of resistance has to be proven durable.

In a discussion of practical breeding for durable resistance to rust diseases in the cereals, Johnson (1978) pointed out that "incomplete" resistance can be race specific as well as race nonspecific, as exemplified by stripe rust (*P. striiformis*) resistance in wheat. Durable resistance usually is incomplete, but on the other hand, incomplete resistance is not necessarily durable. In fact, Johnson noted that in many cases incomplete resistance is specific and unstable. There are now a large number of studies in the literature that back up these assertions. A few of these will be discussed briefly. In diseases such as the rusts and powdery mildews, the relative length of the latent period is commonly regarded as one of the potential components of durable resistance. Thus, it would be expected to be polygenic in inheritance. Parlevliet (1976), in a study of latent period in barley leaf rust (*Puccinia hordei*), found that the very long latent period of barley varieties Minerva and Vada (about 17 days as compared to about 8 days in susceptible varieties) was conditioned by a single recessive gene. In addition, there were four or five minor genes with additive effects involved. Parlevleit (1977) also approached the problem from another angle. He used the barley varieties Vada, Julia, and Berac, which were known to have polygenic resistance to *P. hordei*. His field experiment, in which five monospore rust isolates were used, was designed so that relative disease severity could be assessed by sampling tillers at random. The primary parameter measured was the percentage of leaf area covered by pustules. Greenhouse studies carried out with the same material were used to evaluate latent period and infection frequency. He found that the cultivar × isolate interaction for latent period was highly significant, with one variety × isolate combination being obviously out of line. Others, however, also contributed to the interaction. Parlevleit concluded that differential interactions occur in this polygenic system, suggesting that polygenic systems in the host can function on a gene-for-gene basis (Flor, 1956) with polygenic systems in the pathogen. Ellingboe (1975) and associates (Martin and Ellingboe, 1976) have also reported work that is instructive in this regard. Under the gene-for-gene hypothesis of Flor (1956), there are four possible combinations of alleles of host and pathogen for any individual host–pathogen gene system. This is the so-called quadratic check. Three of the four will result in susceptible infection types. The experimental material consisted of two isogenic wheat lines, one containing the Pm4 gene for powdery mildew (*E. graminis*) resistance and the other containing its allele conditioning susceptibility. Mildew cultures with and without the corresponding virulence alleles resulted in infection types that were resistant or susceptible as expected, i.e., one of the four combinations of alleles conditioned resistance and the other three susceptibility. The three did not condition equal susceptibility, however; one conditioned a slower rate of hyphal development than the other two. Rate of hyphal development is ordinarily regarded as a manifestation of general resistance, and its obvious dependence in this case on single gene differences served to blur further the distinction between general and specific resistance. Thus, it is not surprising that in a recent consideration of the problem, Parlevliet (1979) concluded that there is no simple way that resistances can be classified unambiguously into two groups such as horizontal vs. vertical, general vs. specific, polygenic vs. oligogenic, etc. There is no easy way that stable and unstable or polygenic and monogenic resistances can be discerned. Each case must be studied individually.

At this point, we have strung together a long line of caveats to the central thesis of this section, namely, that the presence of general resistance in our crops will furnish a significant degree of insurance against losses to new pathogens. The caveats notwithstanding, we still believe in general resistance as one, but only one, important form of insurance. The case of common rust (*P. sorghi*) of corn furnishes a convincing example. Hooker (1967) noted that all conditions necessary for the development of corn rust are present in the northern United States. Great acreages of the host are grown, weather usually is favorable for the disease, and the fungus usually is present. Furthermore, the fungus is known to have the capacity to damage severely susceptible corn. All commercial hybrids, however, have polygenic resistance, which has effectively protected field corn against rust ever since the crop has been grown in the region. The varieties that are protected adequately by general resistance to *P. sorghi* are fully susceptible, however, to another, closely related rust fungus, *P. polysora*. Thus, even if *race* specificity is not involved, *pathogen* specificity is, and diversity still is the best insurance.

VII. TOLERANCE TO DISEASE AS INSURANCE

Tolerance to plant disease, as opposed to resistance, is an attractive concept that has particular appeal from the standpoint of insurance against exotic pests. For definitional purposes, we follow Caldwell *et al.* (1958), who clearly differentiated "tolerance" from "resistance" and defined it as the ability of a susceptible plant to endure severe attack by a pathogen without sustaining severe losses in yield. The tolerant variety shows signs and symptoms similar to those of a susceptible variety, but the tolerant variety is damaged less by the infection (Simons, 1969). Tolerance, in theory at least, has an important advantage over resistance. A new form of a pathogen to which a previously tolerant variety was nontolerant would have little selective advantage on the host. Therefore, it would not be expected to show the rapid differential increase that is characteristic of new forms of pathogens that are virulent on previously resistant varieties, and on which they have a very high selective advantage.

Schafer (1971) published a detailed review of the extensive literature on tolerance to plant diseases. Here we will give only a few illustrative examples that are taken largely from our own experience. In his early studies of crown (*P. coronata*) and stem (*P. graminis*) rust on the wild species of oats in Israel, Wahl (1958) showed that some had excellent tolerance to both rusts despite severe infection. This finding was later substantiated experimentally by Simons (1972a), who showed that visually susceptible lines of *A. sterilis* from Israel actually had a degree of tolerance that could be transferred to cultivated oats and then measured in replicated trials. This tolerance, expressed in terms of yield and seed weight reduction attributable to infection, gave significant protection to plants carrying it. Earlier work (Simons, 1969) had shown that the tolerance of certain cultivated oats to crown rust was inherited as a complex quantitative trait. Thus, genetically, it seems safe to assume that it would be expected to show traits that are associated with general resistance, such as effectiveness against a wide variety of pathogenic types and durability over time. Therefore, tolerance deserves a place in any consideration of measures that can be used to insure against severe damage by introduced disease organisms.

Although a good case has been made for tolerance (Schafer, 1971), it now appears that, as more precise methods of measuring disease progress and severity are developed, many effects previously attributed to or described as tolerance may be shown to be due to some low or moderate form of resistance. Parlevliet (1975) made a careful study of what he called the "partial resistance" of barley to leaf rust. He was particularly interested in the latent period and measured this trait very carefully, showing that it was related to the partial resistance of the host. The point here is that if he had not measured this trait so carefully, the superior performance of the cultivars that had the longer latent periods would probably have been attributed to tolerance. For practical plant-breeding purposes, such "resistance" traits in varieties that appear to be susceptible visually are probably ordinarily best handled as if they actually did represent tolerance. We believe, therefore, that the concept of tolerance and the technique of measuring it by measuring its effects on the host plant will have practical application, both in terms of its providing insurance against exotic disease organisms and in terms of handling it in breeding programs.

VIII. GEOPHYTOPATHOLOGY AND INSURANCE

Geophytopathology is a relatively new branch of the science of phytopathology that deals with the geographic distribution patterns of plant diseases, the causes of these patterns, and geographic aspects of disease control in general (Weltzien, 1978). Geophytopathology has important implications for a discussion of insurance against losses from exotic pathogens. As Zadoks and Schein (1979) noted, if we could see the future, we might avoid some serious epidemics and develop management systems before serious problems arose. To make intelligent guesses about the future, we must know the past and present status of specific pathosystems.

Reichert and Palti (1967) described what they called a ''pathogeographical'' approach to the prediction of disease occurrence that is particularly applicable to this discussion. According to them, the basic data needed are (1) the world distribution of the pathogen and its host, (2) the ecological requirements of the pathogen and host, (3) limitation of pathogenicity due to such host and parasite factors as age of the host, presence of young tissue, fruiting, host resistance, etc., (4) climatic conditions with due regard to elevation, day length, topography, etc., and (5) cultural conditions such as irrigation, soil type, etc.

If the geographic distribution of a pathogen and its host is known, and sufficient information on their ecological requirements is available, it should be possible to predict disease occurrence in previously uncontaminated areas, or in places where the host might be introduced as a new crop (Weltzien, 1972).

By using the same reasoning in reverse, we can start with a specific crop-growing area of our country that lacks a certain pathogen, and then locate other areas in the world having the same ecological conditions and also having the pathogen. Such areas might or might not be centers of origin of the crop and of its disease organisms, but valuable information could be obtained in either case. Study of the disease would show its potential destructiveness in the area or the environment in question, and this knowledge would influence the amount of effort we might justify expending to prevent entry of the pathogen or to eradicate a pathogen that gains a foothold. Research on the disease where it is common would be immediately applicable should the disease become established in the United States, because environmental conditions would be similar. Such research would provide valuable information on the potential usefulness of resistance, genetic diversity, specific sources of resistance, chemical control, or the application of cultural methods.

Geophytopathology could show us what areas in the United States might now be regarded as high-risk areas for the introduction of new pathogens. Extra effort could then be concentrated on crops in the high-risk areas in terms of stressing genetic diversity and general application of research from the area where the disease is known to exist.

The work of Smith *et al.* (1973) can be cited as an instructive example of the value of geographical mapping to potential future problems in plant pathology. They mapped *Colletotrichum*-infected weeds in connection with a study on the biocontrol of a serious weed in rice. Potentially useful information could be obtained from mapping the occurrence of weeds that might reasonably be expected to serve as hosts for exotic viruses that might be introduced.

Percy (1982) identified soil properties and climatic conditions conducive to *Phymatotrichum omnivorum* in the United States, and thus established a geographic range within which the fungus was a potential danger in North America. Examination of soil and climate data from other continents showed that similar conducive conditions existed in parts of Africa and the Indian subcontinent. No such areas were found in South America. Such information has obvious value in making decisions on quarantine regulations and other control measures for these areas.

As another example, Drandarevski (1969) made a detailed epidemiologic analysis of environmental conditions associated with epidemics of powdery mildew (*Erysiphe polygoni*) of sugarbeets in Germany. He then determined the degree of risk from powdery mildew epidemics in other major sugarbeet-growing areas of the world. His map of North America showed southern California to be a high-risk area. Only a few years after this was published, Kontaxis *et al.* (1974) reported a severe epidemic of powdery mildew on sugarbeets in the Imperial Valley of California. They noted, incidentally, that about 90% of the 65,000 acres of beets in the area were planted to a single variety, USH9.

A somewhat different application of the principles of geophytopathology is concerned with another disease of sugarbeets, yellow wilt, which is caused by a leafhopper-borne virus (Anonymous, 1981). This potentially destructive disease is now known to occur only in Chile and Argentina. But similar weather, and weeds that are good hosts for both the leafhopper and the virus, exist in many other beet-growing areas. In view of the volume of traffic between those countries and such areas, spread and establishment of the pathogen into new areas, including the United States, seem inevitable. Extensive testing of commercial sugarbeet varieties in Argentina and Chile in past years revealed none that were resistant. With this in mind, a cooperative program to breed resistant varieties for the United States and other areas is now in progress. Wilt-resistant lines, selected in Chile from material that probably resulted from accidental outcrossing to wild Chilean *Beta vulgaris,* were crossed in the United States with agronomically superior United States varieties. Seed was then sent to South America, where the plants were grown under exposure to yellow wilt. Seed of resistant plants was multiplied in the United States for further evaluation, selection, and hybridization necessary to achieve a high level of resistance in an adapted variety.

R. A. Kilpatrick (personal communication) has coordinated an extensive international testing program that is another good example of the practical application of geophytopathological principles in achieving insurance against exotic pathogens. This program involves the testing of breeding lines and potential sources of resistance to rusts and other diseases of small grains. Uniform observation nurseries consisting of such lines have been established throughout the world, providing information on the reaction of breeding lines and sources of resistance to exotic races and pathogens that might become established in new areas. With this program, it is possible to preselect varieties for resistance to exotic pathogens before the pathogen is introduced.

Another specific application of geophytopathology has been suggested by Baker and Cook (1974). They hypothesized that the best place to study control of a given disease may not necessarily be where the disease is now economically important, but rather where the pathogen is indigenous or where the disease once was epidemic but now had subsided. When such areas have been located through the techniques of geophytopathology, they can be searched for antagonists, resistance, etc., that might act as insurance to check an exotic pathogen.

IX. CONCLUDING REMARKS

This chapter has stressed diversity as insurance against loss. This seems justified since diversity is the only hedge against a *future* risk situation, as from an unknown exotic pathogen.

People are accustomed to thinking of diversity if they seek to minimize risk in the world of finance. That is, an investor who desires a well-balanced portfolio will not want all stocks, all bonds, or all real estate. Neither will he want all his stock in, say, shipping companies, or all his real estate in, say, shopping malls. The same principle is true in the world of biology.

We have spent considerable time discussing resistance as insurance. Many pathologists and breeders now labor in the vineyard—and false security—of horizontal resistance, wanting to believe that it really is effective against all races, present and potential. We mentioned that the very effective horizontal resistance of corn in the United States to *P. sorghi* was *no* help when *H. maydis* race T hit in 1970. Nor does it help in combatting a closely related pathogen, *P. polysora*. Thus, even if horizontal resistance seems adequate against known strains of a given pathogen, diversity still is the only known protection against unknown strains or pathogens.

The question frequently is asked, "How much diversity is needed to protect a crop from a highly epidemic pathogen?" The answer seems to be, "Not much!" We cited three-way barley mixtures in the United Kingdom and oat multilines in the United States in which populations with as little as one-third resistance gave dramatic protection. It is true also that, in the United States, the major epidemics of corn have occurred since the switch from double-cross hybrids to single-crosses. Thus, there is strong circumstantial evidence that the level of diversity in a double-cross corn hybrid is adequate to protect that crop from corn pathogens in disease-favorable environments across millions of acres, and that the lack of diversity in single crosses is inadequate.

As relatively little diversity is required to protect a plant population, it should be possible to attain that level without undue sacrifices in yield and quality. Thus, the main impediment to utilizing diversity seems to be a philosophical one. The pathologist–breeder team should not ask "How pure can I make this variety?" but rather "How much diversity can I retain without undue sacrifice in yield and quality?"

Diversity is one of nature's ways of utilizing the many different types of resistance available, of keeping selection pressure low, and of assigning to a given ecosystem much of the capacity for its own self-regulation. Seemingly for these sound ecological reasons, agroecosystems with only minimal diversity of resistance genotypes provide not only protection against *known* pathogenic strains but, more importantly, insurance against future disease risks from as yet *unknown* exotic pathogens.

REFERENCES

Aiyer, A. K. Y. N. (1949). Mixed cropping in India. *Indian J. Agric. Sci.* **19,** 439–543.

Anderson, E. (1952). "Plants, Man and Life." Little, Brown, Boston, Massachusetts.

Anonymous (1970). Southern corn leaf blight special issue. Part II. *Plant Dis. Rep.* **54,** 1099–1136.

Anonymous (1972). "Genetic Vulnerability of Major Crops." Natl. Acad. Sci., Washington, D.C.

Anonymous (1981). Yellow wilt-preparaing for a sugarbeet disease invasion. *Agric. Res.* **29,** 6–8.

Baker, K. F., and Cook, R. J. (1974). "Biological Control of Plant Pathogens." Freeman, San Francisco, California.

Borlaug, N. E. (1953). New approach to the breeding of wheat varieties resistant to *Puccinia graminis tritici. Phytopathology* **43,** 467 (abstr.).

Borlaug, N. E. (1959). The use of multilineal or composite varieties to control airborne epidemic diseases of self-pollinated crop plants. *In Proc. Int. Wheat Genet. Symp., 1st, 1958* pp. 12–27.

Borlaug, N. E. (1965). Wheat, rust and people. *Phytopathology* **55,** 1088–1098.

Brinkman, M. A., and Frey, K. J. (1977). Growth analysis of isoline-recurrent parent grain yield differences in oats. *Crop Sci.* **17,** 426–430.

Browning, J. A. (1971). Corn, wheat, rice, man: Endangered species. *J. Environ. Qual.* **1,** 209–211.

Browning, J. A. (1974). Relevance of knowledge about natural ecosystems to development of pest management programs for agroecosystems. *Proc. Am. Phytopathol. Soc.* **1,** 191–199.

Browning, J. A. (1980). Genetic protective mechanisms of plant-pathogen populations: Their coevolution and use in breeding for resistance. *In* "Biology and Breeding for Resistance to Arthropods and Pathogens in Agricultural Plants" (M. K. Harris, ed.), pp. 52–75. Tex. Agric. Exp. Stn., College Station.

Browning, J. A., and Frey, K. J. (1969). Multiline cultivars as a means of disease control. *Annu. Rev. Phytopathol.* **7,** 355–382.

Browning, J. A., and Frey, K. J. (1981). The multiline concept in theory and practice. *In* "Strategies for the Control of Cereal Diseases" (J. F. Jenkyn and R. T. Plumb, eds.), pp. 37–46. Blackwell, Oxford.

Browning, J. A., Simons, M. D., and Torres, E. (1977). Managing host genes: Epidemiologic and genetic concepts. *In* "Plant Disease: an Advanced Treatise" (J. G. Horsfall and E. B. Cowling, eds.), Vol. 1, pp. 191–212. Academic Press, New York.

Caldwell, R. M. (1968). Breeding for general and/or specific plant disease resistance. *Proc. Int. Wheat Genet. Symp., 3rd, 1968* pp. 263–272.
Caldwell, R. M., Schafer, J. F., Compton, L. E., and Patterson, F. L. (1958). Tolerance to cereal leaf rusts. *Science* **128,** 714–715.
Cournoyer, B. M. (1970). Crown rust epiphytology with emphasis on the quantity and periodicity of spore dispersal from heterogeneous oat cultivar-rust race populations. Ph.D. Thesis, Iowa State Unversity Library, Ames.
Davidse, L. C., Looijen, D., Turkensteen, L. J., and Van Der Wal, D. (1981). Occurrence of metalaxyl-resistant strains of *Phytophthora infestans* in Dutch potato fields. *Neth. J. Plant Pathol.* **87,** 65–68.
Dekker, J. (1976). Acquired resistance to fungicides. *Annu. Rev. Phytopathol.* **14,** 405–428.
Dinoor, A. (1969). The role of the alternate host in amplifying the pathogenic variability of oat crown rust. *Res. Rep. Sci. Agric.* **1,** 734–735.
Drandarevski, C. A. (1969). Untersuchungen über den echten Rübenmehltau *Erysiphe betae* (Vanha) Weltzien. III. *Geophytopathologische Untersuchungen. Phytopathol. Z.* **65,** 201–218.
Dubin, H. J., and Torres, E. (1981). Causes and consequences of the 1976–1977 wheat leaf rust epidemic in northwest Mexico. *Annu. Rev. Phytopathol.* **19,** 41–49.
Ellingboe, A. H. (1975). Horizontal resistance: An artifact of experimental procedure? *Aust. Plant Pathol. Soc. Newsl.* pp. 44–46.
Flor, H. H. (1956). The complementary genic systems in flax and flax rust. *Adv. Genet.* **8,** 29–54.
Frey, K. J., and Browning, J. A. (1971). Association between genetic factor for crown rust resistance and yield in oats *Crop Sci.* **11,** 757–760.
Glass, E. H., and Thurston, H. D. (1978). Traditional and modern crop protection in perspective. *BioScience* **28,** 109–115.
Groenewegen, L. J. M., and Zadoks, J. C. (1979). Exploiting withinfield diversity as a defense against cereal diseases: A plea for "poly-genotype" varieties. *Indian J. Genet. Plant Breed.* **39,** 81–94.
Harlan, J. R. (1956). Distribution and utilization of natural variability in cultivated plants. *In* "Genetics in Plant Breeding," pp. 191–208. U.S. Brookhaven Natl. Lab., Upton, New York.
Harlan, J. R. (1971). Genetics of disaster. *J. Environ. Qual.* **1,** 212–215.
Hartley, C. (1939). The clonal variety for tree planting: Asset or liability? *Phytopathology* **29,** 9 (abstr.).
Hooker, A. L. (1967). Genetics and expression of resistance in plants to rusts of the genus *Puccinia. Annu. Rev. Phytopathol.* **5,** 163–182.
Jensen, N. F. (1952). Intra-varietal diversification in oat breeding. *Agron. J.* **44,** 30–34.
Johnson, R. (1978). Practical breeding for durable resistance to rust diseases in self-pollinating cereals. *Euphytica* **27,** 529–540.
Johnson, R., and Allen, D. J. (1975). Induced resistance to rust diseases and its possible role in the resistance of multiline varieties. *Ann. Appl. Biol.* **80,** 359–363.
Johnson, T. (1961). Man-guided evolution in the plant rusts. *Science* **133,** 357–362.
Kiyosawa, S. (1977). Development of methods for the comparison of utility values of varieties carrying various types of resistance. *Ann. N.Y. Acad. Sci.* **284,** 107–123.
Kontaxis, D. G., Meister, H., and Sharman, R. K. (1974). Powdery mildew epiphytotic on sugarbeets. *Plant Dis. Rep.* **58,** 904–905.
McGregor, R. C. (1978). People-placed pathogens: The emigrant pests. *In* "Plant Disease: An Advanced Treatise" (J. G. Horsfall and E. B. Cowling, eds.), pp. 383–396. Academic Press, New York.
Marshall, D. R. (1977). The advantages and hazards of genetic homogeneity. *Ann. N.Y. Acad. Sci.* **287,** 1–20.
Martin, T. J., and Ellingboe, A. H. (1976). Differences between compatible parasite/host genotypes

involving the Pm4 locus in wheat and the corresponding genes in *Erysiphe graminis* f. sp. *tritici*. *Phytopathology* **66,** 1435–1438.

Murphy, H. C., (1952). Problems involved in breeding oats for disease resistance. *Phytopathology* **42,** 482–483 (abstr.).

Pady, S. M., Hansing, E. D., and Johnston, C. O. (1947). Kansas mycological notes: 1946. *Trans. Kans. Acad. Sci.* **50,** 45–54.

Parlevliet, J. E. (1975). Partial resistance of barley to leaf rust, *Puccinia hordei*. I. Effect of cultivar and development stage on latent period. *Euphytica* **24,** 21–27.

Parlevliet, J. E. (1976). Partial resistance of barley to leaf rust, *Puccinia hordei*. III. The inheritance of the host plant effect on latent period in four cultivars. *Euphytica* **25,** 241–248.

Parlevliet, J. E. (1977). Evidence of differential interaction in the polygenic *Hordeum vulgare-Puccinia hordei* relation during epidemic development. *Phytopathology* **7,** 776–778.

Parlevliet. J. E. (1979). Components of resistance that reduce the rate of epidemic development. *Annu. Rev. Phytopathol.* **17,** 203–222.

Parlevliet, J. E. (1981). Disease resistance in plants and its consequences for plant breeding. *Plant Breed. 2* [*Two*] [*Proc. Plant Breed. Symp.*], *2nd, 1979* pp. 309–364.

Paxman, P. J. (1963). Variation in *Phytophthora infestans*. *Eur. Potato J.* **6,** 14–23.

Percy, R. G. (1982). A theoretical range potential for *Phymatotrichum omnivorum* (Shear) Duggar. *Proc. Beltwide Cotton Prod.—Res. Conf.* p. 22 (abstr.).

Politowski, K., and Browning, J. A. (1978). Tolerance and resistance to plant disease: An epidemiological study. *Phytopathology* **68,** 1177–1185.

Priestley, R. H., and Bayles, R. A. (1980a). Varietal diversification as a means of reducing the spread of cereal diseases in the United Kingdom. *J. Natl. Inst. Agric. Bot.* (*G.B.*) **15,** 205–214.

Priestley, R. H., and Bayles, R. A. (1980b). Factors influencing farmers' choice of cereal varieties and the use by farmers of varietal diversification schemes and fungicides. *J. Natl. Inst. Agric. Bot.* (*G.B.*) **15,** 215–230.

Priestley, R. H., and Byford, P. (1980). Yellow rust of wheat. *U.K. Cereal Pathog. Virulence Surv., 1979 Annu. Rep.* pp. 15–23.

Priestley, R. H., and Wolfe, M. S. (1980). Evidence for the effectiveness of cultivar diversification in reducing the spread of yellow rust and mildew in cereals. *U.K. Cereal Pathog. Virulence Surv., 1979 Annu. Rep.* pp. 71–74.

Reichert, I., and Palti, J. (1967). Prediction of plant disease occurrence: A patho-geographical approach. *Mycopathol. Mycol. Appl.* **32,** 337–355.

Schafer, J. F. (1971). Tolerance to plant disease. *Annu. Rev. Phytopathol.* **9,** 235–252.

Schmidt, R. A. (1978). Diseases in forest ecosystems: The importance of functional diversity. *In* "Plant Disease: An Advanced Treatise" (J. G. Horsfall and E. B. Cowling, eds.), Vol. 2, pp. 287–315. Academic Press, New York.

Segal, A., Manisterski, J., Fischbeck, C., and Wahl, I. (1980). How plant populations defend themselves in natural ecosystems. *In* "Plant Disease: An Advanced Treatise" (J. G. Horsfall and E. B. Cowling eds.), Vol. 5, pp. 75–102. Academic Press, New York.

Sherf, A. F. (1954). The 1953 crown and stem rust epidemic of oats in Iowa. *Proc. Iowa Acad. Sci.* **61,** 161–169.

Simons, M. D. (1969). Heritability of crown rust tolerance in oats. *Phytopathology* **59,** 1329, 1333.

Simons, M. D. (1972a). Crown rust tolerance of *Avena sativa*-type oats derived from wild *Avena sterilis*. *Phytopathology* **62,** 1444–1446.

Simons, M. D. (1972b). Polygenic resistance to plant disease and its use in breeding resistant cultivars. *J. Environ. Qual.* **1,** 232–240.

Simons, M. D. (1979). Influence of genes for resistance to *Puccinia coronata* from *Avena sterilis* on yield and rust reaction of cultivated oats. *Phytopathology* **69,** 450–452.

Simons, M. D., Luke, H. H., Chapman, W. H., Murphy, H. C., Wallace, A. T., and Frey, K. J. (1957). Futher observations on races of crown rust attacking the oat varieties Landhafer and Santa Fe. *Plant Dis. Rep.* **41,** 964–969.

Smith, R. J., Daniel, J. J., Fox, W. T., and Templeton, G. E. (1973). Distribution in Arkansas of a fungus disease used for biocontrol of northern jointvetch in rice. *Plant Dis. Rep.* **57,** 695–697.

Stevens, N. E. (1948). Disease damage in clonal and self-pollinated crops. *J. Am. Soc. Agron.* **40,** 841–844.

Tozzetti, G. T. (1767). True nature, causes, and sad effects of the rust, the bunt, and other maladies of wheat, and of oats in the field. *In* "Phytopathological Classics No. 9," pp. 1–139. Phytopathol. Soc., St. Paul, Minnesota.

Trenbath, B. R. (1977). Interactions among diverse hosts and diverse parasites. *Ann. N.Y. Acad. Sci.* **27,** 124–150.

Ullstrup, A. J. (1972). The impacts of the souther corn leaf blight epidemics of 1970–1971. *Annu. Rev. Phytopathol.* **10,** 37–50.

van der Plank, J. E. (1963). "Plant Diseases: Epidemics and Control." Academic Press, New York.

van der Plank, J. E. (1968). "Disease Resistance in Plants." Academic Press, New York.

Vavilov, N. I. (1957), "World Resources of Cereals, Leguminous Seed Crops, and Flax, and Their Utilization in Plant Breeding." Acad. Sci. USSR, Moscow.

Wahl, I. (1958). Studies on crown rust and stem rust on oats in Isreal. *Bull. Res. Counc. Isr., Sect. D* **6,** 145–166.

Wahl, I. (1970). Prevalence and geographical distribution of resistance to crown rust in *Avena sterilis. Phytopathology* **60,** 746–749.

Watson, I. A. (1970). Changes in virulence and population shifts in plant pathogens. *Annu. Rev. Phytopathol.* **8,** 209–230.

Watson, I. A., and Singh, D. (1952). The future for rust resistant wheat in Australia. *J. Aust. Inst. Agric. Sci.* **18,** 190–197.

Way, M. J. (1979). Significance of diversity in agroecosystems. *Proc.: Opening Sess. Plenary Sess. Symp., Int. Congr. Plant Prot., 9th, 1979* pp. 9–12.

Weltzien, H. C. (1972). Geophytopathology. *Annu. Rev. Phytopathol.* **10,** 277–298.

Weltzien, H. C. (1978). Geophytopathology. *In* "Plant Disease: An Advanced Treatise" (J. G. Horsfall and E. B. Cowling, eds.), Vol. 2, pp. 339–360. Academic Press, New York.

Wolfe, M. S. (1979). Pathogen population control in powdery mildew of barley. *Proc.: Opening Sess. Plenary Sess. Symp., Int. Congr. Plant Prot., 9th, 1979* pp. 145–148.

Wolfe, M. S., and Barrett, J. A. (1980). Can we lead the pathogen astray? *Plant Dis.* **64,** 148–155.

Zadoks, J. C., and Parlevliet, J. E. (1979). The integrated concept of disease resistance: A modeler's view of general and specific resistance. *Proc.: Opening Sess. Plenary Sess. Symp., Int. Congr. Plant Prot., 9th, 1979* pp. 187–190.

Zadoks, J. C., and Schein, R. D. (1979). "Epidemiology and Plant Disease Management." Oxford Univ. Press, London and New York.

CHAPTER 17

International Cooperation on Controlling Exotic Pests

ALVIN KEALI'I CHOCK

Region II, International Programs
Plant Protection and Quarantine
Animal and Plant Health Inspection Service
United States Department of Agriculture
The Hague, The Netherlands

I. INTRODUCTION

Two centuries ago, when the United States was founded, international cooperation in controlling exotic pests[1] was not necessary, for it took months for a

[1]The term "pest," as used in this chapter, has the same definition as that in Article II, Section 2, of the International Plant Protection Convention (IPPC), revised text, which states: "the term 'pest' means any form of plant or animal life, or any pathogenic agent, injurious or potentially injurious to plants or plant products; and the term 'quarantine pest' means a pest of potential national economic importance to the country endangered thereby and not yet present there, or present but not widely distributed and being actively controlled."

EXOTIC PLANT PESTS AND
NORTH AMERICAN AGRICULTURE

ISBN 0-12-757880-3

vessel to reach it from Europe, the only area of trade for the new country. Any fruits and vegetables which were brought or carried by the passengers and crew would have already been eaten. If they had spoiled, they were thrown overboard. Cargo of plants and plant products was limited, thus reducing the possibility of new exotic pests arriving and establishing themselves in the new country.

Even one century ago, with the expansion of the United States to the western part of North America, the increased speed of travel (relatively speaking), and trade with several different continents—Europe, Africa, Central America, and Asia—contact with the outside world was still limited because of the large size and diversity of the United States. Although immigrants were welcomed from most parts of the world (at least from Europe), international trade was still small and not significant. The European immigrants adapted the Indian crops of corn, cotton, potatoes, and tobacco, and eventually made monocultures as we know them today. But they also brought fruits, vegetables, and grains which were familiar to them, and with these importations of nursery stock came the inadvertent importation of new pests, including weeds. Some established themselves in the absence of their natural enemies, whereas others did not thrive in their new habitat. Until the last quarter of the nineteenth century, most of the exotic pests were from Europe (McGregor, 1973). Some exotic pests, such as the European corn borer, boll weevil, spotted alfalfa aphid, green bug, alfalfa weevil, pea aphid, Hessian fly, wheat stem sawfly, brown wheat mite, apple mites, and cabbage worm, were accidentally introduced, whereas the gypsy moth was introduced for scientific research and then escaped, becoming a pest.

To some degree the United States was like a giant island, surrounded by two large bodies of water, the Pacific Ocean on the west and the Atlantic Ocean on the east. These natural barriers slowed down the establishment of new exotic pests. As with any insular environment, however, as demonstrated by the ecological history of Hawaii, when a newly arrived pest managed to overcome the barrier, it established itself rapidly in the new ecosystem, and in many cases its spread was difficult to retard and control. The surrounding oceans also created in the minds of most Americans an illusion of protection and isolation, which was perpetuated in national policy and budgeting to some degree. This illusion of isolation was only slightly disturbed when America participated, as a late entrant, in World War I. But it was "over there" and we were "over here." And we did, after all, have those two great oceans to protect us from potential enemies (human as well as agricultural) in Europe. The Pacific Ocean was even more expansive than the Atlantic, with twice the distance from the nearest continent, and that in itself was sufficient protection from all kinds of foes, including agricultural pests. After the conclusion of World War I, we retreated into our protective shell.

With the advent of World War II in 1939, there was a slight increase in our international involvement. But this was a European war, and the occupation of parts of China by Japan in 1931 was just a problem with a few Japanese warlords. We still had the protection of our two oceans, and this illusion continued for 2 more years, until the Japanese attack on Pearl Harbor on December 7, 1941. The illusion of isolation was finally shattered. Although some efforts were made to return to isolationism, these were by and large unsuccessful because of worldwide developments and the resulting erosion of our former geographical barriers. These barriers had already been breached by increased commerce and the subsequent introduction of more exotic pests, both from the East and the West.

The war brought vast improvements in transportation, both surface and air. These changes meant growth in trade between countries. Not only were they faster, they were also cheaper. After the war, trade and transport between different parts of the world continued to increase, and the United States was no longer isolated or "protected" by two large bodies of water, as in the past.

In the meantime, the *Insect Pest Act* of 1905 was enacted "to prohibit importation or interstate transportation of plant pests." This was supplemented by the *Plant Quarantine Act* of 1912, which provided for federal legislation to restrict and control the entry of plants and plant products in order to prevent the entry of plant pests. The legislation was mainly concerned with the importation of nursery stock, its certification inspection in the country of origin, and its subsequent inspection in the United States, mostly at that time by state collaborators. It was not until May 1, 1928, that an amendment to the Plant Quarantine Act provided authority to stop and, without a warrant, inspect, search, and examine persons, vehicles, receptacles, boats, ships, or vessels—and to seize and destroy or otherwise dispose of plants and plant products or other articles found to be moving or to have moved in interstate commerce, or to have been brought into the United States. The *Mexican Border Act* of 1942 gave authority to regulate, inspect, clean, and disinfect railroad cars, vehicles, baggage, and other materials from Mexico. The *Organic Act* of 1944 gave the Secretary of Agriculture authority to cooperate with farmers' associations, individuals, and Mexico to detect and control plant pests. This act also gave authority to certify plants and plant products for export certification. Thus began the involvement of the United States Department of Agriculture (USDA) in international programs. The *Federal Plant Pest Act* of 1957 gave authority to regulate the movement of "plant pests" into or through the United States. This term covers not only insects but also diseases, viruses, nematodes, and mollusks. The act also provided for emergency actions and inspection, without a warrant, of persons and means of conveyance because of the possible threat of spread of a plant pest. The act repealed the *Insect Pest Act* of 1905 and the *Mollusk Act* of 1951, and strengthened plant protection and quarantine activities.

The expansion of the plant quarantine force, or agricultural quarantine inspection, as we know it today, did not begin until after World War II. In 1956, the

Plant Quarantine Training Center was established in New York City to train uniformly the new inspectors. It began with a 6-month training program, which included 3 months of academic and observation training and 3 months of supervised on-the-job training in maritime and airport activities in the Port of New York. Since then, the training center has moved to Battle Creek, Michigan, and is now located in Frederick, Maryland. It is now known as the Professional Development Center (PDC), with many training programs, including an 8-week New Officer Training (NOT) course.

More international trade, growth in international travel, expansion of trade in agricultural commodities, containerization of cargo, and an increase in opportunities for pest hitchhikers on nonagricultural cargo have resulted in changes in the traditional means of inspection at ports of entry at U.S. borders. Procedures had to be modified, policies had to be changed, and new methodology and technology was created to meet the rapidly changing times. This is a continuing and ongoing activity in the United States.

The organization of the regulatory arm of the USDA has also changed. Formerly part of the Agricultural Research Service (ARS), the Animal and Plant Health Inspection Service (APHIS) came into being on April 2, 1972, with three major components, now known as:

Plant Protection and Quarantine (PPQ)
Veterinary Services (VS)
Administrative Management (AM)

PPQ itself was a merger, the year before, of the Plant Quarantine Division (PQD) and the Plant Protection Division (PPD), which had been known for many years as the Plant Pest Control Division (PPCD). The former was concerned with inspection at ports of entry at U.S. borders and territorial overseas areas, and had responsibility for enforcing foreign and territorial quarantines, whereas the latter was responsible for domestic quarantines and control programs. Both divisions were plant protection organizations, so it was natural that they should be one rather than two separate organizations. In addition, PPQ is responsible for regulating the entry of animal by-products.

A. Principles of Plant Quarantine

The U.S. plant quarantine system is now 70 years old, and is intended to prevent the introduction or spread of plant pests. Legal restrictions are made on the movement of commodities and articles in order to prevent or inhibit the establishment of plant pests in areas where they are not known to occur (Spears, 1974).

The principles of plant quarantine, as practiced by U.S. federal and state governments, were formulated by the National Plant Board in 1931 and slightly modified in 1936. The establishment of a quarantine must be based on the following premises: (1) the pest must offer an actual or expected threat to substantial interests; (2) the proposed quarantine must be a necessary or desirable measure for which no other substitute, involving less interference with normal activities, is available; (3) the objective of the quarantine must be reasonable; and (4) the economic gains expected must outweigh the costs of administration and interference with normal activities.

When it is determined that a quarantine against a particular pest is necessary to prevent the introduction or spread of the pest, a notice of intent is published in the *Federal Register* and a public hearing is scheduled to allow interested parties to present their views. If the biological evidence for the quarantine is affirmed and the need for it still exists, the quarantine document is prepared by PPQ and reviewed for legal aspects by the Office of General Counsel (OGC), USDA. It is then signed by the APHIS Administrator and published in the *Federal Register*.

Revision of the quarantine may take place (1) when requests for additional commodities are made by the importer; (2) when additional or alternate commodity treatments are developed through research; (3) when less stringent requirements can be made, and yet retard the pest; (4) when more stringent requirements must be made to prevent introduction due to the establishment of the pest in a foreign country; (5) when the pest has been eradicated in a regulated area and restrictions are no longer necessary; (6) when treatment facilities are no longer available in a specifically approved port of entry; and when (7) entry can be made in certain ports without need for treatment.

Thus, quarantines must keep abreast of plant protection practices. These practices change as new developments in technology and methodology are discovered and implemented, such as biological controls (parasites, viruses, bacteria), pesticides (new uses of existing pesticides, different dosages, or even new pesticides), breeding (for resistant varieties), cultural methods, pest distribution, and ecological conditions. In other words, there must be means of revising quarantines as new developments occur.

Quarantines should be established only to prevent the entry or to control the distribution of a pest, and not to restrict trade or to serve as a nontariff barrier. Quarantines are for plant protection and *not* for trade protection.

B. Mutual Benefits of Cooperation

It has been said that plant pests do not recognize political boundaries, and the erection of fences, immigration controls, etc., will not deter their movement. Geographical barriers, on the other hand, such as a high mountain range or an expansive body of water, may deter and retard the natural spread of some plant pests.

The need for cooperation with a neighboring country has often led to bilateral discussions, both informally and formally, on ways in which to cooperate. This may have been to control the spread of an agricultural pest, such as the pink bollworm with Mexico; to cooperate in the inspection and dissemination of information concerning khapra beetle-infested ships; or to establish similar foreign quarantines to prevent the introduction of a pest which could potentially damage the crops of both neighboring countries.

The sharing of methodology and control developments reduces possible research costs on the part of both countries and eliminates duplicative efforts.

The further need for cooperation among several countries has led to the organization and development of regional plant protection organizations (see the appendix to this chapter), most of which have been formed under the umbrella of the International Plant Protection Convention. If there are common crops, and potentially damaging pests not known to occur in those countries, then there is a need for common phytosanitary requirements to protect those crops. Similar quarantine strategies can also be adopted by these countries for mutual protection and more effective action.

The sharing of pest distribution information is essential in order to plan effective and accurate quarantine actions. However, many countries are reluctant to share this information because of possible trade embargoes. Although this remains a possibility, a country's pursuit of an up-to-date and accurate reporting system on the distribution of plant pests will win it the respect and confidence of its neighbors. In addition, cooperative efforts can be made to prevent the further spread of the plant pest by all parties concerned. Unless accurate information is distributed, there remains the further danger that the importing country, because of its lack of confidence in the exporting country's surveys, reporting system, and plant protection service, will invoke general prohibitions rather than prohibitions against specific pests and crops, plants, plant products, and other commodities. It is thus incumbent upon the exporting country to provide as accurate information as possible on the distribution and control of its plant pests, along with timely data on pest outbreaks.

II. INTERNATIONAL PLANT PROTECTION CONVENTION

A. Background

The International Plant Protection Convention (IPPC) was approved under Article XIV of the Constitution of the Food and Agriculture Organization of the United Nations (FAO) by the Sixth Session of the FAO Conference in November 1951 and came into force on April 3, 1952, when three signatory governments ratified the Convention.

The revised text of the IPPC was approved by the Twentieth Session of the FAO Conference in November 1979. It will come into force 30 days after the acceptance of the revised text by two-thirds of the contracting parties, of which there are 82.

1. *History*

In May 1950, The Hague Conference convened by FAO and the Kingdom of the Netherlands considered and accepted a plant protection agreement which had been drafted by FAO. This draft was a synopsis of recommendations which had been submitted to FAO by its member states. A month later, a panel of plant protection officials from Canada and the United States met informally in Rome to consider the draft, which covered the principles and administration of international plant quarantine. Comments were solicited from a number of specialists, and a revised text was presented at a special session of the FAO Conference in 1950. Some member countries wanted more time to review and consider the draft text. Their views and comments were summarized and circulated by the FAO Secretariat. A meeting of plant protection specialists met in September 1951 to coordinate the views expressed in the comments, and agreed upon a final draft, which was approved by the FAO Conference in November 1951 (Chock, 1976, 1979).

The principles and administration of international plant quarantine, rather than technical matters, were discussed at The Hague Conference in 1950. Many different kinds of phytosanitary certificates were required by importing countries, and a standard model was sorely needed. In addition, it was recognized that the certificate was only as good as the capability of the issuing agency, and that there were certain elements which should be required for national plant protection organizations. A world reporting service to provide information about plant pests and changes in phytosanitary regulations was also needed, and this led to the concept of the *FAO Plant Protection Bulletin*. There was also a need to provide for the standardized establishment of regional plant protection organizations. Some of these were later established under Article VII of the IPPC and are FAO statutory bodies, whereas others fall under the aegises of this same article but are independent of FAO, while still cooperating with FAO (Mulders, 1977).

2. *Previous Agreements*

When the IPPC came into force on April 3, 1952, it replaced three other conventions (Article X of the IPPC):

The International Convention on measures to be taken against the *Phylloxera vastatrix* of November 3, 1881

The additional Convention signed at Berne on April 15, 1889

The International Convention for the Protection of Plants signed at Rome on April 16, 1929

The 1929 Convention was the result of preparatory work by the International Institute of Agriculture in Rome since 1923. Rapidly changing conditions in plant protection and World War II did not allow it to be an effective instrument. Although 46 countries participated at the 1929 meeting, only 12 ratified the Convention, and another 5 adhered to it later (Mulders, 1977).

B. Convention Provisions

The International Plant Protection Convention has as its aims:

Strengthening international efforts to prevent the introduction and spread of pests of plants and plant products
Securing international cooperation to control pests and to promote measures for pest control
Adoption by each country of the legislative, technical, and administrative measures to carry out the Convention's provisions

The Convention has 15 articles. Its more important provisions, which serve to protect the world's crops, are summarized below.

1. *National Plant Protection Organization*

Article IV of the Convention requires each contracting party to establish as soon as possible, and to the best of its ability, an official plant protection organization which can satisfactorily do the following:

Carry out surveys to report the existence, outbreak, and spread of plant pests, and to control these pests
Inspect plants under cultivation, and plants and plant products in storage or transportation
Inspect consignments of plants and plant products moving in international trade, as well as other articles which may act incidentally as carriers of pests
Inspect and supervise storage and transportation facilities involved in international trade to prevent the dissemination of plant pests
Disinfest or disinfect plant materials, and their containers, storage places, or transportation facilities, moving in international trade
Inspect commodities and issue phytosanitary certificates relating to their phytosanitary condition and origin
Strengthen advisory services about pests and means of their prevention and control
Carry out research and investigations in plant protection
Report to FAO (and other contracting parties) information on the outbreak and spread of plant pests

Publish its plant quarantine regulations and requirements and distribute them immediately to FAO (and other governments)

In addition, this article requires each contracting party to submit a description of the scope of its plant protection organization to FAO. In 1976, while serving as FAO's Plant Quarantine Officer, I circulated a questionnaire (prepared with the assistance of John M. Mulders, Canada Agriculture) to all FAO members and parties of the Convention. The response to the questionnaire was excellent and showed the great variety of national plant protection organization capabilities. They varied from an Acting Agricultural Officer (Tuvalu in the South Pacific) who performs plant protection duties only on a part-time basis to an inspectorate of more than 1500 in the United States.

Training, equipment, and facilities, as well as other factors, are diverse, and in many cases are completely lacking. The parties to the Convention are seeking ways to improve and strengthen their national plant protection organizations. Many of the developed countries, through their bilateral and multilateral agencies, and international organizations such as FAO, are helping the developing countries attain this goal.

In 1977, another questionnaire (prepared by Robert P. Kahn, USDA) was circulated to those organizations which indicated that they maintained plant quarantine postentry stations. The object of this questionnaire was to determine existing facilities and capabilities.

Every 2 years, FAO circulates an updated list of the addresses of national and regional plant protection organizations. Although updated information is solicited at any time, the list is usually finalized in July of the odd-numbered years (Pastore-Gagliardi and Karpati, 1981).

2. *Model Phytosanitary Certificate*

Section II,A,1 pointed out the diversity of phytosanitary certificates which existed prior to the IPPC. Although the IPPC model has been universally adopted, it has long been recognized that the certificate is only as good as the capability of the issuing organization, and that this differs from country to country, depending upon the experience and training of the inspecting officers (Mulders, 1977).

Only recently have plant protection training programs been available as a separate discipline at the university level. Heretofore, one had to major in entomology or plant pathology, and instead of receiving a broad background in plant protection, one was forced to specialize. Workshops, seminars, and short courses have been sponsored by various institutions and organizations. The University of Wageningen in the Netherlands offers a 1-year course in plant protection for international participants. The USDA has for many years provided annually a course in plant quarantine for international participants, who are sponsored and funded by FAO, USAID, or their own country. A plant quarantine training center was established by FAO, the United Nations Development Programme (UNDP), and the Government of Nigeria in Ibadan to provide annual courses for national and international participants. More remains to be done in PPQ training, especially in the developing countries, and personnel need to be taught in programs geared to the abilities of the countries (Brown, 1977).

During a 1-week plant quarantine (PQ) workshop sponsored by the South Pacific Commission (SPC), it was found that the specific PQ training background of the participants ranged from nothing at all to the 13-week curricula of the plant quarantine training centers of the United States and Australia (Chock, 1977, 1979).

The training programs enable the contracting parties to comply with Section 1.(a) of Article V of the IPPC, which states that the "inspection shall be carried out and certificates issued only by or under the authority of technically qualified and duly authorized officers and in such circumstances and with such knowledge and information available to those officers that the authorities of importing countries may accept such certificates with confidence as dependable documents."

Thus, the inspectors must be properly trained, with an extensive knowledge of plants pests and available information about the phytosanitary regulations and requirements of the importing countries.

a. Export Certificate. This certification is based upon two items:

Inspection to determine freedom from quarantine pests and relative freedom from other injurious pests

Conformance with the current phytosanitary requirements of the importing country

The issuance of a certificate does not allow carte blanche entry into the importing country. It must, in any event, be in accordance with the provisions of the Convention, especially with respect to the qualifications of the inspectors as described above. It must be pointed out that, regardless of the documentation, the importing country has every right to reinspect the plants, plant materials, or other commodities if it wishes. The degree of reinspection will depend to some extent upon the importing country's confidence in the exporting country's plant protection service, and also on the particular pests which may be present in the consignment (depending to some degree upon the importing country's agricultural crops). The value of the exporting country's certification depends upon (1) the qualifications of its inspectors; (2) the plant quarantine restrictions of the importing country, and their effectiveness; (3) the acceptance of plant protection and quarantine measures by the populace of the importing country; (4) the effectiveness of its plant protection programs; and (5) the control programs for containing or eradicating important agricultural pests (Chock, 1977, 1979).

One point must be emphasized about the phytosanitary certificate. It is a document from one plant protection service to another, attesting to freedom from quarantine pests. The certificate is *not* a commercial document, nor does it guarantee the quality of the commodity. Plant protection services are now refusing to provide a certificate for commodities which seldom provide an opportunity for the spread of plant pests, such as canned goods, chocolate bars, and similar items.

b. Reexport Certificate. At the 1973 Ad Hoc Consultation on the International Plant Protection Convention held at FAO, it was pointed out that there is now an extensive trade in some countries involving importation and reexport, i.e., breaking down of the original consignment into a number of smaller ones. This requires assurances that infestation did not take place since the issuance of the original phytosanitary certificate.

The Netherlands Plant Protection Service prepared a paper for the 1976 Government Consultation on the International Plant Protection Convention, also held at FAO, which explained two types of transit through a country:

> Direct transit—the consignment passes from the country of origin through the transit country in the same means of conveyance to the country of destination
>
> Indirect transit—here the means of conveyance is changed, or the consignment is broken up into several consignments or unpacked and rehandled.

The purposes of the reexport certificate are (1) to declare that the importation was accompanied by a phytosanitary certificate in the country of origin; (2) to declare that during repacking or storage, the requirements of the country of destination have been met; (3) to perform any additional inspection if necessary; (4) and to indicate the basis for issuance of the reexport certificate (based on the original certificate only, or also on an additional inspection by the reexport country's service) (Chock and Mulders, 1976a,b, 1977).

3. *World Reporting Service*

Articles VI and VII provide that contracting parties are to inform FAO and regional and national organizations about their quarantine requirements and pest information. To this end, the *FAO Plant Protection Bulletin* was established as a scientific journal to provide this information about the incidence of plant pests of economic importance (Johnston, 1979). To this end, the *Bulletin* has endeavored to provide this information on a timely basis. However, many countries continue to delay the issuance of this information. The cooperation of all countries is essential in order to make this program a success.

In addition, the *Bulletin* is expanding its coverage by providing solutions for practical crop protection problems, particularly in the developing world. It will be an effective source of information for plant protection workers to improve their knowledge (Brader, 1977).

4. *Digest of Plant Quarantine Regulations*

Originally published in 1949 and revised in 1952, the *Digest* soon went out of print and also out of date, although Supplement I was published in 1954. Supplements II and III were strictly regional in content. Changes in legislation and regulations are published regularly in the *FAO Plant Protection Bulletin* (Chock, 1979). The need for a central handbook has long been recognized, and the beginnings of a revised summary were initiated in late 1978, with a letter to all national plant protection services, supplemented by an article in the *Bulletin* (Chock and Mulders, 1978). The new summary was assembled in mid-1982, with publication in a looseleaf format under a contract arrangement with the European and Mediterranean Plant Protection Organization (EPPO). The first issuance will not be complete, and all countries are urged to submit their contributions so that the future issuances will be complete and up-to-date.

5. *The 1979 Amendments to the IPPC*

The first firm initiation for the IPPC revisions was the 1973 Consultation (Reddy *et al.,* 1973), which had been preceded by other meetings over the decade. This was followed by the 1976 Consultation, which recommended adoption of a revised text, together with a revised model phytosanitary certificate and a new certificate for reexport. This was presented before the Nineteenth Session of the FAO Conference in 1977. Although there was general agreement on the proposed amendments, there were some differences of opinion on a few points, resulting in a postponement of final approval for the revised text until the Twentieth Conference in 1979. In the interim, an ad hoc subsidiary body of the Committee on Agriculture (COAG) reviewed the amendments and adopted a revised text for consideration by the Twentieth Conference.

The major changes to the IPPC are as follows:

a. Preamble. Clarification of the recognition of cooperation to prevent the spread of pests across and within national boundaries.

b. Article II. The definition of plant products has been broadened; in addition, throughout the IPPC, whenever the word "plants" appears, it is followed by "plant products." The practical effect is that in making the decision to carry out an inspection or to take other phytosanitary measures, the overriding factor is whether that article creates or constitutes a pest risk.

This article uses two new terms which are of significance: "pest" and "quarantine pest." These apply only to the interpretations of the provisions of the IPPC, and national legislation is not required to conform to them.

c. Article IV. The term "containers" has been expanded to include packing material and anything else which accompanies plants or plant products.

d. Article V. The revised certificates should be used for both plants and plant products whenever a certificate is required by the importing country. In addition, certificates must have the wording of the model certificates. There is also a restriction on the importing country's right to require additional declarations, since the wording of the certificate states that the consignment meets the specified requirements of the importing country. This article also provides for the reexport certificate.

e. Article VI. This article contains a recommendation for the listing of pests whose introduction is prohibited or restricted. This listing is now limited to pests of potential economic importance to the country concerned. Information on plant quarantine requirements must be submitted to FAO, the regional plant protection organization of which the country is a member, and other plant protection services which are directly concerned. This change alleviates the tasks of small countries. A new section (4) is added which recognizes FAO's practice of providing information on plant quarantine regulations.

f. Article VIII. An addition to the text recognizes the current practice of most regional plant protection organizations of providing their members with information on phytosanitary matters.

g. Model Phytosanitary Certificate. Of prime significance is the revision of the certifying statement, which now states that the inspection was performed according to "appropriate procedures," and introduces the concept of "quarantine pests" (considered to be free from) vs. "other injurious pests" (practically free from). This part involved the most discussion, both at the 1973 and 1976 Consultations, the 1979 COAG consultative group, and the 1977 Conference. It was the most difficult part of the IPPC to be approved.

h. Model Phytosanitary Certificate for Reexport. This is a new certificate.

III. REGIONAL PLANT PROTECTION ORGANIZATIONS

Article VII provides for international cooperation, and obligates IPPC members to provide FAO with information about plant pests, pest outbreaks, and plant quarantine requirements on a timely basis. FAO, on the other hand, is responsible for distributing this information. The individual responsible for administering the IPPC is the Plant Quarantine Officer, who is a member of the Plant Protection Service (AGPP), Plant Production and Protection Division (AGP), Agriculture Department (AG), Food and Agriculture Organization of the United Nations (FAO) in Rome. The present incumbent is Dr. Joseph F. Karpati, who is loaned to FAO by PPQ-APHIS-USDA.

Article VIII provides for regional plant protection organizations to carry out the provisions and aims of the IPPC, and also to give information on plant protection to their members. Three of the regional organizations are FAO statutory bodies: Plant Protection Committee for the South East Asia and Pacific Region, Caribbean Plant Protection Commission, and Near East Plant Protection Commission (see the appendix). The others are independent organizations set up by the member countries in the region.

Most of these organizations hold periodic meetings, which average at least once every 2 years. They have a secretariat, which is funded on a pro rata basis by the member countries for the independent organizations, and by FAO for the statutory bodies. In the case of the FAO organizations, the secretariat consists of a Regional Plant Protection Officer who is responsible for the plant protection activities in the region, some of which may be directly related to the regional organization, and some which may not be directly connected. He is usually assisted by at least a half-time secretary, clerk, administrative assistant, or combination thereof. The independent organizations tend to be better staffed, with one to three professionals and supporting administrative/clerical personnel, but some, due to difficulties in financing and organization, may become inactive. All of them generally issue newsletters, journals, papers, or some other means of information to keep their members cognizant of the latest developments in plant protection activities within and without the region. They form a cohesive bond between the member countries. The amount of activity generated is usually related to the funding of the organization. In general, all of them have top-quality plant protection specialists in the organization's secretariat.

Most of these organizations seek to identify the pests which could cause considerable harm to the region's agricultural crops and to determine means by which these pests could establish themselves in the region. The organizations also seek to provide common legislative and regulatory measures which would best protect the region's agricultural interests (Reddy, 1976).

Descriptions of these organizations may be found in the FAO Plant Protection Bulletin, Vol. 25(4) (Mathys, 1977; Reddy, 1977b; Berg, 1977; Addoh, 1977), the "Proceedings of Symposia, IX International Congress of Plant Protection" (Mathys and Smith, 1980; Berg, 1980; Hopper, 1980; Reddy, 1980; Addoh, 1980), and elsewhere (Helms, 1980; Reddy, 1977a, 1981; Rohwer, 1979; Shannon, 1980; and Smith, 1979).

IV. INTERNATIONAL PROGRAMS, PPQ, APHIS, USDA

One of the recommendations of the McGregor Report (1973) was that PPQ-APHIS develop a source inspection system, which was termed the Agricultural

Source Inspection and Surveillance Technique (ASIST) in that document. The advantages are as follows:

Inspection can be performed while shipments are being assembled, which is important with increased containerization.
The cost of inspection and treatment is borne by the commercial exporter.
The procedures provide for less governmental interference with commerce.

It is pointed out that for ASIST to succeed, there must be strong cooperation by the exporter and his government's plant protection service. The purpose of this system is to have the exporter ship pest-free commodities.

This program began in the Netherlands with shipments of bulbs to the United States in 1951. This program has also been undertaken in Chile, Australia, New Zealand, Japan, Belgium, France, England, South Africa, and other countries.

To expand this system, and also to develop a pest information network which would move America's first line of defense from the U.S. border ports of entry to the source of the pests and commodities, APHIS implemented International Programs in 1981. The Mexico Region was expanded to include all of Latin America. New regions were established in the rest of the world, as follows:

Region I (Latin America)—headquartered in Monterrey, Mexico, (to be moved to Mexico City), with Ed L. Ayers, Jr., as Regional Director. Area offices are located in Mexico City, Mexico; Guatemala City, Guatemala; Santiago, Chile; and Lima, Peru.
Region II (Europe and Africa)—headquartered at The Hague, the Netherlands, with Alvin K. Chock as Regional Director. Two offices are presently in the Region: Lisse, the Netherlands; and Mannheim, Federal Republic of Germany. An additional office is planned for location in Africa as soon as possible, and additional offices later in the Near East as well as other locations in Africa.
Region III (Asia and the Pacific)—headquartered temporarily in Agana, Guam, with Stanley S. Miyake as Regional Director. This office will be moved to a more central location within the region. One area office is now located in Tokyo, Japan, another will soon be established in New Delhi, India, and other area offices will be established.

V. CONCLUSIONS

One country cannot do it alone, nor is it sufficient to have only two or three do it together. Without international cooperation, plant protection cannot be fully effective and efficient. This has been recognized for many decades, and has become increasingly emphasized with the growth in international trade. This was the objective of the founding fathers of the International Plant Protection Convention more than three decades ago.

The preamble of the IPPC states that the "contracting parties, [recognize] the usefulness of international cooperation in controlling pests of plants and plant products and in preventing their spread." This Convention obligates its members to undertake certain responsibilities. We must do our best to meet these obligations, for only through international cooperation will we be able effectively to retard the spread of plant pests to North America.

APPENDIX: Regional Plant Protection Organizations

Caribbean Plant Production Commission (CPPC)	
Member governments	Barbados, Colombia, Cuba, Dominica, Dominican Republic, France, Grenada, Guyana, Haiti, Jamaica, the Netherlands, St. Lucia, Surinam, Trinidad and Tobago, United Kingdom, United States of America, Venezuela
Year of establishment	1967
Address	c/o FAO Regional Office for Latin America[a] Casilla 10095 Santiago Chile
Technical Secretary	Mario A. Vaughan (pro tempore)
Comité Interamericano de Protección Agrícola (CIPA)	
Member governments	Argentina, Bolivia, Brazil, Chile, Paraguay, Uruguay
Year of establishment	1965
Address	Luís Saenz Peña 1940 Buenos Aires Argentina
Scientific Executive Director	Mario Carlos Zerbino, Argentina (last recorded director) At present, this organization is inactive
European and Mediterranean Plant Protection Organization (EPPO)	
Member governments	Algeria, Austria, Belgium, Bulgaria, Cyprus, Czechoslovakia, Democratic Republic of Germany, Denmark, Federal Republic of Germany, Finland, France, Greece, Guernsey, Hungary, Ireland, Israel, Italy, Jersey, Luxembourg, Malta, Morocco, the Netherlands, Norway, Poland, Portugal, Rumania, Spain, Sweden, Switzerland, Tunisia, Turkey, Union of Soviet Socialist Republics, United Kingdom, Yugoslavia
Year of establishment	1950
Address	1, rue Le Nôtre 75016 Paris France
Director General	G. Mathys

APPENDIX *Continued*

Interafrican Phytosanitary Council (IAPSC)	
Member governments	Algeria, Angola, Arab Republic of Egypt, Benin, Botswana, Burundi, Cameroon, Central African Republic, Chad, Comores, Congo, Djibouti, Equatorial Guinea, Ethiopia, Gabon, Gambia, Ghana, Guinea, Guinea-Bissau, Ivory Coast, Kenya, Lesotho, Liberia, Libya, Madagascar, Malawi, Mali, Mauritania, Mauritius, Morocco, Mozambique, Niger, Nigeria, Rwanda, São Tomé and Principe, Senegal, Seychelles, Sierra Leone, Somalia, Sudan, Swaziland, Tanzania, Togo, Tunisia, Uganda, Upper Volta, Zaire, Zambia
Year of establishment	1956
Address	P.O. Box 4170 Yaoundé Cameroon
Scientific Secretary	Abdel Lebrun Mbiele
Junta del Acuerdo de Cartagena (JUNAC)	
Member governments	Bolivia, Colombia, Ecuador, Peru, Venezuela
Year of establishment	1969
Address	Casilla de Correo 3237 Lima Peru
Executive Secretary	Jaime Rodríguez
Near East Plant Production Commission (NEPPC)	
Member governments	Bahrain, Cyprus, Egypt, Iran, Iraq, Jordan, Kuwait, Lebanon, Libya, Pakistan, Qatar, Saudi Arabia, Somalia, Sudan, Syrian Arab Republic, United Arab Emirates, Yemen Arab Republic
Year of establishment	1963
Address	FAO Regional Office for the Near East FAO, Rome Italy
Technical Secretary	Vacant
North American Plant Protection Organization (NAPPO)	
Member governments	Canada, Mexico, United States of America
Year of establishment	1976
Address	Dirección General de Sanidad Vegetal Secretaría de Agricultura y Recursos Hidráulicos Guillermo Pérez Valenzuela No. 127 México 21, D.F. México
Executive Secretary	Felipe Romero Rosales
Organismo Internacional Regional de Sanidad Agropecuaria (OIRSA)	
Member governments	Costa Rica, El Salvador, Guatemala, Honduras, Mexico, Nicaragua, Panama
Year of establishment	1955
Address	Apartado Postal 1654 San Salvador El Salvador

(*continued*)

APPENDIX *Continued*

Executive Director	Carlos Meyer Arévalo
Plant Protection Committee for the South East Asia and Pacific Region (PPC/SEAP)	
Member governments	Australia, Bangladesh, Burma, Democratic Kampuchea, Fiji, France, India, Indonesia, Lao People's Democratic Republic, Malaysia, Nepal, New Zealand, Pakistan, Papua New Guinea, Philippines, Portugal, Socialist Republic of Vietnam, Solomon Islands, Sri Lanka, Thailand, United Kingdom, Western Samoa
Year of establishment	1956
Address	FAO Regional Office Bangkok 2 Thailand
Executive Secretary	Huang Ke-xun
South Pacific Commission (SPC)	
Member governments	American Samoa, Australia, Commonwealth of Northern Marianas, Cook Islands, Federated States of Micronesia, Fiji, France, French Polynesia, Guam, Kiribati, Marshall Islands, Nauru, New Caledonia, New Zealand, Niue, Norfolk Island, Palau, Papua New Guinea, Pitcairn Islands, Solomon Islands, Tokelau, Tonga, Tavalu, United Kingdom, United States of America, Vanuatu, Wallis and Futuna Islands, Western Samoa
Year of establishment	1947
Addresses	*Headquarters* B.P. D5 Noumea Cedex New Caledonia *Plant Protection Office* Box 2119 Suva Fiji
Plant Protection Officer	Vacant

[a] FAO is in the process of reestablishing the CPPC secretariat in the Caribbean.

REFERENCES

Addoh, P. G. (1977). The International Plant Protection Convention: Africa. *FAO Plant Prot. Bull.* **25,** 164–166.

Addoh, P. G. (1980). Regional plant protection in Africa: Programs and problems. *In* "Proceedings of Symposia, IX International Congress of Plant Protection: Fundamental Aspects," Vol. I, pp. 358–360. Entomol. Soc. Am., College Park, Maryland.

Berg, G. H. (1977). The International Plant Protection Convention: Central America and the Caribbean. *FAO Plant Prot. Bull.* **25,** 160–163.

Berg, G. H. (1980). CPPC/OIRSA, plant protection/quarantine problems. *In* ''Proceedings of Symposia, IX International Congress of Plant Protection: Fundamental Aspects,'' Vol. I, pp. 347–350. Entomol. Soc. Am., College Park, Maryland.

Brader, L. (1977). 25 years of FAO contributions to plant protection. preface. *FAO Plant Prot. Bull.* **25,** 145–146.

Brown, A. L. (1977). Training in plant quarantine. *FAO Plant Prot. Bull.* **25,** 201–203.

Chock, A. K. (1976). ''Background Paper on the International Plant Protection Convention,'' Working Pap. AGP/IPPC/76/1. FAO, Rome.

Chock, A. K. (1977). ''Principles of Plant Quarantine and the International Plant Protection Convention. SPC Workshop and Training Course in Plant Quarantine, Suva, Fiji.'' SPC, Suva, Fiji.

Chock, A. K. (1979). The International Plant Protection Convention. *In* ''Plant Health, the Scientific Basis for Administrative Control of Plant Diseases and Pests'' (D. L. Ebbels and J. E. King, eds.), pp. 11–22. Blackwell, Oxford.

Chock, A. K., and Mulders, J. M. (1976a). ''Suggested Amendments to the International Plant Protection Convention,'' Working Pap. AGP/IPPC/76/2. FAO, Rome.

Chock, A. K., and Mulders, J. M. (1976b). ''Amendments Proposed and Incorporated in the Revised Draft of the Convention, Working Pap. AGP/IPPC/76/3. FAO, Rome.

Chock, A. K., and Mulders, J. M. (1977). ''Report of the Government Consultation on the International Plant Protection Convention held in Rome, Italy, 15–19 November 1976,'' AGP:1976/M/13. FAO, Rome.

Chock, A. K., and Mulders, J. M. (1978). Digest of plant quarantine. *FAO Plant Prot. Bull.* **26,** 1–4.

Devlin, D., Chock, A. K., and Dobbert, J. (1978). ''Revision of the International Plant Protection Convention.'' FAO, Rome.

Food and Agriculture Organization (1951). ''International Plant Protection Convention.'' FAO, Rome.

Food and Agriculture Organization (1980). ''Revised Text of the International Plant Protection Convention.'' FAO, Rome.

Helms, W. F. (1980). ''North American Plant Protection Organization.'' PPQ, AHPIS, USDA, Hyattsville, Maryland.

Hopper, B. E. (1980). Organization, objectives, and quarantine problems of NAPPO. *In* ''Proceedings of Symposia, IX International Congress of Plant Protection: Fundamental Aspects,'' Vol. I, pp. 350–353. Entomol. Soc. Am., College Park, Maryland.

Johnston, A. (1979). Information requirements for effective plant quarantine control. *In* ''Plant Health, the Scientific Basis for Administrative Control of Plant Diseases and Pests'' (D. L. Ebbels and J. E. King, eds.), pp. 55–62. Blackwell, Oxford.

McGregor, R. C. (1973). ''The Emigrant Pests, a Report to Dr. Francis J. Mulhern, Administrator, APHIS.'' PPQ, APHIS, USDA, Washington, D.C.

Mathys, G. (1977). European and Mediterranean Plant Protection Organization. *FAO Plant Prot. Bull.* **25,** 152–156.

Mathys, G., and Smith, I. M. (1980). Harmonizing quarantine regulations in the European and Mediterranean Plant Protection Organization. *In* ''Proceedings of Symposia, IX International Congress of Plant Protection: Fundamental Aspects,'' Vol. I, pp. 344–346. Entomol. Soc. Am., College Park, Maryland.

Mulders, J. M. (1977). The International Plant Protection Convention—25 years old. *FAO Plant Prot. Bull.* **25,** 149–151.

Pastore-Gagliardi, G., and Karpati, J. F. (1981). ''List of National Plant Quarantine Services,'' AGPP/MISC/35. FAO, Rome.

Reddy, D. B. (1976). ''Recommended Measures for Regulation of the Importation and Movement of Plants,'' PPC/SEAP Inf. Lett. 111, pp. 1–17. FAO, Bangkok.

Reddy, D. B. (1977a). "Plant Protection Committee for the South East Asia and Pacific Region. Twenty Years of Progress. Summary of Activities, 1956–1976." FAO-RAFE, Bangkok.

Reddy, D. B. (1977b). The International Plant Protection Convention: Plant Protection Committee for the South East Asia and Pacific Region. *FAO Plant Prot. Bull.* **25,** 157–159.

Reddy, D. B. (1980). Plant protection in Asia and the Pacific: Review of progress and future orientation. *In* "Proceedings of Symposia, IX International Congress of Plant Protection: Fundamental Aspects," Vol. I, pp. 353–358. Entomol. Soc. Am., College Park, Maryland.

Reddy, D. B. (1981). "Plant Protection Committee for the South East Asia and Pacific Region, 25 Years of Service, 1956–1980." FAO, Bangkok.

Reddy, D. B., Mulders, J. M., Gonzales, R. H., and Whittemore, F. W. (1973). "Report of the ad hoc Consultation on the International Plant Protection Convention held in Rome, Italy, 9–13 July 1973," Meet. Rep. AGP:1973/M/7. FAO, Rome.

Rohwer, G. G. (1979). Plant quarantine philosophy of the United States. *In* "Plant Health, the Scientific Basis for Administrative Control of Plant Diseases and Pests" (D. L. Ebbels and J. E. King, eds.), pp. 23–34. Blackwell, Oxford.

Shannon, M. J. (1980). "International Cooperation in Pest Prevention." PPQ, AHPIS, USDA, Hyattsville, Maryland.

Smith, I. M. (1979). EPPO: the work of a regional plant protection organization, with particular reference to phytosanitary regulations. *In* "Plant Health, the Scientific Basis for Administrative Control of Plant Diseases and Pests" (D. L. Ebbels and J. E. King, eds.), pp. 13–22. Blackwell, Oxford.

Spears, J. F. (1974). "A Review of Federal Domestic Plant Quarantines." PPQ, AHPIS, USDA, Hyattsville, Maryland.

Epilogue

Dr. Horsfall introduced our treatise by documenting some of the social effects pest species have had on man. He presented a very convincing argument that pest species have had and always will have a profound effect on man's behavior. How then have we dealt with our introduced pest problems? It appears from a historical glance that pests have fared very well while man has suffered great economic losses. Where did we fall short? The answer to this question is "Know thine enemy." It is evident from past records that in dealing with new pest species we knew too little too late. The result was the establishment of a flourishing pest species before some form of control could be implemented.

Considering the ecogenetic factors that come into play when a pest species is introduced into a new environment, it is a wonder that so many pests have actually been able to move about as they have. While it is impossible to intercept all pest introductions through quarantine inspections, only a very few of those that get through have any chance of actually becoming established. Rapid detection and accurate identification of a newly introduced pest is essential if we are to deal with it effectively. This type of reaction can only come about if we have a network of agencies staffed by well-trained individuals. These agencies must start at a grass roots level and go through state governments to the federal government. Communication must be prompt and action decisive. An example of what we currently have is clearly shown by the medfly infestation that occurred in 1981 in California. We believe it is accurate to say that, because of communication problems and an inability to implement effective control measures, many millions of dollars were wasted and an economic disaster was barely avoided. Can we long walk this narrow path without serious economic consequences?

Certainly a substantial effort should be directed at research on exotic pests. Many potential exotic pests, in the form of arthropods, pathogens, and weeds, are biding their time until the right combination of factors results in their introduction. It is impossible for any research organization to do in-depth studies on so many pests. However, many of these pests, because of their economic importance, have undergone in-depth investigations in the areas of the world in which they now occur. In some cases sufficient information should exist for making some kind of judgment on their risk potential and probable economic impact should they be introduced. Not all but many of our more serious introduced pests came from environments quite similar to the new ones they now inhabit. This type of information should play a major role in anticipating which pest species will strike next. It seems reasonable that the more of these organisms we can scrutinize, the more chance there is that some overall pattern may develop identifying instrinsic factors important in determining high-risk pest species. The cost effectiveness of such a program would be directly proportional to the numbers of pest species scrutinized. On the other hand, in-depth investigations of several pest species over 15–20 years would probably miss completely any such pattern.

What do we do about foreign pests that have not "flagged" themselves in their native habitat as potentially destructive if introduced? Sailer indicates that we would not have expected two-thirds of the important introduced insects to have become major pests based on their behavior in their native habitat. One approach to this problem is to take our plant material to the pests. Important crop plants could be planted in foreign countries to screen out potentially hazardous exotic insects and plant pathogens. Present worldwide plantings of wheat and soybeans already provide an opportunity to look at potentially dangerous exotics that may cause us a problem if introduced.

There are also new research tools available for dealing with pests already introduced. Rare allele studies can identify the exact parent population of the pest, thereby making the selection of biocontrol agents more precise and in all probability more effective. This approach to the biocontrol of a pest would involve a predictive method rather than the empirical or shotgun method used in the past. Another tool to be considered is the predictive model for population dynamics, epidemiology, and crop loss that might occur should a new pest be introduced. The ability to predict what would happen, how long it would take to happen, and what types of strategies are needed to avert an economic disaster would provide the advantage needed to deal with an ancient enemy. It is unfortunate that models of this type, for the more complex pest systems, are many years in the future.

How then do we acquire some sort of insurance against the ravages of these pests? Diversity can and does appear to work in nature. How do we achieve diversity in our agromonocultures? Multicropping, crop rotation, trap crops, and genetic manipulation all contribute to diversity within the monoculture system. A very low level of resistance, about one-third, in multiline plantings of small grains gives very good control of certain pathogens.

Much of what we have discussed makes even more sense when coupled with the idea of international cooperation. The movement of pests about the world is an international problem. All countries are directly dependent on the production of food worldwide and as such should be committed to stopping plant pest movement.

Charles L. Wilson
Charles L. Graham

Index

B

D

J

K

L

S

Y

Z